GRAVITY AND TECTONICS

A WILEY-INTERSCIENCE PUBLICATION

GRAVITY AND TECTONICS

Edited by
KEES A. DE JONG
Department of Geology
University of Cincinnati
Cincinnati, Ohio

ROBERT SCHOLTEN
Department of Geosciences
The Pennsylvania State University
University Park, Pennsylvania

JOHN WILEY & SONS, New York · London · Sydney · Toronto

Copyright © 1973, by John Wiley & Sons, Inc.

All rights reserved. Published simultaneously in Canada.

No part of this book may be reproduced by any means, nor transmitted, nor translated into a machine language without the written permission of the publisher.

Library of Congress Cataloging in Publication Data:

De Jong, Kees A., 1939–
Gravity and tectonics.

"A Wiley-Interscience publication."
Includes bibliographies.
1. Geology, Structural—Addresses, essays, lectures.
2. Gravity—Addresses, essays, lectures.
I. Scholten, Robert, 1923– joint author.
II. Title.

QE601.S33   551.8   73-1580
ISBN 0-471-20305-X

Printed in the United States of America

10 9 8 7 6 5 4 3 2 1

Stones have begun to speak, because an ear is there to hear them. Layers become history and, released from the enchanted sleep of eternity, life's motley, never-ending dance rises out of the black depths of the past into the light of the present.

From *Conversation with the Earth*, by Hans Cloos, Alfred A. Knopf, Inc., 1953.

Rein W. Van Bemmelen
June 1972

To

REIN W. VAN BEMMELEN

"The earth heaves and
the mountains are hurled
into the sea" (*Psalms* 46:2)

"Thou shalt not push" (*Caveat* 32:4)

Preface

The Role of Gravity in Crustal Deformation

Even though this book is not centered exclusively around the concept of gravity tectonics but rather around the manner and extent to which gravity may have played a role in the deformation of the lithosphere, a useful start may nevertheless be made by briefly reviewing what the term "gravity tectonics" has meant to different authors. First, the concept is scale-dependent. In its *broadest sense* it states that all mass displacements in the earth are ultimately the result of gravitational forces. Contraction or expansion of the earth or large parts of the earth, thermal convection in the mantle, hot mantle plumes, and plate drifting—all of these processes may be seen in light of the potential gravitational energy of the global masses involved and the expenditure of that energy. This is justified as long as one chooses to consider the energy system in a large enough perspective. In this global perspective, however, the concept may also cease to be of much practical use to the geologist who seeks answers to regional structural problems.

Gravity tectonics in the restricted sense implies only that a given, regionally integrated tectonic system at the crustal level has lost potential gravitative energy in the course of deformation, even though parts of the system may have gained energy. Seen in this light the concept is much sharpened and takes on greater potential significance in the field of tectonics. At the same time it becomes clear that we may also restrict the term gravity tectonics too much by applying it only to the loss of energy suffered by rock masses in a small part of the integrated system, while failing to see the system in its entirety. This is not a plea for the application of the theory of gravity tectonics to any given complete mountain belt, but it is nevertheless true that failure to see the whole picture has clouded understanding of the theory since its inception.

The simple example of a salt diapir may

serve as a useful illustration of the notions of scale and system. If a diapir rises to a height less than about 1 km below ground level, its top part will be heavier than the surrounding sediments. The continued rise of the top part is now no longer due to the gravitational body force within it, but rather to the buoyant driving force of the salt column below the 1-km level. The top is actually gaining potential energy while the system as a whole is losing energy. Thus it is justifiable to consider all salt diapirs as gravitational tectonic phenomena. The same reasoning may be applied to the intrusion of plutons and, more controversially, perhaps to mantled gneiss domes and the initiation of basement nappes (see Ramberg, this volume).

In this example, the gravitational instability involved and the expenditure of that energy were mainly vertical. The other type of instability is mainly in a lateral sense, as is the resulting movement. Thus Van Bemmelen (1960, 1966) named the Austroalpine nappes gravitational in spite of objections raised because the nappes clearly moved updip on the overthrust surface and thus against gravity, presumably indicating crustal compression. Van Bemmelen, however, never intended to deny this, but he did deny any contradiction between "compressional" and "gravitational" tectonics. His purpose was to indicate that the Austroalpine nappes represent merely the northern part of a system that includes the hinterland of the nappes—a system that lost free potential energy during the northward movement of the nappes. The increase in potential energy in the Alps proper was regarded as compensated regionally by a decrease in the southern part of moving mass, and in these terms the gravity tectonics model was considered justified as applied to the Austroalpine nappes. Thus the acceptance or rejection of this model depends again upon the spatial extent given to the system.

Finally, not only the areal extent, but also the level at which tectonic processes occur within the crust must be taken into account. It is by no means impossible to have crustal compression occur at a lower level and extension by gravitational tectonics at a higher one, either simultaneously or at different times (e.g., see the paper by Choukroune and Séguret on the Pyrenees in this volume).

Growth of the Concept of Gravity Tectonics

The concept of gravity tectonics has been developed along three lines of thought. One line was based on direct observations or space-time reconstructions of structural features that appeared to indicate a loss of potential energy of the rock mass involved, as in the case of downslope gliding. The second line was more indirect and sprang from considerations of rock mechanics. The great "plasticity" of rocks was thought to exclude external forces as the cause of observed deformations, which could therefore have been formed only by an internal body force, the force of gravity. This reasoning is reinforced in cases where very superficial masses have moved across an essentially undeformed substratum. The third line followed the conceptual truism that, in the final analysis, there is no force other than gravity that could be responsible, no matter how indirectly, for all tectonic deformations observed at the earth's surface. Haarmann (1930) gives a clear account of the writings of the early proponents of gravity tectonics: Gillet-Laumont in 1799, Kühn in 1836, Naumann in 1849, Herschel in 1856—all of whom explained folding as the result of the downward movement of rock masses along an inclined plane.

The second line of thought that led to the concept of gravity tectonics was followed by Bombicci (1882) and later Italian geologists, who based their ideas on the structure of the Apennines. A brief summary of the articles of these authors, published in periodicals with very restricted circulation, is given by Dal Piaz (1943); among them are Bonarelli (1901), Anelli (1923, 1935), Migliorini (1933), and Signorini (1936). The Apennine structures were initially considered of local importance only, not involving large tectonic transport, and a comparison was often drawn between these structures and landslides. Then, in 1907, Steinmann recognized overthrusts in the Northern Apennines, and in subsequent years it was realized that the upper nappe, the Ligurides (see Elter and Trevisan, this volume) had moved at least 100 and perhaps 200 km. This change in interpretation did not alter the view that the cause of deformation was gravitational, and in 1934 the first integrated tectonic synthesis of a single mountain belt fully based on the concept of gravity tectonics was given by De Wijkerslooth de Weerdesteyn. In papers defending the gravitational origin of the Ligurian nappe and the other structural features in the Northern Apennines, emphasis is laid on the highly incompetent character of the rocks (mainly the so-called argille scagliose), which "by no stretch of imagination can be thought able to transmit the stress needed to transport the whole as one coherent mass" (De Sitter, 1964, p. 252). Since an external force could not have been the cause of the observed deformations, gravitational forces had to be invoked.

Reyer's publications (1888, 1892, 1894) represent the third approach to gravity tectonics. At the time when he wrote his books the contraction theory was used to explain all tectonic structures. Reyer rejected the notion of contraction and stressed instead the occurrence of contemporaneously formed compressional and distensional phenomena, capable of being produced only by the force of gravity. Surprisingly, he did not give any actual examples of gravitational tectonic phenomena but tried to demonstrate the reality of his theory by experiments that were extremely sophisticated for his time. In so doing, he became the first to formulate, in qualitative terms, the theory of scale models,[1] which led him to make use of clay–water mixtures. Until then the concept of gravity tectonics had been applied exclusively to the explanation of local structures, but Reyer saw that it should operate also on a much larger scale, as in the process of mountain building. This point of view, that gravity is the only force available that will ultimately produce all tectonic phenomena, was revived several decades later by Haarmann (1930) and Van Bemmelen (1931). Reyer was a predecessor of these two workers in another aspect as well: he searched for the cause of the vertical uplift that induced the horizontal sliding. His theory was that the thick pile of sediments in the geosyncline created a gradual increase in their temperature and that the resulting expansion would lead to uplift (compare the paper by Schuiling in this volume).

Although the notion of downslope gliding on a restricted scale is now accepted by many geologists, the broader concept of gravity tectonics is not, and it is recognized that even in its more limited application the theory still lacks a firm quantitative underpinning, particularly with regard to the problems of mechanics and slope.

---

[1] Reyer (1892, p. 4) wrote, as translated: "When a deformation is produced in weak materials through the application of small forces and in a short time period, we may conclude that we should also be able to deform, as does Nature, materials with greater competence by the use of larger forces applied over longer periods of time.

"With plastic substances and with material which ruptures easily we can study, at a small scale, the successive and final stages of the deformation process; such a material even permits us to make, with little force and in a short time, a typical fold mountain belt, and we may justifiably conclude that material of greater competence will behave analogously at a larger scale and over a longer time.

"When we are successful in creating fold mountain belts . . . under experimental circumstances that are analogous to natural circumstances, then we must attach a great significance to the long disregarded geologic experiments."

The Slope Problem

Lemoine gives in this book a review of the "slope problem" in downslope gliding and shows how complex this apparently simple problem actually is. The concept of slope is so important because it helps the student of the structure of the earth's crust visualize that particular type of gravitational mass movement that reduces the free potential energy of a system by dominantly lateral displacement (as opposed to dominantly vertical adjustments by diapirism). Such lateral movements may be of three different types and may be upslope as well as downslope, as shown in Fig. 1. Particularly the upslope type of movement has not been widely recognized by students of gravity tectonics, although it was stressed repeatedly by Haarmann and Van Bemmelen.

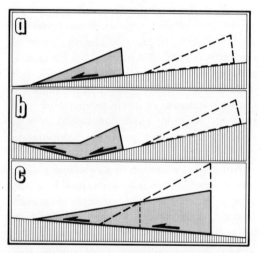

FIG. 1 Three types of mass movement that may result from gravitational instability. (a) Movement is downslope; (b) part of the movement along the fault plane is downslope, the other part is upslope; (c) all movement along the fault plane is upslope. a and b are commonly named gravitational sliding, and c gravitational spreading (and flattening).

The situation depicted in Fig. 1a, where the entire slope is known to dip in the direction of transport, is least controversial, but even here two problems may arise: (1) the slope may have been created in its present form after the mass movement occurred, and (2) the angle of slope may be so small as to lead many to reject the notion of gravity tectonics. The first problem must be solved within the context of the regional structural evolution. The second was partly met by Goguel (1948; also see Kehle, 1970) who discussed mass movement over a zone of low viscosity (rather than a discrete surface), and Hubbert and Rubey (1959), who analyzed the problem in terms of high pore pressures along the thrust surface. However, even with the very low slopes permitted by these mechanical analyses, the excessive height of the source area may still pose a problem in cases of long-distance transport.

Situations such as those shown in Fig. 1b present the additional difficulty that upslope movement is involved. If the rearward portion of the slip surface has been eroded or has subsided, the only part still visible today will be the one that slopes in a direction opposite to that of the movement, and gravity tectonics may be rejected on that basis alone (see Laubscher on the Jura Mountains, this volume). In the spreading model of Fig. 1c the problem is different; it lies mostly in the difficulty of proving that vertical flattening has indeed occurred. The advantage of this model is that (1) it eliminates the slope problem for movement may be upslope, and (2) no area of tectonic denudation needs to be looked for (a frequent difficulty in the gliding models) (see Price on the Canadian Rocky Mountains, this volume).

In summary, it can be said that in the past 25 years we have acquired a much better understanding of the problems of slope and the mechanics of mass movement, although neither problem has yet been eliminated.

About this Book

This book is divided into three sections. In the first section the role of gravity is discussed from a general point of view. The second and third parts deal with "case histories" of specific areas, evaluating the role gravity has played in deformations observed in various mountain belts. Part 2 is concerned with the Alpine mountain chains around the Mediterranean Sea, and Part 3 contains the views of North American authors regarding large segments of the Phanerozoic orogenic belts in that continent.

Regional studies lie at the very root of geological science, but at the same time they represent the danger of introducing a certain provinciality and bias, since it is extremely difficult to remain fully informed about recent studies in parts of the world other than one's own. Parts 2 and 3 of this book will serve an additional purpose if they enable the reader to become more fully acquainted with the structure and evolution of mountain belts in a different continent. A comparison between the two continents of Europe and North America may reveal that some geologic phenomena are better represented in one continent than in the other, and that geologic thought has at times run along parallel, and at other times along divergent lines.

The Alpine Systems in the Mediterranean Region

One interesting fact that emerges from several papers on this area is the continuous gradation in size that appears to exist from the smallest sedimentary particle via large blocks to mountain-sized sedimentary klippen (olistolites). Elter and Trevisan (this volume)[2] give examples of such transitions into major allochthonous masses in the Adriatic and Po basins, and they underline the importance of olistolites and olistostromes in the tectonic history of the Northern Apennines. Caire describes the same size gradations in the Calabro-Sicilian arc and shows how the advancing nappes supplied an influx of "orogenic" sedimentary klippen to the subsiding basins. These are the "precursory" olistolites of Elter and Trevisan, which constitute one of three types in their classification, the other two being derived from a seafloor that was slightly tilted during sedimentation.

The precursory olistolites and olistostromes are generally intercalated in well-dated sediments, and they provide testimony to the long time involved in the gravitational sliding of allochthonous masses. Elter and Trevisan demonstrate that the average velocity of the Subligurian nappe that moved into the 50-km-wide Macigno basin was about 1 cm/year. Olistostromes and olistolites also formed in front of allochthonous units in the Pyrenees according to Choukroune and Séguret, who calculate the velocity of sliding on the basis of the age of synsedimentary deposits, and find it to be in the order of 0.5 cm/year.

To explain the formation of the Pyrenean structure Choukroune and Séguret discuss the probability of several different models in their paper. They combine a crustal compression and a gravity tectonics model. The crustal compression is responsible for basement-involved features, such as the Gavarnie nappe and the North Pyrenean fault. This model is supported by the orientation of cleavage and other small-scale features, and by the absence of corresponding tectonic denudation areas over the axial part of the chain. Upward vertical movements of the core of the mountain belt caused southward displacement of the South Pyrenean unit over a maximum distance of 50 km by gravity gliding, as is suggested by extension features in the glide mass and a denudation area to the rear of the nappe.

A different argument in favor of free gliding is used by Clar in discussing the Eastern Alps. He states that the final movement of the overthrust units of the Northern Calcareous Alps in the mid-Tertiary must have been gravitational because foreland sediments prove that the central part of the

---

[2] In subsequent references to articles in *Gravity and Tectonics* the words "this volume" are omitted and the author's names are not followed by the year of publication.

Eastern Alps was already attacked by erosion long before. There were thus no rocks that could transmit a northward push by this time. Furthermore, before their final emplacement the nappes of the Northern Calcareous Alps had already moved in the Late Cretaceous, and the sediments of that period suggest to Clar the existence of a northward slope which may have caused earlier submarine sliding of the nappes into the continuously subsiding Penninic trough.

The gravitational origin of the Ultrahelvetic nappes in the Prealps of France and Switzerland has been postulated on still different grounds, namely the observation that the nappes represent stratigraphic units that occur in reverse sequence. This is taken to indicate that they have successively slipped off the raised source area, a phenomenon termed "diverticulation" by French authors (see Debelmas and Kerckhove, Fig. 5, and Lemoine, Fig. 4).

The superficial nature of some thrusts is indicated by the fact that they appear to have overridden an eroded land surface, and thus represent *epiglyptic* movement—a phenomenon described by Debelmas and Kerckhove for the Subbriançonnais and Piemont nappes in the Western Alps. It is also beautifully displayed in the Southern Alps and mentioned in the paper by Clar. Still another indication of shallow thrusting is the replacement of the sedimentary cover by an allochthonous cover of more internal origin. The type area for this *remplacement de courverture* is also in the French Alps (Ellenberger, 1958), but a description of the same process in the Northern Apennines (replacement of the Tuscan sedimentary cover by the Ligurian nappe) is given in this volume by Elter and Trevisan.

Debelmas and Kerckhove argue that the extreme thinness and large width of the Autapie nappe (approximately 0.5 and 100 km, respectively) point to the conclusion that this nappe was gravitationally emplaced. Its movement must have been exceptionally fast, a few centimeters per year at least, since some of the allochthonous rocks are of approximately the same age as the autochthonous sediments in front of the nappe.

Trümpy also discusses the rate of the nappe movements. According to him vertical and gliding movements tend to be relatively slow and continuous, whereas the great overthrust movements during the late Eocene to early Oligocene in the Central Alps were episodic in nature and proceeded at the rapid rate of at least 4 and perhaps 6 cm/year. This is faster than the calculated velocities of the gravity nappes in the Northern Apennines and Pyrenees, but some other gravity nappes such as the Autapie nappe may have moved at a comparable velocity. The question may be raised whether individual sliding movements may not be even more rapid during moments of gravitational instability, but alternate with prolonged periods of rest so that only the average velocity would be low.

The entire question of periodicity versus continuity of orogenic activity is raised by Wunderlich. His data from Crete and the Peloponnese appear to indicate more or less continuous orogenic movements during the Tertiary. Gravitational nappes are believed to be carried forward on top of shifting uplifts according to the "surf-riding" principle. Wunderlich also approaches gravity from a point of view distinctly different from that of other contributors to this volume. He discusses the present gravity anomalies in the various Mediterranean mountain chains and relates these to the stage of orogenic evolution that each chain has currently achieved, the Peloponnese–Crete island arc being the youngest and the only one still undergoing orogeny.

In the last paper in the section on Mediterranean geology Turner describes how the oldest rocks of the autochthonous series on the island of Cyprus consist of an immense pile of pillow lavas, intruded by a large gabbroic pluton and innumerable dikes, and often compared to a fossil midoceanic ridge. The ultrabasic rocks in the center of this sequence, at Mount Olympus, have commonly been explained as part of the upper mantle at the base of this fossil ridge, but Turner presents arguments showing that Mount Olympus may instead be an erosional remnant of a large overthrust mass, the Cyprian gravity nappe.

Mountain Belts in North America

One phenomenon displayed far better in North America than in Europe is crustal stretching during orogenic stages. In Europe, crustal stretching appears to be minor by comparison, more difficult to document, and characteristic more of the geosynclinal than the orogenic stage (Trümpy). That extension has occurred in the western Cordillera of North America is beyond question, but the degree and the mechanics of the extension are controversial. Wright and Troxel describe how, in the southwestern Great Basin, an upper crustal zone has been stretched 30 to 50 percent by low-angle normal faulting and how this zone may be underlain by one in which extension is accomplished by both the emplacement of igneous bodies and solid flow of the host rock. Hose and Daneš, on the other hand, give evidence and a mathematical analysis of stretching in the upper part of the crust only, accomplished by younger-over-older low-angle faults in the rear of very large masses whose front is marked by older-over-younger overthrusts. Thus, in their view, frontal compression in the east compensates for stretching in the rear, the whole mass moving under the force of gravity.

Whereas Hose and Daneš discuss exclusively the suprastructure of the eastern Great Basin, the infrastructure of the same area is also taken into account by Roberts and Crittenden, who describe it in terms of penetrative deformation, imbricate thrusts, and recumbent folds. These tectonic features indicate crustal shortening in the Jurassic to Early Cretaceous, which preceded the uplift that permitted the Late Cretaceous to Early Tertiary sliding of the suprastructure. This provides a good example of *Stockwerktektonik*, a term introduced by Wegmann in 1930 and commonly used in Europe to designate different types of tectonic behavior at different levels of the crust.

Tectonic denudation does not occur in the eastern Great Basin, as it does in parts of the Apennines and Pyrenees, and the question remains whether extension of the supracrustal mass in the rear was sufficient to account for the amount of overthrusting in front. Tectonic denudation is, however, classically displayed in the Middle Rocky Mountains of Wyoming because of the fortuitous "fossilization" of the Heart Mountain detachment surface by a cover of volcanic rocks very soon after tectonic denudation occurred (Pierce). Although there is no doubt about the gravitational origin of the Heart Mountain detachment mass, many questions can be raised, and are raised by Pierce, about the mechanics of its emplacement.

The paper by Moench on Appalachian tectonics in western Maine is one that would have immensely pleased Haarmann, one of the foremost proponents of gravitational tectonics. Haarmann (1930) was of the opinion that folding in the Rheinisches Schiefergebirge of northwest Germany was produced by gravity-impelled movements of newly deposited sediments toward the center of the geosynclinal trough (*Volltroggleitung*). Other workers in that area have not accepted Haarmann's interpretation, partly because axial-plane cleavage was generally believed to have formed long after diagenesis of the sediments. Still, the concept may not have to be abandoned altogether, partially in view of Maxwell's (1962) interpretation of cleavage as a soft-sediment phenomenon. Van Bemmelen (1949) applied the term "compressive settling" to a similar phenomenon in Indonesia, but the geological evolution of the Merrimack synclinorium as interpreted by Moench has more in common with the Rheinisches Schiefergebirge, even though the particular structural manifestations are somewhat different. During its deposition the wedge of clastic sediments in the Merrimack trough became divided into slump units or "cells" with a relation between faulting and folding so intimate that Moench coins the term "fault-fold." Each cell was folded and compacted under its own weight against the barrier of sediments in the trough. Moench also draws an interesting comparison between the Rangeley fault-fold structures and the great Taconic klippen of Vermont.

One other problem that may be encountered in the interpretation of allochthonous masses as sliding features is that the source area of the nappes may no longer lie directly behind them but may have been displaced far away by subsequent movement along strike-slip faults. This is admirably illustrated in Newfoundland by Kay. The strike-slip faults may in turn be related to drifting of lithospheric plates, and a manifestation of such drifting may be the Cambro-Ordovician Dunnage melange regarded by Kay as having formed in a trough above a subduction zone along which oceanic crust was underthrust in a northwesterly direction. The melange is also said to contain slides of graywacke and mudstones with boulders, and one gains the impression that it may not only be a mixture of lithologies but also have a mixed origin.

Hsü comes to this conclusion in discussing the Franciscan rock sequence in California. There, sedimentary processes including submarine sliding may have been largely responsible for the introduction of exotic elements; the sequence later became a melange when it was dragged down along a shifting Benioff zone. Hsü also gives a mechanical analysis showing that the "toe effect" can lead to the densely spaced thrust surfaces typical of a melange.

The papers by Price, Foose, Viele, and Root may suggest that the notion of gravitational spreading (as opposed to gravitation gliding) is at present a matter of interest in North America only. Although Wunderlich presented a paper in 1953 on the structure of the Harz Mountains in northern Germany, with a tectonic evolution quite comparable to that of the Beartooth Mountains block described by Foose in this volume, good examples of gravitational spreading appear to be restricted to North America.

Gravitational flattening and spreading indicate the continuous deformation of a mass under its own weight, whereby flattening relates mainly to the vertical and spreading mainly to the lateral movement. As Price stresses in his paper on the southern Canadian Rockies, the view whether a deformation is of the "continuous" or "discontinuous" type depends on the scale of observation: displacements along a series of discrete slip surfaces seen at the scale of outcrop may be regarded as continuous flow at the scale of an entire mountain belt. If spreading occurs over a discrete slip surface, movement may be downslope or upslope; upslope gravitational spreading is possible if the upper boundary of the deformed mass possessed a sufficiently large gradient in the direction of tectonic transport.

Spreading need not be confined to supracrustal rocks, as shown by Foose's discussion of the structure of the Middle Rocky Mountains in Wyoming and Montana. The uplift of crustal blocks gave rise to dilation of macrojoints and development of microjoints in the crystalline rocks, furthering the bending of initially vertical faults around the blocks into reverse faults. The continued rise and free expansion of the block created graben within it and locally bent the bounding faults into low-angle positions. The response of the uplifted block to the force of gravity resulted in a deformation which strongly suggests gravitational spreading if the scale of observation is large enough. The discontinuous extensional movements at the top of the uplifted mass and the discontinuous compressional movements at its margin then fit into a single picture of continuous deformation by flattening and spreading.

Viele presents in his paper the view that the rocks of the Benton-Broken Bow uplift constitute a pile of nappes which slid into a flysch basin during the Mississippian. Small-scale gravity slides and olistostromes from the backs of these nappes added to the load in the flysch basin, "the combined weight of all acting as a plunger that squeezed the basin fill northward and upslope."

The slope of the basement over which the sedimentary cover of the folded Appalachians was displaced toward the northwest is discussed by Root in his paper. He concludes that there are no indications for the former existence of a basement uplift of 9 km, required to generate a basement slope for gravity sliding of the cover. On the other hand, as the papers by Roberts and Crittenden and by Scholten point out, there is not necessarily a direct relationship between basement slope and the slope of a décollement surface, unless it can be demonstrated that the two surfaces are the same.

Scholten's paper leads to the conclusion that different tectonic processes may dominate adjacent segments of a mountain chain. Contrasted are the U.S. Rockies east of the Idaho Batholith with gravitational sliding and cascading down the slope of a décollement zone, and the Rockies farther north with deformation by buoyant upwelling and gravitational spreading according to Price's model. Price stresses that the latter process leads to isostatic subsidence of the cratonic margin in front of the advancing masses, the sedimentation in this migrating foredeep giving a record of the intensity of deformation in the whole orogenic belt.

Summary

The contributions in this volume underline several fundamental points:

1. Crustal deformation may have multiple causes, leading to multiple structural manifestations.
2. The force of gravity has been a factor in the tectonic evolution of many mountain belts.
3. The importance of its role appears to have varied considerably from place to place.
4. The role assigned to gravity in tectonics depends in large part on the scale of observation.

KEES A. DE JONG
ROBERT SCHOLTEN

University Park, Pennsylvania
Cincinnati, Ohio
June 1972

References

Dal Piaz, Giambattista, 1943, L'influenza della gravità nei fenomeni orogenetici: *Atti Reale Accad. Sci. Torino*, v. 77, p. 1–52.

De Sitter, L. U., 1964, *Structural geology*, 2nd ed.: New York, McGraw-Hill, 551 p.

De Wijkerslooth de Weerdesteijn, P. J. C., 1934, *Bau und Entwicklung des Apennins, besonders der Gebirge Toscanas*: Amsterdam, Geologisch Instituut, 426 p.

Ellenberger, F., 1958, *Etude géologique du Pays de Vanoise*: Mém. Serv. Carte Géol. France, 561 p.

Goguel, J., 1948, *Introduction à l'étude mécanique des déformations de l'écorce terrestre*: Mém. Expl. Carte Géol. France, 530 p.

Haarmann, Erich, 1930, *Die Oszillationstheorie, eine Erklärung der Krustenbewegungen von Erde und Mond*: Stuttgart, Ferdinand Enke, 260 p.

Hubbert, M. King, and W. W. Rubey, 1959, Role of fluid pressure in mechanics of overthrust faulting. I: Mechanics of fluid-filled porous solutions and its application to overthrust faulting: *Geol. Soc. America Bull.*, v. 70, p. 115–166.

Kehle, R. O., 1970, Analysis of gravity sliding and orogenic translation: *Geol. Soc. America Bull.*, v. 81, p. 1641–1664.

Maxwell, J. C., 1962, Origin of slaty and fracture cleavage in the Delaware water gap area, New Jersey and Pennsylvania, *in* Engel, A. E. J. and others, eds., Petrologic studies: A volume in honor of A. F. Buddington: New York, Geol. Soc. America, p. 281–311.

Reyer, Eduard, 1888, *Theoretische geologie*: Stuttgart, 868 p.

———, 1892, *Geologische und geographische Experimente*: I. Heft: *Deformation und Gebirgsbildung*: Leipzig, Engelmann, 52 p.

———, 1894, *Geologische und geographische Experimente*: III. Heft: *Rupturen*; IV. Heft: *Methoden und Apparate*: Leipzig, Engelmann, 32 p.

Steinmann, G., 1907, Alpen und Apennin: *Monatsber. Deut. Geol. Ges.*, v. 59, p. 117–185.

Van Bemmelen, R. W., 1931, Kritische beschouwingen over geotektonische hypothesen: *Natuurk. Tijdschr. Ned.-Indie*, v. 91, p. 93–117.

———, 1949, *The geology of Indonesia*: v. Ia: *General geology of Indonesia and adjacent archipelagoes*: Den Haag, Government Printing Office, Martinus Nijhoff, 732 p.

———, 1960, New views on East Alpine orogenesis: *Intern. Geol. Congr.*, 21st, Copenhagen, v. 18, p. 99–116.

———, 1966, On mega-undations: a new model for the earth's evolution: *Tectonophysics*, v. 3, p. 83–127.

Wegmann, C. E., 1930, Über Diapirismus (besonders im Grundgebirge): *Bull. Com. Géol. Finl.*, v. 92, p. 58–76.

Wunderlich, Hans Georg, 1953, Bau und Entwicklung des Harznordrandes bei Bad Harzburg: *Geol. Rundschau*, v. 41, p. 200–223.

Contents

Prologue   *xxiii*

Rein W. Van Bemmelen—
An Appreciation
by Richard M. Foose   *xxv*

*Major Publications by
R. W. Van Bemmelen*   *xxix*

PART 1   GENERAL PAPERS

1. Introduction to Part 1
   by Kees A. De Jong and
   Robert Scholten   3
2. The Gravitational Instability
   of the Lithosphere
   by Frank Press   7
3. Gravitational Instability and
   Plate Tectonics
   by Wolfgang R. Jacoby   17
4. Active Role of Continents in
   Tectonic Evolution—
   Geothermal Models
   by R. D. Schuiling   35
5. Model Studies of Gravity-
   Controlled Tectonics by the
   Centrifuge Technique
   by Hans Ramberg   49
6. Incremental Strains
   Measured by Syntectonic
   Crystal Growths
   by David W. Durney and
   John G. Ramsay   67
7. The Mechanics of
   Retrogressive Block-Gliding,
   with Emphasis on the
   Evolution of the Turnagain
   Heights Landslide,
   Anchorage, Alaska
   by Barry Voight   97

PART 2   PAPERS ON THE ALPINE SYSTEM IN THE MEDITERRANEAN REGION

1. Introduction to Part 2:
   Mountain Building in the
   Mediterranean Region
   by Kees A. De Jong   125

2. Tectonics of the Pyrenees: Role of Compression and Gravity
by P. Choukroune and M. Séguret — 141
3. The Calabro-Sicilian Arc
by A. Caire — 157
4. Olistostromes in the Tectonic Evolution of the Northern Apennines
by P. Elter and L. Trevisan — 175
5. Large Gravity Nappes in the French-Italian and French-Swiss Alps
by Jacques Debelmas and Claude Kerckhove — 189
6. About Gravity Gliding Tectonics in the Western Alps
by M. Lemoine — 201
7. Jura Mountains
by H. Laubscher — 217
8. The Timing of Orogenic Events in the Central Alps
by Rudolf Trümpy — 229
9. Review of the Structure of the Eastern Alps
by E. Clar — 253
10. Gravity Anomalies, Shifting Foredeeps, and the Role of Gravity in Nappe Transport as Shown by the Minoides (Eastern Mediterranean)
by Hans G. Wunderlich — 271
11. The Cyprian Gravity Nappe and the Autochthonous Basement of Cyprus
by William M. Turner — 287

PART 3 PAPERS ON THE OROGENIC SYSTEMS IN NORTH AMERICA

1. Introduction to Part 3: Tectonic Divisions of North America
by A. J. Eardley — 305
2. Tectonic Evolution of Newfoundland
by Marshall Kay — 313
3. Down-Basin Fault-Fold Tectonics in Western Maine, with Comparisons to the Taconic Klippen
by Robert H. Moench — 327
4. Structure, Basin Development, and Tectogenesis in the Pennsylvania Portion of the Folded Appalachians
by Samuel I. Root — 343
5. Structure and Tectonic History of the Ouachita Mountains, Arkansas
by G. W. Viele — 361
6. Mesozoic Evolution of the California Coast Ranges: A Second Look
by K. Jinghwa Hsü — 379
7. Shallow-Fault Interpretation of Basin and Range Structure, Southwestern Great Basin
by L. A. Wright and B. W. Troxel — 397
8. Orogenic Mechanisms, Sevier Orogenic Belt, Nevada and Utah
by Ralph J. Roberts and M. D. Crittenden, Jr. — 409
9. Development of the Late Mesozoic to Early Cenozoic Structures in the Eastern Great Basin
by Richard K. Hose and Zdenko F. Daneš — 429
10. Vertical Tectonism and Gravity in the Big Horn Basin and Surrounding Ranges of the Middle Rocky Mountains
by Richard M. Foose — 443
11. Principal Features of the Heart Mountain Fault and the Mechanism Problem
by William G. Pierce — 457

12. Gravitational Mechanisms in the Northern Rocky Mountains of the United States
by Robert Scholten   473

13. Large-Scale Gravitational Flow of Supracrustal Rocks, Southern Canadian Rockies
by Raymond A. Price   491

Prologue

The degree to which the force of gravity has played a role in causing or modifying tectonic processes at different levels and scales of the Earth's crust and mantle has been much debated for more than a century. At times concepts in which gravity was regarded as a major cause of tectonism significantly influenced geological thinking; at other times such concepts have been minimized. In the past decade the emphasis on major horizontal movements involved in plate tectonics have, perhaps unjustly, caused the pendulum to swing away from discussions of gravitational mechanisms in the crust. Yet in many respects these ideas remain useful and capable of further development and of sharpening our thinking about tectonics. It seems appropriate, therefore, to re-evaluate the role of gravity in crustal deformation at this time.

The most comprehensive concepts of tectonic deformation at all levels and scales have been put forward since 1931 in a series of articles and books by R. W. Van Bemmelen. He developed the more vaguely delineated ideas embodied in the oscillation theory of Haarmann into his more concrete undation theory. The primary "geotumors" ascribed to "cosmic causes" by Haarmann became undations attributed by Van Bemmelen to energy sources internal to the Earth. Secondary tectogenesis was seen as the result of gravitational adjustments away from these large uplifts. The importance of scale and level was increasingly emphasized and finally embodied in a classification of the primary uplifts into local, minor, meso-, geo-, and mega-undations, with the deep-seated mega-undations giving rise to continental drift and the opening of ocean basins by lateral sliding on a global scale. Thus, to Van Bemmelen, all of tectonics may be seen as the expression of a single dynamic system of vertical equilibrium-disturbing and subhorizontal equilibrium-seeking mass displacements, the former causing the buildup of potential gravitative energy and the latter involving the expenditure of that energy. It is perhaps no exaggeration to say that this has, until now, been the most complete and perspective view of tectonics, and it is for this reason that the present volume is dedicated to the man who developed that view over four decades of acute field work, extensive reading within a wide range of the geological literature, and imaginative conceptual thinking.

The purpose of this book is not, however, to present a collection of contributions by "true believers," as the reader will find. Nothing could have been farther from the spirit of Van Bemmelen himself, who was ever open to new data and new thoughts that enabled him to refine, enlarge, or even profoundly modify his own concepts as they had evolved up to that time, and who would rather plant seeds than freeze systems. Many new fundamental data and ideas concerning such matters as plate tectonics and the mechanisms of deformation have become available in recent years. The time is ripe to take stock and see how different authors view the role of gravity in tectonics according to current knowledge and understanding of tectonic principles and the evolution of specific mountain ranges.

THE EDITORS

June 1972

Rein W. Van Bemmelen—
An Appreciation

We stood on the front deck of my home overlooking the Coast Ranges and San Francisco Bay at Palo Alto, California, and talked about mountain-building processes. The easy manner of our conversation and the unusual breadth and flow of ideas marked our guest as clearly as his published work had already done. This was 1961. I was seeing Professor Rein Van Bemmelen for the first time since the International Geologic Congress meetings in 1952 and 1960. During his three-month tour of the United States under auspices of the American Geological Institute that year Van Bemmelen made a deep impression on many American geologists with his quiet charm and the scope and stimulus of his geologic thinking. It was apparent that he was sensitive and receptive to the ideas of others and enjoyed hearing them and debating alternatives.

Our discussion that day, as I remember it, dealt largely with the tectonic framework and structural mechanisms of the Basin and Range of Nevada, which he had just visited on a field trip, and with the Middle Rocky Mountains, where I had been working in the 1950s, but he drew extensively on his own pertinent field observations in the Alps and in Indonesia. Emergent from our discussions that day were two sharp personal impressions. The first was the remarkable way in which Van Bemmelen attempted to integrate a wide variety of geologic data and to synthesize them. It was clear to me that he was strongly moved to understand the *entire* story! While having the courage not to stumble on minute details, nonetheless he was interested in those details, wishing not to ignore them. While answering the easier geologic questions—"What?" "Where?" "When?"—he was obviously also seeking answers to "Why?" and "How?" The second impression from that day was his lucid exposition of the world-wide importance of vertical movements within the crust and mantle as a major cause for horizontal transport of major crustal segments powered by the force of gravity.

Two years later, when I spent a year in Europe, I had occasion to meet several times with Van Bemmelen under widely differing circumstances: once at the meetings of the Geologische Vereinigung in Bern where we were surrounded by professional colleagues; once at the University of Utrecht where I had gone to lecture; and also on a field trip conducted by him and Professor M. G. Rutten through the Alps and Dolomites for their students. From these occasions I have at least three other vivid personal impressions.

In the field Van Bemmelen proved to be an extraordinarily keen and perceptive observer, often accurately predicting the nature of the next outcrop even though it was in new territory. And many times in the field he revealed his interest in attempting to hypothesize a "whole" geological story—thereby sharpening the attention of all of us to search for yet other clues that might confirm or modify the hypothesis.

A second impression stems from a long personal conversation I had with him while at the meetings in Bern. He was outlining his current thoughts on the undation theory and gravity tectonics and "worrying" about the nature of some of the new geologic evidence from studies along the mid-Atlantic rift and in the ocean basins which seemed to support the concept of continental drift. He expressed his concern that some of his earlier ideas may, in fact, have been too "fixistic" and that it might be necessary to develop a more "mobilistic" concept with which these new data would be compatible. Indeed, his articles published during the 1960s reveal an evolution of thought regarding global tectonics that marks on the one hand a vigorous mental tenacity with respect to the core of his earlier ideas and on the other hand a great conceptual flexibility that enabled him to accept new data and new ideas, and to abandon parts of earlier hypotheses, unfettered by any sense of false pride.

My third impression was entirely on the personal side. I enjoyed observing his friendly relationships with students—neither condescending nor soft, but rather stimulating, open, and helpful. From others, and from Rein himself, I had known something of his deep devotion to his wife and of conscious efforts by him to limit his professional activities so as to spend as much time with her as he could. One day in The Netherlands, when I visited their home, I met his charming "Pop," a sensitive and delightful lady, and felt something of the unusual warmth between them and the deep and satisfying relationship they have long enjoyed. One couldn't help but sense the important force she has been in all his professional life.

*   *   *

Rein Van Bemmelen was born in Djakarta, capital of the Dutch East Indies, in 1904. He attended the schools there until his enrollment in 1921 at the Technical University of Delft, where he assisted Professors H. A. Brouwer and G. A. F. Molengraaff. His doctoral degree in 1927 was based on his investigations of the Betic Cordillera in the province of Granada, Spain.

Van Bemmelen returned to Java as a geologist and mining engineer. In 1934–1935, while on European leave, he studied soil mechanics with Professor Karl Terzaghi at the Technical University in Vienna. He was named Chief of the Volcanological Survey of the Dutch East Indies in 1940.

During the Japanese invasion of Java Van Bemmelen served in the homeguard. He was a prisoner of war from March 1942 until the end of 1945. Following his release, he traveled alone through war-ravaged Java seeking his wife and son and finally finding them in another camp whence he led them to safety.

Van Bemmelen left Java in 1946 for the Hague, where he was commissioned by the Dutch government to prepare a comprehensive report on the general and economic geology of Indonesia. An original manuscript of this treatise, with all illustrations, was lost during the war, necessitating a completely new writing and preparation of maps and sections. This gigantic effort resulted in the well-known three-volume treatise, *The Geology of Indonesia*, which appeared in 1949, marking the centenary of the Dutch East Indian Geological and Mining Survey.

From 1950 until his retirement in 1969 Van Bemmelen was Professor of Geology at the University of Utrecht, where he lectured in economic geology and profoundly influenced generations of young geologists through his research, lectures, discussions, and guidance in the field, to say nothing of the stimulus he provided to geologic thinking through his professional colleagues in Europe and throughout the world.

The remarkable scientific productivity of Van Bemmelen, manifested by more than 180 publications written in Dutch, German, French, and English during the 45-year period of 1927–1972 has dealt chiefly with three subjects: volcanoes and volcanism, gravity tectonics, and the undation theory. Other subjects that have commanded his attention are economic geology, methods and philosophy of geology, and the impact on human history of volcanic activity. In addition, Van Bemmelen produced six major geologic maps at a scale of 1:100,000 of Sumatra and Java and the monumental treatise on the geology of Indonesia.

When, as an 8-year-old child, Rein was taken by his meteorologist father to the summit of a volcano near Bandung, Java, he decided to become a volcanologist so as to understand its origin and history. Some 20 years later he described this volcano in a paper that revealed his interest in both volcanology and gravity tectonics (5).[1] The volcano Prahoe had, he thought, become too steep for its height and collapsed. He described the Toba depression as a great caldera in Sumatra (6) and, with Rutten in 1955 (9), he described calderas that had developed in Iceland. Recently (16) he has discussed the geologic and human effects of caldera formation at Santorini volcano in the eastern Mediterranean.

Although his interest in volcanology has continued throughout his career, more significant contributions have been made by Van Bemmelen with his studies of gravity tectonics and his syntheses expressed by development of the undation theory. As a young man in Java he was deeply influenced by Erich Haarmann's ideas expressed in his book *Die Oszillationstheorie* published in 1930. Haarmann's idea that vertical movements within the crust are of fundamental importance rang true for Van Bemmelen. Much of his fine work and great conceptualization evolved from that time and those ideas. Gravity could cause tectonic structures not only at centers of active volcanism (5), he reasoned, but also in the larger scale of mountain belts such as the Alps (3, 8). Since the early 1960s his ideas have evolved to include explanations of still larger scale processes such as continental drift (11, 13, 17). A striving for completeness, for explaining the entire structural evolution of specific areas (e.g., the Southern and Southeastern Alps; 14, 15) and for clarifying the causes and mechanisms of crust-mantle processes as well as the details displayed in a kinematic analysis, have always characterized Van Bemmelen's work.

During the 40 years since Van Bemmelen first sketched the outlines of his synthesizing theory on crustal undation, as it evolved from a rather fixistic to a more mobilistic concept and from first considerations of small-scale phenomena to very large-scale features, he has considered earth processes to be cyclic: equilibrium-disturbing processes cause equilibrium-seeking processes which, in turn, are also equilibrium-disturbing processes, and so on. Vertical movements of the crust must

---

[1] Numbers in parentheses refer to the reference list at the end of this paper.

result; these disturb the gravitational equilibrium and, in consequence, horizontal movements in crust and mantle are caused; these "try" to re-establish equilibrium. The motor for these cyclic phenomena was sought by Van Bemmelen in physico-chemical processes, first in the crust (2, 8) and later in the mantle (11, 17). The vertical movements are undations (primary tectonics) and the horizontal readjustments are gravity (secondary) tectonics. In all his published work on the theory he has sought to present a comprehensive interpretation of the structural evolution of the earth's crust (13, 17).

Van Bemmelen is keenly aware of the fact that knowledge evolves and theories may change. As he said at the end of his introduction to a recent summary of his own work by Hans Havemann,[2] "Ich hoffe, dass die folgende Generation der Erdwissenschaftler daraus besonders die positiven und brauchbaren Ansichten fur eine synthetische Konzeption der Erdgeschichte herauslesen wird." [I hope that the coming generation of Earth scientists will accept from these ideas the positive and useful aspects and will apply them toward a synthesizing concept of the Earth's history.] Through his lucid presentation of imaginative concepts that have been directed toward the goal of understanding earth processes, Rein Van Bemmelen has made a great contribution to geologic thought in our century.

RICHARD M. FOOSE
Department of Geology
Amherst College, Massachusetts

---

[2] Havemann, Hans, 1969, Die Entwicklung der Undationstheorie R. W. Van Bemmelens: *Geologie*, v. 18, p. 775–793.

Major Publications by R. W. Van Bemmelen

1. (1927) *Bijdrage tot de geologie der Betische Ketens in de Provincie Granada*: Delft, Waltmann, 176 p. (Doctoral thesis, Department of Mining, Technical University, Delft, The Netherlands: advisor, H. A. Brouwer.)
2. (1932) De Undatie-theorie (hare afleiding en toepassing op het westelijk deel van den Soendaboog) (Engl. summary): *Natuurk. Tijdsch. Ned. Indie*, v. 92, p. 85–242. (The undation theory and its application to western Indonesia.)
3. (1933) Die Anwendung der Undationstheorie auf das Alpine System in Europa: *Proc. Kon. Akad. Wetensch. Amsterdam*, v. 36, p. 686–694. (In the Mediterranean mountainbelts a distinction is made between vertical primary and horizontal secondary movements.)
4. (1933) The undation theory of the development of the earth's crust: *Proc. 16th Intern. Geol. Congr. Washington, D.C.*, v. 2, p. 965–982. (First description of the undation theory in an easily accessible periodical.)
5. (1934) Ein Beispiel für Sekundärtektogenese auf Java: *Geol. Rundschau*, v. 25, p. 175–194. (The collapse of a volcano in Indonesia.)
6. (1939) The volcano-tectonic origin of Lake Toba (North Sumatra): *De Ing. Ned. Indie*, v. 6, p. 126–140. (Lake Toba is the type-example of a volcano-tectonic depression.)
7. (1949) *The geology of Indonesia*: Vol. Ia, *General Geology* (732 p.); Vol. IB, *Portfolio*; Vol. II, *Economic Geology* (265 p): The Hague, Staatsdrukkerij/Martinus Nijhoff. (Monograph on the geology of Indonesia; reprinted in 1970.)
8. (1954) *Mountainbuilding*: The Hague, Martinus Nijhoff, 177 p. (Discusses the undation theory and its application to the geology of Indonesia.)
9. (1955) *Tablemountains of Northern Iceland*: Leiden, Brill, 217 p. (With M. G. Rutten.)
10. (1960) New views on East-Alpine orogenesis: *Rep. 21st Intern. Geol. Congr. Copenhagen*, v. 18, p. 99–116. (The great compressional nappes are the result of crustal spreading.)

[1] More extensive lists of publications by R. W. Van Bemmelen can be found in the bibliographies of publications 7 and 17. His approximately 180 publications can be grouped into 24 papers on volcanics, 62 papers on tectonics (especially the undation theory), 24 papers on regional geology in Indonesia, 8 papers on the structure of the Alps, 35 papers on petrologic subjects, 10 papers on economic geology, and 18 articles on miscellaneous subjects.

11. (1961) The scientific character of geology: *J. Geol.*, v. 69, p. 453–463. (Of the several papers by R. W. Van Bemmelen on this subject, this is the only one in the English language.)
12. (1962) Geotektonische Stockwerke (eine relativistische Hypothese der Geotektonik): *Mitt. Geol. Ges. Wien*, v. 55, p. 209–232. (Introduction to the concept of relativism in structural geology.)
13. (1964) Phénomènes géodynamiques, à l'échelle du globe (géonomie), à l'échelle de l'écorce terrestre (géotectonique) et à l'échelle de l'orogénèse alpine (tectonique): *Soc. Belge Géol. Paléontol. Hydrol.*, Mem. 8, 127 p. (Comprehensive article; continental drift is considered as an exclusively secondary phenomenon, the megaundations being primary.)
14. (1966) The structural evolution of the Southern Alps: *Geol. Mijnb.*, v. 45, p. 405–444. (This paper incorporates the results of field work by geology students from the University of Utrecht in the Southern Alps.)
15. (1970) Tektonische Probleme der Östlichen Südalpen: *Geol. Ljublijana*, v. 13, p. 133–158. (The fourth of a series of discussions concerning the structure of the southeastern Alps.)
16. (1971) Four volcanic outbursts that influenced human history (Toba, Sunda, Merapi, Thera): *Acta 1st Intern. Sci. Congr. on Thera*, Sept. 1969, p. 136–151.
17. (1972) *Geodynamic models, an evaluation and a synthesis*: Amsterdam, Elsevier, 267 p. (Reprint of 5 papers, 1964–1970, on the undation theory, with introduction and conclusion.)
18. (1972) Driving forces of Mediterranean orogeny (Tyrrhenian test case): *Geologie en Mijnbouw*, v. 51, p. 548–573. (A critical review of new data on the Tyrrhenian region, leading to a model of active mantle diapirism as the driving force of orogenesis.)

GRAVITY AND TECTONICS

PART 1

General Papers

ROBERT SCHOLTEN

Department of Geosciences
Pennsylvania State University
University Park, Pennsylvania

KEES A. DE JONG

Department of Geology
University of Cincinnati
Cincinnati, Ohio

Introduction to Part 1

The articles in Part 1 all deal with models and techniques that are of general significance to geodynamics. Continental drift and plate tectonics are the main subjects of the first two papers by Press and Jacoby. The importance of vertical movements is emphasized by Schuiling in light of geothermal considerations and by Ramberg in light of experimental studies on the effect of gravity. Next a paper by Durney and Ramsay shows how the study of small-scale objects can give us insight into structural processes of a larger scale, and, finally, Voight discusses gravitational gliding on the basis of landslide studies.

All of these authors deal with gravitational instability at various levels and scales in the upper layers of the earth. Gravitational instability has been related to continental drift since Taylor (1910) and Wegener (1912) postulated this concept. Taylor suggested that creep of the continents away from the poles produced the arcuate mountain belts in the Tethys zone, and Wegener called upon the gravitational attraction of the equatorial bulge. Daly (1926) and Haarmann (1930) also stressed the importance of gravitational forces—they related the process of mountain building in island arcs to the sliding of the continents over an inclined surface.

After the continental drift theory received a new impetus in the past decade, Van Bemmelen (1964) was the first to suggest that the plates were not passively reacting to movements in the upper mantle. According to him, they were not dragged by convection currents but slid actively over their base as the result of the free potential energy of the continents. Upwarps within the mantle are seen as the primary movement, and the secondary and mainly horizontal movement is performed by large crustal plates in which the continents are embedded.

Seismological evidence that the lithosphere is more dense that the asthenosphere is presented by Press, and petrological models are given for this system. Press demonstrates that there is enough potential energy in the lithosphere–asthenosphere system to drive the movements of the crustal plates. Jacoby also reviews the structure of the earth's crust and mantle and shows that the gravitational instability of the lithosphere with respect to the underlying asthenosphere leads to the diapiric upward rise of the latter. The resulting slope of the upper boundary of the asthenosphere would be sufficient to set the plates of the lithosphere into motion. The other expression of the gravitational instability of the lithosphere–asthenosphere system is the sinking of the frontal parts of the plates into the upper mantle. The sinking lithosphere pulls the plates toward the trenches.

The mechanical energy available in the lithosphere–asthenosphere system is several orders of magnitude less than the energy delivered to the surface by heat flow. This is the basis Schuiling uses to outline a geothermal model of crustal evolution. In his model the gravitational instability is explained as the result of the heating of the upper mantle below the continents, which act as blankets for the mantle. Schuiling also offers a geothermal model for the development of orogenic belts. He calculates the effect of heat production in linear belts or centers of crustal thickening and shows that great vertical uplift should occur there. He then explains the structural evolution of the Cyclades, orogenic center of a bend (the Aegean arc) in the Mediterranean mountain chain, by means of this model.

Voight describes how new, quantitative data can be obtained by studying in detail a well-known landslide in Alaska. The largely unexpected result is that the subhorizontal component of the force of gravity acting on a 4° slope is less than the lateral pressure from "plastic wedges" against the rear part of the landslide. Again it is shown how necessary it is to be careful when a mechanical interpretation is given of the emplacement of allochthonous masses, whether landslides or tectonic overthrust sheets. It was inevitable that Voight's study should have led him to a landslide interpretation of one of America's showpieces of gravity tectonics, the Heart Mountain overthrust (also refer to the paper by Pierce in Part 3 of this volume).

The extent and manner in which gravity has played a role in the emplacement of the Helvetic nappes in Switzerland is open to some question. Ampferer started the discussion in 1934 by stating that the Helvetic nappes in Glarus had slid from the massifs in the south. In a recent analysis of the mechanics of the Glarus overthrust, Hsü (1969) distinguished between a main movement which was not, and a later movement which probably was, induced by gravity. It may be expected that the new technique in structural geology presented by Durney and Ramsay in their paper will eventually lead to a better understanding of the directions and the mechanics of transport of the Helvetic overthrusts. Their study is concerned with the geometry of crystals that grew during the movement of the nappes. With this information, a chronological correlation can be made between the deformation history from nappe to nappe.

Ramberg's experimental approach to structural geology leads to a comparison of deformational structures in very small objects (5–10 cm) with those in large geologic bodies. Experiments have often played a fundamental role in the explanation of tectonic phenomena. It is very likely that the concept of gravity tectonics would have been accepted earlier if Hall, the first experimentalist in structural geology, had tilted his table so that the tablecloth had slid into folds by its own free potential energy, rather than pushing the tablecloth over the table into folds. Reyer (1895) realized that the reduction of the time factor in experiments had to be compensated by a drastic change in the properties of the materials in his models. He used a mixture of clay and water similar to the mixture used in later experiments by Hans and Ernst Cloos. The mathematical scale model theory developed by Hubbert (1937) showed that the materials used should have very low viscosities, such as that of honey. This makes a precise study of the deformations produced during the experiment virtually impossible. To overcome this difficulty, Ramberg employs a centrifuge model and can therefore use materials with much higher viscosities. His results, obtained with strict application of the scale model theory, are strikingly similar to such major geologic phenomena as batholiths, nappes, gneiss domes, and mid-oceanic ridges.

Taken together, the articles in this section demonstrate that much has been learned in recent years concerning the effect of gravitational forces in geodynamic processes, and that new approaches have been developed that shed light on the problem.

References

Ampferer, O., 1934, Über die Gleitformung der Glarner Alpen: *Sitzungsber. Akad. Wiss. Wien, mathem.-naturw. Kl.*, v. 143, p. 109–121.

Daly, R. A., 1926, *Our mobile earth*: New York, London, C. Scribner's Sons, 342 p.

Haarmann, E., 1930, *Die Oszillationstheorie, eine Erklärung der Krustenbewegungen von Erde und Mond*: Stuttgart, Ferdinand Enke Verlag, 260 p.

Hsü, K. J., 1969, A preliminary analysis of the statics and kinetics of the Glarus overthrust: *Eclog. Geol. Helv.*, v. 62, p. 143–154.

Hubbert, M. K., 1937, Theory of scale models as applied to the study of geologic structures: *Geol. Soc. America Bull.*, v. 48, p. 1459–1520.

Reyer, E., 1892–1894, *Geologische und geographische Experimente*, v. I, III, IV: Leipzig, Engelmann, 52 p., 15 p., 17 p.

Schardt, H., 1893, Sur l'origine des Préalpes romandes: *Eclog. Geol. Helv.*, v. 4, p. 129–142.

Taylor, F. B., 1910, Bearing of the Tertiary mountain belts on the origin of the earth's plan: *Geol. Soc. America Bull.*, v. 21, p. 179–226.

Van Bemmelen, R. W., 1964, Phénomènes géodynamiques à l'échelle du globe (géonomie), à l'échelle de l'écorce terrestre (géotectonique) et à l'échelle de l'orogenèse alpine (tectonique): *Soc. Belge Géol. Paléontol. Hydrol.*, Mém. 8, 127 p.

Wegener, A., 1912, Die Entstehung der Kontinente: *Peterm. Mitt.*, v. 58, p. 185–195, 253–256, 305–309; *Geol. Rundschau*, v. 3, p. 276–292.

FRANK PRESS

Department of Earth and Planetary Sciences
Massachusetts Institute of Technology
Cambridge, Massachusetts

The Gravitational Instability of the Lithosphere[1]

The kinematics of plate tectonics is now reasonably well understood, especially as regards the relative motions of the larger plates. This success is due primarily to the development of paleomagnetic stratigraphy and the remarkable feat of drilling through the sedimentary cover in the deep oceans. Although many details remain to be worked out, especially concerning the motions of smaller plates and the record of plate motions prior to about 200 million years ago, the main problem to be attacked in the next decade concerns the dynamics of plate motions.

As is the case with any new subject, many hypotheses have been advanced to explain the driving system for plate tectonics, but none have been substantiated to the point of widespread acceptance. At this stage of the development it seems appropriate to summarize recent geophysical results that pertain to the properties of the plates and the underlying asthenosphere. In particular, we review the evidence indicative of the gravitational instability of the lithosphere with respect to the underlying asthenosphere. Petrological schemes for making the lithosphere more dense than the asthenosphere are outlined.

Whether gravitational instability is a major factor in producing plate motions cannot be decided on the basis of presently available data or theory. The possibility that a gravitationally unstable lithosphere will be a factor in tectonics has been proposed by several investigators including Ringwood and Green (1966), Elsasser (1967), Hales (1969a), Press (1969), Isacks and Molnar (1969), Ringwood (1969), Ito and Kennedy (1970b), and Jacoby (1970). Van Bemmelen (1964) was of course an early advocate of the important tectonic role played by gravitational instability.

[1] This research was sponsored by the Office of Naval Research under Contract Nonr 1841 (74).

Geophysical Evidence Bearing on Gravitational Instability in the Lithosphere

The density distribution in the upper mantle is not easily determined. In principle, the phase and group velocities of surface waves and free oscillation periods depend on the density distribution. However, because of the scatter in the data and the lack of a complete data set, a unique density distribution cannot be obtained from the inversion of these data. This is not to say that certain conclusions about the density distribution cannot be reached.

Figure 1 presents a large number of models of the upper mantle which give the distribution of shear velocity and density consistent with the preceding data as well as the seismic refraction velocity at the M-discontinuity, the travel time of shear waves, the mass, and moment of inertia of the earth. These models (Press, 1970) were all selected randomly by a Monte Carlo procedure and they fit the data equally well. A number of interesting features emerge. Every oceanic and shield model shows a low velocity zone, whereas this zone may be present but is not required for the tectonic regions. In the latter case the low velocity may reach to the top of the mantle rather than forming a channel below the M-discontinuity. Another interesting feature is the greater thickness for the high velocity lid under the shields than is the case for oceans. If we arbitrarily define the thickness of the lithosphere as that depth where the velocity falls below 4.5 km/sec, then the suboceanic lithosphere seems to be about 80 km thick, and the subshield lithosphere seems to be twice that thickness. It is tempting to associate the low-velocity zone with a low-density zone because of the well-known correlation of velocity with density. However, it is widely accepted that the low-velocity zone is almost certainly caused by partial melting of the mantle. The appearance of a small amount of melt can drastically affect the shear velocity without necessarily introducing a significant reduction in density.

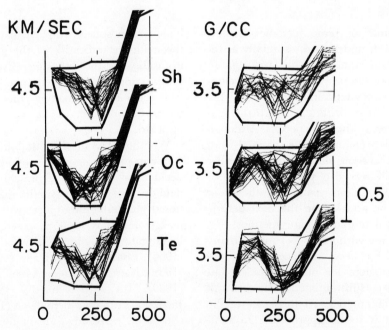

FIG. 1 Upper mantle shear velocity and density models which fit geophysical data for shields (Sh), oceans (Oc), and tectonic (Te) regions (Press, 1970). Vertical bar shows ordinate scale 0.5 km/sec or g/cm^3. Abscissa is depth (km). Heavy lines define bounds within which Monte Carlo models were generated.

The density distributions in Fig. 1 show without exception a rapid increase in density with depth in the range 10–70 km. Every model shows a density exceeding 3.5 g/cm^3 somewhere in the region between 70 and 220 km. Unfortunately, the solutions lack the resolution required to specify where in this depth range the maximum density is to be found. This high value of density exceeds that which we would expect for a normal peridotite; only a highly garnetiferous peridotite or an eclogite would fall in this range. Most of the density models for oceanic and shield mantles show a decrease in density below a high-density cap, the low-density values falling into the peridotite range. Unfortunately, this is not a unique feature because some models show monotonic density distributions. Every successful model for tectonic regions shows the density reversal. On the basis of these results we make a proposal which is nonunique but consistent with the data, that the lithosphere, as defined by the high-velocity cap above the low-velocity zone, contains high-density components in its lower half. Later we will use petrological arguments to select the models with a density inversion in the asthenosphere in a hypothesis for generating a gravitationally unstable lithosphere.

Anderson (1970) and Dziewonski (1971) also find earth models with a high-density lithosphere. Dziewonski's result occurred only with the assumption of a 70-km thickness for the suboceanic lithosphere.

An independent source of evidence for high density in the lower half of the lithosphere is obtained from recent seismic refraction measurements. Hales (1969b) and Hales, Helsley, and Nation (1970) present evidence for an abrupt increase in P-wave velocity at a depth of 80–90 km under the central United States and at a depth of 57 km under the Gulf of Mexico. The velocity increases from 8.05 to 8.3–8.45 km/sec under the shield Under the Gulf the increase is from 8.0 to 8.77 km/sec. Hales interprets this discontinuity as a transition to a higher density phase. Using Birch's relationship $\Delta V/\Delta \rho = 3$ (km/sec)/(g/cm^3), these velocity discontinuities correspond to density increases in the range $0.1 \leq \Delta \rho \leq 0.25$ g/cm^3.

A number of workers (Langseth, LePichon, and Ewing, 1966; McKenzie, 1967; Sleep, 1969) have calculated the thermal history of the lithosphere as it cools in its motion away from the source region in the mid-ocean ridges. Figure 2 shows isotherms for a specific model with assumed spreading rate of 1 cm/yr

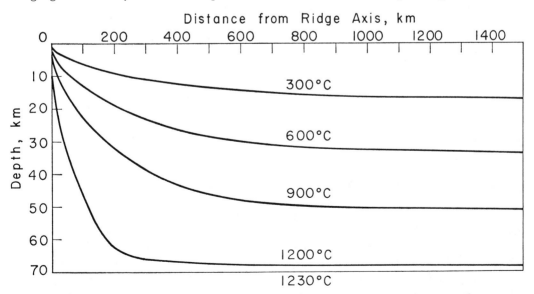

FIG. 2 Isotherms in a lithosphere spreading at the rate 1 cm/yr. Thickness of the lithosphere is assumed to be 70 km. Bottom and source region assumed to be at the solidus for eclogite.

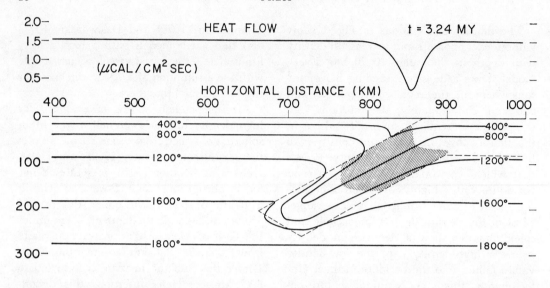

FIG. 3 Isotherms in a sinking slab after 3.24 million years according to Toksöz and others (1971). Spreading rate assumed is 8 cm/yr. Shading indicates zone of phase changes.

and with the boundaries held at the dry solidus for eclogite. Other models with faster spreading rate, with boundary temperatures appropriate for wet or dry peridotite or eclogite, and with different thicknesses of the lithosphere have been calculated by Forsyth and Press (1971). Other studies (McKenzie, 1969; Griggs, 1972; Toksöz, Minear, and Julian, 1971) have provided numerical models for the thermal regime of the lithosphere in the subduction zones. Independent of the preceding discussion these calculations show that the cool, sinking slab is more dense, hence gravitationally unstable with respect to the adjacent asthenosphere.

Figure 3 is an example of isotherms in the model of Toksöz, Minear, and Julian (1971). The densification of the lithosphere implied by the surface wave and refraction data described in the preceding paragraphs would enhance this effect. Griggs (1971) even suggests that the olivine–spinel phase transformation that normally occurs at about 400 km may occur at shallower depths in the sinking lithosphere because of the reduced temperature. Thus thermal contraction as well as phase changes provide a mechanism for raising the density of the lithosphere above that of the adjacent asthenosphere, certainly in the subduction zone where the plate plunges back into the asthenosphere, and probably under the abyssal plains.

Ultimately, the stress system involved in plate tectonics will probably become evident from studies of source mechanisms of earthquakes. These studies based on first motion of P waves and surface wave radiation patterns will reveal the distributions of earthquake-producing stress in the lithosphere. Preliminary results reported by Isacks and Molnar (1969) are consistent with gravitational instability of the sinking lithosphere as a major source of stress. The earthquake mechanisms imply the stress pattern in Fig. 4. Downdip extension predominates where the lithosphere sinks gravitationally through the weak, less dense asthenosphere, exerting a pull on the surface portion of the plate. When the sinking lithosphere reaches the more dense, stronger region below the asthenosphere, downdip compressions seem to be evidenced in the earthquake patterns. More work is needed to substantiate these important results and also to find the pattern of stress release in the ridge-rift source regions.

Is there enough energy in gravitational

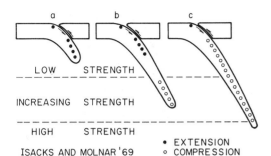

FIG. 4 Stress distributions in slabs of the lithosphere sinking into the asthenosphere and the underlying region of increasing strength. According to Isacks and Molnar (1969) these stress patterns receive experimental support from earthquake mechanism studies. The downgoing slabs pull on the surface portion of the plates until they "hit bottom," decreasing the pull and producing compressive stresses (after Isacks and Molnar, 1969).

instability to drive plate tectonics? Using the expression $4\pi a^2 hd\Delta\rho g/2$ to estimate the energy, where a is the earth's radius, g is gravitational acceleration, $h = 100$ km is the thickness of the lithosphere, $\Delta\rho = 0.2$ g/cm^3 is the density contrast between lithosphere and asthenosphere due to thermal contraction and phase changes, and $d = 200$ km is the thickness of the asthenosphere, we find 10^{35} ergs of available energy. If this is released over 10^8 years, the energy available annually amounts to about 10^{27} ergs, which may be compared to the 10^{25} ergs released annually in earthquakes and the 10^{28} ergs delivered to the surface by heat flow. This comparison with energy release of earthquakes takes on special significance in view of Brune's (1968) result that rates of slip for earthquakes on transform faults and shallow earthquakes in subduction zones agree with rates obtained from magnetic anomalies and geodetic data. It seems as if gravitational instability cannot be neglected as a factor in driving the system. Moreover, the gravitational instability of the sinking plate would exert tensional or compressive stresses (see Fig. 4) acting downdip

of the order of 1 kb in the absence of phase changes and 2 kb if changes have occurred. These values might be reduced by a factor of 2 if viscoelastic support of the slab is taken into account (Smith and Toksöz, 1970). This is more than enough to account for earthquakes.

Physical arguments are often important in eliminating some hypotheses and fostering others. Elsasser (1967), McKenzie (1969), and Jacoby (1970) have made "order of magnitude" calculations testing several hypotheses for the driving forces of plate tectonics. Elsasser and Jacoby argue that the forces exerted by the sinking lithosphere, which are transmitted through the plate, can overcome the viscous drag at the boundary with the asthenosphere. Jacoby also includes diapirism under the ridges as a driving force that is particularly important for plates not associated with a subduction zone. McKenzie admits that gravitational effects cannot be neglected but argues for viscous drag, exerted on the bottom of the lithosphere by convective processes below, as the primary driving force of the largest plate. These differing results warrant a cautious response to calculations that are sensitive to the uncertain composition and rheological and thermodynamic properties of the mantle.

A Petrological Model

The lack of uniqueness of petrological models of the lithosphere at this stage of the development of the subject is best evidenced by the number of different models advanced in recent years by highly qualified investigators (e.g., see Wyllie, 1970). The most that can be done in advancing a particular model is to start with assumptions based on one's own estimate of the most significant and best established experimental data. We select the following as a basis for a petrological model of the lithospheric plate:[1]

---

[1] We define the lithosphere as the strong, high Q, high elastic velocity, relatively young plates which cap the upper mantle, in which continents are embedded and which respond primarily as rigid blocks to the stresses responsible for their relative motions.

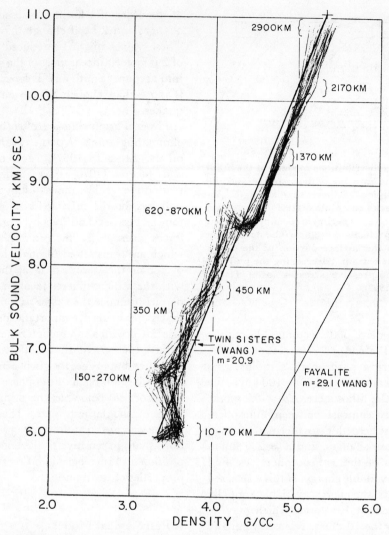

FIG. 5 Bulk sound velocity-density graph of earth models that fit geophysical data. Solid line connecting crosses is Wang's (1968) reduction of shock data for Twin Sisters dunite for which mean atomic weight $m = 20.9$. Note that for the lithosphere (10–70 km) the models imply $m > 21$ and for the region 150–620 km, $m < 21$ (Press, 1970).

1. The boundary between the suboceanic lithosphere and asthenosphere is characterized by an abrupt reduction in shear velocity and Q. This boundary is a solidus. The asthenosphere below is in a state of partial melting (Press, 1959; Birch, 1970; Anderson, 1970; Lambert and Wyllie, 1970).
2. There is insufficient resolution to definitely associate a density change with the lithosphere–asthenosphere boundary. However, a bulk sound velocity-density graph (Fig. 5), based on seismological data (Press, 1970), indicates that $21 \leq m \leq 23$ (m is the mean atomic weight) for the mantle above 70 km and $20 \leq m \leq 21$ for the depths 270–620 km. Since m reflects the Fe/Mg ratio, the lithosphere shows a

higher value for this ratio than most of the upper mantle.
3. The lithosphere is formed in a source region coincident with the oceanic ridge-rift structures from materials coming from below.
4. The M-discontinuity outside of the ridge-rift system is characterized by seismic refraction velocities near 8.0 km/sec. These velocities are appropriate for peridotite, or dunite, but not for eclogite or highly garnetiferous peridotite.
5. An anomalous low-density wedge extends laterally into the lithosphere from the source region. The shape and extent are uncertain (compare Keen and Tramontini, 1970, and Talwani, LePichon, and Ewing, 1965).
6. The unmetamorphosed igneous rocks dredged from the sea floor are a complex suite. For our purposes the rough categorization of basalts, gabbros, and peridotites is sufficient.

We identify the molten fraction of the asthenosphere as basalt because it is the main effusion of oceanic volcanoes and forms the igneous terrain below the sediments of the sea floor. Partial melting from a parent which is predominantly peridotitic produces a magma with augmented Fe/Mg ratio. A mush consisting of basaltic magma dragging residual peridotite rises buoyantly in the ridge-rift source region. The mush contains a larger fraction of basalt than the asthenosphere because the mobility of the liquid is so much greater than the residual peridotite. The liquid reaching the top of the column forms the basaltic ridges and the gabbroic suboceanic crust. The mush in the column adheres to the lithosphere. Large pockets of basalt entrapped in the mush cool as closed systems. The cooling magma forms gabbro and goes through a series of transitions as the plate spreads from the source region, forming pyroxene granulite, garnet granulite, and eclogite. The peridotite fraction undergoes transition to spinel peridotite and garnet peridotite in the cooling, spreading plate. The sequence of the transitions depend on pressure and temperature (hence spreading velocity and thickness) of the slabs. The process may also be rate-dependent and certain transitions may not occur in the cooler, upper lithosphere in the time available before the slab is destroyed in the subduction process. In a personal communication Kennedy (1970) also proposed that rate process and equilibrium boundaries may both be important in the lithosphere. The sequence of lateral variations produces lateral density (and topographic) changes and seismic velocity changes, which may be detected by future geophysical experiments.

Figures 6 and 7 are examples of specific models based on compositions of gabbro and peridotite, respectively (Forsyth and Press, 1971). The phase diagrams for these compositions (Ito and Kennedy, 1971; MacGregor, 1968) are mapped on isotherm models (as in Fig. 2) in constructing the sequences of phase assemblages. The density fronts formed by the phase boundaries may correspond to the anomalous wedges referred to earlier. According to Forsyth and Press the combination of peridotite and eclogite and the use of the wet solidus as the base of the lithosphere offer the advantage of fitting the topographic relief, explaining the peridotite dredged from the sea floor and accounting for the high m, high density, and P-velocity of the lower lithosphere. That the P-velocity at the M-discontinuity matches peridotite is also explained if the transition to eclogite does not proceed at this shallow depth because of rate limitation.

Thus phase transitions transform the lower lithosphere into a more dense phase assemblage than the parent material in the underlying asthenosphere. According to this hypothesis gravitational instability of the lithosphere is a component of a complex convection system involving chemical fractionation and liquid–solid and solid–solid phase transitions. The convective circuit includes rigid body motion of the lithospheric

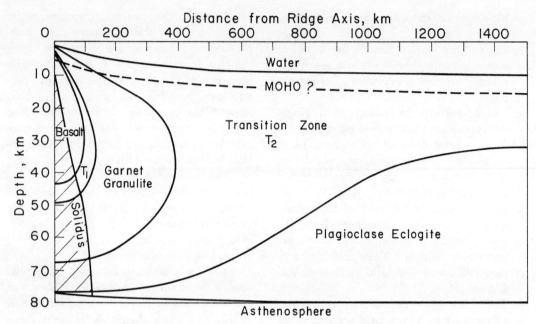

FIG. 6 Model of phase assemblages in a lithosphere based on gabbroic composition, spreading rate 1 cm/yr, and dry solidus. Topographic relief is too large, peridotite dredges and seismic velocities at M-discontinuity are not explained by this model (Forsyth and Press, 1971).

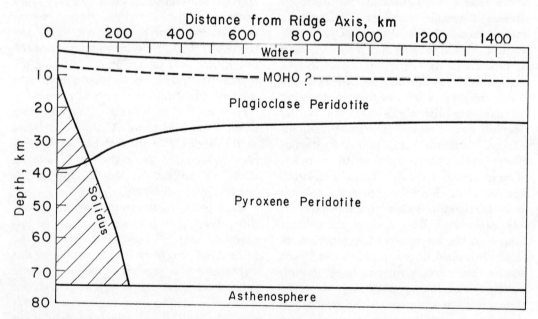

FIG. 7 Model of phase assemblages in a peridotitic lithosphere for spreading rate 1 cm/yr and dry solidus temperature at boundary. High density, high velocity, high m indications for lower lithosphere are not explained by this model (Forsyth and Press, 1971).

plates and fluid flow and creep in the partially molten asthenosphere. It is an interesting question whether the process is a transient one or a steady-state phenomenon.

Hales, Helsley, and Nation (1970) agree with Birch (1970) that the appearance of garnet is important in increasing the density of the lithosphere, and not whether the other principle component is pyroxene as in eclogite, or olivine as would be the case for a garnet peridotite.

Ito and Kennedy (1970a, 1970b) argue for an eclogitic lithosphere and asthenosphere, the boundary between the two regions being set by the solidus of eclogite. In this way they explain those earth models which show a high-shear velocity, high-density lithosphere, and high-density, low-shear velocity asthenosphere. This structure is consistent with geophysical data if the suboceanic mantle immediately below the M-discontinuity is a transition phase between garnet granulite and eclogite. However, there remains the problem of accounting for the topographic change between the ridges and abyssal plains if the lithosphere is predominantly eclogite. The density changes in the basalt–eclogite transformation would require a much larger change in relief than the 4 km ± which is observed.

Ringwood and Green (1966) proposed the basalt–eclogite transformation as the principal driving force for sea floor spreading. This is similar to our hypothesis in many respects but differs in that only crustal basalt transforms to eclogite, and this occurs only in the subduction zone. The gravitational instability of the dense blocks of eclogite drives the advective flow system in the Ringwood–Green mechanism.

Van Bemmelen (1964) anticipated an essential feature in all of these proposed systems: physicochemical processes play a key role in gravity tectonics. He pointed out that convection in the geophysical sense must include compositional and phase changes in a current system involving materials in fluid and crystalline states.

References

Anderson, D. L., 1970, Geophysical evidence on the petrology of the mantle: *Mineral. Soc. Am. Spec. Paper 3*, p. 85–93.

Birch, F., 1970, Interpretations of the low velocity zone: *Phys. Earth Planet. Internat.*, v. 3, p. 178–181.

Brune, J. N., 1968, Seismic moment, seismicity and rate of slip along major fault zones: *Jour. Geophys. Res.*, v. 73, p. 777–784.

Dziewonski, A. M., 1971, Upper mantle models from "Pure Path" dispersion data: *Jour. Geophys. Res.*, v. 76, p. 2587–2601.

Elsasser, W. M., 1967, Convection and stress propagation in the upper mantle: *Princeton Univ. Tech. Rep.*, v. 5, June 15.

Forsyth, D. W., and Press, F., 1971, Geophysical tests of petrological models of the spreading lithosphere: *Abst. 52nd Annual Meeting Am. Geophys. Un.*

Griggs, D., 1972, The sinking lithosphere and the focal mechanism of deep earthquakes, in Robertson, E. C., ed., *The nature of the solid earth: Symposium in honor of Francis Birch:* New York, McGraw-Hill, p. 361–384.

Hales, A. L., 1969a, Gravitational sliding and continental drift: *Earth Planet. Sci. Lett.*, v. 6, p. 31–34.

———, 1969b, A seismic discontinuity in the lithosphere: *Earth Planet. Sci. Lett.*, v. 7, p. 44–46.

———, Helsley, C. E., and Nation, J. B., 1970, P travel times for an ocean path: *Jour. Geophys. Res.*, v. 75, p. 7362–7381.

Isacks, B., and Molnar, P., 1969, Mantle earthquake mechanisms and the sinking lithosphere: *Nature*, v. 223, p. 1121–1124.

Ito, K., and Kennedy, G. C., 1970a, personal communication, June.

———, 1970b, The fine structure of the basalt-eclogite transition: *Mineral Soc. Am. Spec. Paper 3*, p. 77–83.

———, 1971, An experimental study of the basalt-garnet granulite-eclogite transition, in Heacock, J. G., ed., *The structure and physical properties of the earth's crust:* Am. Geophys. Union, Geophys. Mon. 14, p. 303–314.

Jacoby, W. R., 1970, Instability in the upper mantle and global plate movements: *Jour. Geophys. Res.*, v. 75, p. 5671–5680.

Keen, C., and Tramontini, C., 1970, A seismic refraction survey on the Mid-Atlantic Ridge: *Geophys. Jour.*, v. 20, p. 473–492.

Kennedy, G. C., 1970, personal communication, June.

Lambert, I. B., and Wyllie, P. J., 1970, Low-velocity zone of the earth's mantle: Incipient melting caused by water: *Science*, v. 169, p. 764–765.

Langseth, M. G., LePichon, X., and Ewing, M., 1966, Crustal structure of mid-ocean ridges: *Jour. Geophys. Res.*, v. 71, p. 5321–5356.

MacGregor, I. D., 1968, Mafic and ultramafic inclusions as indicators of the depth of origin of basaltic magmas: *Jour. Geophys. Res.*, v. 73, p. 3737–3746.

McKenzie, D. P., 1967, Some remarks on heat flow and gravity anomalies: *Jour. Geophys. Res.*, v. 72, p. 6261–6274.

———, 1969, Speculations on the consequences and causes of plate motions: *Geophys. J.*, v. 18, p. 1–32.

Press, F., 1959, Some implications on mantle and crustal structure from G waves and Love waves: *Jour. Geophys. Res.*, v. 64, p. 565–568.

———, 1969, The suboceanic mantle: *Science*, v. 165, p. 174–176.

———, 1970, Regionalized earth models: *Jour. Geophys. Res.*, v. 75, p. 6575–6581.

Ringwood, A. E., 1969, Composition of the crust and upper mantle: *Geophys. Mon. 13, Am. Geophys. Union*, p. 1–17.

———, and Green, D. H., 1966, An experimental investigation of the gabbro-eclogite transformation and some geophysical implications: *Tectonophysics*, v. 3, p. 383–427.

Sleep, N. H., 1969, Sensitivity of heat flow and gravity to the mechanism of sea-floor spreading: *Jour. Geophys. Res.*, v. 74, p. 542–549.

Smith, A. T., and Toksöz, M. N., 1970, Stress distribution in a downgoing slab: *Am. Geophys. Union Trans.*, v. 51, p. 823.

Talwani, M., LePichon, X., and Ewing, M., 1965, Crustal structure of the mid ocean ridges, 2. Computed model from gravity and seismic refraction data: *Jour. Geophys. Res.*, v. 70, p. 341–352.

Toksöz, M. N., Minear, J. W., and Julian, B. R., 1971, Temperature field and geophysical effects of a downgoing slab: *Jour. Geophys. Res.*, v. 76, p. 1113–1138.

Van Bemmelen, R. W., 1964, The evolution of the Atlantic mega-undation (causing the American continental drift): *Tectonophysics*, v. 1, p. 385–430.

Wang, C., 1968, Constitution of the lower mantle as evidenced from shock-wave data: *Jour. Geophys. Res.*, v. 73, p. 6459–6476.

Wyllie, P. J., 1970, Ultramafic rocks and the upper mantle: *Mineral. Soc. Am. Spec. Paper 3*, p. 3–32.

WOLFGANG R. JACOBY

Seismology Division, Earth Physics Branch
Department of Energy, Mines and Resources
Ottawa, Canada[1]

Current address: Institut für Meteorologie
und Geophysik
Johann Wolfgang Goethe Universität
Frankfurt a.M., W. Germany

Gravitational Instability and Plate Tectonics[2]

The earth's surface is made up of a few large and several smaller plates that are generated at mid-ocean ridges and move away toward deep-sea trenches; here they dip into, and are absorbed by, the mantle (Isacks and others, 1968; LePichon, 1968; Morgan, 1968). In this picture the horizontal motions are predominant; tectonic activity and related vertical motions appear to be simply the consequences of interactions at the plate boundaries.

The question of what drives the plates remains one of the most intriguing problems of the new global tectonics. The following considerations may direct our search for an answer.

It appears ultimately to be gravity that orders and stratifies the earth with the densest materials at greatest depth. Beside gravity, tidal forces and electromagnetic forces appear negligible. On the other hand, thermal expansion, chemical differentiation, and mineralogical phase changes can result in strong forces, capable of disturbing the gravitational equilibrium of the mass distribution within the earth. The total heat flow of about 3×10^{13} W through the earth's surface (Jessop, 1970) may be taken as an indicator of the amount of energy available from these sources. If conduction and radiation cannot efficiently transfer this amount of heat, gravitational instability of the mass distribution and some mechanism of convective heat transfer will result. The actual pattern of motions, however, will depend, not only on gravity and the thermal, chemical,

---

[1] Contribution of the Earth Physics Branch No. 420.
[2] Many discussions with colleagues at the Earth Physics Branch are gratefully acknowledged. The comments and criticism of the manuscript, particularly by C. Wright, R. P. Riddihough, and M. R. Dence, have been very helpful for the presentation of this paper.

and mineralogical energy release, but also on the internal structure and the boundary conditions of the system.

Our present knowledge of the system of motions is restricted to those of the plates. The complete system must involve motions in the mantle apart from the plates. To obtain clues about these unknown motions we must analyze the structure of the mantle.

The analysis attempted in this paper is mainly a mechanical one from the point of view of gravitational instability. After a discussion of the concept of gravitational instability and a review of the structure of the crust and mantle, a description of a mechanism is given that appears to be the consequence of the gravitational instability of the upper mantle and the structural constraints to the possible motions. This analysis is a further development of ideas presented earlier (Jacoby, 1970a, 1970b). In addition, various hypotheses of mantle convection are assessed.

Gravitational Instability

A mass distribution is unstable under two necessary conditions: (1) it has free potential energy, that is, its potential energy is larger than the minimum possible one (Van Bemmelen, 1965a, 1965b); and (2) it is capable of directional movement and rearrangement toward a distribution of lower potential energy (Jeffreys, 1970, p. 425). Both conditions together are sufficient for instability, and motions will necessarily result.

This is illustrated in Fig. 1, showing a relief surface with four identical spheres, instantaneously at rest. Sphere 1 is gravitationally stable because its free potential

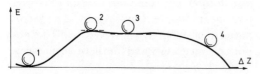

FIG. 1 An illustration of gravitational equilibrium and instability. The four spheres can move freely over the relief surface (thick line).

energy E is zero. Although sphere 3 has an energy E greater than zero, it is stable because a small displacement ΔZ will enlarge rather than reduce E. Only spheres 2 and 4 satisfy both conditions of instability, though in a different way. Sphere 2 has a potential energy E greater than zero, which will be reduced upon a displacement ΔZ in either direction; this is the case of an indeterminate equilibrium, which is commonly considered as unstable. If sphere 4 is not fixed in its position it also clearly satisfies both conditions of instability; generally, this is the type of instability considered in this paper.

It is crucial to keep the *two* conditions of instability in mind when analyzing motions within the earth. The first condition for instability is obvious; there must be, for instance, density reversals with depth or lateral density variations. The second condition is often forgotten; but without mobility, a density reversal or lateral variation is stable. On the other hand, boundary conditions and structural constraints to the possible motions may result in guided motions, whose direction may be very different from vertical.

It is important to ask also where and at what depth gravitational instability may arise or may be inherent in the structures. If the plates are moved by the drag exerted by convection cells in the mantle, the decisive gravitational instability lies in the mantle. If the plates are driven by gravity acting on them and move against drag from below, the decisive gravitational instability lies in the system of the plates and the underlying asthenosphere.

Of great geological importance is the rate of energy release or the "power" of the motions resulting from instability. The *magnitude of the instability* may therefore be defined as the rate at which the free potential energy E is released, or as E/T where T is a typical time constant or the inverse of the "mobility." The magnitude of the instability can, by definition, also be computed as the sum or integral of the scalar product of the inherent forces and the velocities of the particles. Note that the definition is not concerned with the energy source generating the instability, but only with given mass distributions. The magnitude of instability may be a useful concept in comparing various hypotheses of the driving mechanisms of plate motions (see below).

The Structure of the Crust and Mantle

The question of the degree of instability of the crust and mantle can be answered only when their structure—their density distribution and mobility—are known. Although we still lack much needed quantitative information, we do know a few qualitative and even quantitative facts about the upper few hundred kilometers of the mantle; our knowledge about the mantle below about 700 km depth is much more scanty.

The Lithosphere

The moving plates of the lithosphere are about 100 km thick. They are thicker (110–130 km) under continental shields (Walcott, 1970b; Wickens, 1971), thinner (70–80 km) under the ocean basins (Press, 1973), and much thinner in tectonically active regions. The strength of the lithosphere is estimated from strain release in earthquakes and from laboratory measurements to be around 10^8–10^9 N/m^2 or 1–10 kb (Griggs and others, 1960). The lithosphere consists of crystalline rocks and transmits seismic waves with relatively high velocities and low attenuation. Loaded by sediments or ice sheets, it bends like an elastic sheet over a fluid substratum, but with time the bending increases by viscous response to the stresses, the viscosity being about 10^{23} N-sec/m = 10^{24} P (Walcott, 1970b; Artyushkov, 1971). The crust–mantle boundary or Mohorovičić discontinuity is a dominant petrological boundary only within the lithosphere.

The lithosphere appears on average to be denser than its substratum, particularly under the oceans and tectonic regions. In some areas the density inversion at the bottom of the lithosphere seems to be quite sharp. This topic is discussed in detail by Press (1973).

The Asthenosphere

The asthenosphere forms a layer of about 200 km thickness below the lithosphere. Its bottom is poorly defined and certainly not sharp. The temperature probably approaches or surpasses the solidus in this depth range so that the asthenosphere may be partially molten. The asthenosphere is "fluid" in a general sense, if compared to the lithosphere. It attenuates S waves and is generally a low-velocity layer for S waves and in places for P waves as well (Press, 1973). It appears to yield under small, long-term loads. The stress (σ)-strain rate ($\dot{\epsilon}$) relation is difficult to determine. On theoretical grounds and from laboratory experiments with ultrabasic rocks (Carter and Avé Lallemant, 1970), a power law for the steady-state creep ($\dot{\epsilon} = A\sigma^n$, where A = constant and $n \approx$ 3–5) is favored. For simplicity, a linear law, that is, Newtonian viscosity ($\dot{\epsilon} = 1/\eta$, where η = viscosity) is commonly assumed which may fit the observations reasonably for a limited range of stresses. The most recent values of the effective viscosity η of a 200-km asthenosphere below a lithosphere loaded by ice are of the order of 10^{19} N-sec/m^2 or 10^{20} P (Artyushkov, 1971; Lliboutry, 1971). As pointed out earlier, the mean density of the asthenosphere, or part of it, appears to be lower than that of the overlying oceanic lithosphere.

The Mesosphere

The mesosphere below the asthenosphere is defined in this paper to extend to about 700 km depth. Two or more transition layers with rapid velocity increases of P and S waves occur between 300 and 700 km. The transitions appear to be sharp enough to reflect seismic waves (Whitcomb and Anderson, 1970). The transitions may be related to mineralogical phase changes (Anderson and others, 1971). Seismic waves appear to be less attenuated than in the asthenosphere above; the creep strength and the density appear to increase with depth in the mesosphere.

The Transition Zone and the Lower Mantle

The transition zone, centered at about 700 km depth, is considered the boundary between the mesosphere and the lower mantle extending down to the core. It is likely that the lower mantle has a chemical composition different from that of the upper mantle (Press, 1970). The effective viscosity of the lower mantle has been estimated to be about 10^{25} N-sec/m^2 (10^{26} P) on the basis of the nonhydrostatic oblateness of the earth's shape (Kaula, 1967). However, the estimate may be two orders of magnitude too high if it is true that the rotational axis moves through the earth's body in response to variations of density inhomogeneities in the mantle, as suggested by Goldreich and Toomre (1969).

Ocean Ridges

The crust and upper mantle structure of the ocean ridges is anomalous. It is probable that the lithosphere thins toward the ridge crests, and that the asthenosphere consequently forms an upward-pointing wedge. This is plausible if the lithosphere is generated at the ridge by cooling and solidification of the rising asthenosphere, and then moves away. This interpretation of the ridge structure is supported by the rise of the crustal discontinuities toward the crest (Ewing, 1969; Talwani and others, 1965), by the strong attenuation of S_n waves traveling in the uppermost mantle across or along ridges (Molnar and Oliver, 1969), and by the delay of seismic arrival times at ridge stations such as in Iceland. The topography and high heat flow are also consistent with this picture. Although the model of a thin lithosphere at the ridge crests is perhaps most plausible, other models have been proposed, such as purely thermal expansion of the lithosphere at the ridges with dikes near the crest (McKenzie and Sclater, 1969).

Lateral density variations must occur within the upper mantle at the ridges since the topographic mass surplus of the ridges is nearly compensated by a mass deficiency at depth. If we accept the existence of an upward-pointing wedge of asthenosphere, then it probably represents the dominant mass deficiency with a density contrast of about -0.05 g/cm^3 (Jacoby, 1970b). This supports Press's (1973) conclusions regarding a possible density inversion at the bottom of the lithosphere, although it does not necessarily prove them. The cause of the density contrast is probably the temperature difference between the rising asthenosphere and the plates, resulting not only in thermal expansion or contraction but also in chemical reactions and phase transformations, as discussed by Press (1973). Free air gravity anomalies of the order of $+10$ mgal appear to occur over the crestal areas of the ridges, whereas the anomalies over the deep sea basins are usually some -10 to -40 mgal (Kaula, 1972).

Deep Sea Trenches and Island Arcs

The crust and upper mantle structure is also anomalous near the deep sea trenches, where lithospheric plates bend down and dip into the deeper mantle to about 700 km maximum depth or the bottom of the mesosphere. On average the dip is about 45°, but there are considerable variations between 10 and 90°. As the cool dipping plates are heated by the hotter surrounding mantle, they are weakened but remain brittle enough to fracture along internal faults under apparent compressive or tensional stresses, mostly aligned with the direction of dip. Some plates appear to be under tension or compression nearly throughout, others appear to be under tension in the higher parts and under compression in the deeper parts with a relatively stress-free zone between (Isacks and Molnar, 1971).

The density of the cool dipping plate is likely to be higher than that of the surrounding mantle, particularly if oceanic lithosphere is generally denser initially than the asthenosphere. Phase transformations such as the basalt–eclogite transformation will enhance the density contrast and are therefore important for the sinking of the dipping portions of the plates, as proposed by Ringwood (1969) and Ringwood and Green (1966).

Above the area where the descending plates intersect the asthenosphere, a hot low-density zone seems to rise to close to the surface. It is marked by an extremely high attenuation of S waves (Barazangi and Isacks, 1971). Some secondary sea floor spreading may occur here (Karig, 1970, 1971). Volcanism and high heat flow in island arcs are related to this phenomenon. The free air gravity anomalies are generally positive in a broad band around the trenches, while a narrower gravity low follows the trenches themselves. This picture is compatible with the high-density dipping plates and the overlying low-density material (Hatherton, 1969a, 1969b; Karig, 1971).

Lithosphere–Asthenosphere "Convection"

If the preceding description of the upper mantle structure is basically correct, both conditions for gravitational instability—density variations and mobility—are satisfied, particularly for the mass distributions at ridges and near trenches. The most important density contrast appears to be that between the plates and the asthenosphere. The horizontally layered structure of the upper mantle, where not disturbed near ridges and trenches, by the rigidity of the plates and by the fluidity of the asthenosphere may guide predominantly horizontal motions. Downward components of motion of the dense plates at the ridges and the trenches and rising of the low-density asthenosphere at the ridges and behind island arcs must be inferred. It appears therefore inevitable to conclude that the gravitational instability of the lithosphere–asthenosphere system must be at least a partial cause for the motions.

In this chapter a special kind of convection of the lithosphere–asthenosphere system is described, based on the foregoing assumptions regarding the structure of the crust and mantle. This convection appears to develop from the mechanical, thermal, chemical, and mineralogical conditions and constraints of the system. It is shown that this convection may explain many features of plate tectonics satisfactorily.

Diapirism and Gravity Sliding

The gravitational instability of the lithosphere–asthenosphere system expresses itself in active diapirism of the "fluid" low-density asthenospheric wedge into the receding and thickening dense plates of the lithosphere at the ridges and behind island arcs, and in gravity sliding of the plates away from the ridges (Ramberg, 1967, 1973; Seyfert, 1967; Maxwell, 1968; Wilson, 1969; Hales, 1969; Jacoby, 1970a, 1970b). These diapiric phenomena are not the same as the deeper mantle plumes or diapirs proposed by Morgan (1971) and Wilson (1972).

The diapirism envisaged here will occur, as long as sufficient energy is available to maintain the volume and the low density and viscosity of the asthenosphere, and as long as the mechanical constraints to the plates allow any lateral spreading. If the constraints do not change drastically, the process has a tendency of self-perpetuation; it takes place where the lithosphere is weakest, that is, preferably below the existing ridges, whether they migrate or not. In other words, the process does not depend on anomalous conditions at certain places in the asthenosphere or deeper. The question how and where a ridge may originate cannot yet be answered on the basis of definite observations.

The supply of light asthenosphere everywhere and similar conditions of cooling on either side of a mid-ocean ridge will tend to generate a symmetric ridge structure. If anomalous conditions do occur at normal depths of, or below, the asthenosphere, such as unusual amounts of heat stored or produced, or an unusual chemical composition of the material, then they will influence and modify the asthenospheric diapirism. Similarly, differences in the plate properties, such as exist between oceanic and continental parts, will also influence the process.

By the cooling and freezing of the asthenosphere the lithospheric plates thicken away from the ridge and mass is transferred from the asthenosphere to the lithosphere. No lasting static equilibrium is reached, in contrast to salt or granitic diapirs which rise only as long as the supply lasts or remains mobile. The tendency for an equilibrium at the ridge is in competition with the effects of cooling, differentiation, and freezing to the plates. An episodic, cyclic, or pulsing behavior is likely to arise, as frequently observed in nature. It may involve periods of relative quiescence and predominant cooling, differentiation, and freezing of the crestal parts of the asthenospheric diapir (Fig. 2, stage 2). In this period the conditions for a renewed buoyant rising of the diapir will grow until it commences again to rise. This will raise the crestal parts of the plates above their own position of equilibrium (Fig. 2, stage 3). To restore the equilibrium the plates will slide down over the relatively steep slope of the upward-pointing wedge of the asthenosphere (Fig. 2, stage 4), as suggested by Hales (1969) and Jacoby (1970a, 1970b). The last two stages are accompanied by earthquakes, earthquake swarms, and volcanic eruptions.

The length of the cycles is difficult to estimate; it is probably variable and may be evident in the periodicity of earthquake swarms and of volcanic eruptions. The cycles are probably much shorter than the lengths of the reversals of the earth's magnetic field of the order of 10^6 years, since there appears to be no obvious interference of the spreading irregularities in the record of the linear magnetic anomalies of the oceans. This fact suggests 10^5 years as an upper limit for the lengths of the spreading cycles. Over many cycles the process of spreading may remain in a relatively steady state on the average for some 10^8 years, modified only by the changing conditions at all plate edges and by the changing size of the plates. These conditions change slowly until the collision of continents may cause an abrupt and drastic change.

The episodic nature of the diapirism and gravity sliding makes the process most effective because intermittent sliding can occur over the steep slope of the asthenospheric wedge, whereas in a truly steady state the plates would move only a few kilometers down across a distance of about 1000 km, a very gentle slope. "Gravity megatectonics" (Ranalli, 1971)—the gravity sliding of the plates—can be effective only in this intermittent way, interrupted by episodes of uplift and growth of the plates, renewing the possibility of steep sliding. A fixed undulation of the plate bottom with sufficient amplitude cannot be used for sliding over large distances, because the surface of the plate would move

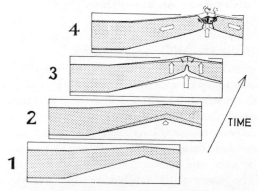

FIG. 2 A cycle of ridge activity in four stages: (1) assumed starting point; (2) cooling, thickening, and densification of the lithosphere; (3) increased diapirism, uplift of the plates; (4) gravity sliding back into or beyond equilibrium position, earthquakes, volcanism. The movements are schematic and their magnitude is exaggerated.

the same way with large vertical displacements.

The cyclical nature of the diapirism and the gravity sliding is proposed here mainly for mechanical reasons. A similar mechanism of sea floor spreading has also been considered by Knopoff (1969); however, the present proposal is different by considering the momentum of the moving plates as completely negligible (Jacoby, 1970b). The cyclical nature is supported by the evidence for ultramafic layered intrusions under the ridge crest that require relatively quiet conditions of formation and differentiation (Aumento and Loubat, 1971).

A discussion of a number of special aspects may shed light on the mechanical, thermal, chemical, and mineralogical conditions, determining the spreading of the plates away from the asthenospheric upwelling under the ridges.

The ridges are composed of rather straight segments, offset along transform faults which are up to several hundred kilometers long. The ridge segments and transform faults are usually perpendicular to each other and the spreading direction is parallel to the transform faults unless some internal deformation of the plates takes place. If a change in spreading direction is forced upon the plates, the transform faults and the ridge segments adjust their direction, as demonstrated in the northern part of the East Pacific Rise (Fig. 3). This way also the gravitational forces on the sliding plates become again aligned with the spreading direction. Another way in which transform faults influence the plate motions is through resistance by friction and intermittent locking.

If the lateral spreading of the plates from a ridge is stopped forcefully, for example, at their opposite edges, the diapirism must stop also; the asthenospheric wedge must cool in this case and will be transformed into normal oceanic lithosphere. This may have happened in the Labrador Sea and in Baffin Bay, while the spreading of the North Atlantic commenced toward the end of the Cretateous (Dietz and Holden, 1970). Anomalous conditions of the asthenosphere or the meso-

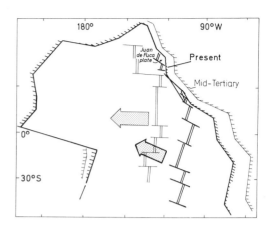

FIG. 3 Spreading in the Pacific. Lighter lines: mid-Tertiary; heavier lines: present. The change in spreading direction since mid-Tertiary time was perhaps caused by the shifting of the active spreading center to the south upon destruction of the northern part of the East Pacific Rise. The direction of the transform faults was changing accordingly.

sphere, or both, may have been important, too, as suggested by the large amounts of basalt that erupted along an arc from Davis Strait to Scotland. A general westward drift of the plates might have brought the anomalous mantle from a position west of Greenland to east of it, thus influencing the rearrangement of the spreading pattern (Fig. 4).

If the plates can move apart but are driven only by the diapirism and gravity sliding from the ridge, the spreading is relatively slow and the lithosphere will be relatively thick at the crest. If, on the other hand, the plates on either side of the ridge, or one of them, can move away easily or may be pulled away (see below), the spreading is rapid and the crestal thickness of the plates is small because there is less time for cooling; in this case the filling of the space between the receding plates may occur not so much by forceful diapirism as by passive emplacement. Median rift valleys typical only of slowly spreading ridges, such as the mid-Atlantic and Indian Ocean ridge systems, but not of rapidly spreading ridges, such as the East Pacific Rise, probably reflect the differences in crustal lithospheric thickness.

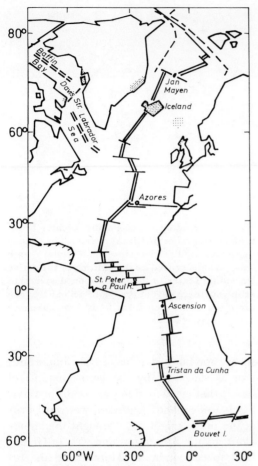

FIG. 4 Spreading in the Atlantic. Between Greenland and Canada spreading started in pre-Tertiary time; spreading between Europe and Greenland commenced during Tertiary time and the mid-Atlantic ridge moved westward while the ridge in Baffin Bay and the Labrador Sea became extinct. The stippled areas indicate Tertiary (in Iceland also Quaternary) basalts; if Europe remained fixed with respect to the deeper mantle, Davis Strait with its Tertiary basalts moved away from the present location of Iceland, suggesting the presence of anomalous mantle at this location. Other less distinct mantle anomalies are suggested by the volcanic islands shown.

The diapirism may also vary along a single ridge from a very active, powerful one to a more passive, filling ridge. Major volcanic islands along a ridge (Jan Mayen Island, Iceland, the Azores, St. Peter and Paul Rocks, Ascension Island, the Tristan da Cunha Group, and Bouvet Island on the mid-Atlantic Ridge; see Fig. 4) may be the expression of stronger diapirism and uplift than along the intervening ridge. The variation could be imposed on the lithosphere–asthenosphere system by variations of mantle temperature or composition or by deeper mantle plumes or diapirs (Morgan, 1971; Wilson, 1972).

An asthenospheric wedge can rise buoyantly only into denser, that is, oceanic lithosphere and is ineffective in continental crust. The East African rift system is an example; it has obviously been less active than the mid-Atlantic and Indian Ocean ridges on either side. A continental rift will develop into a true oceanic ridge preferably if the rift is the extension of an active oceanic ridge, as the Red Sea is an extension of the Carlsberg Ridge and the Gulf of California is an extension of the East Pacific Rise. The history of East Africa, with rifting episodes as far back as Precambrian time (Dixey, 1956) but without appreciable lateral spreading, must be contrasted with the relatively short history of spreading and subsequent extinction of some oceanic ridges, such as the Labrador Sea and Baffin Bay. This difference can probably be attributed to the different thermal and compositional conditions under oceans and continents. A continent with its concentrated heat sources (Roy and others, 1968) may act like a heating blanket, particularly if it remains over the same part of the mantle for a long time; that is, it may keep a hot asthenospheric wedge in existence without much active rising. Under the ocean the asthenosphere will cool more rapidly if it is prevented from rising.

Subduction

Sinking or subduction of the heavy plates near the trenches into the deeper mantle is the other major expression of the gravitational instability of the lithosphere–asthenosphere system (Elsasser, 1967; Jacoby, 1970b;

Isacks and Molnar, 1971). The mode of sinking is controlled by the mechanical, thermal, and compositional conditions of the system. The temperature, the mineral phases, the density contrast, and the rate of weakening and absorption of the plate depend on the rate of heating, and therefore on the velocity of sinking. Mechanical constraints to the motions are guiding them.

The deeper a plate sinks into the increasingly rigid mesosphere, the stronger are the resistances to the motion. The maximum depth appears to lie at about 700 km. Plates that extend to this depth are always under downdip compression, some nearly completely, others only in their deeper parts (Isacks and Molnar, 1971); they are more or less supported by the mesosphere and can therefore add only part of their sinking force to the driving of the horizontal plate motions. Lateral motion of the descending plates is inhibited even more than motion in the dipping direction. This may result in contortions of the deeper parts of the plates (Fig. 7).

The sinking force also depends on the original thickness, and hence on the age of the plate and the distance from the ridge. Fast sinking may decrease the distance from the ridge and the force of sinking. This seems to have happened when the East Pacific Rise met the trench that existed during the Tertiary along the west coast of North America (Fig. 3; see also Atwater, 1970); the small Juan de Fuca plate relict is hardly sinking beneath North America now; this plate seems to grow as the active Gorda Rise, Juan de Fuca Ridge, and Explorer Ridge migrate northwestward (Berry and others, 1971).

Continental lithosphere cannot sink into the mantle, since the light continental crust riding on a descending plate will create a buoyant uplift that will stop the sinking. Thus the collision of a continental block with a trench will lead to a major readjustment of the plate motions (McKenzie, 1969), involving changes of spreading directions and the creation of new trenches and ridges.

The plates appear to remain essentially coherent during sinking instead of breaking into many individually moving pieces. This is suggested by their remarkably regular, gently curved geometry, which is close to that of an intact, elastically downward-bent spherical shell (Fig. 5a). This is also the explanation for the straightening of the plates in dip after bending beyond the elastic limit. Sinking of individual blocks with regular velocities is therefore not required as an explanation for the straightening (Fig. 5b; see also Lliboutry, 1969). At the edges of, or in the corners between, some arcs, deviations from the foregoing simple geometry of the dipping plates result in distortions of the stress pattern; however, the seismicity seems only slightly influenced by them (Isacks and Molnar, 1971).

The coherency of, and the stresses in, the dipping plates suggest that they are coupled with their horizontal parts. This means that the sinking can partly drive the horizontal motions. If uncoupled, the sinking parts could only indirectly and less efficiently aid the plate motion by displacing the substratum and creating a flow that would rise again under the ridges. The rapid spreading of the East Pacific Rise, however, indicates that the sinking of the plates along most margins of the Pacific indeed adds to the driving force directly.

The coherent dipping plates cut off and channel the flow of the asthenosphere, thus

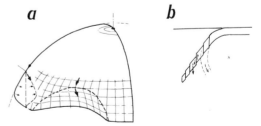

FIG. 5 The shape of the descending plate: (a) three-dimensional model of a downbent spherical elastic shell (arrows indicate plate motion); (b) two-dimensional model of blocks, individually sinking with uniform velocity (Lliboutry, 1969). Model a is the more likely explanation for the shape of the descending plates.

constraining the motions of the asthenosphere and the plates. Without this constraint gravity would pull some plates more steeply down than observed, for example, below the Andes (see below and Fig. 9), or the inclined plates would sink vertically (Fig. 6a) causing the trench to migrate oceanward and the asthenosphere and mesosphere to flow in the same direction. Instead, an asthenosphere that is restricted in its flow may act like a more or less fixed "roller" around which the plate bends (Fig. 6b). The guiding force from the roller on the bending plate is reflected in the depression of the surface below the equilibrium level, and in a corresponding negative free air gravity anomaly over the trench (Hatherton, 1969a, 1969b), superimposed on a broad positive anomaly.

At the lateral edge of a sinking plate, asthenospheric and perhaps also mesospheric matter may be free to flow around the plate edge, thereby reducing the constraint against trench migration. Diapirism may create new crust behind the arc. The flow may contort the edge and bend the plate. There are several examples for such a behavior of lateral edges: the northern end of the Tonga arc (Fig. 7), the eastern end of the Sunda arc, and perhaps the northern ends of the plates sinking beneath South America (Colombia), the Caribbean, and the South Sandwich arc (Isacks and Molnar, 1971).

Return Flow

The description of the lithosphere–asthenosphere "convection" is incomplete without a discussion of the return flow that must occur to balance the mass transport by the moving plates. We do not know, however, how deep the flow may reach and can only speculate about it at present.

Extent to great depth could lead to mantle convections, very similar to those proposed previously by many authors (discussed later), but with a reversed driving mechanism. The driving instability would lie in the lithosphere–asthenosphere system and the deeper currents would be driven by it. In the previous hypotheses exactly the opposite is assumed. Orowan (1969) has proposed a return flow by *en block* movement of the lower rigid mantle; the lithospheric and assumed asthenospheric mass transport toward the shrinking Pacific would be balanced by downward displacement of the lower mantle and core away from the Pacific hemisphere; however, to keep the heavy core centered, flow of mantle matter across the core must be postulated.

The asthenosphere may be the dominant channel for the return flow because of its high fluidity (Wilson, 1969; Jacoby, 1970b). In a simple two-dimensional picture the asthenospheric flow would be opposite to the plate

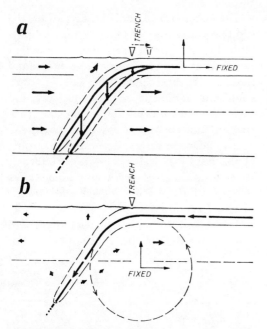

FIG. 6 The sinking of the plates: (a) purely vertical motion of all parts of the inclined plate; it requires a large transport of asthenosphere and mesosphere, plus the plate matter absorbed by them, to the right, but no horizontal motion of the plate (coordinate system fixed in right-hand plate); (b) plate motion around fictitious roller with restricted motions of asthenosphere and mesosphere; only absorbed plate matter is transported (coordinate system fixed in mesosphere). The motion of the coordinate system with respect to the deeper mantle is of physical significance.

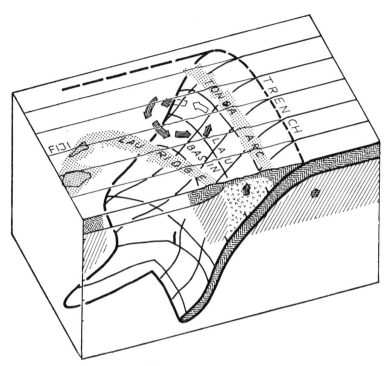

FIG. 7 Block diagram of the northern edge of the Tonga descending plate. Note the bending of its higher parts. The arrows indicate the hypothetical asthenospheric flow from behind (below) to the front of (above) the plate (after Isacks and Molnar, 1971).

movement and balance its mass transport exactly. In the real earth the motions are more complicated; the flow can spread laterally and is influenced by variations of the channel width. At least in the regions of subduction where the plates are absorbed by the mesosphere, subasthenospheric flow must occur. Because of the increased rigidity of the mesosphere the flow is probably very slow and distributed over a broad area.

Figure 8 illustrates qualitatively the suggested flow.

Flow of the asthenosphere over large horizontal distances requires a negative pressure gradient in the direction of the flow. Neither the topography of the ocean basins nor the gravity field reflects such a pressure gradient in an obvious manner, except perhaps in the vicinity of the subduction zones where broad positive gravity anomalies occur

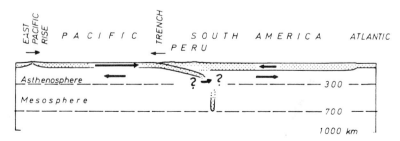

FIG. 8 The plate descending beneath Peru. Restrictions of transport of the asthenosphere and westward advance of South America may require influx of asthenosphere from Pacific into Atlantic regime of motion (after Isacks and Molnar, 1971).

(Kaula, 1972). It is suggested that a gradual thickening of the aging lithosphere away from the ridge and a corresponding widening of the asthenospheric channel toward the ridge may provide the pressure gradient (Fig. 8).

The flow in the channel and the plate motions meet resistance because of the shear strain in the marginal zones of flow (Fig. 8, center). Thus in the present model the plates move against the drag from below, whereas in the deep-mantle convection hypotheses the flow is dragging the plates.

The cutting-off and channeling of the asthenospheric flow by the dipping plates must also influence the global pattern of plate motions. For example, the opening of the Atlantic and the advance of South America toward the closing Pacific require flow of sublithospheric matter toward the Atlantic. This may explain the unusual geometry of the plates dipping beneath South America (Fig. 9; see also Isacks and Molnar, 1971), the eastward migration of the East Pacific Rise relative to South America, and the eastward drift of parts of the eastern Pacific plate between the Americas and South America and Antarctica (Deuser, 1970).

Forces Acting on the Moving Elements

A component of gravity acts in the direction of the plate motions at the ridges and in the subduction zones. The plates are under compression from the ridges and under tension from their sinking ends. Gravity acts against the upward motion of the asthenospheric wedge at the ridges and of mesospheric and asthenospheric matter around the sinking plates. This, of course, does not contradict their buoyancy; because of the density contrast there is a net driving force in the direction of the motions. In principle the system is in a state of convection.

The differential motions in the system meet resistances increasing with the relative velocities. The sources of resistance are the return flow in the asthenosphere and mesosphere, the dipping of the plates into the increasingly rigid mesosphere, the bending of the plates near the trenches, and the friction along transform and other faults. If the dipping plate carries continental crust, the buoyant uplift of the light mass in the subduction zone can resist the sinking of the plate completely.

A quantitative estimate of the gravitational driving forces and the resistances, with reasonable model parameters (as discussed previously), is an important test of the significance of the proposed lithosphere-asthenosphere convection. Such an estimate (Jacoby, 1970b) has shown that the model is physically possible. The forces acting on the assumed model appear to be balanced when the plate velocities are of the order of several centimeters per year, as observed.

The model may also explain the differences in the spreading rates of the ridges qualitatively. The plates on either side of the Atlantic and Indian ridges are spreading slowly (order of 1 cm/yr), driven almost exclusively by the diapirism and gravity sliding at the ridges. The plates on either side of the East Pacific Rise are spreading fast, driven not only from the rise but also by the sinking

FIG. 8 Qualitative illustration of proposed return flow through the asthenosphere and mesosphere (near the descending plate). The arrows indicate the direction and the velocity. The central part of the figure shows the velocity distribution of the motions with depth.

nearly all around the Pacific. Ultimately, not only the mechanical driving forces and resistances but also the thermally and mineralogically controlled rates of plate generation and destruction will influence the velocity of spreading. The importance of these rates grows with the velocity of spreading.

The compressive force from the ridges on the plates is demonstrated by a number of observations: the downdip compression of some of the descending plates throughout; the east–west compression, perpendicular to the East Pacific Rise, indicated by seismic records of an earthquake in the middle of the Nazca Plate west of Peru (Mendiguren, 1971); the occurrence of chains of marine volcanoes perpendicular to the ridge which tend to form perpendicularly to the minimum compression, hence parallel to a compressive stress (Walcott, 1970a, 1970b; Jacoby, 1970b); perhaps the seismic anisotropy with the maximum P velocity perpendicular to the ridges (Keen and Tramontini, 1970; Keen and Barrett, 1971); and the change of the spreading direction from westerly to northwesterly in the Pacific since the mid-Tertiary, when part of the northern East Pacific Rise ceased to exist on meeting North America and was replaced by the San Andreas transform fault (Atwater, 1970), thus shifting the center of the compressive force toward the south (Fig. 3).

Energy Available

Thermal and other forms of energy to maintain the process seem to be amply available. The mechanical energy required by the present mechanism is about one-tenth of the total heat energy flowing through the earth's surface, and one hundred times the global energy release of earthquakes (Jacoby, 1970b). Press (1973) found the same relation by assuming that the potential energy of the dense plates above the less dense asthenosphere is released in about 100 million years.

If the mechanism is a kind of convection, it is to be expected that it should remove a considerable portion of heat from the earth's interior. This seems indeed to be the case, since the heat loss in the process of plate generation may be as high as 45 percent of the total average heat flow (Sclater and Francheteau, 1970).

The ultimate source is not important in the present discussion. However, we may assume that is is partly thermal and chemical energy stored in the earth, partly radiogenic energy, and partly rotational energy, dissipated by the action of the tides (Shaw, 1970; Christoffel, 1969).

Lithosphere–Asthenosphere "Convection" Versus Deeper Mantle Convection

Whole-mantle convection currents of some regular pattern are assumed to govern the surface features of the earth. The hypothesis was clearly expressed as early as 1889 by Fisher (p. 375–381) and was strongly supported by Holmes (1931) to explain Wegener's (1922) continental drift. A possible regular pattern of whole-mantle convection is illustrated in Fig. 10. Upon the recognition of moving rigid lithospheric plates, similar mantle convections were called upon to drag the plates from below. The deep sea trenches and the dipping plates are supposed to form with the downgoing mantle currents. The ridges are not necessarily considered to be the places of upwelling deep-mantle currents,

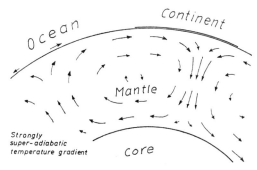

FIG. 10 One possible regular pattern of mantle convection. The regions of upwelling would tend to be surrounded by polygonal belts of sinking flow. The locus of instability is the mantle below the crust.

but they are regarded by some as merely the weakest zones of the plates where they continue to break when pulled apart. In some recent convection models (e.g., Torrance and Turcotte, 1971) variable viscosity in the mantle is assumed resulting in flow concentration and a depth limitation of the convection cells; the plates define rigid boundary conditions in these models. The driving gravitational instability lies in the "overheated" mantle, which is capable of creep.

A different kind of convection would arise if hot, low-density fluids rose as plumes through narrow channels while the relatively rigid mantle compensated the upward flow by slow general sinking (Fig. 11). Morgan (1971) proposed that such plumes rise from the lower mantle and flow out radially in the asthenosphere, where they drag the plates along. Wilson (1972) suggested diapirs in the upper mantle from 300 to 400 km depth, raising the plates and triggering gravity sliding. In Morgan's scheme the driving instability lies in the lower mantle; in Wilson's scheme the driving instability is confined to the upper mantle, so that the localized diapirs only control the lithosphere–asthenosphere system.

What drives the plates? Can we eliminate one or several hypotheses?

Although the effective viscosity of the lower

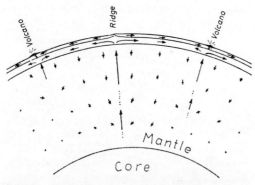

FIG. 11 Deep mantle plumes (Morgan, 1971) or mantle diapirs (Wilson, 1972) are localized vertical structures flowing out radially into the asthenosphere. The return flow would be a very slow sinking along the qualitatively indicated lines.

mantle (see above) may permit convection, the compositional change at about 700 km depth and the sharp phase transition zones make whole-mantle convection highly unlikely (Knopoff, 1967a). There is also little direct evidence for deep-mantle instability; most features of the topography and gravity field are closely related to the crust, the plates, and the asthenosphere. If the magnitude E/T (see above) of deep instabilities is at all comparable to those of the lithosphere–asthenosphere system, the free potential energy E would have to be large because of the large time constant T; the causative density variations and the topographic effects should be detectable. Furthermore, the probable transparency of the lower mantle to thermal radiation may make the development of unstable structures unlikely.

Moreover, whole-mantle convection seems incapable of explaining the migration of the Atlantic and Indian ridges that seems required by the absence of trenches and the presence of the intervening East African rift; this must reduce the area of the Pacific in spite of the spreading of the East Pacific Rise and therefore also requires some of the trenches to migrate. If drag from below is the dominant driving force, one must either postulate a rapid change in the pattern of the mantle convections, which would contradict the estimated time constant of about 10^9 years to set up steady convection (Knopoff, 1967b), or else postulate essential decoupling of the deep-mantle convection from the plate motions in order to allow migration of ridges far from the centers of upwelling mantle. However, the latter postulate defeats the argument for predominance of drag from below and in any case would not explain the migration of trenches. For the same reason, whole-mantle convection, though possibly able to explain the present land–sea distribution (Runcorn, 1969), does not explain how only 250 million years ago all continents could have formed one large Pangea (Wegener, 1922).

Nearly the same arguments are valid against the proposal that lower-mantle plumes drag the plates by their radial outflow into the asthenosphere. The drag, originating from several centers, would, indeed, be less effective than regular mantle convection. Hence lower-mantle plumes do not appear to be important for the driving of the plates by drag.

If localized diapirs exist below the asthenosphere, it is probable that they exert an influence on the gravitational instability of the lithosphere–asthenosphere system, but their importance is difficult to estimate. It seems at least unlikely that a few local subasthenospheric diapirs could fix the position of the ridges over the deeper mantle. More probably, the plates would move so that the ridges would migrate off the deeper diapirs, which would bring about asymmetry of the ridge structure. The symmetry is an argument for the general diapirism of an asthenospheric wedge occurring wherever the ridge migrates. Volcanic islands near junctions of three ridges have been assumed to indicate deep plumes that drive the three plates apart. However, the volcanic activity seems to migrate with the triple junctions, as around Africa, and is therefore perhaps more readily explained as a mechanical consequence than as a cause of the plate motions near the triple junctions.

A driving mechanism of plate motions that lies in the instability of the lithosphere–asthenosphere system appears to explain the plate motions without encountering the difficulties just discussed. If the forces inherent in the lithosphere–asthenosphere system have been estimated correctly and are indeed sufficient to maintain the plate motions, there is no need to postulate deeper-mantle convection. This notion depends crucially on the decoupling of the plates from the deeper mantle and hence on the assumption of a low effective viscosity (about 10^{19} N-sec/m^2 or 10^{20} P) of the asthenosphere. The weak coupling has been supported by Elsasser (1971). On the other hand, McKenzie (1969) has argued for the dominance of drag from below over gravitational forces on the plates, but his conclusion is true only for the model of upper-mantle structure and deeper-mantle convection that he assumed; in this model the thinning of the lithosphere and the diapirism under the ridges as well as the sinking plates were neglected.

Conclusions

The instability of lithosphere–asthenosphere system seems to be the most plausible mechanical cause for the plate motions. We probably know enough about its gross features to make realistic, though crude, models. The assumptions underlying the present model are probably more realistic than those underlying deep-mantle convections.

It may appear strange that the vertical force of gravity may bring about the largely horizontal plate motions. However, the plates and the asthenosphere can only move essentially horizontally, except near ridges and trenches. The gravitational forces on the small parts of the plates and the asthenosphere directly affected by them are apparently sufficient to drive the large parts that move horizontally. In everyday experience we observe something similar: although the winds are predominantly horizontal, they are also driven by gravity and heat and they are guided by the structure of the atmosphere.

References

Anderson, D. L., Sammis, C., and Jordan, T., 1971, Composition and evolution of the mantle and core: *Science*, v. 171, p. 1103–1112.

Artyushkov, E. V., 1971, Rheological properties of the crust and upper mantle according to data on isostatic movements: *Jour. Geophys. Res.*, v. 76, p. 1376–1390.

Atwater, T., 1970, Implications of plate tectonics for the Cenozoic tectonic evolution of Western North America: *Geol. Soc. Am. Bull.*, v. 81, p. 3513–3536.

Aumento, F., and Loubat, H., 1971, The mid-Atlantic ridge near 45°N. XVI. Serpentinized ultramafic intrusions: *Can. Jour. Earth Sci.*, v. 8, p. 631–663.

Barazangi, M., and Isacks, B., 1971, Lateral variations of seismic wave attenuation in the upper mantle above the inclined earthquake zone of the Tonga island arc: Deep anomaly in the upper mantle: *Jour. Geophys. Res.*, v. 76, p. 8493–8516.

Berry, M. J., Jacoby, W. R., Niblett, E. R., and Stacey, R. A., 1971, A review of geophysical studies in the Canadian Cordillera: Can. Jour. Earth Sci., v. 8, p. 788–801.

Carter, N. L., and Avé Lallemant, H. G., 1970, High temperature flow of dunite and peridotite: Geol. Soc. Am. Bull., v. 81, p. 2181–2202.

Christoffel, D. A., 1969, Earth tide mechanism for sea-floor spreading (abs.): Am. Geophys. Union Trans., v. 50, p. 672.

Deuser, W. G., 1970, Hypothesis of the formation of the Scotia and Caribbean Seas: Tectonophysics, v. 10, p. 391–401.

Dietz, R. S., and Holden, J. C., 1970, Reconstruction of Pangea: Breakup and dispersion of continents, Permian to Present: Jour. Geophys. Res., v. 75, p. 4939–4956.

Dixey, F., 1956, The East African rift system: Colonial Geol. Min. Resources Suppl. Ser., London, 63 p.

Elsasser, W. M., 1967, Convection and stress propagation in the upper mantle: Princeton University Tech. Rep. 5, p. 223–246.

———, 1971, Sea-floor spreading as thermal convection: Jour. Geophys. Res., v. 76, p. 1101–1112.

Ewing, J., 1969, Seismic model of the Atlantic Ocean, in Hart, P. J., ed., The earth's crust and upper mantle: Am. Geophys. Union, Geophys. Monogr. 13, p. 220–225.

Fisher, O., 1889, Physics of the earth's crust, 2nd ed. London, Macmillan, 391 p.

Goldreich, P., and Toomre, A., 1969, Some remarks on polar wandering: Jour. Geophys. Res., v. 74, p. 2555–2569.

Griggs, D. T., Turner, F. J., and Heard, H. C., 1960. Deformation of rocks at 500 to 800° C: Geol. Soc. Am. Mem. 79, p. 39–104.

Hales, A. L., 1969, Gravitational sliding and continental drift: Earth Planet. Sci. Lett., v. 6, p. 31–34.

Hatherton, T., 1969a, Gravity and seismicity of asymmetric active regions: Nature, v. 221, p. 353–355.

———, 1969b, Similarity of gravity anomaly patterns in asymmetric active regions: Nature, v. 224, p. 357–358.

Holmes, A., 1931, Radioactivity and earth movements: Geol. Soc. Glasgow Trans. 18, p. 559–606.

Isacks, B., and Molnar, P., 1971, Distribution of stresses in the descending lithosphere from a global survey of focal mechanism solutions of mantle earthquakes: Rev. Geophys. Space Phys., v. 9, p. 103–174.

Isacks, B., Oliver, J., and Sykes, L. R., 1968, Seismology and the new global tectonics: Jour. Geophys. Res., v. 73, p. 5855–5899.

Jacoby, W. R., 1970a, Active diapirism under mid-oceanic ridges (abs.): Am. Geophys. Union Trans., v. 51, p. 204.

———, 1970b, Instability in the upper mantle and global plate movements: Jour. Geophys. Res., v. 75, p. 5671–5680.

Jeffreys, Sir H., 1970, The earth, its origin, history and physical constitution, 5th ed.: London, Cambridge University Press, 525 p.

Jessop, A. M., 1970, How to beat permafrost problems: Oilweek, v. 20, p. 22–25.

Karig, D. E., 1970, Ridges and basins of the Tonga-Kermadec island arc system: Jour. Geophys. Res., v. 75, p. 239–254.

———, 1971, Origin and development of marginal basins in the western Pacific: Jour. Geophys. Res., v. 76, p. 2542–2561.

Kaula, W. M., 1967, Geophysical implications of satellite determinations of the earth's gravitational field: Space Sci. Rev., v. 7, p. 769–794.

———, 1972, Global gravity and tectonics, in Robertson, E. C., ed., The nature of the solid earth: Symposium in honor of Francis Birch: New York, McGraw-Hill, p. 385–405.

Keen, C. E., and Barrett, D. L., 1971, A measurement of seismic anistropy in the northeast Pacific: Can. Jour. Earth Sci., v. 8, p. 1056–1064.

Keen, C. E., and Tramontini, C., 1970, A seismic refraction survey on the mid-Atlantic ridge: Geophys. Jour., v. 20, p. 473–491.

Knopoff, L., 1967a, On convection in the upper mantle: Geophys. Jour., v. 14, p. 341–346.

———, 1967b, Thermal convection in the earth's mantle, in Gaskell, T. F., ed., The earth's mantle: New York, Academic Press, p. 171–196.

———, 1969, Continental drift and convection, in Hart, P. J., ed., The earth's crust and upper mantle, Am. Geophys. Union, Geophys. Monogr. 13, p. 683–689.

LePichon, X., 1968, Sea-floor spreading and continental drift: Jour. Geophys. Res., v. 73, p. 3661–3697.

Llibourty, L. A., 1969, Sea-floor spreading, continental drift and lithosphere sinking with an asthenosphere at melting point: Jour. Geophys. Res., v. 74, p. 6525–6540.

———, 1971, Rheological properties of the asthenosphere from Fennoscandian data: Jour. Geophys. Res., v. 76, p. 1433–1446.

Maxwell, J. C., 1968, Continental drift and a dynamic earth: Am. Scientist, v. 56, p. 35–51.

McKenzie, D. P., 1969, Speculations on the consequences and causes of plate motions: Geophys. Jour., v. 18, p. 1–32.

———, and J. G. Sclater, 1969, Heat flow in the eastern Pacific and sea-floor spreading: *Bull. Volcanol.*, v. 33, p. 101–117.

Mendiguren, J. A., 1971, Focal mechanism of a shock in the middle of the Nazca Plate: *Jour. Geophys. Res.*, v. 76, p. 3861–3879.

Molnar, P., and Oliver, J., 1969, Lateral variations of attenuation in the upper mantle and discontinuities in the lithosphere: *Jour. Geophys. Res.*, v. 74, p. 2648–2682.

Morgan, W. J., 1968, Rises, trenches, great faults and crustal blocks: *J. Geophys. Res.* v. 73 p. 1959–1982.

———, 1971, Convection plumes in the lower mantle: *Nature*, v. 230, p. 42–43.

Orowan, E., 1969, The origin of the oceanic ridges: *Sci. Am.*, v. 214, p. 103–119.

Press, F., 1970, Regionalized earth models: *Jour. Geophys. Res.*, v. 75, p. 6575–6581.

———, 1973, The gravitational instability of the lithosphere (this volume).

Ramberg, H., 1967, *Gravity, deformation and the earth's crust, as studied by centrifuged models*: London, New York, Academic Press, 214 p.

———, 1973, Model studies of gravity-controlled tectonics by the centrifuge technique (this volume).

Ranalli, G., 1971, The expansion-undation hypothesis for geotectonic evolution: *Tectonophysics*, v. 11, p. 261–285.

Ringwood, A. E., 1969, Composition of the crust and upper mantle, *in* Hart, P. J., ed., *The earth's crust and upper mantle*: Am. Geophys. Union, Geophys. Monogr. 13, p. 1–17.

———, and Green, D. H., 1966, An experimental investigation of the gabbro-eclogite transformation and some geophysical implications: *Tectonophysics*, v. 3, p. 383–427.

Roy, R. I., Blackwell, D. D., and Birch, F., 1968, Heat generation of plutonic rocks and continental heat flow provinces: *Earth Planet. Sci. Lett.*, v. 5, p. 1–12.

Runcorn, S. K., 1969, Convection in the mantle, *in* Hart, P. J., ed., *The earth's crust and upper mantle:* Am. Geophys. Union, Geophys. Monogr. 13, p. 692–698.

Sclater, J. G., and Francheteau, J., 1970, The implications of terrestrial heat flow observations on current tectonic and geochemical models of the crust and upper mantle of the earth: *Geophys. Jour.*, v. 20, p. 509–542.

Seyfert, C. K., 1967, Dilatational convection as a mechanism for sea-floor spreading (abs.): *Geol. Soc. Am. Program, 1967 Annual Meeting*, p. 200.

Shaw, H. R., 1970, Earth tides, global heat flow, and tectonics: *Science*, v. 168, p. 1084–1087.

Talwani, M., LiPichon, X., and Heirtzler, J. R., 1965, East Pacific rise: The magnetic pattern in the fracture zones: *Science*, v. 150, p. 1109–1115.

Torrance, K. E., and Turcotte, D. L., 1971, Structure of convection cells in the mantle: *Jour. Geophys. Res.*, v. 76, p. 1154–1161.

Van Bemmelen, R. W., 1965a, Mega-undations as cause of continental drift: *Geol. Mijnb.*, v. 44, p. 320–333.

———, 1965b, Der gegenwärtige Stand der Undationstheorie: *Mitt. Geol. Ges. Wien*, v. 57, p. 379–399.

Walcott, R. I., 1970a, Flexure of the lithosphere at Hawaii: *Tectonophysics*, v. 9, p. 435–446.

———, 1970b, Flexural rigidity, thickness and viscosity of the lithosphere: *Jour. Geophys. Res.*, v. 75, p. 3941–3954.

Wegener, A., 1922, *Die Entstehung der Kontinente und Ozeane:* Braunschweig, F. Vieweg & Sohn, 140 p.

Whitcomb, J. H., and Anderson, D. L., 1970, Reflection of $P'P'$ seismic waves from discontinuities in the mantle: *Jour. Geophys. Res.*, v. 75, p. 5713–5728.

Wickens, A. J., 1971, Variations in lithospheric thickness in Canada: *Can. J. Earth Sci.*, v. 8, p. 1154–1162.

Wilson, J. T., 1969, Aspects of the different mechanics of ocean floors and continents: *Tectonophysics*, v. 8, p. 281–284.

———, 1972, Mantle diapirs as a principal driving mechanism and unifying concept in geonomy (in preparation).

R. D. SCHUILING
Department of Geochemistry
Vening Meinesz Laboratory
Utrecht, Netherlands

Active Role of Continents in Tectonic Evolution—Geothermal Models[1]

Much of the present-day attention in tectonics is focused on the exciting developments in global tectonics. This emphasis on plate tectonics tends to obscure the fact that within individual plates a smaller-scale tectonism takes place; this centers around orogenic nodes, where the primary tectonic movement is vertical uplift and subsidence. The strong vertical uplift causes gravity slides from the orogenic centers toward their surroundings.

In this paper a geothermal model is used to explain the formation of such nodes, which are only indirectly linked to former geosynclines and orogenic belts. The preferential location of such nodes in the concave sides of bends in an orogenic belt follows logically from the model.

Before introducing this model we will outline some ideas on the movements of plates, which can also be considered in terms of heating of mantle columns under continents and cooling of oceanic mantle columns.

General Considerations

Orogeny takes energy. The source of this energy and the global distribution of its production should therefore be of prime interest to tectonics. It has been shown (Nieuwenkamp, 1956; Verhoogen and others, 1970) that even in tectonically active regions the energy required for orogeny is only a fraction of the amount of energy produced by radio-

---

[1] I wish to thank Professor R. W. Van Bemmelen for spending so much of his time and thought on improving the ideas set forward in this paper, although large differences of opinion still remain. I also wish to thank my colleagues J. E. Meulenkamp, T. Kwak, and B. J. H. Jansen for their patient help in developing the ideas on the evolution of the Cyclads, as well as Professor H. N. A. Priem, of the ZWO Laboratory of Isotope Geology in Amsterdam, for putting the first geochronological measurements on the Cyclads at my disposal.

active decay of U, Th, and K. The same holds true for volcanism and metamorphism. There is therefore no need to look for any other sources of energy to drive these processes.

An average continental crust of 35 km in a state of thermal equilibrium, consisting of 17.5 km of "granitic" material and 17.5 km of lower crustal material (granulites, anorthosites, charnockites, or similar rocks), produces about 38 cal/cm$^2 \cdot$ yr (see Table 1). This is only slightly less than the total observed heat flow on the continents, which means that the present surface heat flow on the continents is maintained almost exclusively by heat produced in the crust itself. The continents act as "thermal blankets." In oceanic areas the heat flow is very variable, but its average value seems to be only a little less than the average heat flow on land. This is usually explained by theories which maintain that the mantle under continents is differentiated and depleted in U, Th, and K, whereas the mantle under the oceans is undifferentiated and still contains the original amounts of the radioactive elements. Such a hypothesis is in direct conflict with the concept of continental drift and sea floor spreading, and the near-equality of continental and oceanic heat flows has in fact been used as a major argument against continental drift. If continents could move, some would come to rest on top of former oceanic mantles, and ocean floor would form from former, differentiated continental mantles. The differences between oceanic and continental heat flows would then become even greater than they would be if a stable thermal regime were established under continents and oceans situated on top of mantle columns with similar compositions of radioactive elements.

This apparent dilemma was resolved (Schuiling, 1966) when the thermal consequences of removing a continent by sea floor spreading were considered. From a consideration of the cooling curves for this situation, it can be seen that even several hundred million years after a continent has been removed by continental drift and a new ocean has been formed, the surface heat flow in this ocean must still be of the same order of magnitude as on the continents. This is due to the fact that unsteady-state cooling of the mantle contributes to the oceanic heat flow, whereas most of the heat generated in the mantle under the continent at their new locations is used to heat up the mantle and so does not contribute to the surface heat flow.

A Geothermal Model for Horizontal Movements of Plates

Some promising recent models for the driving force of plate tectonics include gravitational

TABLE 1 Heat Production of a Crust Consisting of 17.5 km of "Granitic" Material and 17.5 km of Lower Crustal Material (granulites, charnockites, anorthosites, etc.)[a]

	Upper Crust
U	$17.5 \times 10^5 \times 2.7 \times 4.10^{-6} \times 0.73$ cal = 13.8 cal
Th	$17.5 \times 10^5 \times 2.7 \times 15.10^{-6} \times 0.20$ cal = 14.2 cal
K	$17.5 \times 10^5 \times 2.7 \times 3.10^{-2} \times 27.10^{-6}$ cal = 3.8 cal
	Lower Crust
U	$17.5 \times 10^5 \times 2.9 \times 4.10^{-7} \times 0.73$ cal = 1.5 cal
Th	$17.5 \times 10^5 \times 2.9 \times 2.10^{-6} \times 0.20$ cal = 2.0 cal
K	$17.5 \times 10^5 \times 2.9 \times 2.10^{-2} \times 27.10^{-6}$ cal = 2.7 cal
	Heat production of the crust 38.0 cal/cm$^2 \cdot$ yr

[a] Average content of radioactive elements in granites and heat production per gram of ordinary U, Th, and K, from Clark (1966). Average content of U, Th, and K in lower crustal material from Lambert and Heier (1968).

instability of the lithosphere (e.g., Hales, 1969; Jacoby, 1970, 1973; Press, 1973). These proposals are similar to some earlier ideas by Van Bemmelen (1964). In all these models the plane of separation of lithosphere and deeper mantle is the low-velocity channel, the asthenosphere. Following Press (1973) we may set the average thickness of the oceanic lithosphere at 80 km and that of the continental lithosphere at 160 km. If it is assumed that the low-velocity zone represents the beginning of melting of the mantle, it is clear that at present the temperature in the upper mantle under the oceans must be higher than under the continents, since the melting curve for peridotite (or of any other likely mantle material) has a positive slope on a P–T plot (Fig. 1). It has been argued (Hales, 1969) that a slope of 1:3000 of the upper boundary of the asthenosphere (i.e.,

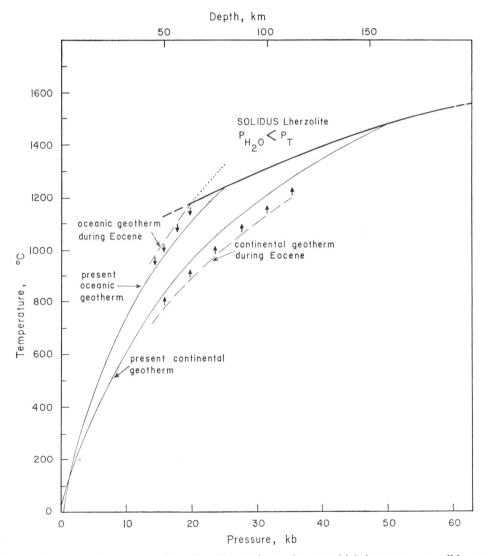

FIG. 1 Inferred positions of continental and oceanic geotherms which intersect a possible mantle solidus (the low-velocity channel) at 160 and 80 km, respectively. If most of the movement of continental lithospheric plates occurred before the middle Eocene, the geotherms must have diverged even more at that time.

the solidus of mantle material) would be enough to set continental drift in motion. Taking the average distance of the middle of a continent to its nearest mid-ocean ridge as 8000 km, we see that the slope of the upper surface of the asthenosphere is 1:100 from the oceans toward the continents. Let us assume therefore that a slope of 1:100 (30 times greater than the minimum given by Hales) is sufficient to provide gravitational instability necessary to move the lithosphere. We will now present a geothermal model which can cause such gravitational instability.

We saw that the annual continental heat flow of 40 cal/cm^2 is accounted for almost entirely by heat produced in the crust itself; this means that heat produced in the underlying mantle is used almost exclusively to heat up the mantle. Most estimates of mantle heat production are between one-half and one-fourth that of the total heat flow. By taking a figure of 20 cal/cm^2 · yr we can calculate that a 200-km mantle column below continental crust will heat up at a rate of about 165° C in 100 million years.[2] Oceanic crust, by contrast, does not account for an annual heat flow of nearly 40 cal/cm^2, and we may estimate that approximately 20 cal/cm^2 · yr is contributed through unsteady-state cooling under the ocean. Consequently, a 200-km column of oceanic mantle will, by a similar calculation, cool at a rate of about 165° C in 100 million years.

If, at any time, the slope of the upper surface of the asthenosphere dips away from the oceans toward the continents, as it does today, heating of the mantle under the continents and cooling of the mantle under the oceans will cause the isotherms to rise under the continents and to descend under the oceans. The slope of the upper surface of the asthenosphere will decrease and will eventually reverse sign. Once a slope of 1:100 of the solidus surface of the mantle has been reached from the continents toward the oceans, a new cycle of continental drift will set in, and the continents will again split apart. In such a way we can visualize that the horizontal movements of the plates are caused primarily by the distribution of the heat sources in the crust, and especially by the difference in heat production of continents and oceans. Since major orogenies in continent-sized belts are almost certainly related to the movement of plates, we see that it is the inhomogeneous distribution of the radioactive elements in the crust that in the final analysis causes and perpetuates the tectonic mobility of the crust.

If it is assumed that a plate moves 5000 km during one cycle and that the average velocity is 5 cm/yr, the average period of active continental drift will last roughly 100 million years. Sufficient heating under the continents and cooling under the oceans to cause inversion of the slope of the upper boundary of the asthenosphere will take another 140 million years, after the convective motion away from the mid-oceanic ridges has stopped. This last figure is obtained on the basis of an apparent temperature difference of the order of 150° C between the suboceanic and subcontinental mantle at a depth of 80 km (Fig. 1) and by assuming rates of heating and cooling of 165° C/100 million years, as discussed earlier. Since the major part of the movement during the last period of continental drift had been accomplished some time during the Eocene ($\sim$ 50 million years ago), it will take another 90 million years (= 140 − 50) to reverse the present slope of the asthenosphere. The general sequence of these events is schematically pictured in Fig. 2.

Many uncertainties enter into this kind of generalized approach. One obvious objection

---

[2] This value is arrived at as follows: taking the specific heat of peridotite as 0.6 cal/cm^3, a 200-km column of peridotite with a cross section of 1 cm^2 will need 120 · 10^5 cal to increase its temperature by 1° C. If the annual rate of cooling under the oceans or of heating under the continents is set at 20 cal/cm^2 · yr, the rate of cooling or heating of a 200-km column will be of the order of 165°C in 100 million years.

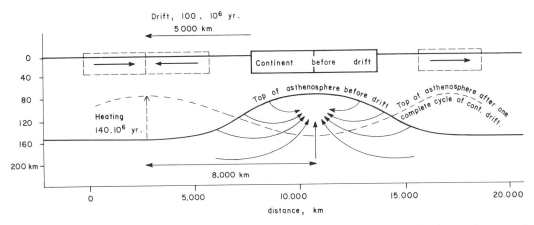

FIG. 2 Schematic view of one complete cycle of continental drift. Heating under the continents and cooling under the oceans takes place during periods of relative immobility, until the gravitational instability built up by these processes is released again in a period of drift.

would be that the latent heat of melting of mantle material in the asthenosphere has been neglected. It can be shown, however, that if the percentage of melt in the asthenosphere is of the order of 1 percent, heating or cooling will be slowed down by only slightly more than 1 million years for a shift of the upper boundary of the asthenosphere by as much as 80 km. We feel therefore justified in neglecting such refinements.

Bearing in mind all uncertainties, we may estimate that a complete tectonic cycle, including the period of active drift with associated orogeny in the frontal parts of drifting plates as well as the subsequent period of stagnation until drift is reversed, will have a period of roughly 240 million years (= 100 + 140). Such a value seems reasonable in the light of the periodicity of orogeny seen in the geological past. The geothermal processes just outlined may constitute the driving mechanism for what has been termed the "pulse of the earth" (Umbgrove, 1947).

Before considering a geothermal model for uplift and subsidence within plates, one obvious question comes to mind. Even if one grants that continents might split apart according to the foregoing geothermal model and that mid-ocean ridges might form along these splits, one cannot help but wonder how certain oceanic ridges came into being at places where there probably was no continent before. The foremost example of this kind is the East-Pacific Rise. Within the framework of the model given here, one might visualize their formation as follows. As a continental lithospheric slab slides down the dip of the low-velocity zone, the mantle at its front end is depressed, and a countercurrent movement of mantle material must take place in order to make up for the mass loss near the existing mid-ocean ridge (the hind end of the slab). This will cause a downwarping of the isotherms below the frontal part of the moving lithospheric slab and therefore a depression of the surface of the asthenosphere. In this way a downward slope of the upper surface of the asthenosphere is also created from the ocean side. This in turn may cause sufficient gravitational instability to set a new sliding system in motion, this time, however, of an oceanic lithosphere. The prediction in this case would be that such oceanic ridges would have a tendency to be asymmetrically arranged with respect to the continents in both directions, and that they generally would originate later than the beginning of the corresponding period of continental drift. There are evidently not enough oceanic ridges to carry out any meaningful statistical analysis, but the prediction seems to fit the general picture.

A Geothermal Model for Vertical Movements within Plates

We have discussed the horizontal movement of plates away from oceanic ridges, caused by the heating up of the mantle under continents, but we have not yet discussed smaller-scale vertical tectonic phenomena that take place within the plates, which may again be caused by differences of heat production.

If we look at the Alpine fold belt in the Mediterranean area, we may distinguish *linear parts* (Central-Eastern Alps, most of the Dinarides, Atlas Mountains, Betic Cordillera) and *arcuate parts* (Western Alps, Carpathians, Circum-Tyrrhenian mountains and the Aegean arc, where the Alpine system swings from a predominantly NW–SE direction into an easterly direction). We will consider first the thermal situation that develops in a linear fold belt, then the geothermal model for the arcuate parts of a fold belt.

Uplift in a Linear Fold Belt

If, as a result of crustal shortening, the continental crust in a linear fold belt has become thicker than in the adjacent cratonic areas, the total heat production in a column under the fold belt will be larger than under adjacent stable areas of normal crustal thickness. Without going into numerical details (see Schuiling, 1969), it can be said that the temperature at the bottom of a continental crust of 40-km thickness is about 200° C higher than at the bottom of a crust of 30-km thickness. The attainment of thermal equilibrium will take time. The radiogenic heat evolved in the crust will be used in part to heat up the sediments and cause the metamorphism of their deeper units, and in part also to heat up the basement and upper mantle. If we restrict the depth of significant heating of the mantle to 100 km,[3] the attainment of thermal equilibrium will take between 10 and a few tens of millions of years, depending on the adopted value for the thermal diffusivity of the mantle. It is mainly the following three effects that cause uplift during this period:

1. Thermal expansion of the column.
2. Phase transitions and anatexis accompanying high-grade metamorphism in the crust.
3. Phase transitions in the mantle and rise of the low-velocity zone.

As erosion proceeds and the column is unloaded, we may add to these an expansion of the material in the column through decompression. For the purpose of calculation we will arbitrarily restrict this decompression effect also to a mantle column of 100 km, although of course decompression will essentially affect everything under the area of unloading.

If we approximate the crust by assuming that it consists mainly of an intermediate plagioclase (the numbers are not very sensitive to great changes in assumed composition), then a heating of 200° C of this crust will lead to a volume expansion of 0.4 percent. By approximating the mantle by an olivine with about 90 percent of the forsterite molecule, a heating of 200° C will lead to a volume expansion of 0.8 percent. Thermal expansion therefore amounts to a vertical movement of 160 m in the crust and 480 m in the mantle.

The contribution of phase transitions in the crust, including anatexis, is difficult to assess but is probably the most important. It is a common observation that in many Alpine systems an earlier low-temperature, high-pressure metamorphic stage (glaucophane schist facies) is followed in time by low-pressure, high-temperature metamorphism.

---

[3] If the time scale of a geological process is of the order of 100 million years, the depth to which a thermal anomaly will make itself felt is of the order of 200 km. If, as in the case now under discussion, the time scale is of the order of 10 million years (or a few tens of million years), thermal effects are significant up to a depth of about 100 km.

Considering that mineralogical transformations from glaucophane schist assemblages to chemically equivalent higher temperature assemblages of the greenschist facies involve a volume increase between 7 and 11 percent (Ernst, 1963), and that melting of granitic material also involves a volume increase of the same order, we will estimate the total contribution of phase transformations in the crust at a 5 percent volume increase. This amounts to a vertical uplift of 2 km in 40-km crustal material.

The phase changes in the mantle are probably restricted in the upper 60 km of the mantle to some transformation of basalt into eclogite near the base of the crust, and transformation of garnet pyrolite into pyroxene pyrolite (Green and Ringwood, 1970). An additional volume increase upon heating is provided by the rise of the surface of the asthenosphere, associated with the beginning of melting of peridotite. Considering that the eclogite–basalt transition is restricted to eclogite pockets in the upper mantle, that the garnet pyrolite toward pyroxene pyrolite transition proceeds with a negligible volume change, and that the amount of melting in the asthenosphere is a small percentage, these factors are unlikely to contribute more than 500 m to the uplift.

We see that the total amount of uplift provided by the preceding effects can be estimated at 3 km or slightly more. If the densities of crust and mantle are set at 2.8 and 3.3, respectively, the combined effects of erosion and in some cases tectonic denudation and compensatory isostatic rise will eventually result in the removal of a total of 17 km of the column. By decompression in this column due to unloading, an additional uplift of 400 m is achieved, which will result in an additional erosion of 2.2 km. (All data on thermal expansion and compressibility used in these calculations can be found in Clark, 1966.)

The total amount of erosion from a linear mountain chain with an original thickening of the crust of 10 km will reach something of the order of 20 km, provided erosion is effective and is much more rapid than thermal equilibration. The interplay of unloading and cooling provides that the normal leveling of a mountain chain will reach to depths of 20 km in its central part by the time it has reached isostatic and thermal equilibrium.

It is clear that this effect, an amount of erosion larger than the original thickening of the crust, is due to the fact that pressure effects are transmitted virtually instantaneously (on the geological time scale), whereas temperature effects are transmitted slowly. Uplift therefore does not stop at the moment the crust has reached the same thickness as the surrounding cratonic areas but *proceeds further*, because in the mantle the heating effects caused by the thickened crust are still felt even after the crust has regained its original "normal" thickness.

Uplift in an Arcuate Fold Belt

We will now consider the particular case in which an otherwise linear mountain belt shows a fairly pronounced bend. The cores of these bends, which we will refer to as orogenic nodes, have subsided in some cases to oceanic depth, even though it can be proved that these same areas were once elevated land masses, in some cases as recent as Miocene in age. The Tyrrhenian Sea, for example, was once a source of gravity slides and a source of sedimentary detritus for its surrounding areas (for a review of these problems see Pannekoek, 1969). The Tyrrhenian Sea is now an ocean-type basin with a maximum depth of 3000 m. The Po basin has been a strongly subsiding area since the Miocene, and thicknesses of the Neogene sediments are of the order of 5 km. The Pannonian basin also contains thick piles of Miocene and younger sediments, whereas it was an elevated area in Alpine time. Similarly, the Cyclad area in the Aegean Sea was until recently a center of uplift from which material was eroded and deposited in its surroundings (Meulenkamp, in press), whereas it is today a drowned mountain area, still undergoing rapid subsidence.

Such arcs in otherwise more or less linear fold belts are favorable locations for the development of centers of younger orogenic movements. It must be remarked here that the bends themselves may owe their origin to the presence of a preexisting tectonic node, already active when the main folding phase took place. In other instances arcs may form as a result of tangential movements along major wrench folds during the main folding phase. In any case, orogenic movements associated with tectonically active nodes are characterized primarily by vertical movements, in contrast to the major fold belts in which vertical movement is preceded by large-scale horizontal translation within the crust.

At the concave side of a curved fold belt the heating effects of both segments of the arc will affect the same column and thus each tends to reinforce the other. At the convex side, on the other hand, the underlying column will be less heated than under a straight fold belt, as the same amount of radiogenic heat is here dissipated over a larger area. The effect is that a thermal blister will form in the node at the concave side, whereas at the convex side the foredeep will be more pronounced than along the straight fold segments. This is in part due to thermal effects directly, but also to the fact that there will be strong foredeep subsidence under depositional load as a consequence of its extensive filling by erosion and gravity slides from the orogenic node. This situation produces its own orogeny, at a later date and of smaller dimensions than the orogeny which produced the linear belt.

Tectonic denudation by gravity slides from the center of the uplift into the foredeep will cause in some cases a fold-arc, which can be more strongly curved than the original curvature in the mountain belt. Since the heating-up and subsequent uplift, erosion, and tectonic denudation of this orogenic node are more pronounced than in the straight segments, which possess otherwise similar crustal thicknesses, it is clear that nodes will become eroded to deeper structural and metamorphic levels than the adjacent straight parts of the mountain chain. For this same reason, however, they will subsequently subside more than the surrounding parts, due to the fact that more radioactive heat sources in the upper crust have been eroded away, thus allowing the upper part of the mantle and the remaining part of the lower crust to cool and shrink more. Deeper parts of the mantle column, on the other hand, are only then reaching their temperature maximum, or their temperature may still be rising, due to the slow transmission of heat in rocks. It is therefore not surprising that the foundering stage of such orogenic nodes is accompanied by the emplacement of shallow intrusions and strong volcanic activity. Both decompression and the slow passage of a "heat wave" will favor the formation of melts. In the Mediterranean we have abundant evidence of volcanic activity accompanying the foundering stage of the nodes, as witnessed by Vesuvius, Stromboli, and Liparic Islands in and around the Tyrrhenian Sea, the many buried sea-mounts in the westernmost part of the Mediterranean Sea, and Milos and Thira (Santorini) in the Aegean Sea.

Once this stage is reached, subsidence can proceed at an accelerated pace, for heat is now lost at a high rate not only because of the loss of radioactive crustal matter, but also owing to the rapid upward convection of magmas. If the process of erosion and tectonic denudation has been very active in an orogenic node, essentially the whole crust may have been stripped off before subsidence sets in; in such a case the result will be a small oceanic basin. This process has been called Mediterranean oceanization (Schuiling, 1969) and may also hold for a number of cases outside the Mediterranean area. Menard (1967) has listed a number of these Mediterranean-type ocean basins, and gives a short description of their properties.

Application of the Geothermal Model to the Cyclad Islands

The Cyclad Islands are situated between the mainland of Greece and Turkey, in the Aegean Sea, north of Crete. Although in recent years their geology has become reasonably well known, much of this knowledge is still available only in unpublished reports of the Department of Geochemistry of the Vening Meinesz Laboratory. Some of these data are summarized here and in Fig. 3.

We can divide the islands into three types, according to their dominant rock types:

1. Islands with glaucophane schists or glaucophanitic greenschists. These are found on Syros, Siphnos, Folegandros, a small part of Milos (?), northern and southern Ios, and the southeastern part of Naxos.
2. Islands with intermediate facies metamorphic rocks (Miyashiro, 1960), or even high-temperature, low-pressure type rocks, associated with granitic intrusives and migmatites. These islands include Paros, part of Antiparos, Mykonos, Delos, the main part of Naxos, and central Ios.
3. Islands with volcanic rocks of Tertiary or Quaternary age, or high-level subvolcanic

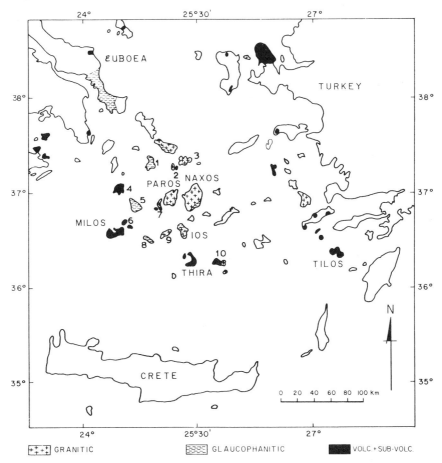

FIG. 3 The Cyclad area. A subdivision has been made into areas of high-temperature metamorphism and granitic rocks, areas with glaucophane-bearing rocks, and areas of Tertiary and Quaternary volcanism and subvolcanic activity. Islands referred to in the text are (1) Syros, (2) Delos, (3) Mykonos, (4) Serifos, (5) Siphnos, (6) Kimolos, (7) Antiparos, (8) Folegandros, (9) Sikinos, and (10) Anaphi.

intrusions. The volcanic rocks occur on Milos, Kimolos, parts of Antiparos, and Thira (Santorini), whereas the subvolcanic intrusions occur on Serifos and Anaphi.

Notwithstanding the fact that some authors (e.g., Wunderlich, 1973) believe that the metamorphism is of Alpine age, there is evidence that in fact the metamorphic basement and the associated intrusives (excluding those on Serifos and Anaphi) are of Prealpine age. Without going into a detailed discussion (see an earlier paper; Schuiling, 1962), we may summarize some of the arguments as follows:

1. All of the metamorphic complexes in the Eastern Mediterranean area show very persistent N–S to NNE–SSW lineations and fold axes. This holds for the Attic-Cycladic Massif and, in Turkey, for the Menderes Massif, the Kaz-Dağ complex, the gneisses south of Bursa, and, as far as is known, for the Rhodopes as well. This N–S direction of the fold axes and lineations is at approximately right angles to the Alpine folding directions in this area.
2. In several places the crystalline basement is unconformably overlain by nonmetamorphic Permian, or only weakly metamorphosed Permian (Papastamatiou, 1963; Marks and Schuiling, 1965). Locally this Permian contains a well-preserved fauna.
3. Very few age determinations have as yet been made on rocks of the crystalline basement. A whole-rock Rb-Sr determination of the central migmatite of Naxos gave 355 million years, a whole-rock determination of an augen gneiss from the Menderes Massif gave 529 million years, and a U-Pb determination on definitely postmetamorphic uraninite collected from the same massif gave 268 million years.

Mica ages of these same rocks, however, are very young. A number of Rb-Sr determinations on biotites, both from the metamorphic complex on Naxos and from a granodioritic intrusion, gave ages between 11.9 and 12.1 million years, whereas a muscovite gave 13 million years. Since it is believed that micas, especially biotites, lose their radiogenic Sr at temperatures around 300° C, it seems that a young Alpine thermal event affected the basement or, alternatively, that these rocks have remained buried fairly deeply ever since the main stage of Hercynian metamorphism and were only recently uncovered by rapid erosion.

From the arguments just given most evidence seems to point to a Prealpine age for the Attic-Cycladic Massif; the high-pressure metamorphic rocks (with jadeite, lawsonite, glaucophane, and pseudomorphs of calcite after aragonite, or kyanite-sillimanite sequences in other places) testify to the fact that at some point during their history the Cyclads must have lost an overburden 20–25 km thick.

The stratigraphic record of the Neogene on Crete (Meulenkamp, 1971) gives an interesting clue to the late Tertiary history of the southern Aegean area. The Neogene overlies unconformably a folded and overthrust basement which comprises rocks ranging in age from Paleozoic to Eocene. The Neogene starts with continental deposits up to several hundred meters thick. It has been proved that the sediments were derived from the north and northeast, that is, from the southern part of the present Aegean Sea. North of Crete this sea is now everywhere deeper than 1000 m, and it may locally reach more than 2000 m. On top of the continental deposits Tortonian marine sediments are found, and thereafter marine sedimentation in Crete continues almost uninterruptedly into Pliocene time.

From these data it appears that, in a schematic way, Crete has undergone the following movements since the main Alpine folding phase:

1. Uplift from Eocene through early Miocene time, during which time Crete formed part of a much greater land mass extending toward the north and northeast.

2. Subsidence in Miocene and Pliocene times, during which, at least at first, Crete received sediments from a northern land mass which must have undergone fairly rapid erosion, as indicated by conglomerates and sandstones in the thick continental deposits.
3. Uplift again since the beginning of the Pleistocene, as marine sedimentation on Crete ceased at the end of the Tertiary, and the Miocene and Pliocene deposits are uplifted above sea level.

The presence of a South Aegean land mass north of Crete, to which the central Cyclads probably belonged, shows that vertical movements in that area took place simultaneous to the movements on Crete, but in the opposite sense. Uplift and erosion took place during Miocene and Pliocene time, followed by rapid subsidence in more recent times, as Crete emerged again.

There is also some indirect evidence for rapid erosion during and after the Miocene from radioactive dating of micas. As stated, biotites from Naxos have Rb-Sr ages of only 11.9–12.1 million years, whereas muscovite is only slightly older (13 million years). It is commonly held that the biotite system becomes closed for loss of radiogenic strontium at temperatures around 300° C, whereas the muscovite system closes at a somewhat higher temperature. Hence the biotites must have cooled through the 300° C isotherm in late Miocene time. Even if we assume the unlikely geothermal gradient of 60°/km, which is twice as high as the world's average, this means that a 5-km overburden has been removed by erosion or tectonic denudation since the late Miocene. This means that the rate of erosion must have been 4 cm/100 yr over the last 12 million years. Differential movements through block-faulting have certainly taken place in the Cyclads, so that this figure may be higher than the average rate of erosion in the Cyclads, but it gives at least an indication of strong relief, high erosion rates, and rapid uplift. This uplift and erosion have since changed into a general subsidence of the Cyclads, which must be regarded as a drowned mountain area. Geomorphologically this can be seen especially well on the smaller islands, where valleys descend into the sea without a trace of a coastal plain. The high positive gravity anomaly over the southern part of the Cyclads, up to +120 mgal, shows that much subsidence is still to follow before isostatic equilibrium will have been reestablished. The volcanism on various islands, so typically associated with the foundering stage of orogenic nodes, is linked in time with the subsidence.

It seems therefore that we have in the Cyclad area an example of an orogenic node, where uplift and erosion as well as the subsequent subsidence have probably been less violent than in the Tyrrhenian Sea, and where the development is younger and has not yet proceeded to such an advanced stage as in most of the other alpine orogenic nodes. We can see here into the Prealpine basement, because the entire Alpine cover has been stripped off by erosion in the recent geological history.

Conclusions

As long as continents and oceans remain fixed, their distribution will definitely produce lateral temperature variations in the underlying mantle. The heat generation in core and mantle, on the other hand, probably approaches radial symmetry, since otherwise geophysical observations on the properties of the deeper earth would show a much more marked lateral variation than in fact has been found (Toksöz and others, 1969). Therefore, if we want to look at the location and causes of processes like orogeny (including uplift and subsidence), volcanism, and metamorphism, we must first examine the effects of the inhomogeneous distribution of the heat production in the *crust*. This is contrary to the usual approach by tectonophysicists, who look for the causes of tectonic processes in the *mantle*.

It seems that the energy needed for the movement of plates is provided by a gravitational instability of the lithosphere. At present, the slope of the upper surface of the asthenosphere is from the oceans toward the continents. This is the situation to be expected at or near the end of a period of continental drift. During a period of relative immobility of the continents, the heating up of the mantle under the continents will cause a reversal of the slope of the upper surface of the asthenosphere. As soon as the gravitational instability caused by this slope becomes large enough, the continents will split apart, and a new period of large plate movements sets in.

A geothermal model has also been applied to certain second-order orogenic features, as contrasted to the first-order orogenic features associated with plate boundaries. A consideration of the thermal situation in a bend of an orogenic system shows that the concave parts of such bends are favorable locations for subsequent, smaller-scale orogenies of a nodal type, in contrast to the major, more or less linear fold belts. The direction of tectonic movement in these nodal-type orogenies is toward the convex side of such bends. In a later, foundering stage these nodes will become basins, accompanied by volcanic activity, and with a structure which tends toward an oceanic-type crust. This model has been applied to the Cyclads, which seem to have reversed their vertical movement only recently. It explains in a general way the geologic history of Crete and the Cyclads since the main stage of the Alpine orogeny. It gives a satisfactory explanation for the location of the South Aegean as a nodal point in an Alpine orogenic belt, and it explains the distribution of the Miocene and Pleistocene sediments on Crete, as well as the erosional history of the Cyclads during those same periods, and their present subsidence. Although not explored in any detail, the geothermal model seems also promising to explain the type, location, and timing of the associated younger volcanism.

The nodal-type orogenies that have been described in this paper can be equated with Van Bemmelen's "meso-undations" (see, e.g., Van Bemmelen, 1969). Although there is still divergence of opinion between his views and the present author's, Van Bemmelen (1969) also arrives at the conclusion that "in the case of arcuate mountain belts and island arcs the direct cause of the orogenic evolution has to be sought in geodynamic processes at a meso-tectonic scale, occurring in the upper mantle, directly underneath the mobile belts." It is felt that the focusing of heat sources at the concave side of a bend in an orogenic system and the defocusing of heat sources at the convex side provides an adequate mechanism for such geodynamic processes.

References

Clark, S. P., Jr., ed., 1966, Handbook of physical constants: *Geol. Soc. Am. Mem. 97*, 587 p.

Ernst, W. G., 1963, Petrogenesis of glaucophane schists: *Jour. Petrol.*, v. 4, p. 1–30.

Green, D. A., and Ringwood, A. E., 1970, Mineralogy of peridotitic compositions under upper mantle conditions: *Phys. Earth Planet. Int.*, v. 3, p. 359–371.

Hales, A. L., 1969, Gravitational sliding and continental drift: *Earth Planet. Sci. Lett.*, v. 6, p. 31–34.

Jacoby, W. R., 1970, Instability in the upper mantle and global plate movements: *Jour. Geophys. Res.*, v. 75, p. 5671–5680.

———, 1973, Gravitational instability and plate tectonics (this volume).

Lambert, I. B., and Heier, K. S., 1968, Estimates of the crustal abundances of thorium, uranium and potassium: *Chem. Geol.*, v. 3, p. 233–238.

Marks, P., and Schuiling, R. D., 1965, Sur la présence du Permien Supérieur non-métamorphique à Naxos: *Prakt. Akad. Athens*, v. 40, p. 96–99.

Menard, H. W., 1967, Transitional types of crust under small ocean basins: *Jour. Geophys. Res.*, v. 72, p. 3061–3073.

Meulenkamp, J. E. (in press), The Neogene in the southern Aegean area: *Botaniska Notiser, Lund.*

Nieuwenkamp, W., 1956, Energy in orogenesis and metamorphism: *Geol. Mijnb.*, N. S., v. 18, p. 128–130.

Pannekoek, A. J., 1969, Uplift and subsidence in and around the Western Mediterranean since the Oligocene: A review: *Verh. Kon. Ned. Geol. Mijnb. Genootschap*, v. 26, p. 53–77.

Papastamatiou, I., 1963, Sur la présence de roches sédimentaires d'âge prétriassique à Mykonos (archipel des Cyclades, Grèce): *Comptes Rendus Acad. Sci.*, v. 256, p. 5167–5169.

Press, F., 1973, Gravitational instability of the lithosphere (this volume).

Schuiling, R. D., 1962, On petrology, age and structure of the Menderes migmatite complex (SW-Turkey): *Bull. M. T. A. Inst., Ankara*, v. 58, p. 71–84.

——— , 1966, Continental drift and oceanic heat-flow: *Nature*, v. 210, p. 1027–1028.

——— , 1969, A geothermal model of oceanization: *Verh. Kon. Ned. Geol. Mijnb. Genootschap.*, v. 26, p. 143–148.

Toksöz, M. N., Arkani-Hamed, J., and Knight, C. A., 1969, Geophysical data and long-wave heterogeneities of the Earth's mantle: *Jour. Geophys. Res.*, v. 74, p. 3751–3770.

Umbgrove, J. H. F., 1947, The pulse of the earth, 2nd ed.: The Hague, Martinus Nijhoff, 358 p.

Van Bemmelen, R. W., 1964, The evolution of the Atlantic mega-undation: *Tectonophysics*, v. 1, p. 385–430.

——— , 1969, The Alpine loop of the Tethys zone: *Tectonophysics*, v. 8, p. 107–113.

Verhoogen, J., Turner, F. J., and Weiss, L. E., 1970, *The earth: An introduction to physical geology*: New York, Holt, Rinehart and Winston, 748 p.

Wunderlich, H. G., 1973, Gravity anomalies, shifting foredeeps, and the role of gravity in nappe transport as shown by the Minoides (Eastern Mediterranean) (this volume).

HANS RAMBERG

Division of Mineralogy and Petrology
University of Uppsala
Uppsala, Sweden

Department of Geology and Geography
University of Connecticut
Storrs, Connecticut

Model Studies of Gravity-Controlled Tectonics by the Centrifuge Technique[1]

The advantages of the centrifuge technique in model studies of gravity-driven and gravity-controlled tectonic deformations have been pointed out on several occasions by the author, and the instrumental setup and the model materials used have been described in some detail (Ramberg, 1963, 1967, 1972). In the strong body-force field affecting centrifuged models, substances with high viscosity and considerable strength will yield under their own "weight," thus greatly increasing the choice of model materials and the possibility of studying composite and complex structures. The materials most frequently used in the experiments at the tectonic laboratory in Uppsala include (1) silicone putty or bouncing putty, pure or mixed with fine powder of magnetite or tungspar (or both) to obtain the desired density, (2) painter's putty of a special plastic-base type, (3) mixtures of collophony and ethylene phthalate, (4) plasticine or modeling clay, (5) oil–wax mixtures, and (6) $KMnO_4$-solutions.

Dome Model Theory

Inverted density stratification often occurs in the Earth's crust, caused, for example, by

---

[1] Several students and research assistants have aided in the laboratory preparation and running of the models. For this valuable assistance I wish to extend sincere thanks to O. Stephansson, who also played a major role in constructing the apparatus, and to H. Sjöström, H. Berner, and R. Häll. Thanks are also due Mrs. A. Kaljusaar, who made the drawings, Mrs. A. M. Ekström, who typed the manuscript, and Mr. A. Wallner, who took some of the photographs. Financial support was provided by the Swedish Natural Science Research Council and the Swedish Board for Technical Development. Last but not least, I wish to extend my sincere thanks to Prof. R. Scholten for editing the manuscript and suggesting alterations which improved its readability greatly.

precipitation of heavy sediments on top of less dense ones, or by lava that solidifies to dense rocks pouring over sediments or other rocks with less density. In the Earth's mantle low-density strata may also form by phase transition from high-density minerals to low-density assemblages (e.g., eclogite → gabbro), or by partial fusion, as in the low-velocity layer in the upper mantle.

The mechanical instability of such structures gives rise to the development of a wavy pattern which gradually changes to a set of domes or diapirs, the top part of the latter spreading laterally and ultimately producing a stable layering with normal density stratification. An evolution of this kind is shown in Fig. 1.

Application of fluid dynamics to the problem of instable density stratification enables one to determine the dominant wavelength and the rate at which the rise occurs in its early stages. The crucial equation in the theory is

$$y_i = y_{0i} \exp{(\kappa_1 q_1 t)} \qquad (1)$$

where y_i is the amplitude at time t of the deflections on interface i in a multilayer complex, y_{0i} is the initial amplitude at time $t = 0$ of the same deflections, $q_1 \equiv \frac{1}{2}(\rho_1 - \rho_2)gh_2/\mu_2$ in which ρ_1 and ρ_2 are the densities of the top layer and the second layer from the top, g is acceleration due to gravity, h_2 is the thickness of layer number 2, and μ_2 is the viscosity of the same layer. Here κ_1 is the numerically largest eigenvalue of a certain matrix which characterizes the dynamics of the system in question, the so-called *coupling matrix*. This matrix is a function of the following dimensionless parameters of the system: (1) the viscosity ratios of the various strata; (2) the density ratios; (3) the ratios between the thicknesses; and (4) the ratios between the thicknesses and the wavelength λ of the deflections. (For detailed explanation see Ramberg, 1968a, 1968b, 1972.)

The significance of the quantity κ_1 for our problem is that κ_1, when plotted as a function

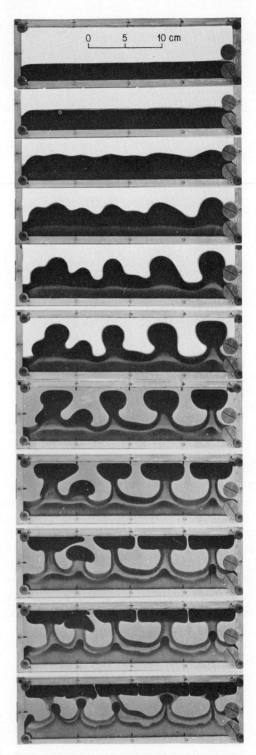

FIG. 1 Successive stages of dome evolution of oil beneath heavy syrup overburden.

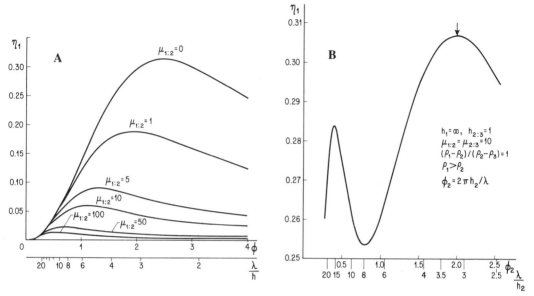

FIG. 2 (A) The eigenvalue κ_1 plotted versus ϕ_2 and λ/h_2 for single buoyant layer with infinite half-space as overburden. $\mu_{1:2} = \mu_1/\mu_2$ is the ratio between viscosities of overburden and buoyant stratum. Maxima give the dominant wavelength. (B) The eigenvalue κ_1 plotted versus ϕ_2 and λ/h_2 for a double layer with infinite half-space as overburden. The dimensionless parameters are $\mu_{1:2} = \mu_{2:3} = 10$; $h_{2:3} = 1$; $(\rho_1 - \rho_2)/(\rho_2 - \rho_3) = 1$; $\rho_1 - \rho_2 > 0$; $\rho_2 - \rho_3 > 0$; $h_1 = \infty$ and the base is rigid. The two maxima coincide with the two distinct wavelengths of simultaneously developing waves or domes.

of the wavelength of the interfacial deflections, goes through a maximum, or sometimes several maxima. The wavelength coinciding with maximal κ_1 is the dominant wavelength, which determines the position of gradually ascending domes and penetrating diapirs.

Figure 2 shows examples of the function $\kappa_1 = (\lambda/h_2)$ for structures developing into one (Fig. 2A) or two (Fig. 2B) sets of domes. The dominant wavelength determines the spacing between the domes. Models with two distinct maxima are particularly interesting because they indicate the possibility of

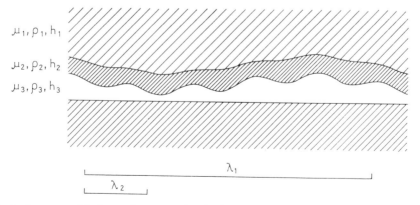

FIG. 3 Part of an unstable layered system developing two sets of waves. See also Fig. 2B.

simultaneous formation of "domes within domes" in geologic structures. In other words, the theory shows that a huge dome with a diameter of the order λ_1 may actually consist of a cluster of smaller domes with a much smaller diameter of the order λ_2 (Fig. 3). This theoretical conclusion is worth recognizing, for example, in connection with the occurrence of composite batholiths in orogenic belts.

Centrifuged Dome Models

Figure 4A shows a well-developed model dome exhibiting the typical shape of such structures: the rim syncline, the trunk, and the hat. Another common detail of the shape, the substratum bulge, is hardly recognizable in this model. It is, however, distinctly developed in the model of Fig. 4B and extremely well developed in the model shown in Fig. 4C.

The rim syncline forms because of the centripetal flow in the buoyant layer toward the center of the dome. This flow is associated with a pressure gradient in the buoyant or "source" layer, the pressure gradually increasing away from the dome. Hence the low pressure in the source layer just outside the dome results in maximal sinking of the overburden around the contact. Contrary to common belief the material ascending in the trunk is not only taken from underneath the

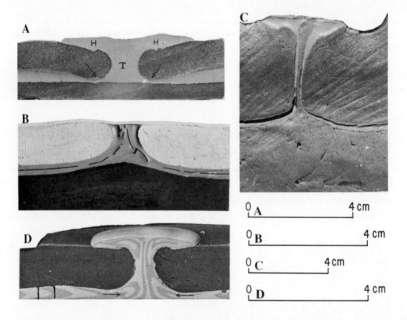

FIG. 4 (A) Dome of silicone putty ($\rho = 1.14$ g/cm^3) having penetrated overburden of painters putty ($\rho = 1.87$ g/cm^3) during centrifugation. Initial structure consisted of horizontal parallel and continuous layers. H = hat of dome, T = trunk. Arrows indicate rim syncline. (B) Silicone-putty dome (light gray, with thin, dark sheet that has broken into boudins) rises through painter's putty overburden. Substratum (black) consists of heavy silicone putty mixed with powder of magnetite. Note bulge on substratum below dome. Original layering horizontal. (C) Layer of light silicone putty that has risen through overburden of painter's putty, spread across the surface, and sucked up the heavy substratum almost to the surface. (D) Model indicating direction of flow in source layer and dome. The initially evenly thick horizontal source layer consisted of alternating strips of dark and light colored silicone with square cross section as indicated by two vertical lines on left side of photo. Present pattern indicates movement. Note that horizontal flow in source layer has passed underneath the rim syncline.

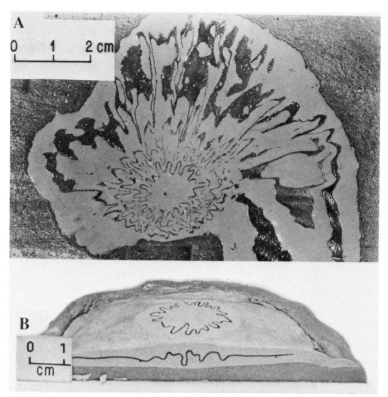

FIG. 5 (A) Horizontal cut through root zone of silicone dome and its source layer, with four thin sheets of competent plasticine strongly buckled during the convergent flow toward center of dome. Fold axes are subhorizontal in source layer some distance away from dome but almost vertical in center of dome. (B) Three-dimensional view of source layer and dome with buckled sheet of plasticine. Overburden is removed and dome is cut close to its root zone.

rim syncline but is supplied from a wider region, the horizontal flow passing underneath the maximal depression as shown in Fig. 4D.

Thin sheets of competent materials embedded in the buoyant silicone layer deform in a pattern that indicates the geometry of the flow. Buckle folds with axes radiating out from the center of ascent are produced during the domal process. Sometimes the buckled sheets are broken by tension fractures oriented normal to the fold axes, and the fragments are separated in boudinagelike fashion.

This evolution of the embedded competent sheets is in harmony with the strain pattern developed during the flow in the source toward the center of ascent. The dome is a dynamic sink at the center of ascent, and the centripetal movement is in the form of a convergent flow toward this sink. Compressive strain normal to incompressible model material (rocks) create so-called *a* tectonites, characterized by fold axes and elongation parallel to the "direction of movement" (Fig. 5A). In the trunk of the dome the flow is essentially vertical and the flow lines are parallel. In this region the fold axes of the buckled embedded sheets are steep to vertical. Examples are shown in Fig. 5B.

Figure 6 illuminates the deformation of embedded competent sheets in convergent and divergent flow of rocks. Figure 6A is a horizontal section cut through a circular ridge which has ascended parallel to the circular

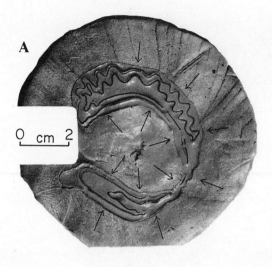

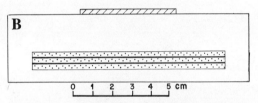

FIG. 6 (A) Horizontal cut through model with circular ridge or anticline of buoyant silicone with two thin sheets of competent plasticine. Note buckling and boudinage structures. Arrows indicate flow in source layer (which is below the section shown). See text and Fig. 6B. (B) Cross section of model shown in Fig. 6A prior to run in centrifuge. Dotted: silicone putty, $\rho = 1.14$ g/cm^3; black: sheets of plasticine; white: painter's putty, $\rho = 1.87$ g/cm^3; inclined hatching: plasticine, $\rho = 1.68$ g/cm^3.

boundary of the model. The initiation of these ridges is due partly to the edge effect mentioned below, partly to a circular heavy layer on top of the overburden (Fig. 6B). In the model two sheets of plasticine were embedded in the buoyant layer of silicone. These layers are now buckled in the outer part of the ridge but stretched and broken into boudins in the inner part. The reason for this contrasting behavior is that during the rise of the ridge the material in its outer part was emplaced by convergent flow, while the material in the inner part of the circular ridge was emplaced by divergent flow (see arrows, Fig. 6A).

In the hat region, which develops when the dome spreads close to the surface or actually across the surface, the flow is again more or less radial, but now in a centrifugal direction, the material moving radially out from the center of the dome. The flowlines diverge when seen in the plane of the surface; as a result, the enclosed competent sheets develop radial tension cracks, and folds formed earlier during the flow in the source layer and in the trunk are likely to flatten out. If the thickness of the hat which spreads laterally over the surface were constant, there would, because of the incompressibility of the materials, be a compressive strain parallel to the radius and thus parallel to the flowlines. This would create buckling about fold axes parallel to the circumference. In other words, so-called b tectonites would be produced in the spreading hat region. However, since the hat actually becomes thinner with distance from the center (Figs. 4A and 4D), there is a vertical component of compressive strain in the spreading hat. Compression parallel to the radius is therefore less than it would have been without the thinning of the hat, and the buckling about axes normal to the flowlines is accordingly weakened.

Strain Pattern Adjacent to Domes

A layered overburden sagging down in the rim syncline and stretching over an early stage dome that has not yet pierced the layers of the overburden takes the shape of a fold which is not produced by lateral shortening. This large-scale fold is generated entirely by vertical movements and vertical forces (Fig. 7).

On a smaller scale, however, exhibiting shorter wavelength and smaller amplitude, true buckle folds form in competent sheets in the overburden because of local lateral compression generated during the sliding down in the rim syncline (Fig. 8). They may also

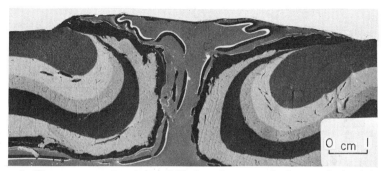

FIG. 7 Dome of silicone putty containing folded and ruptured sheets of plasticine which have penetrated a layered overburden of painter's putty. Note pronounced rim syncline and inversion of strata.

FIG. 8 Model showing boudins above dome and buckling of surficial strata which have been pulled down in the rim syncline.

be produced in the overburden outside the edge of the spreading lobes of the dome (Fig. 9).

Often, the spreading of the hat causes an inversion of the overburden strata, the resulting pattern being a huge recumbent fold following more or less the shape of the rim syncline, the trunk, and the lobes of the hat. The contact is partly conformable and partly discordant (Fig. 10). The strata in the upper limb of the recumbent fold—on the top of the hat—are usually strongly thinned and/or stretched to widely separated boudins (Fig. 8).

In cases where the substratum bulge is pronounced, competent layers embedded in the substratum also are pressed into buckle folds because of the centripetal flow of the

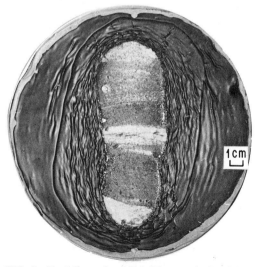

FIG. 9 Buckling of surficial layer of plasticine outside the edges of a spreading dome.

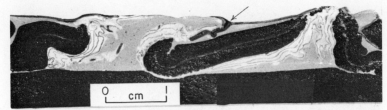

FIG. 10 Cross section through model of silicone putty with thin sheets of plasticine, risen through an overburden of painter's putty and spread in the form of a recumbent fold on the surface.

substratum toward the bulge below the dome. This flowage in the substratum is implemented by the active flow in the buoyant stratum, the movement being transferred by friction coupling. Since the flow in the substratum toward the center of the domal ascent is convergent, the stretching parallel to the flow lines should result in tension fractures normal to the radial flow in competent sheets embedded in an incompetent substratum not too far below the lower boundary of the buoyant layer.

Initiation of Domal Rise

Experiments show that in model materials with vanishing yield strengths it is impossible to prevent the development of waves, and thus the rising of domes, if low-density strata are buried below denser ones. This is true even if the effective viscosity of the materials is very high, though if the viscosity is high, the rise of the waves on a straight interface is correspondingly slow. It seems that even if one tries to make each layer exactly uniform in density, viscosity, and thickness, and to form the interfaces and surfaces in the layered models exactly parallel to the equipotential surface, there are irregularities on the molecular level which initiate domes in our centrifuged models. If carefully prepared, the models develop waves in quantitative accordance with the dynamic theory of the dominant wavelength as embodied in Eq. 1. The best test of the theory, however, is by oil-syrup models run without a centrifuge, as shown in Fig. 1.

Domes are also initiated at predetermined irregularities purposely built into the overburden or buoyant layer, including uneven thickness, fractures and faults, variable density, and initial wavy shape of the layer. There is also a definite boundary effect in the circular models, domes tending to rise parallel to the circumference.

Subsiding Bodies

During the overturn of an unstable density stratification to the stable arrangement characterized by downward increasing density, the conspicuous phenomenon is the ascent of domes and anticlines, such as salt diapirs and ridges, which abound in connection with halite deposits, as well as batholiths, granite-cored anticlines, and the mantled-gneiss domes characteristically encountered in orogenic belts. For incompressible materials the volume of matter that descends to replace the space left by the elevated bodies is the same as the volume it replaces. The descent is, however, distributed over a much wider region than the rise, hence the vertical displacement of the subsiding material is much less than that of the ascending material. Yet the subsiding movement is also significant from the point of view of tectonics, sometimes perhaps more so than the diapiric rise. Probably, the subsidence is often so inconspicuous as to remain unnoticed by field geologists. Whereas the batholiths, diapirs, mantled domes, and so on, move toward the surface where they can be readily observed, the descending masses generally become inaccessible to direct observation.

Let us look at models of subsiding sheets of relatively dense materials. An interesting and perhaps unexpected feature of such sheets is that they assume a bent shape with the convex side facing upward while sinking through the less dense substratum, provided that their lateral extent is between one and two dominant wavelengths (see above). Figure 11 shows a model of a subsiding sheet.

The flow pattern in the substratum and the overburden of the sinking sheet is of interest for tectonic theories. In the substratum the flow points away from the center toward the edges, whereas in the overburden the flow is from the edges to the center. In other words, the adjacent flow wraps around the sheet in two symmetrical branches, starting from beneath the center of the sheet and ending above the center. This movement produces intense lateral compression and buckle folds in layers that happen to exist in the overburden, for example, in the models of Fig. 11. In the course of the subsidence the substratum may ultimately pierce the central region of the sheet and move through in the form of a diapir.

If the width of the heavy sheet is many times the dominant wavelength characteristic for the particular system, only the outer edges will bend down and the substratum will pierce through at many points and form diapirs with an interspacing controlled by the dominant wavelength.

The writer considers it likely that phenomena of this nature occur in geosynclines. It is not unrealistic to assume that sheets of basic volcanic rocks, when thick enough (say of the order 5 km or more) and buried below several kilometers of geosynclinal sediments, do, in fact, sink through a gneissic or granitic substratum. Could it be, for example, that the heavy masses that "seem" to occur below the Jotun nappes in the Norwegian Caledonides (Smithson, 1964) and below the Ivrea zone in the Alps (Berckhemer, 1968; Kaminski and Menzel, 1968) are remnants of thick surficial sheets or huge sills of basic igneous rocks that, after solidification, have sunk through the geosynclinal substratum? If that is so, the Hercynian massifs and the gneissic core of the Pennine nappes in the Alps and the gneissic masses and the Nordland "granites" in the Scandinavian Caledonides would be the more or less mobilized (partly molten?) and recrystallized Prealpine and pre-Caledonian basement that rose into the column of geosynclinal sediments and lavas, thus constituting the buoyant counterparts of the sinking masses. We now find the uplifted buoyant masses at the surface, but most of the heavy masses have sunk to inaccessible levels.

One may also wonder to what extent processes of this kind take place at the bottom of the sea. Quite clearly, much basic lava is extruded across the sediments on the ocean bottom and becomes intercalated with sedimentary strata. Do the low-density subjacent sediments tend to pierce through the igneous rocks, thereby allowing these rocks to sink? What happens along the edges of continents, where Sial underlies the column of sediments

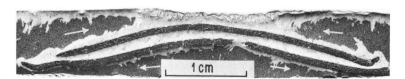

FIG. 11 Two layers (black) of heavy painter's putty that have sunk through a substratum of silicons putty (dark and light gray). Note buckling of thin surficial sheets of plasticine above the sinking layers. Arrows indicate direction of flow.

and lavas? Would a thick layer of basic lava resting directly on the Sial along the continental edge actually descend through the Sial, forcing it upward in the form of domes with nappelike lobes and recumbent anticlines? Experiments and quantitative dynamic calculations (Ramberg, 1972) both indicate that these are possibilities that should be considered.

The Effect of Size on the Velocity of Dynamic Evolution

Geometrically similar structures (but possibly of very different size, as in the case of a small laboratory model compared with a huge orogenic belt) will develop in a similar way provided that dynamically significant parameters meet certain conditions. In layered systems of Newtonian materials with vanishing strength and no lateral stresses in excess of those caused by gravity (or the centrifugal force in centrifuged models) the two conditions for dynamic similarity are simple:

1. The viscosity ratio $\mu_1/\mu_2/\mu_3$, must be the same in all structures.
2. The density ratio, $\rho_1/\rho_2/\rho_3$, must be the same in all structures.

Since geometric similarity is already specified, equal thickness ratio, equal wavelength/thickness ratio, and equal amplitude/wavelength ratio are implied, as are of course equal numbers of layers and similar boundary conditions on top and bottom in the structures compared. The slow rate of tectonic deformation of crystalline rocks permits us to disregard inertial forces.

Layered structures fulfilling the foregoing requirements have identical eigenvalues κ_i (see p. 50) irrespective of their geometric size, their absolute viscosities, and their absolute densities.

The velocity v of the growth of amplitude of anticlines in a set of dynamically similar structures can be expressed by the equation

$$v = \kappa_1 q_1 y \qquad (2)$$

where κ_1 is the eigenvalue, y the amplitude of the uplift, and

$$q_1 \equiv \tfrac{1}{2} \frac{(\rho_1 - \rho_2)h_2 g}{\mu_2}$$

where ρ_1 is the density of the top layer, ρ_2 that of the second layer, h_2 the thickness of the second layer, μ_2 its viscosity, and g the acceleration due to gravity. For a centrifuged model the centripetal acceleration a (which in our models is up to 4000 g) is substituted for g.

If the densities and the viscosities of corresponding layers are the same in all the structures compared, we see that v is proportional to the product $h_2 y$. However, if geometric similarity exists, y is proportional to h_2 and it follows that

$$v = k h_2^2$$

where k is the same constant for all structures compared. Thus the velocity increases with the square of the linear dimension in dynamically similar structures if all consist of materials whose effective viscosities and densities are the same in the various structures. Applied to gravity tectonics, this means that the larger the structure of a given kind, be it a salt diapir, a mantled gneiss-granite dome, a gravity-driven nappe, or a subsiding heavy layer in the mantle, the faster it develops. For materials such as crystalline rocks which possess a finite strength and show a high effective viscosity, only structures larger than a certain minimum size (which varies depending upon the rheological parameters of the rocks involved and the exact geometry of the structure) can possibly have been molded essentially by gravity forces. For smaller structures the strength and viscosity is too high for the body force of gravity to be effective, except indirectly by creating the boundary stresses around a tectonic system.

Models of Orogens

A Theoretical Example

The similarity between certain salt diapirs, mantled gneiss-granite domes, batholiths, and other sialic plutons in orogens, and Pennine-type nappes (which compare strikingly with spreading lobes in some domes) makes it tempting to believe that orogenesis is a process powered by gravitational instability of a kind comparable to that which produces salt diapirs and mantled gneiss-granite domes (Van Bemmelen, 1966; Wegmann, 1930; Ramberg, 1967). If the substratum below at least a part of the geosynclinal column is sialic, its density is a little less than the average density of lavas and metasedimentary rocks in the overlying stack of strata. For example, using published densities of Alpine metamorphic rocks (Wenk and Wenk, 1969), we find that a column of 90 percent schist of various types and 10 percent greenstone plus amphibolite has an average density of 2.77 g/cm^3. A gneissic-granitic basement would have a density of about 2.67 g/cm^3.

To see if this density contrast is sufficient to start and maintain the overturn of the original instable layering toward the stable arrangement, we must know the thicknesses, the effective viscosities of the rocks, and their yield strengths. Although we know practically nothing about yield strengths we can make some reasonable suggestions as to the viscosities. If the basement is buried below, say, 15 km of sediments, the temperature would be 200–300°C, and a "softening" of the solid gneiss to a viscosity $\mu_4 = 10^{20}$ P seems not unrealistic. Let the sialic basement be 7.5 km ($h_4 = 7.5 \cdot 10^5$ cm) thick and let it rest on a simatic sub-basement with density $\rho_5 = 3.1$ g/cm^3, viscosity $\mu_5 = 10^{22}$ P, and thickness $h_5 = \infty$ (infinite). We separate the geosynclinal column above the sialic basement into two main units, an uppermost unit consisting of unmetamorphosed sediments with density $\rho_2 = 2.55$ g/cm^3, viscosity $\mu_2 = 10^{17}$ P, and thickness $h_2 = 7.5 \cdot 10^5$ cm; and a lower unit of metasedimentary rocks and lavas with density $\rho_3 = 2.77$ g/cm^3, viscosity $\mu_3 = 10^{21}$ P, and thickness $7.5 \cdot 10^5$ cm. Layer number 1 on the very top is the sea, which is disregarded in our calculation because of its negligible viscosity and low density.

The preceding data lead to the following pertinent dimensionless parameters, which allow us to calculate the eigenvalue κ_1:

$$\frac{\mu_2}{\mu_3} = 10^{-4}, \quad \frac{\mu_3}{\mu_4} = 10, \quad \frac{\mu_4}{\mu_5} = 10^{-2}$$

$$\frac{(\rho_2 - \rho_3)}{-\rho_2} = \frac{-0.22}{-2.55} = 0.0862745,$$

$$\frac{(\rho_3 - \rho_4)}{-\rho_2} = \frac{0.1}{2.55} = -0.0392157,$$

$$\frac{(\rho_4 - \rho_5)}{-\rho_2} = \frac{-0.43}{-2.55} = 0.1686275$$

$$h_{2:3} = 1, \quad h_{3:4} = 1, \quad h_5 = \infty$$

The dominant wavelength may be computed from the value of ϕ_2 at maximal κ_1, where $\phi = 2\pi h_2/\lambda$. Since, in Fig. 12, maximal κ_1 fixes ϕ_2 at 1.4 and the λ/h_2 ratio at 4.5, and since $h_2 = 7.5$ km, the computed dominant wavelength is 33.7 km. This should then be the approximate spacing between the anticlines or diapirs of the mobilized sialic basement.

We are, however, particularly interested in the velocity of ascent of these protrusions from the basement. Equation 2 supplies an answer to that question when we select a certain stage of the domal rise with a certain amplitude, say, $y = 3$ km, which is less than 10 percent of the dominant wavelength, and employ the appropriate value for q_1, which is

$$q_1 = \tfrac{1}{2}\frac{(\rho_1 - \rho_2)h_2 g}{\mu_2}$$

$$= \frac{-2.55 \cdot 7.5 \cdot 10^5 \cdot 981}{2 \cdot 10^{17}}$$

$$= -9.3808 \cdot 10^{-9} \text{ sec}^{-1}$$

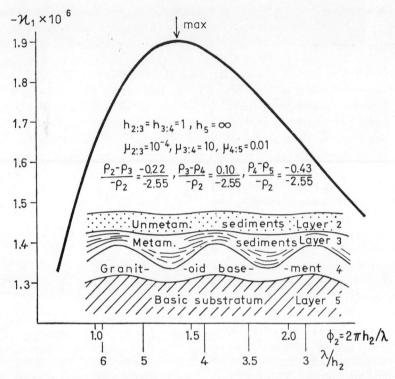

FIG. 12 Cross section of theoretical geosyncline and its basement with plot of the function $\kappa_1 = f(\lambda/h_2)$ and tabulation of the critical dimensionless parameters. Symbols are explained in text.

With $\kappa_1 = -1.8996 \cdot 10^{-6}$ and $y = 3 \cdot 10^5$, the velocity of ascent is
$v = \kappa_1 q_1 y_1 = 1.8996 \cdot 10^{-6} \cdot 9.3808$
$ \cdot 10^{-9} \cdot 3 \cdot 10^5$ cm/sec
$ = 5.346 \cdot 10^{-9}$ cm/sec
$ = 0.168$ cm/yr

Though rising at a very slow rate, the movement amounts to several tens of kilometers in the course of the lifetime of orogens, which is measured in the order 100 million years. We therefore conclude that the model is realistic.

Experimental Models of Orogens and Batholiths

Figures 13 to 17 show selected models of orogens after run in the centrifuge, and Fig. 18 permits a comparison with an actual orogen. Pertinent data are given in the figure captions. The model shown in Fig. 15 was run to obtain a calculated time span for the evolution of a natural orogenic structure which is dynamically similar to the model. The necessary data are found in Ramberg (1967, Table 8). On the basis of density and effective viscosity data selected for the natural geosynclinal column and its sialic basement we obtain 122 million years from the geosynclinal stage to the mature orogenic stage shown in Fig. 15. The data selected for the rocks seem realistic; thus it is satisfying that the time obtained is of the order usually assumed for the evolution of an orogen. Our experiments strongly support Van Bemmelen's (1954, 1966) ideas of orogenic evolution.

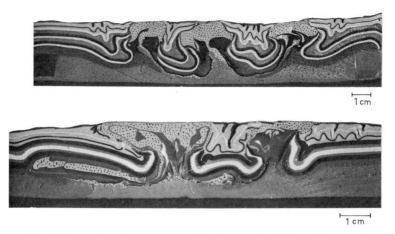

FIG. 13 Two models of domes that have penetrated and strongly deformed an overburden of layered materials. Note significant differences in details of structures between the two models.

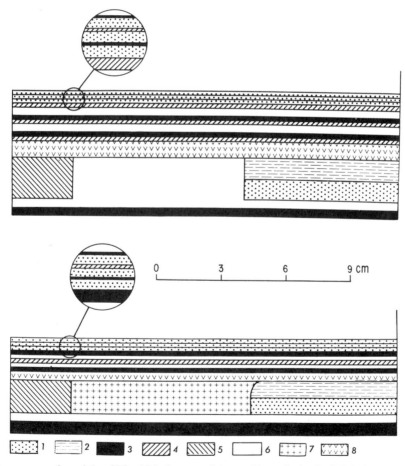

FIG. 14 Structures of models of Fig. 13 before run in centrifuge. Only the left halves of the symmetrical models are shown. (1) Silicone putty, $\rho = 1.14$ g/cm^3; (2) silicone putty, $\rho = 1.58$ g/cm^3 in upper model, $\rho = 1.33$ g/cm^3 in lower model; (3) plasticine, $\rho = 1.68$ g/cm^3; (4) plasticine, $\rho = 1.78$ g/cm^3; (5) plasticine, $\rho = 1.71$ g/cm^3; (6) gray painter's putty, $\rho = 1.85$ g/cm^3; (7) yellow painter's putty, $\rho = 1.85$ g/cm^3; (8) green painter's putty, $\rho = 1.85$ g/cm^3.

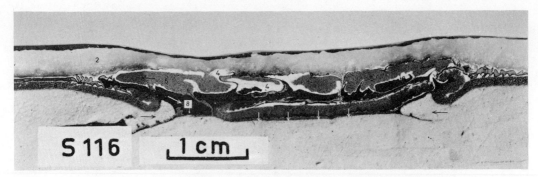

FIG. 15 Fold-mountain model formed from the initial structure shown in Fig. 16.

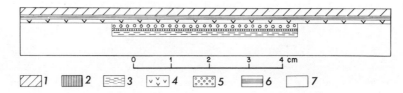

FIG. 16 Initial structure of model shown in Fig. 15. (1) Oil-wax mixture, $\rho = 0.9$ g/cm^3; (2) white silicone putty, $\rho = 1.14$ g/cm^3; (3) gray silicone putty, $\rho = 1.25$ g/cm^3; (4) painter's putty, $\rho = 1.87$ g/cm^3; (5) black silicone putty, $\rho = 1.35$ g/cm^3; (6) silicone with thin layers of plasticine; (7) painter's putty, $\rho = 1.87$ g/cm^3.

FIG. 17 Cross section of model in which a layer of low-density silicone (2) has penetrated an overburden of silicone with higher density (black, 3) and ascended into a top layer of low-density silicone (white, 1) with thin sheets of stiff plasticine. Compare cross section of Appalachian structures in Fig. 18.

Models of Continental Drift

The drifting apart of continents and the opening up of oceans are imitated in centrifuged models in which a diapir of material slightly less dense than the average mantle rises through the mantle and spreads underneath the crust (Fig. 19; see also Fig. 20, which shows the initial structure prior to run in the centrifuge and gives the pertinent parameters). The spreading of the hat of the diapir, whose density is higher than that of the crust, transmits tensile stresses to the overlying crust. Depending on the strength and effective viscosity of the crust, it fractures or thins (or both) and the fragments or boudins become separated by considerable distances above the spreading mushroom.

It is interesting to compare this model with a cross section of the Mid-Atlantic Ridge as interpreted by Talwani and others (1965) (Fig. 21). A "pillow" of low-density material

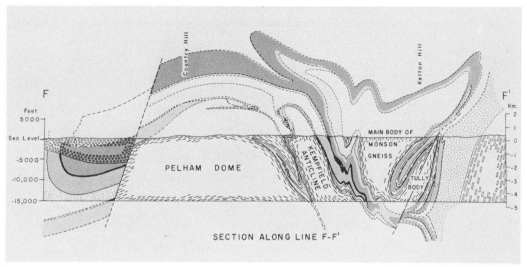

FIG. 18 West-east cross section of Appalachian structure in Massachusetts (from Zen and others, 1969). Horizontal scale same as vertical.

is pictured below the rift zone, exactly as found in the experimental model.

Physical State of Rising Mass

It is significant that our model studies (including a large number of tests not presented in this brief account) exclude almost conclusively any possibility that the rising and spreading mass can be as fluid as a silicate melt when scaled up to natural dimensions. All known silicate melts would be too fluid to form the bulky mushroom shape actually developed in the models. Buoyant model materials whose viscosity is related to that of the model mantle and model crust (viscosity of the order 10^6 to 10^8 P) in the same way that the viscosity of a magma is related to the viscosity of the earth's mantle and crustal rocks will penetrate the overburden as narrow dikes or sills rather than forming bulky, dome-like bodies. The low viscosity of the model magma required by proper dynamic scaling allows it to flow rapidly along the narrow channelways without plastically deforming the adjacent model rocks. At the ends of magma-filled fractures, however, the pressure in the fluid may produce tensile stresses in the host rock sufficient to propagate the fracture, thus allowing the magma to proceed until, perhaps, it reaches the Earth's surface. Even if we consider the most viscous silicate melt—obsidian, with a viscosity of $\sim 10^6$ P at 1400° and 1 atm—and allow for the highest pressure coefficient recorded in the recent data compilation by Clark (1966), we can hardly assume a viscosity greater than 10^{12} P for a true silicate melt at a depth between 20 and 40 km. Yet it is at such a depth that the spreading lens is situated in the profile of the Mid-Atlantic Ridge according to Talwani and others (1965). Assuming 10^{22} P as the viscosity of the surrounding rocks and mantle materials, the viscosity ratio of the melt and its environment is 10^{-10} Now dynamic similarity requires the same ratio between the model magma and the model crust and mantle. The effective viscosity of the stiffer components of the model rocks (the model clay sheets) and of the softer component (painter's putty) is of the order 10^8 and 10^6 to 10^7 P, respectively. For the model magma to be properly scaled it must

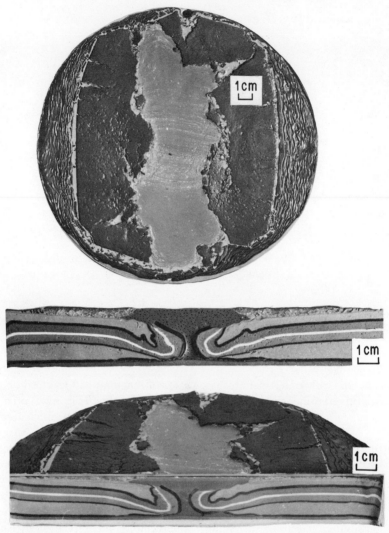

FIG. 19 View from above and profiles of model simulating ocean spreading and continental drift. Compare Figs. 20 and 21.

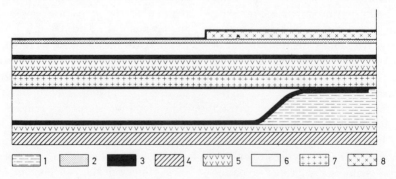

FIG. 20 Cross section of model shown in Fig. 19 before run in centrifuge. Only the left half of the symmetric model is shown. (1) Silicone putty, $\rho = 1.46$ g/cm^3; (2) silicone putty, $\rho = 1.28$ g/cm^3; (3) plasticine, $\rho = 1.68$ g/cm^3; (4) plasticine, $\rho = 1.78$ g/cm^3; (5) green painter's putty, $\rho = 1.85$ g/cm^3; (6) gray painter's putty, $\rho = 1.85$ g/cm^3; (7) yellow painter's putty, $\rho = 1.85$ g/cm^3; (8) mixture of paraffin wax and oil.

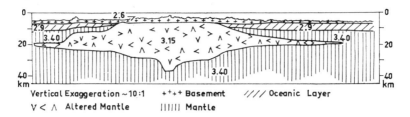

FIG. 21 Cross section of the Mid-Atlantic Ridge (from Talwani and others, 1965).

have a viscosity of the order 10^{-2} to 10^{-4} P ($10^{-10} \times 10^8$ to $10^{-10} \times 10^6$). That is, the model magma should be even less viscous than water.

Based on arguments of this kind we have run a number of models intending to imitate the movement of magma within a matrix of solid rocks and crystalline mantle material. Some of the results are discussed in the author's contribution to the recent Liverpool symposium on Mechanism of Igneous Intrusion (Newall and Rast, 1970). The structures are strikingly different from those formed when the rheological parameters of the buoyant material and the overburden are close. The magma-imitating models (Fig. 22) do not develop a spreading lens similar to the lenticular body of anomalous mantle material believed to exist below the Mid-Atlantic Ridge and the Red Sea rift. Such a lens does develop in models simulating the rise of a crystalline batholithic body in crystalline rocks (Figs. 19 and 23). On the other hand, the faults and the subsided central block of the rift valleys are features consistent with a *local* occurrence of magma just below

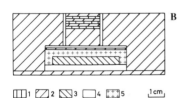

FIG. 22 (A) Model simulating the dynamics of a fluid magma surrounded by solid rocks. Initial structure shown in B. The model magma has found its way to the surface along a narrow fracture on the right-hand side of the sunken central block (from Newall and Rast, 1970). (B) Initial structure of magma-imitation model shown in A. (1) White plasticine, $\rho = 1.78$ g/cm^3; (2) black plasticine, $\rho = 1.68$ g/cm^3; (3) KMnO$_4$ solution; (4) white painter's putty, $\rho = 1.85$ g/cm^3; (5) gray painter's putty, $\rho = 1.85$ g/cm^3.

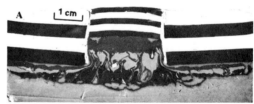

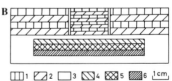

FIG. 23 (A) Model simulating the rise of a crystalline batholithic body surrounded by crystalline rocks. Initial structure shown in B. The "batholith" is too viscous and stiff to penetrate along the narrow zones of weakness on the sides of the central block which therefore has been lifted as a roof above the rising "batholith." Note contrast with the magma-imitation model shown in Fig. 22A. (B) Initial structure of crystalline batholith model shown in A. (1) and (2) Same as in Fig. 22B; (3) painter's putty, $\rho = 1.85$ g/cm^3; (4) silicone putty, $\rho = 1.58$ g/cm^3; (5) silicone putty, $\rho = 1.44$ g/cm^3; (6) silicone putty, $\rho = 1.26$ g/cm^3.

the faulted and subsided block (Fig. 22). It is tempting to suggest that, as a result of decreased pressure, local melting takes place in a rising plastic but crystalline mass, and that the melt penetrates fractures and surfaces as lava while a large crystalline mushroom spreads laterally below the crust, or at least below the uppermost part of the crust.

References

Beloussov, V. V., 1966, Modern concept of the structure and development of the Earth's crust and the upper mantle: *Geol. Soc. London Quart. Jour.*, v. 122, p. 293–314.

Berckhemer, H., 1968, Topographie des "Ivrea-Körpers" abgeleitet aus seismischen und gravimetrischen Daten: *Schweizer. Mineral. Petrol. Mitt.*, v. 48, p. 234–246.

Clark, S. P., 1966, Handbook of physical constants: *Geol. Soc. Am. Mem. 97*, 587 p.

Eskola, P., 1948, The problem of mantled gneiss domes: *Geol. Soc. London Quart. Jour.*, v. 104, p. 461–476.

Gutenberg, B., 1959, The asthenosphere low-velocity layer: *Ann. Geofis.*, v. 12, p. 439–452.

Hubbert, M. K., 1937, Theory of scale models as applied to the study of geologic structures: *Geol. Soc. Am. Bull.*, v. 48, p. 1459–1520.

Kaminski, W., and Menzel, H., 1968, Zur Deutung der Schwereanomalie des Ivrea-Körpers: *Schweizer. Mineral. Petrol. Mitt.*, v. 48, p. 255–260.

Newall, G., and Rast, N., eds., 1970, *Mechanism of igneous intrusion*: *Geol. Jour. Liverpool, Special Issue*, v. 2, 261 p.

Ramberg, H., 1963, Experimental study of gravity tectonics by means of centrifuged models: *Bull. Geol. Inst. Univ. Uppsala*, v. 42, p. 1–97.

———, 1967, *Gravity, deformation and the Earth's crust*: London, Academic Press, 214 p.

———, 1968a, Instability of layered systems in the field of gravity, I: *Phys. Earth Planet. Int.*, v. 1, p. 427–447.

———, 1968b, Instability of layered systems in the field of gravity, II: *Phys. Earth Planet. Int.*, v. 1, p. 448–474.

———, 1972, Inverted density stratification and diapirism in the Earth: *Jour. Geophys. Research*, v. 77, p. 877–889.

Smithson, S. B., 1964, The geological interpretation of the Slidre positive anomaly: *Norges Geol. Undersökelse Skr.*: v. 228, p. 270–283.

Talwani, M., LePichon, X., and Ewing, M., 1965, Crustal structure of the mid-oceanic ridges: *Jour. Geophys. Research*, v. 70, p. 341–352.

Talwani, M., Windisch, Ch. C., and Langseth, M. G., 1971, Reykjanes ridge crest: *Jour. Geophys. Research*, v. 76, p. 473–517.

Van Bemmelen, R. W., 1954, *Mountain building*: The Hague, Martinus Nijhoff, 177 p.

———, 1966, On mega-undations: A new model of the Earth's evolution: *Tectonophysics*, v. 3, p. 83–127.

Wegmann, C. E., 1930, Über Diapirismus: *Bull. Comm. Géol. Finl.*, v. 92, p. 58–76.

———, 1935, Zur Deutung der Migmatite: *Geol. Rundschau*, v. 96, p. 305–350.

Wenk, H. R., and Wenk, E., 1969, Physical constants of Alpine rocks: *Beitr. Geol. Schweiz. kleinere Mitt.*, v. 45, p. 343–354.

Zen, E-an, White, W. S., Hadley, J. B., and Thompson, J. B., Jr., eds., 1968, *Studies of Appalachian geology*: New York, Interscience, 475 p.

DAVID W. DURNEY
JOHN G. RAMSAY

Department of Geology
Imperial College of Science
London

Incremental Strains Measured by Syntectonic Crystal Growths

Those movements that take place in the Earth's crust and mantle as the result of changes in physical and chemical state at depth are expressed near the surface as vertical and horizontal displacements. The continental distribution of marine Mesozoic and Tertiary sediment basins indicates that widespread vertical uplifts take place, and the diapiric rise of masses of igneous rock testify to smaller scale vertical uplifts. The older ideas of continental drift have now been extended by newer hypotheses of sea floor spreading to support the original idea of large-scale horizontal movements of plate-like crustal slabs which are slowly rammed together to produce complex vertical and horizontal displacements in narrow orogenic zones. Geotectonic studies are aimed at deducing the displacement history of the earth on all scales to see if there are any recurrent patterns that might help us to assess the significance of these processes.

The structural geologist can investigate many aspects of this problem, for he is able to study the morphology of structures that have developed from displacements; these structures are now exposed at the surface, but many were formed at considerable depths and were later unroofed by erosion. In recent years there has been a concerted research effort aimed at measuring the amounts of internal distortion in naturally deformed rocks which has been caused by the displacement processes, first to assess the significance of the fold and fault structures seen in these rocks, and second to attempt to measure the large-scale shape changes that have taken place. So far, these investigations have tried to assess the value of the total or finite strain using techniques of measuring the forms of objects of known original shape, such as fossils, oolites, and pebbles embedded in the rock before it was deformed.

To gain a fuller understanding of the evolution and kinematic development of regional and local deformation we need to have information on the incremental strain history. The aim of this contribution is to describe some new techniques for determining the orientations and values of the principal strain increments during a natural progressive deformation, techniques which are based on a study of the growth forms of crystals developed syntectonically in the deforming rock. As a result of these studies Durney (1972a) has been able to determine the evolutionary history of part of the Helvetic nappes of the Alpine mountain chain and evaluate, from geometric evidence gleaned from the deformed rocks, the sequential deformation history of the nappes of central Switzerland, a deformation zone generally believed to have developed by gliding of the upper levels of the Alpine orogenic zone under the influence of gravitational force. As a result of the applications of these new techniques it is becoming possible to set out a more complete picture of the differences in deformational history at different levels in the nappe pile. The use of these new techniques in other orogenic zones lead us to believe that they may have widespread applicability.

Finite and Incremental Strain

Deformation is a process whereby points in a material continuum undergo an orderly displacement or change of position relative to each other; the distance between any two points is generally altered, and the angle between any two nonparallel lines is generally changed. If the displacement gradient is constant or nearly so over a small element of the body, the displacement differences in this element set up a state of homogeneous strain. Any original spherical marker of unit radius embedded in the element is deformed into an ellipsoid generally known as the *finite strain ellipsoid* (Thompson and Tait, 1879). In naturally deformed rocks these conditions often hold over the scale of a rock outcrop, a hand specimen, or a thin section. Many techniques are now available for assessing the shape and orientation of the finite strain ellipsoid using the deformed shapes of objects that may not have been spherical originally (see summary in Ramsay, 1967a, pp. 185–254), and the application of this finite strain ellipsoid concept has been most successful in helping to solve many problems in deformed rocks (Sorby, 1853, 1908; Harker, 1885; Cloos, 1947).

During deformation any originally spherical object embedded in the rock will undergo a series of progressive shape changes, and this process of sequential evolution is known as *progressive deformation* (Flinn, 1962). Figure 1 illustrates a progressive deformation sequence within a two-dimensional surface and depicts a series of evolving strain ellipse shapes as selected "stills" taken from a continuous "movie picture" at intervals $t_{,1}$ $t_2, \ldots, t_9$. At the end of the deformation (t_9) the total distortional effects are recorded by the shape of the finite strain ellipse. The nature of the evolution of the deformation may be best visualized by determining the amount of additional distortion necessary to transform any one ellipse in the series into the next. This additional distortion can be best appreciated by considering the shape changes that would affect a new unit circle emplaced in the material at some stage during the deformation. This circle (stage t_6) will be transformed into an ellipse (stage t_7) known as the *incremental strain ellipse* (Ramsay, 1967b). Because natural strain sequences are inevitably rotational (Ramsay, 1969, p. 52; Ramsay and Graham, 1970, p. 795), the principal axes of each successive incremental strain do not generally coincide, and they are not generally coaxial with the ellipse recording the total strain at the time of superposition of the strain increment. Because the general

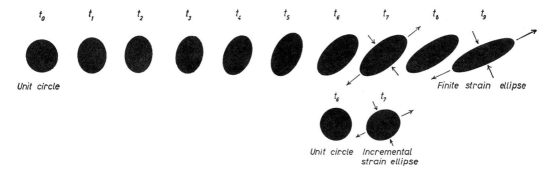

FIG. 1 A progressive deformation sequence in two dimensions. The shape of the strain ellipse at different stages indicates the total strain at successive periods during the deformation. The incremental strain ellipse at t_7 records the additional distortion between stages t_6 and t_7.

progressive deformation sequence in natural processes results from the successive accretion of small, noncoaxial incremental distortions, we have been seeking methods by which it might be possible to determine the nature of these sequential changes. Unfortunately nature does not provide a series of successively introduced circular or spherical objects into progressively deforming rocks, but some evidence of the incremental changes is occasionally preserved by the forms and orientations of folds and fissures which are developing during the deformation.

The studies we have made of the morphology of crystals growing in veins which are progressively opening indicate that the commonly observed fiber habit is controlled by the incremental dilations in these veins. It is very common for tensile and shear fractures to form in rocks deforming at high levels in the crust. In many environments these fractures are opened either as a result of a continuation of the deformation which led to their initiation or as a result of later and perhaps unrelated increments of deformation. These fissures become filled with crystalline material generally derived from the surrounding rock by solution or diffusion. Our studies indicate that many of these tectonic veins are filled synchronously with the fissure dilation and that there are no actual spaces available for vuglike crystal growth. The crystals preserved in such veins generally show an elongate fibrous or needlelike form quite unlike their normal habit. The long axes of these crystals are often curved—sometimes very markedly so—and our studies have shown that this curvature is best interpreted as the result of changes in the direction of progressive opening of the fissure (Ramsay, 1967a, p. 90, 248; Wickham and Elliott, 1970). We have also been able to show that many of the "striated" surfaces of faults are not slickenside grooves caused by the grinding of broken material between the fault walls, but that these are long crystal fibers growing in the direction of differential displacement across the fault.

There appears to be a wide variety of small-scale structures to be seen in naturally deformed rocks in which one finds syntectonic growths of fiberlike crystals. The separation zones of boudins, crinoid columns, and broken belemnites (Daubrée, 1879; Badoux, 1963; Ramsay, 1967a; Durney, 1972b) very commonly contain vein filling of needlelike form. Uneven strain distribution around resistant objects such as pyrite cubes, fossil fragments and detrital particles may lead to the development of pressure shadows around the object with fiber growth in the pressure shadow "eyes" (Mugge, 1930; Pabst, 1931; Durney, 1972b). The geometric forms of syntectonic crystal fibers will now be described, with particular emphasis on the diagnostic features of each type of structure which enable the incremental strain history to be deduced. This will be followed by a description of numerical techniques for calculating the values of the strain increments, and the results of applying these techniques to evaluate the progressive deformation history of part of the Helvetic nappes of Switzerland.

Syntectonic Crystal Fibers in Veins

Elongate growth textures are very commonly found in the crystal components that grew syntectonically in veins and fissures. Often these veins contain no cavities, and the crystal fibers show little or no sign of having grown into a cavity. Although the crystal needles frequently show a very strong shape orientation, it is very common for the degree of preferred orientation of their optic axes to be very poor, or like that of similar mineral species found in the vein walls, and the textures are not always consistent with inward growth from the walls. In many examples of fiber structure that we have examined the long axes of the crystals show a marked curvature, with the curvature changes of the crystals on one side of the vein being a mirror image of those on the other (Figs. 2, 5, 6, 9, 13, 14). The curving crystals do not appear to be significantly deformed; they show little or no sign of fracture or internal strain, and it is deduced that they were born with a curved form. Where a vein displaces some pre-existing structure in the rock so that the

FIG. 2 Siliceous limestone cut by a vein consisting of fiber crystals of intergrown quartz and calcite, Gstellihorn, Canton Bern, central Switzerland. The vein filling has probably been derived from the adjacent wall rock by processes of diffusion and solution transport.

total dilation across the vein can be calculated, the fiber length (straight or curved) always joins points on opposite walls of the vein which were once in contact. We therefore conclude that the fibers are not controlled by the orientation of the vein walls, but that they grow in a direction controlled by the differential displacement across the fissure. A number of different types of fiber growth have been recognized, each showing characteristic geometric forms that appear to be best explained by differing ways whereby each crystal in the vein undergoes progressive growth.

Type 1. Syntaxial Growth

This type of growth occurs in those veins where the crystalline filling of the vein is made of a species that is particularly common in the wall rocks of the vein (such as might occur in quartz veins cutting quartzites or calcite veins cutting limestones), and where the vein walls are irregular on a small scale and controlled by the shapes of the crystal boundaries on the wall. In these cases the initial fracture appears to have been guided along mechanically weak grain contacts in the original rock. Straight fiber growth is generally perpendicular to the walls, and in many occurrences we have been able to show that the vein filling is an overgrowth on the original wall rock grains and takes place in optical continuity with them—hence the term *syntaxial* (Fig. 3). Under these circumstances the degree of preferred optic orientation of the fibers at the vein edge is inherited from

FIG. 3 Quartz pressure shadow growth against euhedral pyrite in a sandstone, Parys Mountain, Anglesey, northern Wales. Note the syntaxial overgrowth of the quartz fibers in optical continuity with the crystals in the matrix (antitaxial against the pyrite).

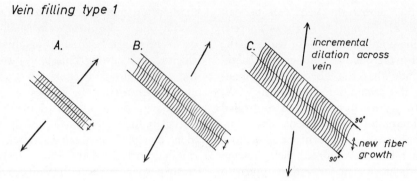

FIG. 4 Interpretation of the curved crystal fibers in veins showing syntaxial crystal growth.

that of the host material along the vein wall.

If the fibers are curved, they show perpendicular relationships with the trend of the wall at the edges of the vein and progressively lose this perpendicularity towards the center of the vein. At the vein center there is a surface which marks a break in optical continuity between the crystals growing from opposite sides (Fig. 4). All the geometric features we have noted are consistent with progressive fiber growth from the edge to the center of the vein as shown schematically in Fig. 4. The crystalline material is progressively added onto the wall rock grains in a direction parallel to the incremental direction of vein dilation at the time of the addition. *The directions of the earliest increments are recorded by the fiber lengths at the margins of the vein, and those of the last increments by the fiber lengths at the center of the vein.*

Type 2. Antitaxial Growth

The vein filling in this type of fiber growth is of a crystal species that is uncommon or absent in the wall rocks of the vein, as might occur where calcite veins have formed in a shale. The crystal fibers are in optical continuity across the vein and appear to grow from a median suture line toward the walls—hence the term *antitaxial*. The median suture is a centrally located surface marked by a line of inclusions of wall rock composition, and it generally shows undulations matching those seen at the vein walls. Curving fibers (Figs. 5 and 6) show symmetry about this median suture line; the fiber lengths at the median suture are subperpendicular to the trace of the suture, whereas at the margin of the vein the fibers are oblique to the vein wall (Fig. 7). All the features observed in veins of this type are consistent with progressive fiber growth along two surfaces at the vein-wall-rock contact, as illustrated in Fig. 8. *The directions of the earliest increments are recorded by the fiber directions at the central median surface, and those of the last increments by the fiber direction at the vein walls.*

Since new material is added along the mechanically weak vein wall, it is common for small fragments of wall rock to be incorporated as inclusions within the crystals. The fragment filled crystals are in marked contrast to the "cleaner" crystals formed by the process of syntaxial growth (cf. Figs. 6 and 3).

Type 3. Composite Growth

Composite veins consist of two or more species of crystalline components arranged in zones parallel to the vein walls. This crystalline material generally consists of components found in nearby rock but where one component is absent from the vein walls, such as might occur where quartz-calcite veins cut limestones (Fig. 9) or quartz-chlorite veins cut sandstone. The crystalline component of the vein which occurs along the vein walls is generally overgrown in optical continuity with crystals of similar species occurring in

FIG. 5 Antitaxial fiber growth of calcite in a vein cutting shale, Tertiary flysch, Tour Sallière, Canton Valais, Switzerland. Note the trace of the median suture and the perpendicularity of the crystal fibers to this trace. Field about 4 mm × 3 mm.

the host rock at the wall, whereas the central component generally shows a median line similar to that characteristic of antitaxial growth with very weak preferred optic orientation. The line of included minerals marking the median suture has a form generally corresponding with that of the adjacent vein walls, and it seems most likely that it originally formed in contact with the walls, and thus marks the first growth site of the central vein filling.

The interface of the two crystalline components that fill the vein is often highly serrated, and its shape is quite unlike that of the walls and the median suture. This shape seems to have originated from minor irregularities in the growth rates of the two crystalline components as they grew toward each other. The optic axes of several hundred quartz and calcite crystals in a composite vein like that illustrated in Fig. 9 were measured on a Universal stage. The preferred orientation of c-axes of calcite fibers is random at the wall where the crystals are welded to those of the limestone matrix, but it increases markedly from walls toward the interface. This phenomenon is analogous to the development of preferred orientation in cavity veins with comb structure (Bateman, 1959, p. 108) and is attributable to the selective growth of suitably oriented crystals (Spry, 1969, p. 162), in this case those with c-axes parallel to the fiber lengths.

There is also a slight development of

FIG. 6 Antitaxial calcite fibers in a vein cutting Cambrian slate, Nantlle, North Wales. The solid-inclusion loaded median suture is traversed by curving crystals in optical continuity across the suture. Field about 3 mm × 4 mm.

preferred orientation of the quartz away from the median suture and toward the quartz-calcite interface. The quartz fibers along the median surface in the areas of quartz together with calcite fibers at the vein edge are all perpendicular to the walls, whereas those near to the quartz-calcite interface are oblique to the wall trend. The composite veins are cut by later fissures, and the fibers in these late fissures are subparallel to the direction of the fibers at the quartz-calcite interface in the earlier veins. All these observations support the conclusion that *the earliest increments are recorded by fiber directions at the median suture and vein walls, and the last extensions are recorded by fiber directions at the interfaces of the two components which fill the vein* as illustrated in Fig. 10. This type of fissure filling is therefore a combination of syntaxial and antitaxial types described previously.

Type 4. "Stretched" Crystals

This type of fiber structure is found in veins with filling of similar composition to those crystals making up the greatest proportion of the host rock. This condition is similar to that of syntaxial growth (type 1), but this growth type of fiber differs in that the fibers are optically continuous right across the vein and there is no central surface separating fibers of differing optic orientation. These fibers appear to have been overgrown with optical continuity on planar surfaces of

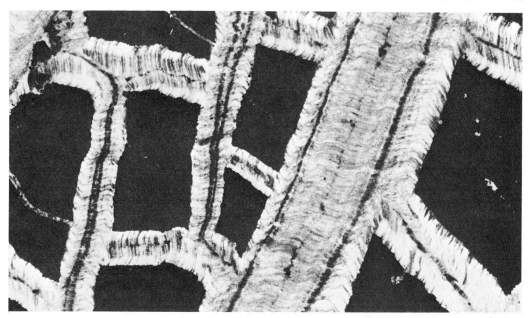

FIG. 7 Complex symmetrical antitaxial calcite and quartz fibers in veins transecting a pyrite layer in Lower Cretaceous marl, Morge valley, Canton Valais, Switzerland. Note well developed median sutures and growth zoning having the same shapes as the walls, indicating fiber growth at the pyrite walls. Field about 7 mm × 10 mm, parallel to lceavage.

original broken crystals along the initial fracture. The initial fracture does not appear to have been guided by the grain contacts in the original rock (Fig. 11 and Fig. 12, type 4). Fibers in veins of this type are generally subparallel and can be either perpendicular or nonperpendicular to the vein walls. The fibers are optically continuous with the separated parts of once single crystals which are now situated on opposite sides of the veins, and there seems to be no unique grain boundary at which grain growth could have occurred. They show a virtually constant width across the vein.

The clue to the origin of this texture is afforded by the presence of abundant planes

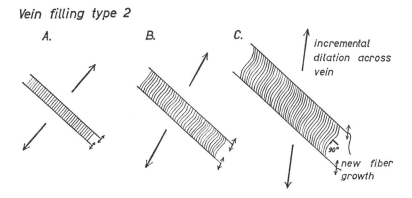

FIG. 8 Interpretation of fiber development in antitaxial growth veins with changing principal strain increment directions.

FIG. 9 Composite veins of quartz and calcite cutting limestone, Leytron, Canton Valais, Switzerland. The pale calcite is overgrown on the calcite of the wall rocks, and the central zone of quartz shows a very marked median suture. Plane subparallel to cleavage.

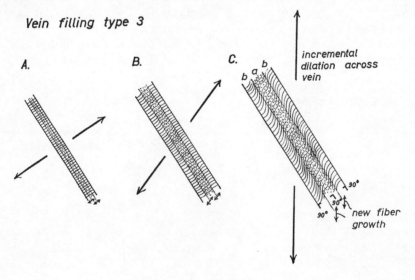

FIG. 10 Fiber formation in a composite vein during progressive opening by changing principal increments of strain. Material b is in optical continuity with the wall rock crystals, and mineral components a and b are added synchronously at their common interface.

FIG. 11 Development of "stretched" crystal dolomite fibers in dolomite rock, Gstellihorn, Canton Bern, Switzerland. Fibers link separated halves of grains in the walls and show the finite displacement vector. Note also the swarms of faint microfractures around the vein tips. Field 2 mm × 3 mm.

of microscopic fluid inclusions representing healed microfractures generally aligned subparallel to the vein walls. It appears that vein dilation took place by repeated microfracturing as individual crystals were "stretched," with the progressive healing by overgrowth of each fracture. It is not possible to predict with confidence the location of the fracture formation at any particular stage during the sequential deformation, and these fibers can only be reliably used for the determination of the total displacement vector across the vein walls.

Because stress concentrations preferentially develop around earlier former fluid inclusions, the microfracturing and overall extension tend to be confined within the vein walls. At times the microfractures could propagate into the host rock or could become initiated independently outside the vein. This would explain the commonly observed random and clustered distributions of this type of vein (Fig. 11), unlike the much more regularly spaced distribution of antitaxial and composite veins.

The characteristic features of these various types of syntectonic fiber growth in veins are summarized in Fig. 12.

Nontectonic Cavity Growths in Veins

In an open cavity crystallization occurs at the fluid-crystal interface, and the crystals tend to grow into the cavity from the walls. The crystals formed in this way generally form parallel or radiating prisms oriented subperpendicular to the growth surface. They usually show an increase in grain size and degree of preferred crystallographic orientation along the direction of progressive growth and are often characterized by euhedral terminations (comb structure) and euhedral growth zones. Clearly the growth directions of these crystals bear no relationship to any displacement by which the cavity may have formed, and they cannot be used to highlight any features of the deformation process.

Syntectonic Crystal Fibers in Pressure Shadows

In deformed rocks containing fragments of composition different from the host material uneven strains often are set up around the fragment. If the fragments are more competent than the host material (e.g., pyrite or crinoidal plates), and if the contact between the fragment and host is mechanically weak, it is common for the matrix to pull away from the fragment in the direction of maximum elongation and so develop a pressure-shadow zone. Crystalline material commonly grows in the pressure shadow and shows a fibrous habit similar to that seen in tensile veins (Cloos, 1971; Durney, 1972a). All the types of crystal growth described previously have been observed, with the exception of the type called "stretched" crystals.

Syntaxial growth is seen in the calcite-filled pressure-shadows around crinoid ossicles which occur at many localities in the Helvetic Alps in strata ranging from the Lias to the flysch (Durney, 1972b, p. 316). The specimens described here were found at Leytron in the Valais and their features seem fairly typical of those we have observed from other localities.

The calcite in these pressure-shadows is a monocrystalline syntaxial overgrowth with tapering and serrated external boundaries seen in cross section, which was built up outward from the crinoid walls. A fibrous texture is often imparted to the calcite growth by acicular inclusions of quartz or calcite, which often have a sigmoidal form. The geometric form of these fibers is illustrated in Fig. 12, 1A and 1B.

Antitaxial growth is most common around pyrite cubes and concretionary bodies made up of aggregates of small pyrite crystals.

Pressure Shadows around Concretionary Pyrite

Well-developed pressure-shadows filled with sigmoidal fibrous quartz or calcite (or a

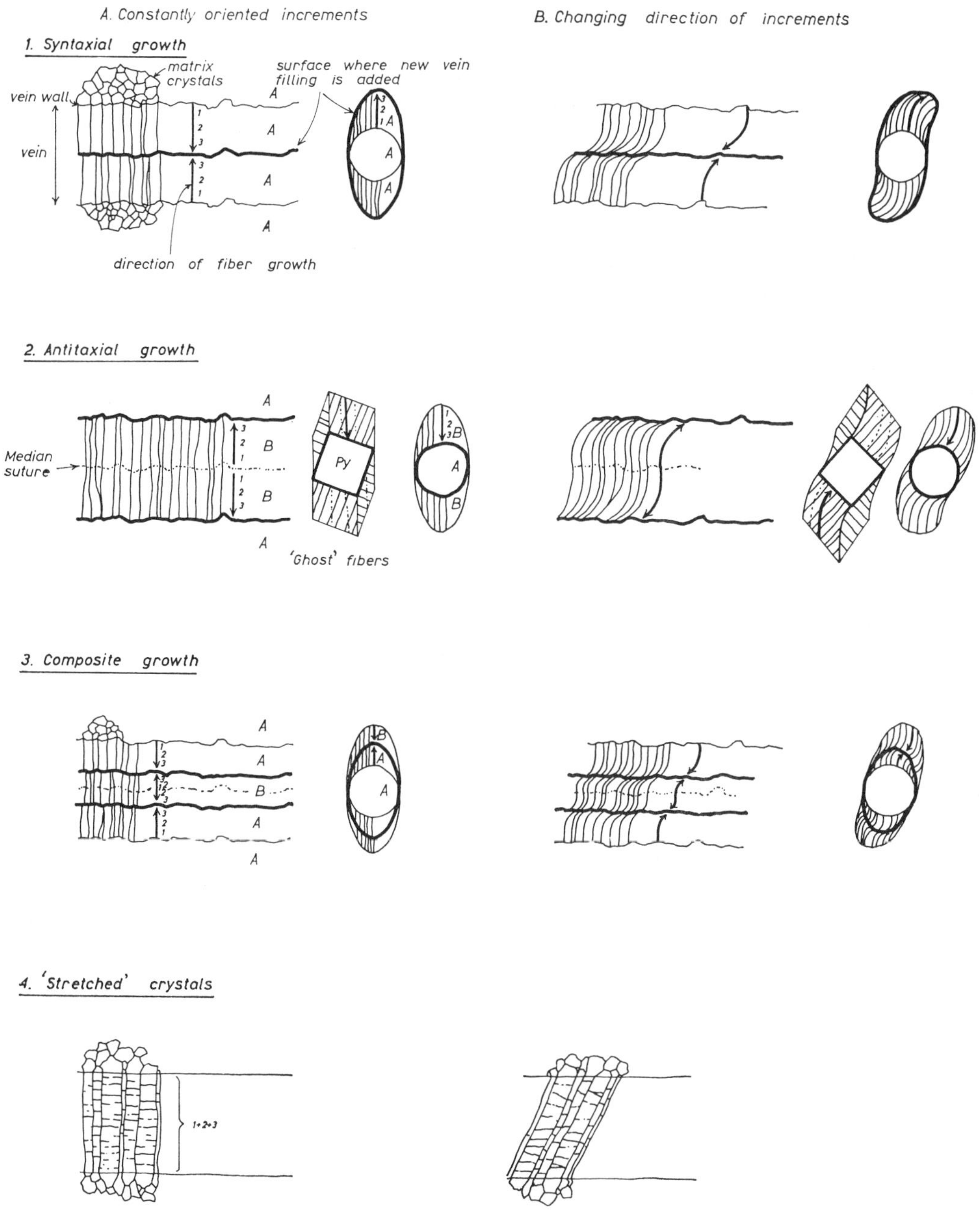

A. Constantly oriented increments

1. Syntaxial growth

2. Antitaxial growth

3. Composite growth

4. 'Stretched' crystals

B. Changing direction of increments

FIG. 12 Synoptic diagram showing the various types of fibrous vein filling and pressure-shadow structure under conditions of constantly oriented principal incremental elongation and of changing orientation of these increments. Materials of differing mineral species are indicated A, B, and Py (a pyrite cube) respectively. Arrows indicate direction of progressive fiber growth, and numbers *1*, *2*, and *3* the sequential development of any one mineral species. The dotted lines in the pressure shadow zones around the cube of pyrite are the "ghost" fibers which form parallel to the principal incremental extension directions.

quartz-calcite intergrowth) frequently occur around small, subspherical concretionary masses of pyrite (less than 0.5 cm diameter). They have been found in all the post-Triassic formations of the Helvetic Alps with the exception of the Urgonian and Nummulitique limestones (Durney, 1972b, p. 316).

The pyrite concretions are composed of an aggregate of fine grains, sometimes in the form of very small cubes and radiating prismatic crystals or, more commonly, small globules of cryptocrystalline material about 10–20 μ diameter (Fig. 13). These globules are known as framboidal pyrite (Love, 1957, 1967; Rickard, 1969; Berner, 1970; Farrand, 1970) and are a characteristic feature of early diagenesis of sediments associated with the activity of anaerobic sulfate-reducing bacteria. When observed under the electron microscope, these globules or framboids are seen to be composed of smaller globules about 0.5 μ diameter, which in turn have a granular texture. This pyrite is therefore exceedingly fine grained.

These concretions are clearly pretectonic in origin and behaved in a completely rigid manner. The pressure-shadows formed around the concretions probably started to grow as soon as ductile extension commenced in the surrounding rock.

The fibrous crystals making up the pressure-shadow fillings are generally parallel and practically undeformed (very weak twinning in calcite, little or no undulose

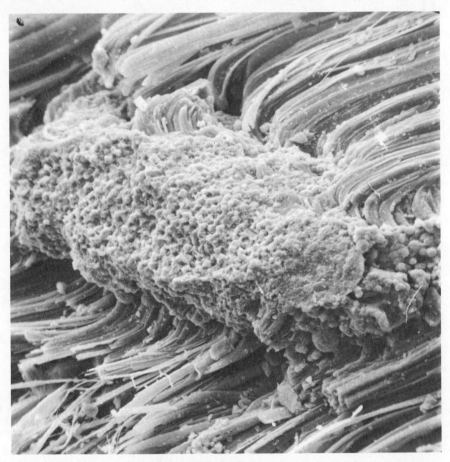

FIG. 13 Electron micrograph of slender sigmoidal quartz fibers in a pressure-shadow structure against a framboidal pyrite aggregate, same locality as Fig. 14. Quartz fibers range from 0.1 to a few microns diameter. Field 150μ, parallel to cleavage.

extinction (Fig. 14)). They show sigmoidal curvature with triclinic symmetry, the center of symmetry being located at the center of the pyrite mass (Fig. 13). The external boundaries of the pressure-shadows are not tapered or serrated; they correspond in shape to those of the host walls, as do any growth zones that may be present within the infillings. This signifies that the external boundaries and growth zones were once in contact with the pyrite, so the material forming the fibers was added at the face of the pyrite. Several other lines of evidence support this conclusion: (1) late microfractures are oriented normal to those fibers closest to the pyrite; (2) the outermost fibers are in optical continuity with grains of the matrix; (3) incipient polygonization is found in the outermost fibers of some specimens; and (4) the sense of fiber curvature is opposite to that seen in pressure-shadows around crinoids (Fig. 12, cf. 1B and 2B) (the two structures occur together in some thin sections).

Pressure Shadows around Euhedral Pyrite

Pressure-shadows around single crystals of euhedral pyrite are similar to those around concretionary pyrite except that the fibers are oriented perpendicular to the pyrite crystal faces, or very nearly so (Ramsay, 1967a, p. 181) and there is generally a marked increase in grain diameter toward these faces. The progression of fiber growth in this type of structure was deduced by Mugge (1930). It is

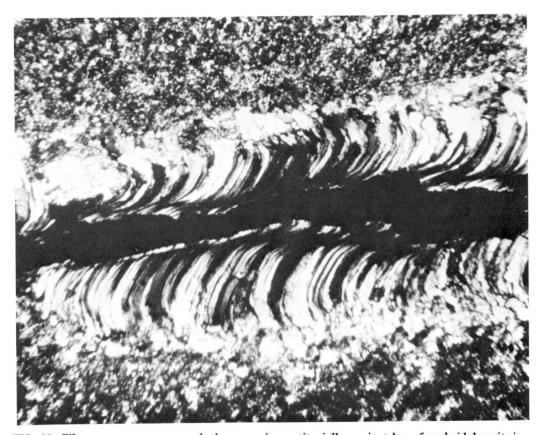

FIG. 14 Fibrous quartz pressure-shadow growing antitaxially against long framboidal pyrite in a siliceous marl, Leytron, Canton Valais, Switzerland. The axis of each crystal of quartz curves through an angle of 90 degrees without change in optical orientation. The fiber crystals are growing with optical continuity on the quartz crystals in the matrix. Field about 3 mm × 4 mm, parallel to cleavage.

the same as that of concretionary pyrite pressure-shadows, with new crystalline material being added at the pressure-shadow-pyrite interface.

A petrofabric investigation was made of quartz-calcite intergrowth fibers in specimens from pyritiferous calcareous slates from Leytron. The quartz c-axes were found to be quite random, both in the small outermost grains and the large innermost grains. This observation agrees with that of Pabst (1931). The c-axes of large calcite grains plotted with respect to the pyrite face showed some slight departure from a random pattern.

In some specimens, parallel sigmoidal "ghost fibers" cross-cut the wall-controlled fibers (Fig. 12, 2A and 2B). The former appear to trace out the displacement path of the external boundary and evidently show the true orientation of crystal growth. It is also apparent that wall-controlled fibers are not necessarily aligned parallel to the growth direction.

The wall-controlled fibers cannot be used to interpret the incremental extension strain because they are not generally parallel to the extension. Where the pyrite faces lie approximately perpendicular to the plane of section it is possible to use either the shape of the partition line (which separates two domains of fiber orientation), the lateral boundaries of the pressure-shadow, or the "ghost fibers," to compute the incremental extensions (Fig. 12, 2A and 2B).

All transitions have been seen from the partition lines of euhedral pyrite pressure-shadows to the fiber traces of framboidal pyrite pressure-shadows, depending on the size of the pyrite crystals.

Summary of Criteria Used to Determine the Growth Sense of Crystal Fibers

If the crystal fibers which originated by "syntectonic growth" are not appreciably modified by recrystallization or internal strain, it is necessary to establish the sense of crystal growth using the following criteria:

1. Crystals attached in optical continuity to grains in the walls grew inward, away from the walls (syntaxial crystals) with the exception of "stretched" crystals.
2. Progressive increase in grain diameter occurs in the direction of growth.
3. Progressive increase in degree of preferred crystallographic orientation occurs in the direction of growth.
4. If there is a difference in orientation of the crystal fiber between its central and marginal zones, the zone in which it is more nearly perpendicular to the vein walls is more likely to be the site of earliest growth (and vice versa) because of kinematic consistency with the extension that initiated the fracture.
5. Within the confines of the two vein walls there is usually an odd number of more or less distinct surfaces approximately parallel to the walls and symmetrically disposed about the vein center.
    a. Surfaces similar to the walls in shape indicate that growth occurred continuously at the walls and therefore in the direction away from the center toward the walls (referred to here as antitaxial crystals). These surfaces may be growth zones, although a more common form is seen as a median suture line down the center of the vein representing the site of the earliest growth.
    b. Surfaces which are significantly more serrated in outline than the vein walls result from the amplification of local growth irregularities during continuously convergent growth from opposite directions. With the exception of serrated growth zoning these surfaces therefore mark the site of latest growth and almost invariably coincide with a mineralogical interface.
6. The veins which are restricted to competent stretched layers or objects sometimes neck down toward the site of latest

growth (provided that it is not situated at the walls).
7. Growth sense in a new structure can be deduced by comparing its fiber curvature with that of another structure from the same locality in which the growth sense is known.

For pressure-shadows analogous criteria may be used, taking "wall" to mean "competent host object contact." Thus all the criteria except perhaps 4 are applicable. Criteria 1, 2, 3, and 7 are fairly straightforward. Pressure-shadow ends which are similar in shape to the host are homologous to the median suture and imply growth from those ends toward the host, whereas those with serrated external ends imply growth away from the host. Pressure-shadows with tapered ends in XZ section or in the section perpendicular to cleavage indicate growth toward the ends, whereas if there is no noticeable thickness change along the pressure-shadow in this section, growth at and toward the host is indicated.

Mechanisms of Syntectonic Fiber Formation

Syntectonic fibers do not possess textures characteristic of free growth in cavities. It seems unlikely that the material of which they are composed was introduced by circulating solutions; it is probable that this material migrated from the adjacent rock by diffusion. Pressure-solution textures are usually associated with these fibers (Durney, 1972b), so the most likely source of the material seems to be pressure-solution activity in the surrounding rock.

Growth will take place at an unfractured grain boundary (probably when the hydrostatic pressure is less than the least compressive stress; otherwise hydraulic fracturing may occur) when the solute concentration (activity) of the grain boundary fluid exceeds the equilibrium value for the particular conditions of normal stress, fluid pressure, and temperature. The crystal growth will tend to be restricted to faces subjected to relatively low normal stress in a differentially stressed aggregate. Under these circumstances, crystal growth causes dilation, a phenomenon usually termed the "force of crystallization" (Correns, 1949), and demonstrated by laboratory experiments (Duvernoy, 1852; Lavalle, 1853; Klocke, 1871; Lehmann, 1904, p. 138; Becker and Day, 1905, 1916; Taber, 1916; Correns, 1926, 1949; Correns and Steinborn, 1939; Brehler, 1951).

The direction of growth is probably controlled by the orientation of the least principal stress at the growth surface, because least work is done against the external stress system when growth occurs in this direction. At grain boundaries the least stress may be oriented obliquely to the growth surface, so growth may occur in almost any direction depending on the orientation of the least stress at any particular time. In contrast to this, free crystal growth in a fluid-filled cavity always takes place perpendicular to the growth surface (which is itself free to be controlled by crystal habit) because the hydrostatic fluid pressure dictates that a principal stress must act perpendicular to the surface and this is generally the least stress.

The most rapid growth should occur at suitably oriented grain boundaries which have the highest coefficient of solute diffusion. According to Spry (1969, p. 14), the rate of grain boundary diffusion (and hence the rate of growth) increases with the degree of misfit between the lattice structures of the two crystals in contact. The following empirical sequence of crystal interfaces arranged in order of increasing preference for growth (increasing mechanical weakness) appears to coincide quite well with a sequence of increasing lattice misfit. This sequence is based on an interpretation of the positions of the growth surfaces of examples described previously using petrographic criteria. Growth occurs at the lowest interface in the sequence in preference to interfaces higher in the sequence:

1. The same mineral in optical continuity (no grain boundary diffusion).
2. The same mineral in optical discontinuity (e.g., a calcite-calcite grain boundary).
3. Two different ionic minerals with related lattice structures (e.g., calcite–dolomite).
4. Two different ionic minerals with unrelated lattice structures (e.g., quartz–calcite).
5. Two different minerals, one ionic and the other semimetallic (e.g., quartz–pyrite or calcite–pyrite).

Furthermore, the fact that the diffusion coefficient is the same for the same mineralogical interfaces explains why the growth at two adjacent growth surfaces is symmetric.

The mechanism of fracturing which initiates most veins, and the mechanism of "stretched" crystal growth differ markedly from the grain boundary growth mechanism just described. They appear to take place under conditions where rate of grain boundary growth cannot accommodate the progressive build-up in effective tensile stress.

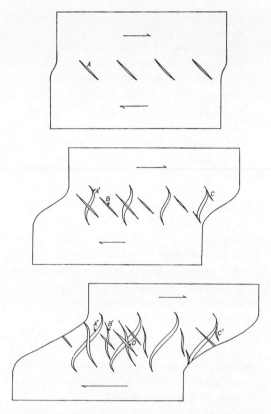

FIG. 15 Sequential development of arrays of en-echelon and sigmoidal tension veins in a progressively evolving shear zone. Arrows indicate the sense of relative shear displacements across the shear zone.

Vein and Fiber Development in Progressive Simple Shear Deformation

To check the previous conclusions on vein and fiber growth during a progressive deformation sequence we examined in some detail the geometric forms of veins and fibers in the en-echelon vein systems formed in shear zones. The general patterns of these vein systems is well known (Shainin, 1950; Wilson, 1961; Ramsay, 1967a; Roering, 1968; Ramsay and Graham, 1970). The initial veins are formed in en-echelon arrays, with individual veins inclined at 45° to the general trend of the shear zone (Fig. 15, A). Increasing shear leads to a rotation of the initial veins to a position which is no longer perpendicular to the directions of the later principal incremental strain extensions; the ends still continue to propagate at 45° to the shear zone, but the central part of the vein now opens in a direction that is no longer perpendicular to the vein walls (Fig. 15, A'). At some stage of rotation the central part of the vein is oriented unfavorably for further dilation, and new vein systems are initiated which cross-cut the earlier formed veins (Fig. 15, B and C). Further shear leads to further rotation of the original veins (Fig. 15, A''), to rotation of the secondary veins (B' and C') and to the formation of new veins (Fig. 15, D).

The geometric forms of the fiber structures developed within these en-echelon and sigmoidal vein arrays depend on the nature of the vein filling, wall rock composition, and strength of grain contacts. In the Upper

Devonian strata of southwest England (west of Ilfracombe) shear zone arrays of quartz veins cut sandstones, and the fiber filling appears to have developed either by syntaxial or by "stretched" crystal growth. Where the shear values are low the zone consists of a series of planar en-echelon veins with fibers perpendicular to the vein walls (Fig. 16A). Where the veins have been further deformed into a sigmoidal form the fiber structure in the central part of the vein shows a curving form; at the vein margin the fibers welded to the quartz grains in the walls show a consistent perpendicular relationship to the walls, whereas those fibers in the central part of the widest veins are not perpendicular to the walls but are oriented at 45° to the direction of the shear zone, parallel to the last direction of maximum incremental extension. These last formed fibers are parallel to those at the tips of the sigmoidal veins (Fig. 16B). At some localities the early formed veins are markedly sigmoidal and cross-cut by late veins, and the fibers in these late veins are aligned parallel to those in the center of the sigmoidal vein. These relationships clearly accord well with the ideas of progressive fiber growth in the direction of maximum incremental extension. Some of the veins show no central surface dividing the quartz filling into parts as is characteristic of syntaxial growth. The most likely explanation of these continuous wall to wall quartz fibers is that they were formed by "stretched" crystal growth mechanisms, with a concentration of late stage microfracturing near the vein center. Fiber patterns characteristic of antitaxial growth at the vein contacts have been described in sigmoidal en-echelon arrays from the Monts d'Ougarta of Algeria by Donzeau (1971).

Fiber Growth along Fault Surfaces

The surfaces of many natural faults often have a smooth polished appearance, and on these slickensided surfaces linear grooves are found. These grooves or striations are generally interpreted as being the result of the gouging action of material broken from the fault walls as the fragments are crushed between the moving fault walls. The linear trends of the striae are generally interpreted as representing the directions of the last differential slip movements of the fault, since any early formed striae are destroyed by later grinding motions. The "roughness–smoothness" principle accepted by geologists for many years as a practical field technique for

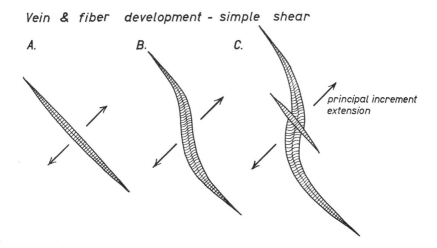

FIG. 16 Progressive evolution of stretched crystal growth fibers in the shear zone illustrated in Fig. 13.

determining the direction sense of the relative displacement of the fault walls (Stočes and White, 1935; Goguel, 1952; Hills, 1953; Billings, 1954) was seriously questioned after work by Paterson (1958) and Riecker (1965) showed that the steps on faults produced during experimental rock deformation were opposed to the sense of motion on the fault surface. Tjia (1964) described field data which appeared to conform with this re-interpretation, and Hills (1963, p. 177) subsequently revised his interpretation of the "roughness–smoothness" principle. Gay (1970) confirmed Paterson's findings for the surface features on initial shear fractures but showed that subsequent movement on these fractures tended to pluck away the early formed steps and produce a new step structure more in accord with the traditional interpretation of fault surface morphology.

Our studies of the surfaces of faults of known differential displacement have consistently confirmed the traditional interpretation of the "roughness–smoothness" technique, but we have been able to demonstrate that the steplike treads that give rise to this feature are not generally formed by cataclasis at the fault surface. In many of the faults we have investigated it is possible to demonstrate that the striae are due to the *growth of crystal fibers along the fault plane*, and that these growth fibers show very close analogies to those seen in tensile veins and in pressure-shadows. A typical example of such striation fibers on a fault surface is shown in Fig. 17. The linear structure consists of long parallel

FIG. 17 Overlapping step-like crystals of fibrous calcite developed on a fault surface, Oviedo, northern Spain.

fibers of quartz with their long axes almost parallel to the fault surface. Individual crystals are sometimes many hundreds of times longer than they are broad and yet show little or no sign of internal deformation The ends of the fibers generally are attached to opposite walls of the fault, and the fiber axes usually cross the fault at a small angle (0–5°). Where we have been able to calculate the total displacement vector across the fault we have found that the length of individual crystal fibers is generally equal to that of the differential displacement vector.

Our explanation for this morphology is shown in Fig. 18A. Since an initial shear fracture is rarely if ever perfectly planar, there will always be a geometric problem posed when the two sides of the fracture move past each other. The protruding high points may be planed away to produce fault breccia and clay gouge; if this does not occur, spaces will be opened up as the high points ride over each other. Since faulting at shallow and medium depths in the crust is commonly associated with conditions of high pore water pressure (Price, 1966), the second of these alternatives is more usual in natural situations. If crystal growth takes place on these surfaces and keeps pace with the progressive separation of the adjacent walls, the morphology of the crystals is likely to be governed by the same principles we have established for those forming in progressive dilating tensile fissures. Because the differential displacement vector across the fault is close to the fault surface, however, the fibers have to

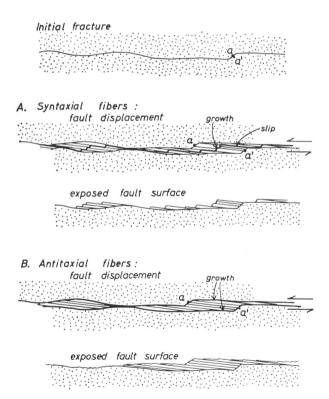

FIG. 18 Formation of crystal fibers along a fault surface by two different mechanisms. The fibers form at a low angle to the fault, and their total length is a measure of the total separation across the fault of two originally adjacent points a and a'. When the fault plane is exposed the fibers formed by both mechanisms have a characteristic step-like structure, a feature which gives rise to the "rough-smooth" effect when the surface is touched with the hand.

develop in this direction and in consequence their long axes will make a small angle with the fault surface. The result is to produce a series of overlapping needlelike fibers (Fig. 18), which, on exposure of the fault surface, have a surface expression as a series of long steps parallel to the fiber axes separated by smaller steep "risers" where individual fibers break across their length (Figs. 17 and 18). This arrangement would clearly account for the "roughness–smoothness" effect when the fault plane is stroked with the hand, and it agrees with the traditional interpretation of this feature. It contradicts the reinterpretation of Hills (1963, pp. 175–177 and his Figure VII-17) of structures he terms "pinnate shears."

The interpretation that slickenside striations give the direction of the last differential displacements on the fault should also be reassessed if these striations represent fiber crystals. The fiber length will be equal to the total differential displacement vector of the fault movements taking place during active growth of the crystal fibers. Any change in the incremental differential displacement vector will be recorded by a change in the orientation of the long axes of a single fiber according to the principles previously established which govern fiber growth in tensile fissures. The interpretation of the directions of fault surface fibers will depend on the type of growth, whether syntaxial, antitaxial, or composite. In most faults we have examined syntaxial growth seems particularly common; the directions of the central parts of curving fibers converging from each wall of the fault indicate the direction of the differential displacement vector during the last stage of fault movement, and those at the fiber ends correspond to the first stages of fault movement. Successively overlapping curved sheaves of fibers give rise to differently oriented fibers at different levels in the crystalline filling along the fault (Fig. 19).

Antitaxial fault fibers show the reverse sense of growth (from center towards walls) and the displacement history is interpreted accordingly, but the same "roughness—smoothness" sense is maintained where the fibers have broken away from the walls (Fig. 18B). "Slip-fiber" asbestos veins found in the Archean of Western Australia, show the characteristics of antitaxial fault growth.

Computation of Incremental Strains from Syntectonic Crystal Growth

Incremental Strains Measured from Veins

It has been demonstrated that curved crystal fibers arise during progressive deformation and that the orientations of the fibers are parallel to the principal incremental elongation. In certain circumstances it is possible to determine the actual values of these incremental extensions by measuring the fiber lengths.

As an example of these techniques we will give an account of measurements made on fibrous veins found in boudinaged limestones in the Liassic slates of Leytron, Canton Valais, Switzerland. The marl slates here occupy the core (root zone) of the Morcles Nappe, the lowest nappe in the Helvetic pile. The limestone layers are generally less than 0.5 cm in thickness and show small-scale boudinage (microboudinage), with the development of fissure veins trending ENE. The fissures are completely filled with sigmoidal fibers of quartz and calcite (Fig. 9), which have developed according to the principles of composite fiber growth (Figs. 10 and 12), the curves lying within the plane of the slaty cleavage.

To compute the extensional strain it is necessary to know the width of the limestone boudins and how much deformation took place within them. Judging from the equidimensional shapes of the original micrite grains the limestone seems to have undergone little internal deformation. The boudin width

FIG. 19 Curved overlapping groups of quartz fibers on a fault surface which are interpreted as being the result of progressive clockwise change in the direction fault movement, Hartland Quay, northern Devon.

appears to have remained approximately constant, and the early fissures (and the earliest fibers) appear to have formed at a very early stage in the deformation. Subsequent extensions in the microboudinage were brought about chiefly by dilation of fissures.

Large drawings of the upper surface of specimens were prepared, each including about 20 complete (wall to wall) fibers. Starting at the sites of earliest growth, sections of these quartz and calcite fibers were identified, having orientations of 0, 10, 20°, and so on, and the total chord length for each 10° interval was recorded. The limestone boudin width corresponding to one wall-to-wall fiber was measured as the sum of half the boudin width at one end of the fiber plus half the boudin width at the other end of the fiber. Such widths were measured for each complete fiber and the total width was recorded.

Two different methods were then used to compute the incremental strains (Fig. 20). These have been called the "standard" and the "convinc" (converted increment) methods, and they give slightly different results. The convinc method is theoretically sound for a single set of fissures, whereas the standard method is best applied to "chocolate tablet" structures (Ramsay, 1967a, p. 113).

In the standard method it is assumed that the axis of incremental X-elongation

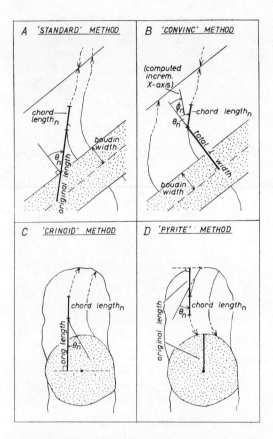

FIG. 20 Methods for determining the values of the principal strain increments using fibrous vein structure (*A* and *B*), syntaxial crystals in pressure-shadows (*C*) and antitaxial crystals around framboidal pyrite aggregates (*D*).

coincides with the fiber axis or chord axis (Fig. 20*A*). The incremental strain for the nth 10° fiber chord is then given by

$$\text{incremental } e_{1_n} = \frac{\text{increase in length}}{\text{previous length}}$$

$$= \frac{\text{chord length}_n \cos \theta_n}{\sum_0^{n-1} \text{chord length}_i \cos \theta_i + \text{boudin width}}$$

where θ is the angle between a particular chord and the normal to the boudin walls and i is a previous fiber chord. Separate calculations were made for the strains associated with the late fissures, using the width of the broken quartz deposits as the original length instead of limestone boudin width. The results of this technique applied to specimens from two localities (71 and 74) are shown in Fig. 21.

In the convinc method the fiber chords are considered to be strain displacements, and it is assumed that no displacements occur simultaneously in any other direction (Fig. 20*B*). The two-dimensional displacement transformation for this is

$$x_1 = ax$$
$$y_1 = cx + y$$

or, in matrix notation,

$$\begin{pmatrix} a & 0 \\ c & 1 \end{pmatrix}$$

where the x-axis is chosen perpendicular to the fissure and the y-axis parallel to the fissure (in this case y is parallel to the length of the boudins), and where

$$a = 1 + \frac{\text{chord length}_n \cos \theta_n}{\sum_0^{n-1} \text{chord length}_i \cos \theta_i + \text{boudin width}}$$

and

$$c = \frac{\text{chord length}_n \sin \theta_n}{\sum_0^{n-1} \text{chord length}_i \cos \theta_i + \text{boudin width}}$$

The incremental strain ellipse for the nth fiber chord is then found by solving the equations

$$(1 + e_1)_n^2 \text{ or } (1 + e_2)_n^2$$
$$= \frac{a^2 + c^2 + 1}{2}$$
$$\pm \tfrac{1}{2}\sqrt{(a^2 + c^2 + 1)^2 - 4a^2}$$

and the orientation of one of the principal incremental strain axes by

$$\tan 2\phi_n = \frac{2ac}{a^2 - c^2 - 1}$$

where ϕ_n is the negative angle between strain axis and normal to the fissure.

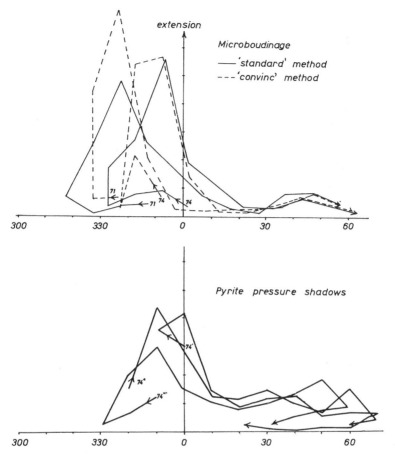

FIG. 21 Changes in the values of incremental extension with change of direction of these principal extensions using the methods described in the text. The data are from Liassic limestones and calcareous slates at Leytron, Canton Valais, Switzerland in the root zone of the Morcles nappe. Increment scale graduated in $e_1 = 0.1$ units.

The results for specimens from localities 71 and 74, using the data as for the standard method, are plotted in Fig. 21. The results are reasonably consistent over the locality; in each graph there is an early reversal of orientation (an inflexion in the fibers) and a major clockwise rotation with a strong early peak between north and northwest and a weaker late peak northeast. There is some spread in orientation of the early peaks, but they all lie between N and NW. It is evident from these graphs that the greatest extension took place between N and NW, even though the stretching lineation parallel to the late peak extension in the slates is oriented in a NE direction.

Incremental strains were computed from the geometry of fibers in pressure-shadow zones of crinoids, assuming the crinoids to have been rigid (they have been neither appreciably distorted nor strongly twinned). The incremental strains were computed by measuring the lengths of fibers as shown in Fig. 20C. The average incremental strains for successive 10° intervals of several fibers in 10 specimens produced results remarkably similar to those derived from microboudinage veins, with the exception that the early NNW peak was rather weak. This difference might be explained by the effects of the high temperature that existed at the time of the NNW extension (homogenization temperature of fluid inclusions in calcite fibers is approximately 400° C). This high temperature would have led to a low calcite solubility and a sluggish rate of calcite solution-transfer.

The pressure shadow fibers around crinoids registered extensions that were less than those of the matrix, presumably because of preferential flow of matrix around the ends of the fragments. The fiber directions are believed to be fairly reliable indicators of the extension directions in the matrix because the equant shapes of the structures would lead to little differential body rotation of the fragments relative to the matrix.

Values of principal incremental strains have also been determined from the pressure-shadow zones around framboidal pyrite.

Fairbairn (1950) and Spry (1969) have suggested that the sigmoidal curvatures of pyrite pressure-shadows result from a "rolling" of the pyrite host by a simple shear couple, rather like the rolling of a ball-bearing between two plates. This interpretation overlooks the fact that an overall relative shear couple cannot be applied to a spherical object by a continuously deforming matrix, and that the matrix only comes into contact with the pyrite over a small percentage of its surface area. Fiber curvatures produced by rolling are nonparallel helices and do not correspond to any pattern observed in our specimens, and the process of simple shear cannot satisfactorily explain the curvatures seen in XY sections, the complexities of curvature (particularly the kinks and inflexions), or the varying radii of curvature seen in both XY and XZ sections. The fibers trace out the displacement path of the external boundary as it was drawn away from the pyrite (Fig. 12, antiaxial growth). As pointed out by Hills (1963, p. 130), the most probable explanation of the fibres is that they are parallel to the extensions in the surrounding matrix.

Incremental extensional strains have been calculated for three specimens from Leytron, using a method that assumes that the fibers were parallel to the incremental extensions (Fig. 20D). By using average successive 10° fiber chord lengths, the nth incremental extension given by this method is

$$\text{incremental } e_{1_n} = \frac{\text{chord length}_n}{\sum_0^{n-1} \text{chord length}_i \cos \theta_i + \text{pyrite radius}}$$

where θ_i is the angle between the ith chord and the nth chord. The results (Fig. 21) are very similar in magnitude and orientation to those calculations based on the crinoid pressure-shadows at the same locality. The fibers around the pyrite masses record an inflexion in the latest stages of extension at about 65° right of N, not seen in the analyses of strain increments from microboudinage. This is attributed to the greater sensitivity of pyrite pressure-shadow growths to incremental deformations.

Regional Extension History of Part of the Helvetic Alps

On the scale of a thin section, different individuals of similar extension structures have consistent fiber curvatures. Consistent progressive deformation data have also been obtained from an analysis of different specimens and different types of fibrous structure. On a regional scale, however, the pattern of progressive deformation varies from one locality to another.

Drawings of concretionary pyrite pressure-shadow fibers from 28 localities in the Helvetic Alps are seen in Fig. 22 to illustrate these regional variations of Alpine extension history. To interpret this map it should be noted that the arrowheads denote the chronological order of extension events and not necessarily the movement direction of the rocks, and that where there are sharp 90° changes of orientation of the fibers the decision as to whether the kink should be drawn

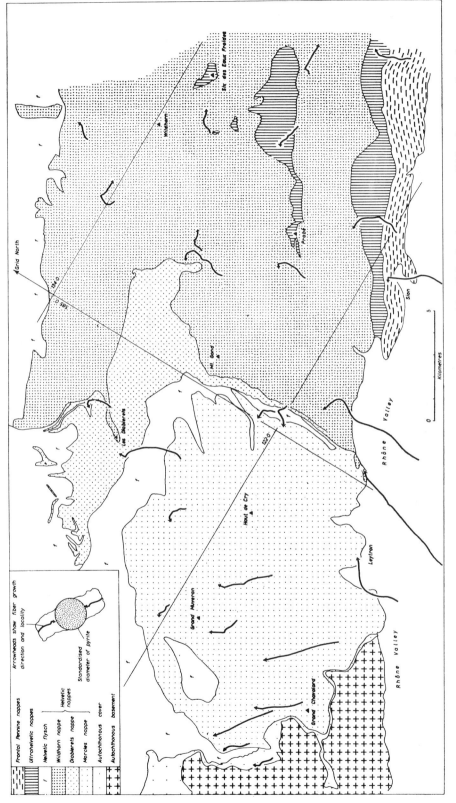

FIG. 22 Synoptic map of the changing principal incremental extension patterns at several localities in the western Helvetic nappes based on measurements made on fibers in pressure shadow zones around framboidal pyrite aggregates.

to the left or to the right was purely arbitrary. These sudden changes of orientation increments probably signify that at some stage the maximum and intermediate incremental strain became equal and then interchanged position during the next increment.

The changes of extension increment orientation in Fig. 22 throughout the area are clearly complex. At some localities the variations are continuous, whereas at others they appear to be abrupt. There appear to be three domains within which the fiber shapes are comparable: (1) the center of the Morcles nappe; (2) the front and normal limb of the Morcles nappe, the Diablerets nappe, and the southern half of the Wildhorn nappe; and (3) Leytron and the northern half of the Wildhorn nappe.

The longest pressure-shadow was found at a locality in the inverted Malm limestone root of the Wildhorn nappe where the finite extension (X) was shown to be more than 17.0 (i.e., >1600 percent). The smallest pressure-shadows and the lowest values of principal maximum extensions occur in the fronts of the nappes.

At corresponding tectonic levels in the nappe pile on opposite sides of the Wildstrubel depression the fibers show consistent predominant orientations and a symmetry about a northwest–southeast line passing through the depression.

A striking feature of the data in Fig. 22 is the extension subparallel to the trend of the Helvetic chain in the central and frontal regions of the Wildhorn nappe; this could be due to longitudinal stretching of the outer arc of the chain after the rocks had been forced through the depression between the Aiguilles Rouges and Aar massifs.

On the basis of characteristic geometric features of fiber curvature, such as inflections and kinks, tentative chronological correlations can be made between adjacent localities to reveal a changing spatial distribution of strain in time. Figure 22 suggests that extension started in the Wildhorn nappe with N–S direction, followed by a NW–SE extension which is especially strongly developed in the Morcles nappe, and finally a small extension oriented NNE–SSW or NE–SW in nearly all nappes. This movement picture agrees with inferences made from major structural and stratigraphic relationships (Lugeon, 1940; Badoux and others, 1959; Badoux, 1960; Trümpy, 1961) that the deformation migrated downward or northwestward through what are now the Helvetic nappes.

Fault fibers have been recorded on large thrusts in the Diablerets nappe, and these directions correlate with the later increments of the regional picture, indicating that thrusts developed fairly late in the history of ductile flow.

Conclusions

To gain a fuller understanding of the evolution and kinematic development of tectonic deformation it is necessary to have information on the way a finite strain state was built up by the superposition of successive increments of deformation. This paper describes the geometric forms of crystals which grew in veins and in pressure-shadow zones around objects resistant to ductile flow (pebbles, fossil fragments, pyrite crystals) during the regional deformation. The form of these crystals (calcite, quartz, chlorite, stilpnomelane, albite) is peculiar in that all are fibrous or needlelike, quite unlike their normal habit. The crystal needles frequently curve; they appear to have grown in this way as a result of changing orientation of incremental deformations during growth. A number of characteristic patterns have been recognized which depend upon the crystal species and composition of the wall rocks, and these have been used to assess the nature of the strain increments. Techniques have been described for converting these general observations on crystal morphology into exact methods for measuring the orientation of and values of the principal axes of incremental strain.

Many striations on slickensided surfaces of natural faults also appear to be the result of growth of needlelike crystal fibers and not due to the mechanical breakup of the walls of the fault and subsequent grinding out of these fragments between the fault walls. Growth of fibrous crystalline material takes place in spaces opened up during movement along an irregular fault surface and leads to the formation of overlapping fiber aggregates. The orientation of these crystals enables the deformation history of the fault to be determined.

The general techniques are applied to a study of part of the Helvetic Alps of Switzerland. Many measurements have been made of crystals found in vein fillings, around fossil fragments, and around pyrite crystals. At any one locality the strain increment pattern that they reveal is consistent, but from locality to locality (and in particular from nappe to nappe) the strain increment patterns change. The uppermost nappes show early phases of deformation not present in the lower nappes, and it appears that the deformation migrated northwestward. The upper nappes show more complicated incremental deformation patterns than the lower ones; the lower nappes were probably confined from the sides and from above, whereas the upper nappes were nearer the surface and were sliding about under the influence of gravity in a less controlled fashion. The largest incremental strains occur in those rocks showing the greatest finite strain and are generally found in the overturned limbs of large recumbent folds, whereas the smallest incremental strains are found in the hinge zones of these folds. At comparable tectonic levels on either side of the Wildstrubel depression the strain increments show a mirror symmetry probably related to preferential flow through this depression. The frontal part of the Wildhorn nappe shows late increments parallel to the nappe front which appear to be best interpreted as the result of divergent flow of the nappe once it had passed through the depression zone.

References

Badoux, H., 1960, Notice explicative, Monthey sheet, *Atlas géologique de la Suisse*.

———, 1963, Les bélemnites tronçonnées de Leytron, Valais: *Bull. Soc. Vaud. Sci. Nat.*, v. 138, p. 1–7.

———, Bonnard, E. G., Burri, M., and Vischer, A., 1959, *Atlas géologique de la Suisse*, St. Léonard, sheet 35.

Bateman, A. M., 1959, *Economic mineral deposits*: New York, John Wiley and Sons, 916 p.

Becker, G. F., and Day, A. L., 1905, The linear force of growing crystals: *Proc. Wash. Acad. Sci.*, v. 7, p. 283–288.

———, 1916, Note on the linear force of growing crystals: *Jour. Geol.*, v. 24, p. 313–333.

Berner, R. A., 1970, Sedimentary pyrite formation: *Am. Jour. Sci.*, v. 268, p. 1–23.

Billings, M. B., 1954, *Structural Geology*: Englewood Cliff, N.J., Prentice-Hall, 514 p.

Brehler, B., 1951, Über das Verhalten gepresster Kristalle in ihrer Lösung: *Neu. Jahrb. Mineral.*, p. 110–131.

Cloos, E., 1947, Oolite deformation in the South Mountain fold, Maryland: *Geol. Soc. Am. Bull.*, v. 58, p. 843–918.

———, 1971, *Microtectonics along the western edge of the Blue Ridge, Maryland and Virginia*: Baltimore, Johns Hopkins University Press, 234 p.

Correns, C. W., 1926, Über die Erklärung der sogenannten Kristallisationskraft: *Sitzungsber. Preuss. Akad. Wiss.*, v. 11, p. 81–88.

———, 1949, Growth and dissolution of crystals under linear pressure: *Disc. Faraday Soc.*, v. 5, p. 267–271.

———, and Steinborn, W., 1939, Experimente zur Messung und Erklärung der sogenannten Kristallisationskraft: *Z. Krist.*, v. 101, p. 117–133.

Daubrée, A., 1879, *Etudes synthétiques de géologie experimentale*: Paris, Dunod, 478 p.

Donzeau, M., 1971, Etude structurale dans le Paléozoique des Monts D'Ougarta: Thèse, 3ème Cycle, Fac. des Sci. Orsay, 55 p.

Durney, D. W., 1972a, Deformation history of the western Helvetic Nappes, Valais, Switzerland: Ph.D. thesis, London Univ., 372 p.

———, 1972b, Solution transfer, an important geological deformation mechanism: *Nature*, v. 235, p. 315–317.

Duvernoy, G., 1852, Die ausdehnende Wirkung der Kristallisationskraft: *Neu. Jahrb. Mineral.*, p. 781–821.

Fairbairn, H. W., 1950, Pressure shadows and relative movements in a shear zone: *Trans. Am. Geophys. Union*, v. 31, p. 914–916.

Farrand, M., 1970, Framboidal sulphides precipitated experimentally: *Mineral. Deposita.*, v. 5, p. 237–247.

Flinn, D., 1962, On folding during three dimensional progressive deformation: *Quart. Jour. Geol. Soc.*, v. 118, p. 385–433.

Gay, N. C., 1970, The formation of step structures on slickensided shear surfaces: *Jour. Geol.*, v. 78, p. 523–532.

Goguel, J., 1952, *Traité de tectonique*: Paris, Masson, 383 p.

Harker, A., 1885, On slaty cleavage and allied rock structures: *Br. Assoc. Adv. Sci. 55th meeting*, p. 813–852.

Hills, E. S., 1953, *Outlines of structural geology*: London, Methuen, 182 p.

———, 1963, *Elements of structural geology*: New York, John Wiley and Sons, 483 p.

Klocke, F., 1871, Beobachtungen und Bemerkungen über das Wachstum der Kristalle: *Neu. Jahrb. Mineral.*, p. 369–392; 571–581.

Lavalle, M., 1853, Recherches sur la formation lente des cristaux à la température ordinaire: *Comptes Rendus, Paris*, v. 36, p. 493–495.

Lehmann, O., 1904, *Flüssige Kristalle sowie Plastizität von Kristallen im Allgemeinen*: Leipzig, Engelmann, 267 p.

Love, L. G., 1957, Micro-organisms and the presence of syngenetic pyrite: *Quart. Jour. Geol. Soc.*, v. 113, p. 429–440.

———, 1967, Early diagenetic iron sulphide in recent sediments of the Wash: *Sedimentol.*, v. 9, p. 327–352.

Lugeon, M., 1940, *Atlas géologique de la Suisse*: Diablerets, Notice explicative.

Mugge, O., 1930, Bewegungen von Porphyroblasten in Phylliten und ihre Messung: *Neu. Jahrb. Mineral. Geol. Palaeontol.*, v. 61, p. 469–520.

Pabst, A., 1931, Pressure shadows and the measurement of orientations of minerals in rocks: *Am. Mineral.*, v. 16, p. 55–61.

Paterson, M. S., 1958, Experimental deformation and faulting in Wombeyan marble: *Geol. Soc. Am. Bull.*, v. 69, p. 465–476.

Price, N. J., 1966, *Fault and joint development in brittle and semi-brittle rock*: London, Pergamon, 176 p.

Ramsay, J. G., 1967a, *Folding and fracturing of rocks*: New York, McGraw-Hill, 568 p.

———, 1967b, *A geologist's approach to rock deformation*: London, Imperial College, 21 p.

———, 1969, The measurement of strain and displacement in orogenic belts, *in Time and place in orogeny*: Geol. Soc. London, Spec. Publ. 3, p. 43–79.

———, and Graham, R. H. 1970, Strain variation in shear belts: *Can. J. Earth Sci.*, v. 7, p. 786–813.

Rickard, D. T., 1969, The microbiological formation of iron sulphides: *Stockholm Contrib. Geol.*, v. 20, p. 49–66.

Riecker, R. E., 1965, Fault plane features: *Jour. Sed. Petrol.*, v. 35, p. 746–748.

Roering, C., 1968, The geometrical significance of natural en-echelon crack arrays: *Tectonophysics*, v. 5, p. 107–123.

Shainin, V. E., 1950, Conjugate sets of en-echelon tension fractures in the Athens limestone: *Geol. Soc. Am. Bull.*, v. 61, p. 509–517.

Sorby, H. C., 1853, On the origin of slaty cleavage: *Edinb. New Philos. Jour.*, v. 55, p. 137–148.

———, 1908, On the application of quantitative methods to the study of rocks: *Quart. Jour. Geol. Soc.*, v. 64, p. 171–232.

Spry, A., 1969, *Metamorphic textures*: London, Pergamon, 350 p.

Stočes, B., and White, C. H., 1935, *Structural geology*: London, Macmillan, 460 p.

Taber, S., 1916, The growth of crystals under external pressure: *Am. Jour. Sci.*, v. 12, p. 532–556.

Thompson, W., and Tait, P. G., 1879, *Treatise on natural philosophy*: Cambridge University Press, 730 p.

Tjia, H. D., 1964, Slickensides and fault movements: *Geol. Soc. Am. Bull.*, v. 75, p. 683–686.

Trümpy, R., 1961, Sur les racines des nappes Helvétiques: *Soc. Géol. France, Livre à la mém. du Prof. Paul Fallot*, p. 419–428.

Wickham, J. S., and Elliott, D., 1970, Rotation and strain history in folded carbonates, Front Royal area, northern Virginia: *Trans. Am. Geophys. Union*, v. 51, p. 422.

Wilson, G., 1961, The tectonic significance of small-scale structures: *Bull. Soc. Géol. Belg.*, v. 84, p. 423–548.

BARRY VOIGHT

Department of Geosciences
Pennsylvania State University
University Park, Pennsylvania
Visiting Lecturer
Delft Technical University
Netherlands

The Mechanics of Retrogressive Block-Gliding, with Emphasis on the Evolution of the Turnagain Heights Landslide, Anchorage, Alaska[1]

A bad earthquake at once destroys our oldest associations; the earth, the very emblem of solidity, has moved beneath our feet like a thin crust over a fluid; one second of time has created in the mind a strange idea of insecurity, which hours of reflection would not have produced.

Darwin: *Voyage of the Beagle*, 1835.

---

[1] I especially thank Rob Scholten for his kind invitation to contribute to this volume, for his encouragement and patience concerning an errant manuscript, but most of all for his contagious and unflagging enthusiasm throughout the years in pursuit, among inanimate objectives, of geologic discovery. My acceptance was automatic; having been born in Nieuw Nederland, having worked for the better part of a decade among the Pennsylvania Dutch and, most recently, among the Delft Dutch, I rather considered the invitation an opportunity to pay homage to the dedication of a countryman. I extend my gratitude to Professor Jacques Dozy for his invitation to join the Geological Laboratory at TH Delft and for his gracious hospitality during my tenure. I am grateful to my other colleagues at Delft, in particular to Frits Koster and Mej. A. Guldemond for assistance in many matters. In the construction of illustrations I was assisted by my daughters Lisa and Barbara and their friends R. Searle and M. Escher; figures were drafted by J. J. Swanink and photographed by A. F. G. Faessen. Helpful discussions of the Turnagain Heights problem with the following gentlemen are acknowledged: S. T. Harding (U.S.C.G.S.), Andrej Werynski (P.A.N., Warszawa), Prof. H. C. Duyster and Ir. G. E. J. S. L. Voitus van Hamme (HBG, Rijswijk), Eppo Viswat (NBB, Den Haag). The Statens Geotekniska Institut, Stockholm, recently forwarded proceedings on the Sköttorp landslide, a report that clearly demonstrated the degree to which many of the arguments presented here were recognized, 20 years ago, by Sten Odenstad. The benefit of stimulating discussions in the field with W. G. Pierce, concerning the Heart Mountain and related overthrust problems, is acknowledged with gratitude. An association of the Geological Field Camp of the Pennsylvania State University with the field station of the Yellowstone-Bighorn Research Association at Red Lodge, Montana, has since 1965 provided the opportunity for me on several occasions to become reasonably familiar with the large-scale detachment faults of that region; I wish to express my appreciation of this fact to fellow members of both organizations. Finally, I am grateful to my family for their extraordinary patience with me as I struggled, with characteristically bad temper, over this manuscript during (what was supposed to be) a Vrolijk Pasen at Scheveningen.

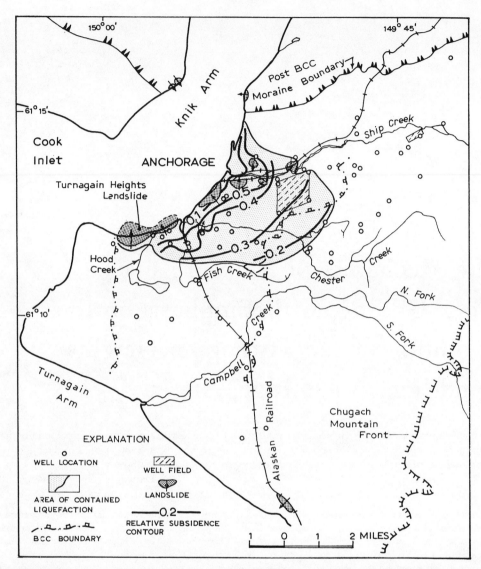

FIG. 1 Map of Anchorage, Alaska, and vicinity, showing locations of the Turnagain Heights slide and other major slides (note shaded pattern). The distribution of ground subsidence (in feet), based on well data (Waller, 1966) and due to compaction, delimits an area (light-shading) of seismically induced subsurface pore pressure enhancement (a so-called low-strength area of contained liquefaction). The extent of Bootlegger Cove Clay is shown by the BCC boundary.

Gravity-induced displacements in geologic media range rather continuously in scale from the downslope motion of a discrete particle to, perhaps, movements on a global scale (Schardt, 1898; Haarmann, 1930; Van Bemmelen, 1936, 1965; Varnes, 1958; Rubey and Hubbert, 1959). The scale of phenomena dealt with in this paper is only slightly more restricted, for gravitational block gliding has been observed in simple laboratory experiments and demonstrated from unequivocal geologic evidence to exist on the scale of mountain ranges. Although the effect of scale can be shown to be of importance, the kinematics and dynamics of block gliding must, irrespective of scale, be approached from the same fundamental considerations.

The purpose of this essay is twofold: (1) to discuss some principles of mechanics applicable to block gliding, as well as to treat those geologic factors essential to an understanding of the role of gravity in the evolution of block-gliding structures; and in this attempt, (2) to provide an example of a well-understood structure produced by retrogressive block gliding.

I have chosen a historic event for an example; rather than groping into the murky depths of geologic time, I thought it wiser to deal with a structure for which some initial and intermediate, as well as final, conditions were known, for which measurements of material properties and material behavior were conducted, and for which eyewitness accounts of the deformational process and of related events were on record. Hence the Turnagain Heights slide is analyzed here, a modest structure by geologic standards, but with a length of 1.6 miles (~ 2.6 km) and volume of 12.5×10^6 yd^3 ($\sim 9.6 \times 10^6$ m^3), a most impressive feature from the viewpoint of students of soil mechanics. Certain features of the slide seem analogous to, and methods of analysis employed here are believed to be applicable to, both smaller and larger structures, and in a subsequent section brief consideration will be given to this possibility.

Turnagain Heights Slide Area

The largest and most devastating landslide in the Anchorage area occurred along coastline bluffs overlooking the Knik Arm head of Cook Inlet, south-central Alaska, a direct result of the Prince William Sound earthquake,[2] March 27, 1964 (Fig. 1).

Approximately 130 acres (~ 53 ha) became involved in a complex, apparently chaotic, displacement pattern characterized by extensive lateral slip and spreading of bluff material into the bay (Figs. 2 and 3). Lateral displacements were as great as 2000 ft (~ 610 m) (Hansen, 1965). Several broad lobes developed (Fig. 4), with headward regression reaching a maximum of 1200 ft (~ 370 m) in the western lobe (Fig. 2); in the east lobe, an area of residential development, 75 homes were destroyed and three lives were lost as the bluff retrogressed 600 ft (~ 185 m) (Fig. 3). Extensive ground fracturing occurred behind the slide escarpment, more especially behind the east lobe (Fig. 4).

Geotechnical Summary

Details of regional and local geology have been given elsewhere,[3] and thus only those features germain to the present discussion need be presented here. The Turnagain area (ground surface elevation approximately 70 ft, or 21 m) is veneered by a surface layer of

---

[2] Revised Richter magnitude 8.5, with an epicenter located 75 miles east-southeast of Anchorage; the energy-release area is assumed to be defined by a band of aftershock epicenters which proceeded southwesterly from Prince William Sound to Kodiak Island, a distance of about 700 miles (Steinbrugge, 1967).

[3] Early important studies of the slide areas in Anchorage were undertaken by an *ad hoc* assemblage of local geologists living in the Anchorage area; a report was released under the authorship of the Engineering Geology Evaluation Group. Subsequent work was conducted under the supervision of the office of the U.S. Army District Engineer, with a major portion of this work contracted to the foundation engineering firm of Shannon and Wilson, Inc., Seattle. Important engineering reports arising from these investigations are available (Shannon and Wilson Inc., 1964; Wilson, 1967; Seed and Wilson, 1967; Long and George, 1967; Seed, 1968). A detailed description of the geology of the Anchorage landslide sites was given by Hansen (1965). Effects of the earthquake on the hydrology of Anchorage have been examined by Waller (1966). Modern work on the surficial geology and groundwater geology of the Anchorage area and on the Quaternary history of the Cook Inlet region had been published at the time of the earthquake (Miller and Dobrovolny, 1959; Cederstrom et al., 1964; Karlstrom, 1964); these papers have provided a firm foundation for more recent geologic studies (e.g., Trainer and Waller, 1965; Grantz et al., 1967; Hansen et al., 1966; Varnes, 1969) and investigations of the slide areas.

FIG. 2 Aerial view, looking eastward, of east lobe of Turnagain Heights slide (USGS photograph); snow-covered ground shown in white. Note large glide-block in foreground and narrower blocks developed as clay ridges, shown in black. Arrowpoint denotes intersection of McCollie Avenue and Turnagain Parkway, referred to in Table 2 and Fig. 6.

fluvioglacial sand and gravel, 5–20 ft ($\sim$ 1.5–6 m) thick, which overlies a regionally extensive marine deposit of silty clay with an average thickness of 100–150 ft ($\sim$ 30–46 m), containing sporadic and discontinuous silt and fine sand seams and lenses (Fig. 1).

This unit, the Bootlegger Cove Clay[4] of Miller and Dobrovolny (1959, p. 35–48), is the source for the damaging landslides in the greater Anchorage area, and its properties have thus been extensively studied. A summary of physical properties is provided in Table 1; a typical boring log is seen in Fig. 5. The unit can be grouped into three zones according to physical property variations. In the central zone, maximum sensitivity[5] coincides with the lowest strength; at Turnagain this zone is located between elevations -10 and $+25$ at the east end of the slide area (with all elevations given in feet); and

---

[4] The abbreviation BCC is used here on some maps and tables, e.g., Fig. 1.

[5] In the jargon of soil mechanics, sensitivity refers to the ratio of the undrained strengths of a cohesive material in "undisturbed" and "remolded" states.

FIG. 3 Aerial view, eastern portion of west lobe of Turnagain Heights slide (USCGS photograph). Snow-covered ground (white); sharp-crested clay ridges (grey to black) well developed near right-center of photograph. Apparent ground striae are shadows cast by morning sun. Arrows point toward a house split and separated by 50 ft (ca. 20 m) due to formation of a clay ridge (location T, Fig. 4).

between +10 and +35 at the west end. Sand lenses occur within clay at the east end, at elevation +10 to +20, and at the west end, above elevation +20 in association with silt seams (Seed and Wilson, 1967). In front of the bluff line, within the tidal zone, a thin, outward-sloping (2.5°) deposit of estuarine silt provided a sliding surface, which enabled the leading edge of the landslide to glide into ice-choked waters of Knik Arm (Figs. 2 and 6).

Deformational Mode and the Geometry of the Sliding Zone

Deformation at Turnagain Heights was characterized by the development of sharp-crested clay ridges (horsts) alternating with collapsed troughs (graben) (Fig. 2); strati-

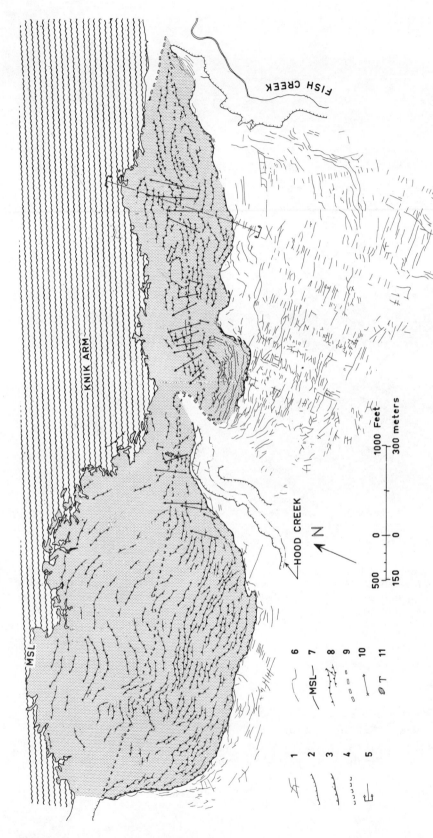

FIG. 4 Map of Turnagain Heights slide area. (1) Fracture pattern behind slide; (2) unaffected bluffs, with "bluff" near Hood Creek of small topographic relief; (3) new bluffline behind slide; (4) position of original bluff within slide area; (5) arrows depict area mapped by Miller and Hansen (Hansen, Pl. 2), with line of cross sections for Fig. 6, this paper; (6) intermittent creek; (7) contact of slide mass and adjacent tidal area with mean sea level; (8) trace of glide block ridges (shown by carets on both sides of line) and scarps of flat-topped glide blocks (carets on one side of line); (9) pressure ridges; (10) displacement vectors; (11) house ruins (T denotes house of L. Thomas, Jr.). [(1) and (10) based on data by Engineering Geology Evaluation Group (1964) and Hansen (1965); other data based chiefly on slide area topography as given by Hansen (1965, Plate 1).]

fication was greatly disturbed within the trough areas but virtually undisturbed within the ridge-shaped glide blocks (Hansen, 1965). Stratification within the displaced ridges remained approximately horizontal, thus requiring nearly rectilinear displacement vectors. Displacements within ridges were chiefly due to translation, with some rotation in the horizontal plane; translational and rotational components in both horizontal and vertical planes were important in the intervening troughs. Ridges were typically aligned more or less perpendicular to the slip direction, with some exceptions (Fig. 4). Detailed mapping by Hansen and Miller along a 120-ft-wide ($\sim$37-m) strip across the eastern slide area (Figs. 3 and 6, section IX; compare Hansen, 1965, Pl. 2) suggest that displacements occurred on a maximum seaward grade of 2.2° (4 percent).

Further indications concerning the sliding zone could be inferred from continuous borehole sampling. These borings revealed within the slide areas a correlation between (1) an approximately horizontal boundary separating low- and relatively high-strength zones (Fig. 5, depth 27–36 ft), (2) positions of clay zones having lower strengths than similar materials at equivalent elevations outside of the slide area (Fig. 5, depth $\sim$30 ft), and

TABLE 1 Geotechnical Data Summary: Bootlegger Cove Clay[a]

Physical, physico-chemical, and mineralogical properties
    Specific gravity: 2.7–2.8
    Particle size $< 2\ \mu$: 26–66%, average, $\sim$ 42%
    Atterberg limits: LL: 22–48%; PI: 4–30% (average in sensitive zone, LL $\simeq$ 32, PI $\simeq$ 12)
    Liquidity index: typically < 1; in sensitive zone, ≥ 1
    Porewater pH: 8.2–10.3
    Equivalent NaCl: 1–6 g/l (34 g/l in seawater)
    Base exchange capacity: 5–11 me/100 g
    Cation abundances: $Ca^{++} > Na^+ > Mg^+$
    Mineralogy: quartz, chlorite, illite, feldspar
    Clay particle orientation: probably "random" in sensitive zone

Strength, stiffness, and related properties
    Undrained strength: variation, 1 tsf (elev. +50) to 0.35 tsf (elev. 0) to 0.6 tsf (elev. -30)
    Remoulded strength: $\sim$ 0.02 tsf
    Cyclic-load strength: $\sim$ 55% of static strength for 30 cycles (elev. +10)
    Sensitivity: < 5–60; 6% of total sample had sensitivity > 20; thickness of sensitive zone: typically 20–30 ft, centered at elev. 0
    Shear modulus: 330 tsf (stiff clay); 100 tsf (soft sensitive clay)
    Seismic velocity: 5000 fps (cf.: 1600 fps, surficial sand and gravel; 7500 fps, underlying till); 2500 fps (heterogeneous, disturbed slide mass)
    Consolidated state: *normally consolidated* in sensitive zone; *overconsolidated* in upper stiff zone

Properties of sand and silt lenses
    Relative density: $\sim$ 60%, estimated from penetration resistance
    Particle size: $\sim$ 0.02–0.20 mm
    Cyclic strength: 30 cycles produces initial pore-pressure rise at a lower value of shear stress than required for failure of clay
    Seismic velocity: 5500 fps

[a] Data from Wilson, 1967; Seed and Wilson, 1967; Kerr and Drew, 1968.

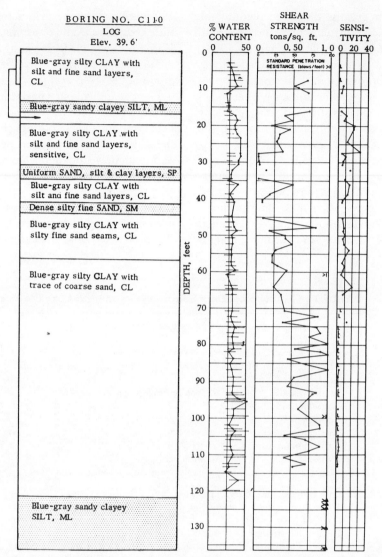

FIG. 5 Typical boring log, Turnagain Heights slide (Wilson, 1967). Note loss of strength within depth range 28–35 ft; the sliding surface is presumed to lie near the base of this zone. Location of boring: in clay ridge associated with block F, 125 ft southwest of block F (Fig. 6, IX). Standard penetration resistance given only for silt at base of log, denoted by triangular symbols. Water contents (w) are given in association with Atterberg limits (LL, PL); zone of high sensitivity roughly coincides with w > LL. Symbol L in sensitivity column refers to values less than 10.

(3) a thin disturbed zone of intermixed clay and silt/sand laminations. Seed and Wilson (1967) concluded that the main surface of sliding was at about elevation +8 near the back boundary of the east lobe (Fig. 2), sloping slightly toward the original toe. A similar surface was noted at elevation +10 in the west lobe (average slope 2.2°), rising toward the west margin to +20 where a quarter-acre "tectonically denuded" area of

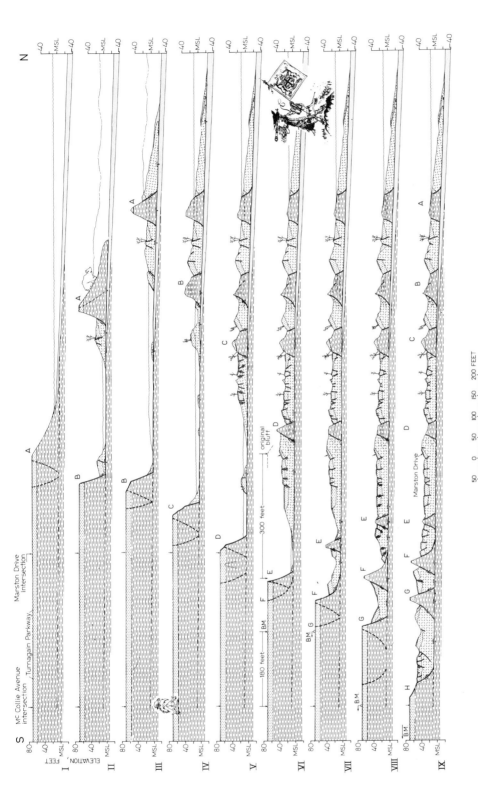

FIG. 6 Sections through east lobe of Turnagain Heights slide, approximately parallel to Turnagain Parkway (see Fig. 4). Sequence of retrogressive events, hypothetical but reconstructed on the basis of section IX (cf. Hansen, 1965, Pl. 2) and eyewitness observations. The symbol B.M. denotes estimated positions of Mr. B. Marston in sections V–IX, at times referred to in Table 2. Heavy and light dotted patterns refer respectively to glacial outwash and to tidal silts; Bootlegger Cove Clay is indicated by a dashed line pattern in relatively undisturbed zones and by cross-hatching in zones of higher strain, e.g., plastic wedges.

105

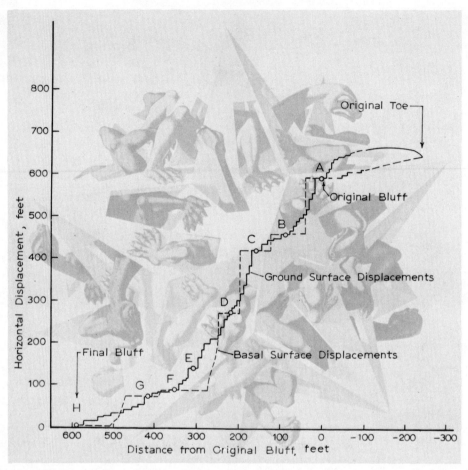

FIG. 7 Horizontal displacements for points within the slide mass at Turnagain Heights plotted versus initial position with respect to the original bluff. Data are given both for points on the ground surface and at the basal contact with the sliding surface, and are derived from the reconstructed sections of Fig. 6.

the slip surface was exposed (see Hansen, 1965, Fig. 43).

Displacement Reconstruction

From the detailed cross section through the slide area published by Hansen (1965, Pl. 2), used in conjunction with information on preslide topography and geometry of the main sliding surface, it was possible from considerations of volume to reconstruct the preslide positions of the major glide blocks (Fig. 6). These are thought to be accurate within about 20 ft ($\sim$6 m).[6] From the initial and final positions, a displacement diagram was constructed (Fig. 7), both for points on the original ground surface and for points within the slide mass along the sliding surface. The "stepped" configuration of the displacement curve, particularly for the basal surface, is

[6] Initial volume reconstructions were made by assuming plane constant-volume deformation and zero postslide erosion; corrections were possible owing to certain points of known displacement, e.g., the Marston Drive intersection with Turnagain Parkway (Fig. 6).

considered to be characteristic of dominant block-gliding motion. The difference between the two curves is an indication that disruption had occurred within certain major portions of the slide mass and suggests that a distinction between "rigid block" gliding and "deformable block" gliding may be useful.[7]

We are not dealing here with a "single block" which collapsed after sliding had been initiated; however, in either case, in any attempt to construct a "hierarchy" of deformational modes, the position of individual blocks *in time* should be considered.

Time Reconstruction

Reconstruction of the chronology of slide events requires the use of eyewitness reports. Such information can be extremely valuable, although objectivity under conditions of earthquake-induced duress is difficult to achieve, and hence the data obtained must be evaluated with careful scrutiny.

At Turnagain, eyewitness accounts indicate that movements began approximately 2 minutes after the start of the earthquake and continued to some extent after the cessation of quake-induced ground motion (Shannon and Wilson, 1964, p. 64; Wilson, 1967, p. 285; Seed and Wilson, 1967, p. 327). A particularly careful and useful eyewitness account was given by Mr. Brooke Marston (see Appendix to this paper); a review of this account is presented in Table 2 together with an estimation of the time required for the events described. This table also provides the locations of the retrogressing bluffline as a function of time, as inferred from the reconstructed cross section (Fig. 6) considered in relation to Marston's account. For example, at estimated time $t_{BM} + 1:30$, Marston describes the bluff to be located 300 ft south of its preslide position; this location very nearly coincides with the reconstructed position of the block at point E (Fig. 6, section VI). Subsequent periods of slumping are described and a time estimate is provided; these are summarized in Fig. 8. Marston's location at various times is indicated by the position BM shown in Fig. 6.

Period t_{BM} represents the minimum time from the start of the quake to the moment when Marston resumed driving. As summarized by Steinbrugge (1967), clock-measured duration of the felt motion varied from 4.5 to 7.0 minutes within greater Anchorage, with duration of heaviest intensity being about 1 minute. The writer estimates t_{BM} to be approximately 3 minutes, based on a recording made by Mr. Robert Pate at a location about 1.5 miles east of Turnagain Parkway.[8] The "probable" time period t'_{BM}, for the Marston sequence, is more likely on the order of 5 minutes (Fig. 8).

The reconstructed displacement-time sequence (Figs. 7 and 8) reveals clearly that the mode of deformation (block gliding) and the initial dimensions of the glide blocks remained about the same throughout the deformational process. Yet the strongest inertial

---

[7] With large motions, it is possible for weak glide blocks to deform by attrition, extrusion, slumping, or other forms of plastic deformation, to the extent that *even at the basal surface* the displacement profile becomes relatively smooth. In such an instance a transition in deformational mode may be considered to have occurred, from "block gliding" to "failure by lateral spreading" (cf. Varnes, 1958; Hansen, 1965, p. 38). At Turnagain Heights lateral spreading appears to have become more important as sliding progressed; deformable block gliding was clearly dominant at the initiation of sliding (Fig. 7, curve for basal surface). Obviously, such phenomenological terms as elastic, plastic, viscous, and their hybrids (e.g., elastic-elastoplastic) remain available for use; but indiscriminant use of these terms leads to confusion. They are not recommended for field use unless due attention is given to their rather restricted meanings in mathematical physics. The suggested terms "rigid" and "deformable" are used here in a purely descriptive sense. Individual glide blocks may be characterized by large strains, depending on the circumstances and materials involved in the deformation, and it is helpful to have a field terminology adequate for their description.

[8] This recording, made available by the U.S. Department of Commerce (see Steinbrugge, 1967), suggests that strong, damaging motions could have lasted 3 minutes and that felt motions continued beyond 4 minutes.

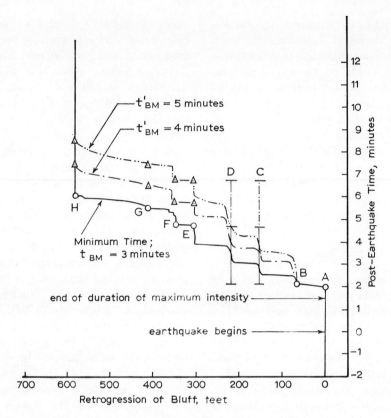

FIG. 8 Postearthquake time versus bluff regression, Turnagain Heights slide (compare Fig. 6 and Table 2).

forces were over within about 1 minute, and the duration of significant inertial forces was over in about 3 minutes. Inertial forces could have been important in early block movements (e.g., blocks A and B; and possibly C and D), but they were not involved in subsequent movements. The main role of the earthquake in the development of the slide was apparently to effect a low-friction surface upon which block gliding could occur.

Velocity Reconstruction

Velocities can be estimated closely from displacement-time reconstructions and other considerations. A calculation for the average velocity ($\bar{V}$) of block A, based on assumptions that the block remained attached to the main slide mass and that forward motion of the block ceased only when bluff retreat stopped, provides a minimum value of about 1–2 ft/sec (~ 0.5 m/sec). An upper-bound estimate of average velocity is 10 times greater, based on the application of principles of conservation of energy [9] to the free gliding of a detached block, and ignoring frictional losses and other forms of resistance which, at Turnagain, were of considerable importance. The front edge of the slide at Turnagain encountered resistance from the pressure of

---

[9] E.g., the instantaneous velocity V at time t is given by $V = V_0 + g(\sin i)t$, where V_0 is initial velocity, g is gravitational acceleration, and i is slope angle. Average velocity $\bar{V} = \frac{1}{2}(V_0 + V)$ over time intervals in which $g \sin i$ is constant. The distance traveled, s, is given by $s = V_0 t + \frac{1}{2}g(\sin i)t^2$.

water throughout the extent of its displacement (Fig. 6); the time estimate of 30 seconds for gliding of the frontal edge, as derived from the free-gliding, instantaneous-stop model, is thus known to be far too short.

Estimates for the movement of interior points provides similar data. For example, an interval of approximately 40 seconds separates bluff retrogressions at points E and F: assuming that the majority of the E block displacement had occurred before F became unstable, then $\bar{V} \simeq 3.5$ ft/sec. The minimum value is obtained if it is assumed that important movements continued over a longer period, that is, until bluff retreat stopped; for this condition $\bar{V} \simeq 1.3$ ft/sec. Although the actual velocities must have varied to some degree as a function of position within the slide mass, their typical values are presumed to be not greatly different from the lower bound values cited above, namely, a few feet per second (~ 1 m/sec; 1 knot).

Limiting Equilibrium

General Application to Gliding Mechanics

A convenient way of evaluating block slide mechanics, at the point in time of incipient motion, is to assess, separately, (1) those force components tending to induce sliding movements, here called driving forces (F_D) and (2) those force components tending to resist sliding, called resisting forces (F_R). For sliding to occur, $F_R \leq F_D$.

As shown in an arbitrary cross section of a block of unit width (Fig. 9), the driving

TABLE 2 Time Estimates for the Retrogression of the Turnagain Heights Bluff

Inferred Position of Bluff	Marston Observations	Estimated Time	
		Minimum	Probable
Original bluffline	Driving westward on McCollie Avenue	—	—
Quake begins	Quake begins	0:00	0:00
Strong vibrations	Auto stopped; N–S shaking	0:00	0:05
Block A begins to move		2:00	2:00
Bluff retreats to B		2:10	2:10
Continued bluff retreat	Quake subsides	t_{BM}	t'_{BM}
Bluff retreats to E	Concludes quake over; decision to drive; starts car	t_{BM} to $t_{BM} + 0:30$	t'_{BM} to $t'_{BM} + 0:30$
	Drives west to Turnagain Parkway, then north 180 ft; notices bluff has retreated 300 ft south of original edge	$t_{BM} + 1:30$	$t'_{BM} + 1:30$
	Leaves auto, runs to driveway	$t_{BM} + 1:30$ to $t_{BM} + 1:45$	$t'_{BM} + 1:30$ to $t'_{BM} + 1:45$
Bluff retreats to F	Additional slumping	$t_{BM} + 1:45$	$t'_{BM} + 1:45$
	Returns to car	$t_{BM} + 1:45$ to $t_{BM} + 1:55$	$t'_{BM} + 1:45$ to $t'_{BM} + 1:55$
	Backs car 180 ft to corner McCollie and Turnagain	$t_{BM} + 1:55$ to $t_{BM} + 2:25$	$t'_{BM} + 1:55$ to $t'_{BM} + 2:25$
Bluff retreats to G; F in motion	Stops at corner; bluff continues to slowly slide northward; continues to back car southward	$t_{BM} + 2:25$	$t'_{BM} + 2:25$
Bluff retreats to H	Backing car south on Turnagain Parkway; bluff slowly breaks away to corner of McCollie and Turnagain Parkway	$t_{BM} + 2:25$ to $t_{BM} + 3:00$	$t'_{BM} + 2:25$ to $t'_{BM} + 3:30$

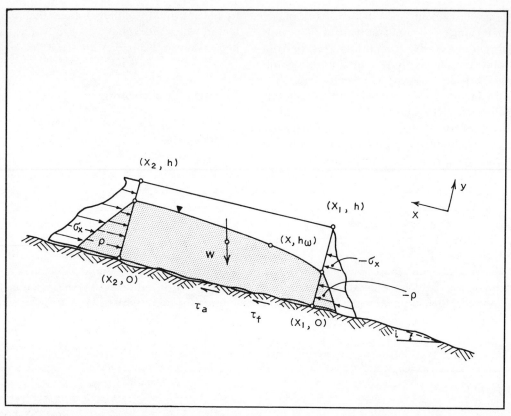

FIG. 9 Simplified diagram of boundary stresses acting on, and body force acting within, a glide block on an irregular inclined sliding surface. Symbols are described in text; dotted pattern refers to the saturated zone and to fluid boundary stresses.

forces are those due to the weight component parallel to the ground slope, the distribution of intergranular pressure σ_x[10] and fluid pressure ρ along the block boundary at $X = X_2$, and seismically induced inertial forces I, if positive.

Resisting forces reflect inertial forces (if negative), the shear strength τ_f capable of being mobilized along the surface of sliding, the apparent strength τ_a developed when a slide block is forced to ride over or shear through local asperities (e.g., Patton, 1966; Rengers, 1970), and the resisting intergranular ground pressure and fluid pressure at the boundary $X = X_1$.

These relationships can be summarized in general terms as follows:

$$\sum F_D = W \sin i + \int_0^h \sigma_x|_{X=X_2} \, dy$$
$$+ \int_0^{h_w} \rho|_{X=X_2} \, dy + I$$

$$\sum F_R = \int_{X_1}^{X_2} \tau_f|_{Y=0} \, dx + \int_{X_1}^{X_2} \tau_a|_{Y=0} \, dx$$
$$+ \int_0^h \sigma_x|_{X=X_1} \, dy + \int_0^{h_w} \rho|_{X=X_1} \, dy$$

Assuming a unit weight of 125 lbf/ft³, $h = 65$ ft, $h_w = 50$ ft, the driving forces at

---

[10] If negative, this term suggests a tensile resistance and should be grouped under "resisting" components. Similarly, the weight component can be resistive with reverse slope.

Turnagain Heights for the static case ($I = 0$) may be reasonably approximated as follows for $X \geq 0$:

$$F_D \simeq 140{,}000 + 810{,}000 \sin i + 8{,}100 \sin i \cdot X$$

where forces are in lbf (pound force units)[11] and X (ft) is measured backward from the bluff line, $X = 0$. The first term represents the force due to an "active" soil and fluid pressure;[12] the second term reflects the weight of the toe in front of the bluff, acting on a slope i; the final term represents the increase in downslope force increments as a function of X. All terms decrease in the direction of the toe for $X \leq 0$ and vanish at the toe.

For $i = 2.2°$ or 4 per cent:
$$F_D \simeq 173{,}000 \text{ lbf} + 325 \text{ lbf/ft} \cdot X \text{ ft}$$

For $i = 0$:
$$F_D \simeq 140{,}000 \text{ lbf}$$

Resisting forces capable of being mobilized at Turnagain include the basal shear strength,[13] here assumed to be chiefly frictional (Coulomb parameters assumed, $C' \simeq 0$, $\mu' \simeq 0.6$; these values are considered reasonable as residual values both for the Bootlegger Cove clay and for sand seams contained within that unit); equivalent roughness is approximated by $C_a \simeq 0$ and $\mu_a = 0.01$–0.10.

For the case with "high" roughness, the increase in resistance beyond the bluff is arbitrarily given by

$$F_R \simeq 3800 \text{ lbf/ft} \cdot X \text{ ft}$$

For "low" roughness:

$$F_R \simeq 3080 \text{ lbf/ft} \cdot X \text{ ft}$$

The rate of increase from toe to bluff is typically nonlinear, dependent upon toe geometry.

The forces F_R and F_D are plotted against distance from the bluff in Fig. 10. To encompass a range of possibilities, three cases of F_D are presented, for slope angles of 8, 0 and -8 percent; two F_R cases are given, curves L and H, representing low (0.01) and high (0.10) "roughness friction coefficients."

Mechanics of Initial Failure

In the initial static condition it is clear that $F_R \gg F_D$, hence sliding is not possible. The driving forces are induced by gravity and cannot greatly change under static conditions. Therefore for failure to occur under static loading, the strength along the surface of sliding must greatly decrease. At Turnagain it must have virtually disappeared; the

---

[11] It has seemed inappropriate to give parenthetic SI or other equivalents after each numerical value involving force or stress terms; for the convenience of the reader: 1 lbf = 4.448 N; 1 tsf $\simeq$ 1 kg/cm² $\simeq$ 1 bar $\simeq$ 10⁵ N/m² $\simeq$ 15 lbf/in.².

[12] This "active pressure" can be considered an effective sustained driving force for gliding blocks that are thick relative to their strength, i.e., for "deformable" glide blocks. The initial state of stress, whether at the "rest" condition or involving residual stress systems, can provide lateral pressure in excess of "active" pressure, and, indeed, can exceed the pressure due to the weight of overburden. However, for blocks on a low-friction base, these pressures will dissipate with small displacements and are incapable of providing a sustained "drive." If extensive soil remolding occurs, the driving forces due to intergranular pressure and pore fluid are changed, and in the limit may be replaced by a single term $\frac{1}{2}\gamma h^2$, where γ is the unit weight of the liquified soil mass. In this instance an increase in the driving force could result. This effect could produce local "bulges" in the driving force curve, which would increase the probability of failure at those positions where strength loss is extensive. The position of the point at which the active pressure reaches its maximum value is discussed in the section on Geometry of Glide Blocks and Plastic Wedges.

[13] Strength parameters are assumed to be reasonably represented by the criteria $\tau_f = C' + \bar{\sigma}\mu'$ and $\tau_a = C_a + \sigma\mu_a$, where $\bar{\sigma}$ is the normal effective stress on the sliding surface, σ is the total stress on this surface, C' and C_a represent strength intercept values at zero normal pressure, μ' is the residual frictional coefficient for sliding on a relatively smooth surface and μ_a is the equivalent additional friction required for sliding over a rough surface.

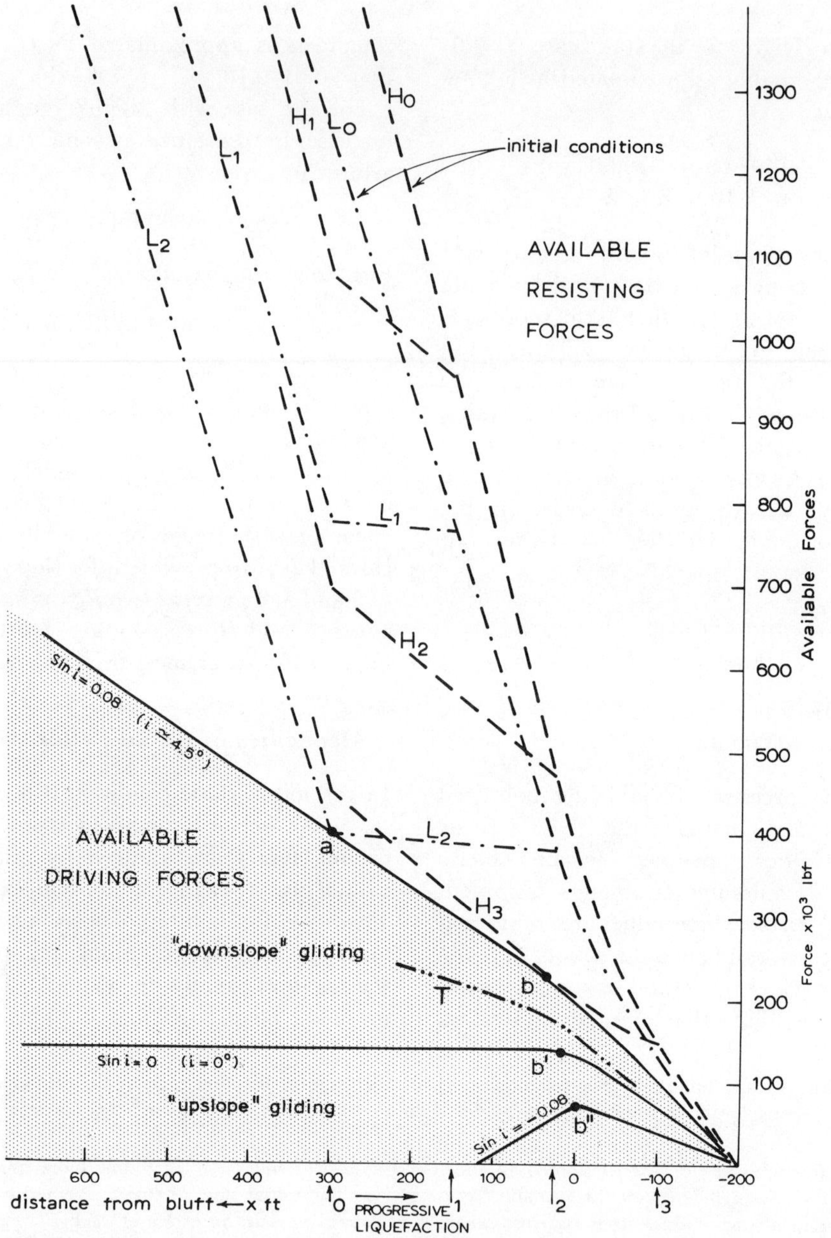

FIG. 10 Resisting forces and driving forces (in shaded region) plotted as a function of distance to the bluff. Symbols are described in text. Curve T represents the estimated driving forces available at Turnagain Heights.

ground in the vicinity of the sliding surface appears to have "liquefied" as a result of earthquake-induced ground motions.[14] The effect of strength loss can be evaluated by examining the change in F_R curves as a result of decreasing strength; in Fig. 10, it is assumed that a thin zero-strength zone begins at an arbitrary (positive) point, $X = +300$, and progresses toward the slope. Discussion is given for three cases of initial block separation: (1) downslope block gliding; (2) block gliding on a zero slope; and (3) upslope block gliding.

Downslope Block Gliding. The F_R curves are profoundly lowered, although the F_R slope remains positive at all positions because roughness exists irrespective of whether strength-loss has occurred (it may be modified); in addition, even "liquefied" soils possess a small but finite strength (e.g., at Turnagain, 40 lbf/ft^2). As liquefaction progresses successively from point 0 to points 1 and 2, curves H_0 and L_0 progress first to positions H_1 and L_1, and then to H_2 and L_2. For the case of low roughness, block separation is predicted[15] at $X = 300$ (i.e., $F_R = F_D$ at point a). "Liquefaction" has progressed to within only 25 ft of the bluff; for sliding to occur, the toe must be ruptured. The available forces are, however, sufficient to cause this rupture. The generalization follows that *for slope $F_R <$ slope F_D within the zone of strength-loss, block separation is predicted at or near the rear boundary of the strength-loss zone*; this is irrespective of the direction of propagation of the strength-loss zone. Some of the larger tabular-shaped glide blocks at Turnagain may be explicable by this mechanism (see Fig. 2).

However, *if slope $F_R >$ slope F_D* (e.g., because of high "roughness" or less profound strength loss), *failure will be predicted at or near the bluff line*. This is shown by H_3 in Fig. 10; $F_R = F_D$ at point b; block separation is predicted at about $X = 40$ ft and occurs when the strength-loss zone has progressed to position 3 at $X = -120$ ft. In this instance the position of the rear boundary of the strength-loss zone is *not* of concern as long as it is located somewhere behind the bluff. The best estimate of conditions at Turnagain Heights is to be found within this category, with an F_D curve lying midway between the F_D curves for slopes of $+4.5$ and $0°$ (Fig. 10, curve T).

Block Gliding on a Zero Slope. The sole driving force component is the gravitationally induced lateral ground pressure.

---

[14] Although the nature of the low-strength zone at Turnagain Heights has been the subject of much interest and some sophisticated experimentation, certain aspects of this subject, which appear not to have been recognized, deserve comment. A most important distinction should be made between the vibratory response of the following materials, both of which undergo compaction and strength loss as a consequence of transfer of overburden load to fluid pressure:

Type I. Materials that undergo strength loss only if subjected to low strains, but which ultimately dilate as shear strains increase and hence develop significant strength.

Type II. Materials that behave as a liquid having the capability to flow indefinitely with vanishingly small resistance. When sufficient water is expelled the particles achieve more stable and dense configuration.

Type II refers to "classical" liquefaction, the "spontaneous" liquefaction of Terzaghi and Peck. Type I materials, in comparison, cannot under ordinary circumstances collapse, even though it is possible for pore pressures to approach overburden pressures if shear strains are kept small. To apply the term "liquefaction" to the materials of Type I may be to apply it out of context. According to Casagrande (1971, p. 202), ordinary sands with relative density > 50 percent are "safe" against liquefaction. The relative density of material in the sand lenses within the Bootlegger Cove Clay is approximately 60 percent (Seed and Wilson, 1967, Fig. 10), which suggests that, in general, the material is probably dilative with large strains, hence of Type I. These materials seem to have been rather widely distributed, as suggested by the regional subsidence curves shown in Fig. 1, and were likely of mechanical significance in providing local low-friction segments which subsequently coalesced to form a surface of sliding. Since these sand and silt lenses comprised closed hydraulic systems under the conditions of seismic loading, however, complete liquefaction of sand is not regarded to have been likely until the adjacent silty-clay structure had collapsed. Hence this author believes liquefaction of clay *preceded* complete liquefaction of sand lenses, rather than the reverse as suggested by Seed and Wilson (1967; see also Seed, 1968), and was essential to its development.

[15] The failure zone is assumed to be contained within two planes disposed at $\theta = \pm(45° - \phi/2)$ about a vertical plane drawn through the predicted point of separation. Thus a "plastic wedge" develops. Here ϕ is the angle of internal friction for the material overlying the sliding surface. Further discussion of the location of the point of failure is given in a subsequent section dealing with The Geometry of the Glide Blocks and Plastic Wedges.

Even with low roughness,

$$\text{slope } F_R > \text{slope } F_D = 0,$$

hence separation is again predicted near the bluff (point b'). A more extensive progression of the strength-loss zone in the direction of the toe is required to achieve the condition $F_R = F_D$.

Upslope Block Gliding. The sole driving force[16] is as in the preceding example, but it now must overcome the upslope weight component in addition to soil strength. As before, separation is possible near $X = 0$ (point b'') with even more extensive strength-loss required toward the toe.

Propagation of the strength-loss zone has been discussed in each of these examples with reference to an interior initial position. If development were to proceed from the toe, Fig. 10 shows that the condition $F_R = F_D$ would immediately obtain, and successive slumping rather than block gliding would result.

Mechanics of Retrogressive Failure

Initial failure is presumed to occur at $X = b$, within a wedge of material more or less symmetrically disposed (at angle θ) about a vertical plane through b [Fig. 11-(1)]. The curve representing the available driving force is then shifted to the left ($F_D' = 0$ at $X = b$) inasmuch as concern is now with stability of the block to the left of point b.

The lateral force developed within the "plastic wedge" serves a dual function at this point in time: it (1) drives block I toward the right at an accelerating rate, and (2) serves as a resisting force for the stable block to the left of b. The lateral force developed within the wedge may undergo an increase due to strength loss, for example, soil remolding associated with failure; hence initially F_R may shift to F_R' [Fig. 11-(1)]. However, as a consequence of displacement u of block I, the average thickness of material within the wedge decreases; thus the lateral pressure will also decrease as sliding progresses.

When the thickness of the plastic wedge decreases to a critical value h_{cr}, the curve of resisting forces F_R' will have decreased to the curve F_R''; F_R'' is tangent at some point c to the curve F_D', and a new plastic wedge develops about c, releasing glide block II [Fig. 11-(2)]. The process is then repeated [Fig. 11-(3)]. Now F_D' shifts to a new position F_D'' relative to the stable block to the left of c, F_R'' shifts initially to F_R''', and subsequently, with sufficient movement of block II, to position F_R'''' [Fig. 11-(4)].

The process can be indefinitely repeated as long as boundary conditions and material properties remain unchanged; these, however, typically change under conditions of natural deformation, and hence the process eventually stops. At Turnagain, several factors may have been involved, including (1) an increase of toe resistance, (2) retrogression to margins of the strength-loss zone, (3) gradual increase of resistance of the strength-loss zone, owing to either "effective stress" increase with dissipation of fluid pressure or increased geometric complexity of the sliding surface, and (4) retrogression toward a domain of lower sensitivity.

Geometry of Glide Blocks and Plastic Wedges

Conclusions can be drawn from the geometry of the blocks and intervening wedges. For example, because the blocks at Turnagain are, for the most part, roughly triangular in shape, it is apparent that failure and plastic wedge development had occurred in the

---

[16] The net gravitationally induced force is plotted in Fig. 10 to facilitate comparison with the two preceding examples. If driving and resisting forces were strictly separated, driving force F_D would remain about as in the second case, and the resistance due to weight acting upon a negative slope would be added as a component of F_R.

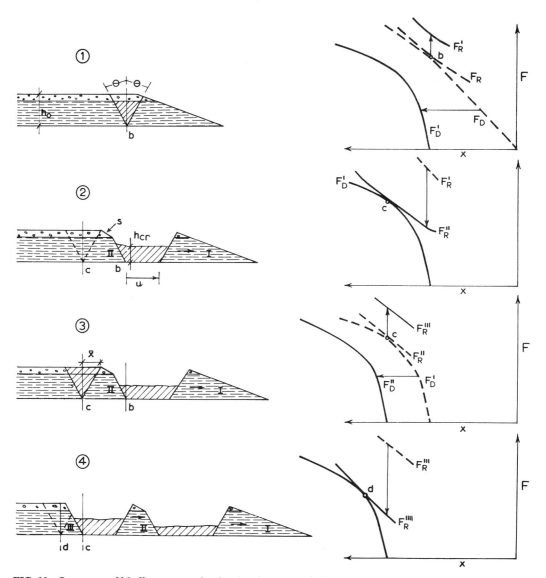

FIG. 11 Sequence of bluff retrogression by development of plastic wedges and block gliding. Resisting and driving forces are plotted (on the right) as a function of distance to the original bluff. The sequence of events associated with each of these force diagrams is shown schematically on the left. Symbols are described in text. Compare sequence with an alternative given in Fig. 6, in which separation of plastic wedges from stable areas has been assumed.[23]

general vicinity of the bluff. This suggests that, according to the ideas elucidated with respect to Fig. 10, in general slope F_R > slope F_D; hence resistance possibly due to a combination of locally incomplete liquefaction and roughness overrides the downslope gravitational component due to the weight of the block. This is not particularly surprising in view of the magnitude of slope ($\sim 2.2°$). However, it is useful to obtain an idea of the actual friction involved in so-called liquefied zones, and it illustrates that *gravitationally induced lateral pressures* were far more important[17] than the downslope weight component so often cited as the cause of block-glide movements. If these blocks were rigid glide blocks rather than deformable blocks, retrogression would not have occurred.[18]

Further, the observed geometry can be used to draw inferences concerning the lateral pressures and material properties. In the simplified example presented here it is assumed that apparent friction and the inclination of the slide plane produce a null net effect, hence analysis is equivalent to that for a frictionless horizontal glide plane. At incipient sliding, the lateral driving force behind an interior glide block must equal the resisting forces; hence in Figure 11-(2), at point c, $F'_D \simeq F''_R$. Employing a total stress analysis,

$$\tfrac{1}{2}\gamma h_0 - 2S_o h_0 = \tfrac{1}{2}\gamma h_{cr}^2 - 2S_r h_{cr}^2$$

where γ is unit weight and S_o and S_r are original and remolded shear strength values (see Odenstad, 1951). This equation can be used to infer the depth of the slip surface if the material strength is measured. Conversely, if the position of the sliding surface is known, it is possible to estimate strength; in this case the strength would fall between limits given by

$$S_o = \frac{\gamma}{4}(h_0 + h_{cr})$$

for an insensitive material, $S_o = S_r$, and

$$S_o = \frac{\gamma}{4h_0}(h_0^2 - h_{cr}^2)$$

for a highly sensitive material, $S_r = 0$. Application of these equations to Turnagain (and ignoring the effect of the surficial sand layer), assuming $h_0 = 60$ ft and $h_{cr} = 40$ ft, leads to

$$S_o \text{ (max)} = 1.56 \text{ tsf}$$
$$S_o \text{ (min)} = 0.52 \text{ tsf}$$

The latter value is in rather good agreement with measured strength values (see Fig. 5; Wilson, 1967; Hansen, 1965, Fig. 8), which suggests that extensive remolding was in fact associated with development of the plastic wedge. The high sensitivity and extremely low remolded strength measured for the Turnagain material (Wilson, 1967) lends credence to this view.

The width of the glide blocks is governed by the slopes of the F_R and F_D curves. It is assumed that full development of active pressure becomes possible at a distance $\bar{X}$ behind each successive bluff, where $\bar{X} = h_0 \tan \theta$ [Fig. 11-(1) and 11-(3)]; θ is governed by the material properties, and a good approximation is given by $\theta = \pm(45 - \phi/2)$ where ϕ is a coefficient of internal friction for effective stresses. Hence the center of the plastic wedge is predicted to be approximately located at $\bar{X}$ in a majority of cases. The initial width of the glide block base (as well as the initial width of the wedge) is predicted to be about $2\bar{X}$, although this would be subject to variation where local slumping [Fig. 11-(2), see point S] has caused modifications of the F_D curve or where inhomogeneities or ground fracturing (see Fig. 4) exists. Hence blocks of width smaller or larger than $2\bar{X}$ (compare blocks II, III in Fig. 11) would not be expected to be uncommon.[19] Subsequent deformation leading

---

[17] A relative assessment of these factors can be readily made by reference to Fig. 10.

[18] Fluid pressures are included in the calculations of lateral pressures within plastic wedges; fluid pressures could, of course, be of equivalent significance with regard to the motion of "rigid" glide blocks separated by fluid-filled fractures.

[19] At Turnagain, assuming $\phi = 30°$, predicted initial block width would be about 70–80 ft, increasing as a function of h_0 toward the front of the original bluff. With slumping outwash at the crest of the bluff, these figures would increase to 90–100 ft; the intersection of the wedge boundaries with local tensile fractures would result in decreased block widths. These values may be compared to the blocks illustrated in Fig. 6.

to extrusion or attrition at the base of the block would decrease glide-block dimensions.

Finally, estimates of displacement can be related to the geometry of the plastic wedge and hence to the pressure distribution and to material strength. For example, assuming planar wedge boundaries, the relative displacement u between two glide blocks is given by

$$u = \left(\frac{h_0^2}{h'} - h'\right) \tan \theta$$

where h' is the height of material within the wedge.[20] At incipient failure, $h' = h_{cr}$, and therefore

$$u_{cr} = \left(\frac{h_0^2}{h_{cr}} - h_{cr}\right) \tan \theta$$

where u_{cr} is the critical displacement required for the development of a new plastic wedge.[21]

The minimum distance between the bases of two glide blocks is given by u_{cr};[22] it would be expected that earlier-developed glide blocks would be, at the end of deformation, separated by greater distance owing to continued spreading within the plastic wedge.[23]

These relationships make refined velocity estimates possible if information is available concerning the timing of the retrogressing bluffline.

Summary and Analogies

Consideration has been given to a general approach for the analysis of block-gliding structures based on rather simple geometrical and mechanical relationships. The Turnagain example was treated in detail; many of the block-glide features that developed at that site appear to be explainable without the invocation of earthquake-induced inertial forces, even though such forces may have played some role in early block movements. Proceeding from the final geometry, it was possible to establish information on the displacement patterns and sequence of block-glide events in time; velocity estimates were given, the retrogressive character of the deformation was established, and a hypothesis concerning the mechanics of this process was presented; finally, deductions were possible concerning block dimensions and the critical values of block displacement required to sustain retrogression.

The essential features of the Turnagain Heights slide include the following:

1. A low-friction basal plane, probably approximately parallel to bedding plane surfaces, but rendered in a virtually "liquefied" state because of earthquake-induced ground motions of unusually long duration.
2. Propagation of the liquefied zone from

---

[20] This is simply a formalization and reversal of the approximation devised by C. Kaye (see Hansen, 1965), according to which the depth of failure is given by the graben area divided by (presumed known) lateral displacement.

[21] This spacing (u_{cr}) should not be confused with the spacing between ridge crests. The latter is approximately equal to $u_{cr} + 2\bar{X}$. At Turnagain, u_{cr} is predicted to be about 30 ft, and predicted minimum spacing between ridges crests is on the order of 100 ft (cf. Fig. 6).

[22] Soil strength can be related to u_{cr}; e.g., for the case of very sensitive material, assuming $S_r = 0$,

$$u_{cr} = \left(\frac{h_0^2}{\sqrt{h_0^2 - (4S_o/\gamma)h_0}} - \sqrt{h_0^2 - (4S_o/\gamma)h_0}\right) \tan \theta$$

[23] In fact, it is possible for a glide block to become separated from its plastic wedge, or for the plastic wedge to become separated from the stable (or more recently stable) block at its rear margin; e.g., a break-in-slope exists at Turnagain between the sliding surface and the steeper tidal slope; it is possible that earlier-formed glide blocks could have moved individually over the steeper tidal slope as a result of the downslope gravity component (Fig. 6, section II).

generally *interior* points, requiring rupture of the toe of the initial glide block.
3. Resistance to movement along the main sliding surface due to local friction nuclei or to surface roughness; although small, this force was generally more important than the driving force produced by the downslope component of the weight of the glide block; if the glide blocks were "rigid" rather than "deformable," retrogression could not have occurred unless fluid pressures within fractures performed a function analogous to "plastic wedges."
4. Development of successive zones of detachment behind the bluffline; at Turnagain these took the form of "plastic wedges" and provided the dominant energy source required to drive the glide blocks forward.
5. Initially, the lateral pressures developed within these wedges were sufficient to support the new bluffline; as the wedge thickness decreased as a result of lateral spreading associated with deformable block gliding, lateral support decreased to the point at which a new plastic wedge could develop behind the bluffline, thus resulting in periodic retrogression.
6. Sensitivity was a factor concerning both the decrease of basal friction and the variation of lateral pressures within plastic wedges, and hence it exerted an important control on the total displacements and on the rate and amount of bluff retrogression.
7. Motion of the frontal glide blocks occurred on a slightly inclined, submerged, erosion-deposition surface veneered by saturated tidal silts; this surface was probably rendered nearly frictionless owing to earthquake vibrations, and thus relatively small basal resistance was encountered here.

Clear analogies are some of the landslides in south Sweden; these slides have also occurred in sensitive, laminated Quaternary silts and clays. Low-friction basal planes appear characteristic, and occurrences of sharp-crested glide blocks alternating with collapsed plastic wedges have been reported (see, e.g., Odenstad, 1951). These slide masses are smaller than at Turnagain Heights, but some are of a similar order of magnitude (e.g., $1-2 \times 10^6$ m^3 at Sköttorp). Retrogressive behavior has been important, and the significant deformational modes appear to be lateral spreading and deformable block gliding. Thus these structures appear to be comparable to Turnagain Heights even in details. Some mechanisms other than earthquake-induced cyclic loading appear to be related to basal strength loss; however, although this subject is poorly understood, sensitivity of the sediments and the development of large fluid pressures seem to have been important factors.

On a far greater scale are the structural elements of the so-called Heart Mountain "overthrust" of northwestern Wyoming. The principal geologic features of the Heart Mountain area are given at length in a paper by Pierce in this volume. The most enigmatic aspect of this structure continues to be the mechanism of emplacement, a problem which has defied resolution for a half-century. In the author's view the Heart Mountain structure is a landslide, perhaps the largest on earth, which, despite differences in scale, shares many features with smaller slides effected by block gliding on low-angle surfaces. Glide blocks at Heart Mountain appear to be of the "rigid" rather than the "deformable" type; nonetheless, in contrast to previously expressed opinions, it appears entirely reasonable to suppose that initial movements could have involved a "push from behind." A "fluid wedge" hypothesis is proposed according to which magma- or water-filled fractures in the rear detachment area provided the thrust necessary to initiate movement on a low-angle sliding surface; the sliding surface is assumed to have been rendered nearly frictionless by enhanced fluid pressures. The proposed "fluid wedge" is analogous in function to the "plastic wedge" described for the slide at Turnagain Heights.

The complete absence of deformation below the basal Heart Mountain décollement, and the presence of fault breccia-derived dikes injected within glide blocks, seem to demand a fluid flotation mechanism for the Heart Mountain structure (Voight, 1973a).[24] The proposed sequence of events is as follows: (1) multiple horizon steam/water injection from either a juvenile source within an active vent (e.g., the Hughes hypothesis), a "thermal contact" source arising from intrusive injection into saturated shales, or both; (2) steam/water sill intrusion by pneumatic/hydraulic uplift and (locally) fracture; (3) local development of steam/water dikes by pneumatic/hydraulic fracture, and consequent development of horizontally-directed "fluid-wedge" forces (cf. Voight, 1972); (4) toe rupture in response to horzontal thrust and pneumatic/hydraulic fracture mechanisms; (5) saturation of land surface by toe water vents; (6) behavior of the upper plate as a flexible slider on a compressible, multiple-phase, largely turbulent squeeze-film lubricant; lubricant escape velocities were sonic, hence "flotation times" for larger blocks were only of a few hours duration; (7) large plate displacements, due principally to the downslope component of gravity and only secondarily to short-lived fluid-wedge impulse; a minimum initial plate length, ~ 10 miles (15 km) on a 2° décollement, was required to overcome the resistance of the transgressive fault riser; about twice this initial length was necessary to produce the observed displacements; (8) movement over the slightly sloping ($\sim 1°$) land surface due to high basal fluid pressures induced by rapid loading; maximum block velocities were likely high, ca. 10^2 K;[25] (9) fragmentation and dispersion of moving glide blocks attributed to pneumatic/hydraulic fracture and fluid wedging (with associated breccia dikes), squeeze film "shear," and/or to local frictional seizing.

Comparison of individual elements associated with the Heart Mountain and Turnagain Heights structures discloses similarities even in detail[26]; there are also a few clear

---

[24] This paragraph was added when the volume was "in press" and hence was not available to Pierce for comment in his paper on Heart Mountain geology; the reader should therefore not suppose that Pierce agrees to the arguments expressed here.

[25] The appropriate velocity unit for geological materials which move in association with a fluid mechanism is clearly the *knot*, abbreviated K; one knot is defined as one nautical (Admiralty) mile (6,076.10 ft; 1.852 km) per hour. Approximate conversions are as follows; 1 K $\simeq$ 1.85 km/hr $\simeq$ 1.15 mph $\simeq$ 1.69 ft/sec $\simeq$ 0.51 m/sec $\simeq$ 1.62 × 10^9 cm/yr. A case can be made, on similar grounds, for listing dimensions of fluid-borne structural elements in knot units; each knot is a division 47 ft 3 in (~ 14.4 m) in extent, having the same relation to one nautical mile as 28 seconds has to one year.

[26] Heart Mountain and Turnagain Heights structures share the following features: (1) a fragmented glide sheet, thin in comparison to length or width dimensions, subjected to large translational displacements; (2) detachment zones which developed as a consequence of block gliding away from relatively stable masses, and within which lateral pressure "wedges" presumably developed; (3) a low angle ($\sim 2°$) basal décollement, approximately parallel to bedding planes, below which surface deformation was negligible; locally this surface was denuded as a consequence of block gliding; (4) a sequence of marine silty clays (or clay shales) located in the vicinity of the sliding surface but not necessarily coincident with it; (5) a ruptured toe, across which several glide blocks transgressed; (6) a frontal glide horizon composed of an exposed surface of erosion and deposition (former land surface at Heart Mountain underlain principally by shales; submerged tidal flat at Turnagain); (7) particulate material at or near the basal décollement which locally display evidence of fluidization (Voight, 1973b); at Heart Mountain, clastic injection dikes within glide blocks[27] of finely-crushed carbonate rock derived from a thin layer of gouge formed on the décollement; at Turnagain, large sand boils within the slide mass (Hansen, 1965, p. A29); (8) a relatively moderate coefficient of sliding friction for glide blocks on the décollement, i.e., $\mu' \simeq 0.6$; (9) a fluid pressure mechanism essential to the mechanism of emplacement, as inferred from (7), (8), and other considerations; (10) a virtually cataclysmic rate of deformation.

[27] Clastic dikes have also intruded into a thick layer of volcanic material which buried the then-exposed surface of *tectonic denudation* soon after gliding had occurred; the age of these dikes therefore post-dates deposition of at least some of these volcanics, although their development is, in general, synchronous with volcanism. Clastic dikes injected into glide blocks are somewhat older; their emplacement seems to have occurred directly in association with block gliding movements.

dissimilarities, e.g., at Heart Mountain the glide mass is much larger and displacements are greater, and a thermal event (volcanism) is closely associated in time with block gliding. Nonetheless, from a mechanical point of view, it is possible to approach both structures from a consideration of similar fundamental considerations. Finally, analysis of the mechanics of these structures suggests an important contrast with regard to the role of gravity; at Heart Mountain, displacements by gliding are attributed principally to the downslope component of weight of overburden, whereas at Turnagain Heights, gravitationally-induced *lateral* pressures acting at the rear of the glide blocks seem most important.

Appendix

An eyewitness account of the Turnagain Heights slide furnished to the Geological Survey by Mr. E. R. Bush, as reported to him by Mr. Brooke Marston, 1900 Turnagain Parkway, Anchorage, Alaska (Grantz et al., 1964):

I was driving my automobile westward on McCollie Avenue when the earthquake occurred. I immediately stopped my automobile and waited until the quake subsided. It appeared to me that the car was rocking from north to south. It rocked so violently that I nearly became seasick. From my car I could observe an earth crack alined north–south and opening and closing from east to west. As soon as the quake subsided I proceeded to drive westward to the corner of McCollie Avenue and Turnagain Parkway. After turning right on Turnagain Parkway and driving approximately 180 feet north, I realized the bluff was gone north of my driveway, which paralleled the bluff in an east–west direction. I got out of the car, ran northward toward my driveway, and then saw that the bluff had broken back approximately 300 feet southward from its original edge. Additional slumping of the bluff caused me to return to my car and back southward approximately 180 feet to the corner of McCollie and Turnagain Parkway. After I stopped at this point, the bluff continued to slowly slide northward as I continued to back my auto southward on Turnagain Parkway. The bluff slowly broke away until the corner of Turnagain Parkway and McCollie had slumped northward. It is my impression that the Turnagain Bluff area slumped northward in segments and that much of the southward receding of the bluff occurred after the major earthquake had subsided. I was never aware of any unusual noise other than the sounds of people calling for help.

References

Casagrande, A., 1971, On liquefaction phenomena: report of lecture, by P. A. Green and P. A. S. Ferguson: *Geotechnique*, p. 197–202.

Cederstrom, D. J., Trainer, F. W., and Waller, R. M., 1969, Geology and ground-water resources of the Anchorage Area, Alaska: *U.S. Geol. Survey Water Supply Paper 1773*, 108 p.

Engineering Geology Evaluation Group, 1964, *Geologic report 27 March 1964 earthquake in Greater Anchorage Area*: Alaska State Housing Authority and the City of Anchorage, Alaska, 34 p.

Grantz, A., Plafker, G., and Kachadoorian, R., 1964, Alaska's Good Friday earthquake, March 27, 1964: A preliminary geologic evaluation: *U.S. Geol. Survey Circ. 491*, 35 p.

Haarmann, E., 1930, *Die Oszillations Theorie*: Stuttgart, Ferdinand Enke, 260 p.

Hansen, W. R., 1965, Effects of the earthquake of March 27, 1964 at Anchorage, Alaska: *U.S. Geol. Survey Prof. Paper 542A*, 68 p.

———, Eckel, E. B., Schaem, W. E., Lyle, R. E., George, W., and Chance, G., 1966, The Alaska earthquake, March 27, 1964: Field investigations and reconstruction effort: *U.S. Geol. Survey Prof. Paper 541*, 111 p.

Karlstrom, T. N. V., 1964, Quaternary geology of the Kenai Lowland and glacial history of the Cook Inlet region, Alaska: *U.S. Geol. Survey Prof. Paper 443*, 69 p.

Kerr, P., and Drew, I., 1968, Quick-clay slides in the USA: *Eng. Geol.*, v. 2, p. 215–238.

Long, E., and George, W., 1967, Turnagain slide stabilization, Anchorage, Alaska: *Am. Soc. Civ. Eng. Proc.*, v. 93, no. SM 4, p. 611–627.

Miller, R. D., and Dobrovolny, E., 1959, Surficial geology of Anchorage and vicinity, Alaska: *U.S. Geol. Survey Bull. 1093*, 128 p.

Odenstad, S., 1951, The landslide at Sköttorp on the Lidan River: *Royal Swedish Geotech. Inst. Proc.*, no. 4, 40 p.

Patton, F., 1966, Multiple modes of shear failure in rock: *Proc. 1st Cong. Int. Soc. Rock Mech., Lisbon*, v. 1, p. 509–513.

Rengers, N., 1970, Influences of surface roughness on the friction properties of rock planes: *Proc. 2nd Cong. Int. Soc. Rock Mech., Belgrade*, v. 1, paper 1-31, 6 p.

Rubey, W. W., and Hubbert, M. K., 1959, Role of fluid pressure in mechanics of overthrust faulting, II. Overthrust belt in geosynclinal areas of Western Wyoming in light of fluid-pressure hypothesis: *Geol. Soc. Am. Bull.*, v. 70, p. 167–206.

Schardt, H., 1898, Les régions exotiques du versant nord des Alpes suisses: *Bull. Soc. Vaudoise Sci. Nat.*, v. 34, p. 114–219.

Seed, H. B., 1968, Landslides during earthquakes due to soil liquefaction: *Am. Soc. Civ. Eng. Proc.*, v. 94 no. SM 5, p. 1055–1122.

———, and Wilson, S. D., 1967, The Turnagain Heights landslide, Anchorage, Alaska: *Am. Soc. Civ. Eng. Proc.*, v. 93, no. SM 4, p. 325–353.

Shannon and Wilson, Inc., 1964, Report on Anchorage area soil studies, Alaska, to U.S. Army Engineer District, Anchorage, Alaska: Seattle, Wash., 109 p.

Steinbrugge, K. V., 1967, Introduction to the earthquake engineering of the 1964 Prince William Sound, Alaska, earthquake, *in* Wood, F. J., ed., The Prince William Sound earthquake, Alaska, earthquake of 1964 and aftershocks: *U.S. Dept. Commerce, Coast and Geodetic Survey Pub. 10-3*, v. 2, p. 1–6.

Thomas, Mrs. L., Jr., 1964, An Alaskan family's night of terror: *Nat. Geog.*, v. 126, no. 1 (July), p. 142–156.

Trainer, F. W., and Waller, R. M., 1965, Subsurface stratigraphy of drift at Anchorage, Alaska: *U.S. Geol. Survey Prof. Paper 525C*, p. 167–174.

Van Bemmelen, R. W., 1936, The undation theory of the development of the earth's crust: *Report 16th Int. Geol. Congr.*, Washington, v. 2, p. 965–982.

———, 1965, Evolution of the Indian Ocean megaundation: *Tectonophysics*, v. 2, p. 29–57.

Varnes, D. J., 1958, Landslide types and processes, *in* Eckel, E. B., ed., Landslides and engineering practice: *Natl. Research Council, Highway Research Board Spec. Rept. 29*, p. 20–47.

———, 1969, Stability of the west slope of Government Hill, Port area of Anchorage, Alaska: *U.S. Geol. Survey Bull. 1258D*, 61 p.

Voight, B., 1972, Fluid-wedge hypothesis and the Heart Mountain and Reef Creek décollements, northwestern Wyoming, U.S.A.: *Geol. Soc. Am. Ann. Mtg. Abs., Minneapolis, Minn.*, p. 698.

———, 1973a, Role of fluid pressure in mechanics of South Fork, Reef Creek, and Heart Mountain rockslides: *Geol. Soc. Am. Northeastern Section Abs., Easton, Pa.*

———, 1973b, Clastic fluidization phenomena and the role of fluid pressure in mechanics of natural rock deformation: *Geol. Soc. Am. Northeastern Section Abs., Easton, Pa.*

Waller, R. M., 1966, Effects of the March 1964 Alaska earthquake on the hydrology of the Anchorage area: *U.S. Geol. Survey Prof. Paper 544B*, 18 p.

Wilson, S. D., 1967, Landslides in the city of Anchorage, *in* Wood, F. J., ed., The Prince William Sound earthquake, Alaska, earthquake of 1964 and aftershocks: *U.S. Dept. Commerce, Coast and Geodetic Survey Pub. 10-3*, v. 2, p. 253–297.

PART 2

Papers on the Alpine System in the Mediterranean Region

KEES A. DE JONG

Department of Geology
University of Cincinnati
Cincinnati, Ohio

Introduction to Part 2: Mountain Building in the Mediterranean Region

The almost continuous Alpine mountain belt in the Mediterranean area can be subdivided into various orogenic systems (Figs. 1 and 2). Most of the attention of this book is directed to the central systems of the Alps and Apennines, since so many classical concepts illustrating the role of gravity in tectonics originated in these regions.

The Rocks

The rocks in the Mediterranean mountain belts can be conveniently grouped into those of the Prealpine geosynclinal basement, the Alpine geosynclinal sequence (divided into a mio- and a eugeosynclinal sequence), and the postgeosynclinal sequence.

Pre-alpine Rocks

A common interpretation of North American geology is that mountain belts, developed at the continental margin, have increased the area of the continent during successive orogenic cycles (Dietz, 1961). At first glance, European geology seems to present a similar appearance. Orogenic belts surround the Russian craton, creating the impression that the stable part of the European continent was greatly enlarged during the Phanerozoic (Fig. 3). In two orogenic cycles Paleo-Europe and Meso-Europe were welded to the Russian craton, whereas Neo-Europe consists of the orogenic systems formed during the Alpine orogenic cycle in the Mesozoic and Cenozoic.

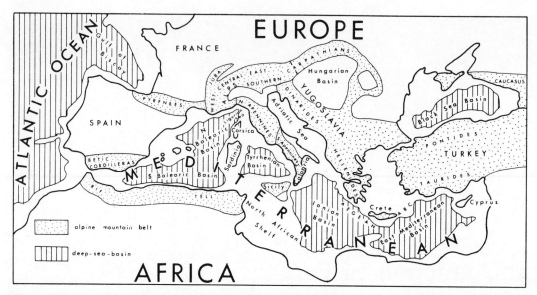

FIG. 1 Names of the major geologic units in the Mediterranean area.

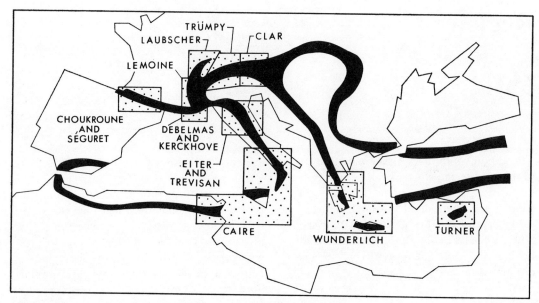

FIG. 2 Locations of the Mediterranean areas described in Part 2.

However, where one can today observe the floor of the geosynclinal sequence of Meso-Europe, it consists of continental (sialic) rocks folded and metamorphosed during earlier orogenic episodes. Extrapolation of the present surface area of Prealpine complexes in Mediterranean mountain belts also shows that the larger part of the Alpine geosyncline was underlain by sialic crust. Thus most of the rocks in the Mediterranean mountain belts were formed not at the continental rise on a floor of sima but on top of a continent

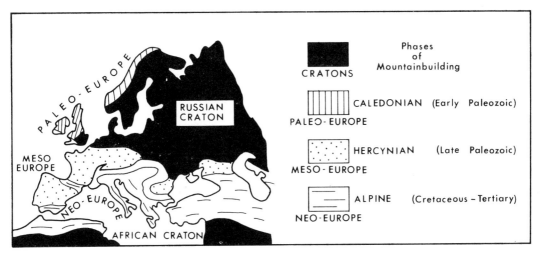

FIG. 3 Positions of the Phanerozoic mountain belts in Europe.

which has not increased its size substantially since the Paleozoic; in fact, a decrease in size is more likely.

The Miogeosynclinal Sequence

At the end of the Paleozoic the Hercynian mountain belts were eroded. Red bed deposits accumulated in fault-bounded troughs in the future Alpine region, and in some areas volcanic rocks were extruded. Marine platform sedimentation progressed from east to west, preceding the deposition of the miogeosynclinal sequence of predominantly carbonate rocks. Evaporites were formed over a wide area during the Late Permian to Early Triassic and in the Late Triassic (Fig. 4).

This evaporite layer, which later determined to a great extent the structural evolution of the Mediterranean orogenic systems, is in places very thick: oil wells have penetrated more than 1500 m of Triassic salt in western Greece (Jenkins, 1972) and locally more than 1000 m in Italy (Cataldi, 1963, in Bortolotti and others, 1970). In northern Africa and close to the Pyrenees the evaporites

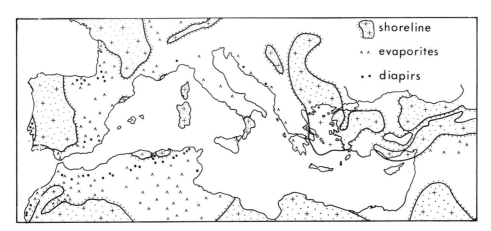

FIG. 4 Distribution of diapirs and Triassic evaporites in the Mediterranean area (after Sander, 1970).

rose diapirically, and in many places they permitted the detachment of the overlying miogeosynclinal sediments, leading to folding and overthrusting above an undeformed basement. The type-locality for this *décollement* is in the Jura Mountains (Laubscher, this volume [1]), but typical décollement areas are also found in the Subalpine chains of the Western Alps (Lemoine) and in the Northern Apennines, where the large Tuscan nappe moved to the northwest over the Upper Triassic evaporites (Elter and Trevisan). In almost all other Mediterranean mountain belts some overthrusts can be related to movements across underlying evaporite layers. The evaporites are commonly associated with dolomites and cellular or cavernous dolomites (*Rauhwacke*) which do not seem to diminish the ability of these horizons to act as detachment surfaces, even where they occur in greater abundance than the evaporites.

In many areas miogeosynclinal sequences are thicker than the eugeosynclinal sequence. For example, Temple (1968) reports thicknesses of about 6 and 3 km, respectively, in the Hellenides, and approximate figures for the Lombardic Basin of the South-Alpine miogeosyncline and the Piemont eugeosyncline are 8 and 2 km, respectively. Flysch sequences in the miogeosyncline are generally younger than those in the eugeosyncline.

In summary, the miogeosynclinal rocks in the Mediterranean area rest on a sialic basement and are characterized by the common occurrence of one or two evaporite layers at the base, an overlying sequence of mainly carbonate rocks, and flysch rocks younger than eugeosynclinal flysch.

The Eugeosynclinal Sequence

Perhaps the best preserved eugeosynclinal sequence in the Mediterranean area is formed by the Ligurian rocks of the Northern Apennines. According to Elter and Trevisan the Ligurian rocks are characterized by their internal derivation, the allochthonous position on the external miogeosynclinal rocks, the presence of ophiolites, and the fact that their floor was oceanic (simatic) rather than sialic. These characteristics apply in varying degrees to the other eugeosynclinal sequences in the Mediterranean mountain belts. In particular, the occurrence of ophiolites induced Stille (1940) to differentiate between mio- and eugeosynclines in the first place, and this still may be the major characteristic not only in the Ligurian domain but also in other Mediterranean mountain belts.

On the other hand, the internal position of the Ligurian eugeosyncline seems to be such a typical feature that Elter and Trevisan prefer to name this trough simply the "internal domain." ("Internal" refers to the distance between the foreland and the paleogeographic zones in the geosyncline, the most internal zone farthest away from the foreland). Similarly, Caire avoids the term eugeosyncline and speaks of rocks of internal origin. Figures 5 and 6 show, however, that in some places the eugeosyncline has an intermediate position between two miogeosynclinal zones and is thus not the most internal zone.

In the Northern Apennines the eugeosynclinal rocks have been moved much farther from their original position than the external miogeosynclinal sequence. In the Southern Apennines the situation is comparable, but in the Sicilo-Calabrian Arc the occurrence of crystalline massifs complicates the tectonic interpretation. According to Caire (see also Caire, 1970), the rocks of the eugeosynclinal trough were transported farthest, over the crystalline massifs of Calabria and Sicily, but Ogniben (1970) concludes instead that the massifs, with an internal miogeosynclinal cover, have undergone the largest tectonic transport in the arc. This situation, an internal miogeosynclinal rock

---

[1] In subsequent references to articles in *Gravity and Tectonics* the words "this volume" are omitted and the authors' names are not followed by the year of publication.

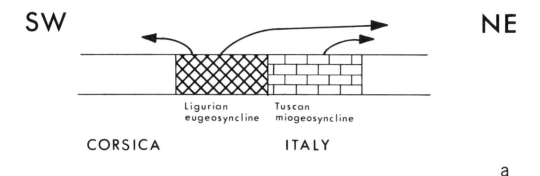

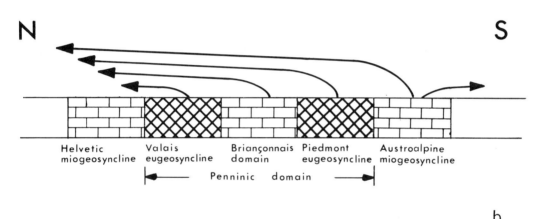

FIG. 5 Position of miogeosynclines and eugeosynclines in the geosyncline of the Northern Apennines (a), and in the geosyncline of the Western Alps (b). The arrows show the direction in which the rocks were transported during orogenesis.

sequence transported across the eugeosyncline, is also common in other Mediterranean mountain belts. In the Eastern Alps the Penninic eugeosyncline is tectonically overlain by the Austro-alpine nappes of miogeosynclinal origin (Clar), and in Crete an allochthonous eugeosynclinal sequence was overridden by miogeosynclinal limestones (Wunderlich). Thus in most of the Mediterranean area the eugeosynclinal rocks have been transported over an intermediate distance, more than the underlying external miogeosynclinal sequence, but less than the overlying internal one.

With regard to rock type, it is noteworthy that high-rank graywackes, regarded by some as the typical eugeosynclinal sediments, are not characteristic of the Mediterranean eugeosynclines, probably owing to the absence of subaerial volcanism. Moreover, metamorphism is not an exclusive characteristic of the eugeosynclinal zone. It is absent, for example, in the Ligurian eugeosyncline, and present in both the eu- and miogeosynclinal

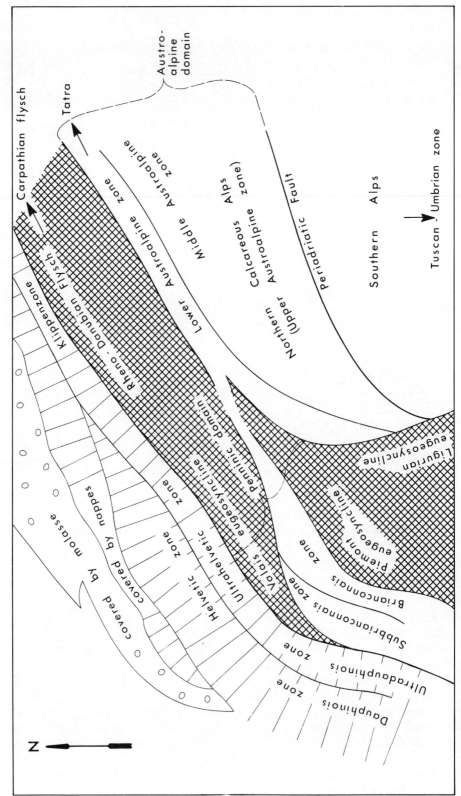

FIG. 6 Facies zones in the Alps (after Trümpy, 1972).

sequences in the Central Alps. The metamorphism in that region is not related to paleogeographic zones, or to the former presence of a heat dome, but is essentially a load metamorphism caused by tectonic burial during the formation of the Alpine nappes (Niggli, 1970).

Ophiolites in the Eugeosyncline

The ophiolitic suite consists of basic rocks, such as basalt and gabbro, and ultrabasic rocks, such as peridotite and serpentinite. Even though Steinmann (1905) recognized long ago that ophiolites and associated rocks represent a particular geosynclinal sequence, different interpretations have abounded, and it is only in the last few years that we have arrived at a certain understanding of ophiolites. The most common occurrences and interpretations of ophiolites are discussed briefly here.

1. Ophiolites Intrusive into Sialic Rocks. In the Pyrenees ultrabasic rocks (lherzolites) (Den Tex, 1969; Choukroune and Séguret) were emplaced as cold intrusives, penetrating the Mesozoic flysch rocks vertically. In the Cyprus gravity nappe the sediments of the Trypa group were contact metamorphosed by ultrabasic magma according to Turner. Basic intrusives in sediments, such as basalt dikes and sills, are commonly so closely related to submarine basalt flows that they are both included in group 2.

2. Ophiolites as Hot Submarine Extrusives. There is no doubt that some occurrences of basaltic ophiolites were submarine lava flows, commonly showing pillow structures. Locally these extrusives are associated with gabbros. However, it now seems unlikely that large, ultrabasic ophiolite masses could have been extruded at the sea bottom, as had been suggested by Brunn (1956) and Aubouin (1965), among others. It is more probable that the fractional crystallization and gravitational layering in these masses took place in the mantle (Bezzi and Piccardo, 1971; Reinhardt, 1969). Their close association with sediments is then secondary, caused either by large-scale thrusting, as in the Vourinos massif of northern Greece (Bortolotti and others, 1969) and the Oman ophiolite massif (Reinhardt, 1969) or by submarine sliding of sedimentary klippen (olistolites: see group 4, below).

3. Ophiolites as Part of the Upper Mantle. These have been emplaced in their present position as part of a large overthrust mass. The basic and ultrabasic rocks in the Apennines and Alps were interpreted by Argand (1924) as oceanic crust or sima (upper mantle material). This idea was taken up more recently by Hess (1962) and applied by Decandia and Elter (1969) to the Northern Apennines and by Peters (1969) to the Alps.

4. Ophiolites as Olistolites. Gansser suggested in 1959 that many ophiolites were not emplaced in the geosynclinal sequence of sediments as intrusives but as sedimentary klippen (olistolites). Italian geologists have shown that this suggestion was correct for a number of occurrences in the Northern Apennines, and also for some in the Dinarides, Hellenides, and the mountain belts of Turkey (Bortolotti and Passerini, 1970). However, this theory does not inform us about the origin of the parent rock of the olistolites, although a submarine extrusive, or an upper mantle origin are likely.

The ophiolites in the Northern Apennines can be classified into olistolites, submarine lava flows, and parts of the upper mantle. The distribution of the upper mantle ophiolites (i.e., the Bracco Ridge ophiolites in Fig. 13 of Elter and Trevisan) is not very extensive, but this might be due to the fact that the detachment surface of the Ligurian nappe was located rather high in the rock sequence and passed only locally through the ophiolites. Therefore the possibility cannot be excluded that the entire Ligurian eugeosyncline of the Northern Apennines was underlain by upper mantle rock (Elter and Trevisan, Fig. 3).

In definitions of geosynclinal realms extrapolations based on the occurrence of a key rock type become extremely important. For example, in spite of the rather restricted occurrence of ophiolites in the Ligurian nappe, it is still common to apply the term eugeosyncline to all Ligurian rocks because many of the characteristics of the sediments are fairly uniform over the whole Ligurian domain, even where ophiolites were absent. These "secondary" characteristics are essential in order to classify geosynclines in the Mediterranean. If the Ligurian-Piemont realm is called eugeosynclinal, the nomenclature is based on a rather limited occurrence of ophiolites, combined with certain common aspects of the sedimentary units.

The Postgeosynclinal Sequence

In places the clastic sediments of the postgeosynclinal sequence (Aubouin, 1965) were deposited before the last tectonic phase with horizontal movements. Two examples from the Alps and two from the Apennines will illustrate this.

The coarse breccias of the mid-Cretaceous to Lower Tertiary Gosau formation in the Eastern Alps, locally transgressive over thrust surfaces, were faulted and transported farther northward during a later tectonic phase (Clar). The Oligo-Miocene Molasse basin of tha Central Alps in Switzerland was filled with up to 6 km of clastic sediments after the late Eocene to early Oligocene Mesoalpine orogeny, but some of the molasse was already being overridden by nappes in late Oligocene time, while sedimentation continued elsewhere (Trümpy). During the Pliocene, the Molasse basin rotated around a vertical axis at its northern end (Laubscher, 1965), at the same time pushing the sedimentary layers of the Jura Mountains over a distance of more than 20 km toward the northwest.

In the Apennines sedimentation at the southern margin of the molassic Po Basin was repeatedly interrupted by olistostromes from the advancing front of the Ligurian nappe (Fig. 8). Olistostromes were also deposited in the postorogenic Mio-Pliocene of Calabria (Caire).

The Structure

When large overthrust sheets (nappes) and recumbent folds were discovered in the Alpine mountain chains around the Mediterranean Sea they were for some time thought to be characteristic of that area. This belief in the uniqueness of the Alps was lost when nappes and large recumbent folds were recognized in many other mountain belts as well; only the impressive scale of this tectonic style within the relatively small area of the Alps proper may still be considered unusual. Recently, a different distinguishing phenomenon has been attributed to the Mediterranean chains, namely the great mobility of relatively small crustal plates or blocks. Another peculiarity of the Alpine mountain belts is the great variety of tectonic transport directions.

Mobility of Crustal Blocks

Rotation of blocks relative to each other in the Mediterranean area was postulated first by Argand (1924) and later by Carey (1958). This notion was not widely accepted until paleomagnetic evidence was found in recent years. The rotation of the Iberian Peninsula is supported by the paleomagnetic work of Van der Voo (1969), the rotation of Corsica by Nairn and Westphal (1967), and the rotation of Sardinia by the work of De Jong and others (1969).

The counterclockwise movement of the Italy–Dinarides block (Fig. 7) is based on paleomagnetic data from the Southern Alps (Zijderveld and others, 1970), and Caire (Fig. 10) presents arguments for a similar rotation of Italy. A single counterclockwise rotation is certainly too simple to reflect the complex structural evolution in this part of the Mediterranean; Clar presents the hypothesis that the frontal part of the Italy–Dinarides block, that is, the Austroalpine unit, performed a clockwise rather than a counterclockwise rotation (Clar, Fig. 9).

In the eastern Mediterranean the Tethys Ocean must have been triangle-shaped and

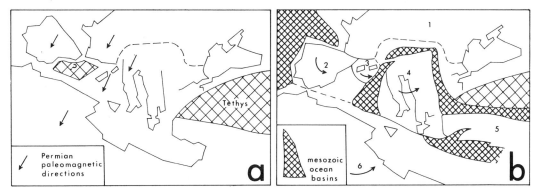

FIG. 7 (a) One possible position of the crustal blocks in the Mediterranean during the Permo-Triassic; the Permian paleomagnetic directions have been aligned parallel in this reconstruction. The position of Spain is based on data in LeBorgne and others (1971). (b) Location of the Mesozoic-Early Tertiary ocean basins (eugeosynclines) in the Mediterranean. The rotations which took place after the Permian are indicated by arrows. Some rotations (e.g., those of Corsica and Sardinia) did not occur until the Late Tertiary. The crustal blocks, between the continents of Europe (1) and Africa (6) are: (2) Iberian Peninsula; (3) Corsica/Sardinia; (4) Italy/Dinarides; and (5) Turkey.

more than 1000 km wide in the Triassic, according to paleomagnetic information (Fig. 7a) and coastline-fit data (Smith, 1971). Brinkmann (1972) considers it very unlikely, however, that such a large oceanic area ever did exist in the eastern Mediterranean. Are the interpretations of the new quantitative data of the past few years incorrect, or are the contradictions regarding the sense of rotation of the Italy–Dinarides block or the existence of a wide ocean in Turkey the result of insufficient knowledge of the geology of these areas?

The mobility of crustal blocks in the Mediterranean is also evident from the formation of small ocean basins during the Jurassic and on into the Early Tertiary (Fig. 7b). These ocean basins have not been preserved, but their former existence can be deduced from the distribution of ophiolites and associated sediments (De Jong, 1972, Fig. 4). The majority of the ophiolites in the Mediterranean area probably may be interpreted as upper mantle rocks, exposed through removal of the overlying sialic rocks. The oldest sediments in the Mediterranean eugeosynclines are invariably of Jurassic or younger age; this suggests that the extensional movements that led to the upward doming of upper mantle material (Maxwell, 1970) and to the formation of eugeosynclines did not start until the Jurassic. Some of the upper mantle rocks were raised sufficiently to be exposed and subsequently emplaced in the form of sedimentary klippen within the sediments (Elter and Trevisan). Finally, during the orogenic phases, ophiolites were thrust across adjoining crustal blocks.

Nappes

Overthrusting in the Mediterranean area stretched over a long time interval (Fig. 8). The first movements occurred during the Cretaceous in the Carpathians and Eastern Alps, whereas the Jura Mountains were folded as recently as the Pliocene. Some movement in the Central Alps may have continued into Pleistocene time (Trümpy).

Although overthrusts of older over younger rocks indicate local compression, this, of course, does not necessarily imply synchronized compression everywhere. Thus, whereas Cretaceous thrusting is evident in the Carpathians and Eastern Alps (Clar), indications for it are vague in the Central Alps (Trümpy).

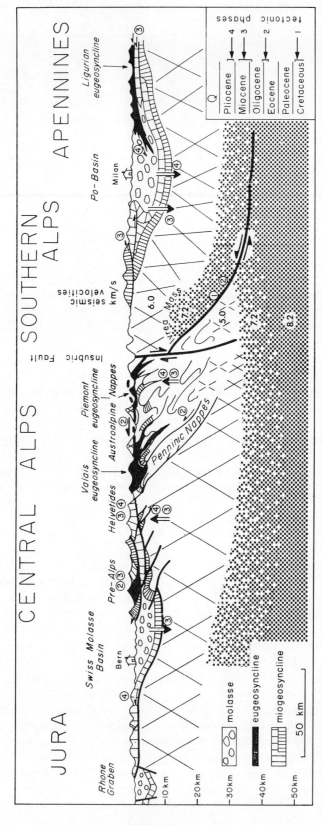

FIG. 8 Schematic composite cross section through Jura Mountains, Alps, and Northern Apennines. Location of section shown in Fig. 9. Seismic velocities after Giese (1968).

In the Ligurian eugeosyncline the evidence only shows vertical movements at this time, and the presence of Upper Cretaceous basalts, plutons, and innumerable dikes in Cyprus (Turner) suggests that the Mediterranean sea floor was actively spreading in that area.

One obvious classification of the enormously diverse group of Alpine overthrust sheets and recumbent folds could be based on the presence or absence of Prealpine basement rocks. Of course, the absence of such rocks does not mean in all cases that the basement rocks were not involved in the thrusting: they might have been eroded away or now be covered by the sea or below higher thrust sheets. An example of a subtle relationship between nappes with and without basement participation is described by Choukroune and Séguret from the Pyrenees. Here, the root zones of the nappes are not covered by younger nappes as they generally are in the Alps. The disappearance in the Alps of the rear parts or root zones of nappes is named "subduction" by Trümpy.

The rocks of the Penninic nappes and recumbent folds were pushed to great depths (as much as 30 km according to Niggli, 1970) by the Austroalpine nappes which constitute the frontal part of the Italy–Dinarides block. The latter nappes consist largely of sialic basement rocks, and it is also possible that the so-called Ivrea zone represents a part of the upper mantle (Fig. 8). There is no direct evidence for basement involvement in the development of the Jura folds or the nappes of the Northern Apennines, but this might be a false impression. The maximal 20-km displacement of the Jura Mountains is related, according to Laubscher, to the emplacement of the Helvetic Morcles nappe. The Prealps and the overthrust masses in the Southern Alps consist exclusively of Alpine geosynclinal rocks, but the fact that the source area of the Prealpine nappes was subsequently covered by the Austroalpine nappes makes it difficult to evaluate the role of the basement during their initial movement. Their final emplacement was almost certainly in the form of superficial slip sheets.

In the Northern Apennines the Ligurian units, derived from the Ligurian eugeosyncline, and a part of the Tuscan nappe, derived from the external miogeosyncline, came from a place which is now covered by the Ligurian Sea. This is not the only place where the source area of overthrust nappes has sunk from direct observation: five of these sunken orogenic centers can be distinguished in the Mediterranean area (Fig. 10). The tectonic transport around these orogenic centers was everywhere radially outward, that is, centrifugal with respect to the centers.

Variety in Tectonic Transport Directions

A striking aspect of Mediterranean tectonics is the great variety of transport directions (Figs. 9 and 10). In the past a common "explanation" was to deny one of the transport directions. For example, the southward movements in the western part of the Southern Alps (De Jong, 1967) were either refuted or explained as the effect of the northward movement of the crystalline basement, so that it became possible to explain the structure of the whole Alps in simple terms: all tectonic structures in the Alps would indicate the relative northward movement of the African continent (Staub, 1923, 1949). When it was realized that the structure of the Alps could not have been caused by a single mechanism, it also became clear that the great variety in transport directions is at least partly related to a succession of separate tectonic phases, commonly of quite different nature. Trümpy shows how the tectonic movements in the Central Alps can be classified into separate orogenic phases, and more than 15 years ago Fallot (1955) and later Van Bemmelen (1960) made a clear distinction in the Alps between older tectonic phases with northward tectonic transport and a younger phase with southward transport as well.

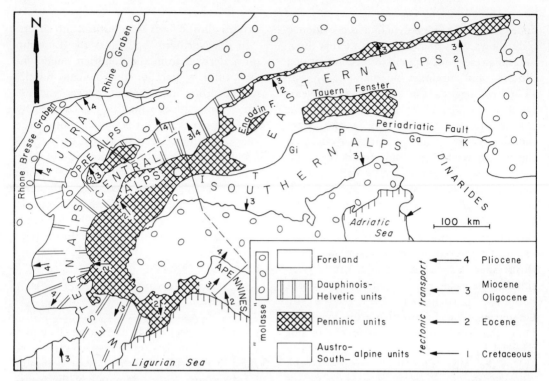

FIG. 9 Schematic geologic map of the Alps. Arrows indicate the directions, and the numbers the approximate times of tectonic movement. The Periadriatic fault system consists of (C) Canavese fault, (I) Insubric fault, (T) Tonale faule, (Gi) Giudicaria fault, (P) Pusteria fault, (Ga) Gail fault, and (K) Karawanken fault.

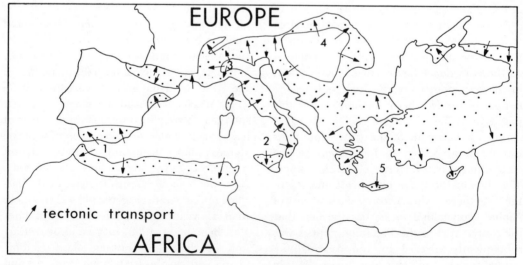

FIG. 10 Directions of tectonic transport in the Mediterranean area. (1) Gibraltar arc; (2) Calabro-Sicilian arc; (3) West Alpine arc; (4) Carpathian arc; (5) Aegean arc.

The centrifugal arrangement of overthrust sheets around orogenic centers in the Mediterranean was explained in an elegant way by Van Bemmelen in 1933 and afterward. According to his model the centers of the orogenic activity were uplifted so that radially outward gravity gliding could occur, followed by a collapse of the central part. Caire develops a similar model for the Calabro-Sicilian arc, but in addition he takes into account the movements of the adjoining crustal blocks. A different kind of interpretation is given by Andrieux and others (1971), and Laubscher (1971), who explain the arc structures of Gibraltar and the Western Alps exclusively in terms of crustal block movements.

Conclusions

The following statements summarize the major characteristics of the Mediterranean mountain belts:

1. Most Alpine geosynclinal sediments are underlain by sialic continental rocks.
2. Evaporite and *Rauhwacke* layers in the lower reaches of the miogeosynclinal sequences determined, to a great extent, the formation of detachment and overthrust surfaces during subsequent orogenesis.
3. The eugeosynclinal rocks, generally thinner than those of the miogeosynclines, and with less carbonate and older flysch rocks, were deposited on a partially simatic (ocean) floor.
4. Some of the ophiolites in the mountain belts are extrusive, others intrusive in origin, and still others may represent displaced parts of the eugeosynclinal ocean floor. These basic and ultrabasic rocks of upper mantle origin were locally overlain by submarine basalts, frequently emplaced as sedimentary klippen and finally thrust over the adjoining continent together with the surrounding sediments.
5. The great mobility of a few crustal blocks has been the major structural theme during the evolution of the Mediterranean region. This mobility may have been part of a larger-scale event, the drifting apart of Europe/Africa and North America.
6. The major movements of the crustal blocks in the Mediterranean region were extensional prior to the Late Cretaceous, continuing into the Early Tertiary in the eastern part (Turkey). The formation of eugeosynclines was related to these movements.
7. The earliest compressional movements in the geosyncline occurred in the Cretaceous and were contemporaneous with the formation of new oceanic crust in other parts of the Mediterranean area.
8. Overthrust sheets can be divided into two types: those with and those without basement participation. In the Central Alps, in which both types occur, a crustal block with internal miogeosynclinal sediments overrode an external miogeosynclinal block, with nappes of eugeosynclinal origin between them.
9. The great diversity in tectonic transport directions must be seen in light of a number of successive tectonic phases with different mechanisms of transport.
10. The arcs of the Mediterranean area have been described as the result of a sequence of uplift, centrifugal sliding, and collapse of the orogenic center, but also as the end product of the collision between two or more plates. It is likely that a combination of these processes can explain the majority of the observed structures.

References

Andrieux, J., Fontboté, J. M., and Mattauer, M., 1971, Sur un modèle explicatif de l'Arc de Gibraltar: *Earth Planet. Sci. Lett.*, v. 12, p. 191–198.

Argand, E., 1924, La Tectonique de l'Asie: *Comptes Rendus 13ème Congr. Géol. Int.*, p. 171–372.

Aubouin, J., 1965, Geosynclines: Developments in geotectonics, v. 1: Amsterdam, Elsevier, 335 p.

Bezzi, A., and Piccardo, G. B., 1971, Structural features of the Ligurian ophiolites: Mem. Soc. Geol. Italiana, v. 10, p. 53–63.

Bortolotti, V., and Passerini, P., 1970, Development of the Northern Apennines geosyncline: Magmatic activity: Sed. Geol., v. 4, p. 599–624.

Bortolotti, V., Passerini, P., Sagri, M., and Sestini, G., 1970, Development of the Northern Apennines geosyncline: the miogeosynclinal sequences: Sed. Geol., v. 4, p. 341–344.

Brinkmann, R., 1972, Mesozoic troughs and crustal structure in Anatolia: Geol. Soc. Am. Bull., v. 83, p. 819–826.

Brunn, J. H., 1956, Contribution à l'étude géologique du Pinde septentrionale et d'une partie de la Macédoine occidentale: Ann. Géol. Pays Helleniques, v. 7, p. 1–358.

Caire, A., 1970, Sicily in its Mediterranean setting, in Alvarez, W., and Gohrbrandt, K. H. A., eds., Geology and history of Sicily: Tripoli, Petroleum Expl. Soc. Libya, p. 145–170.

Carey, S. W., 1958, A tectonic approach to continental drift, in Carey, S. W., ed., Continental drift, a symposium: Hobart, University of Tasmania, p. 177–355.

Decandia, F. A., and Elter, P., 1969, Riflessioni sul problema delle ofioliti nell' Apennino settentrionale (Nota preliminare): Atti. Soc. Toscana Sci. Nat., Mem., Ser. A, v. 76, p. 1–9.

De Jong, K. A., 1967, Tettonica gravitativa e raccorciamento crostale nelle Alpi Meridionali: Boll. Soc. Geol. Ital., v. 86, p. 749–776.

———, 1972, Mediterranean geology—A review: Kagaku, v. 42, p. 283–291.

———, Manzoni, M., and Zijderveld, J. D. A., 1969, Palaeomagnetism of the Alghero trachyandesites: Nature, v. 224, p. 67–69.

Den Tex, E., 1969, Origin of ultramafic rocks, their tectonic setting, and history: A contribution to the discussion of the paper "The origin of ultramafic and ultrabasic rocks" by P. J. Wyllie: Tectonophysics, v. 7, p. 457–488.

Dietz, R. S., 1961, Continent and ocean basin evolution by spreading of the seafloor: Nature, v. 190, p. 854–857.

Fallot, P., 1955, Les dilemmes tectoniques des Alpes Orientales: Ann. Soc. Géol. Belgique, v. 78, p. 147–170.

Gansser, A., 1959, Ausseralpine Ophiolithprobleme: Eclog. Geol. Helv., v. 52, p. 659–680.

Giese, P., 1968, Die Struktur der Erdkruste im Bereich der Ivrea-Zone, Ein Vergleich verschiedener seismischer Interpretationen und der Versuch einer petrographisch-geologischen Deutung: Schweiz. Mineral. Petrol. Mitt., v. 48, p. 251–284.

Hess, H. H., 1962, History of the ocean basins: in Engel, A. E. J. et al., eds., Petrologic studies: A volume in honor of A. F. Buddington: New York, Geol. Soc. America, p. 599–620.

Jenkins, D. A. L., 1972, Structural development of Western Greece: Am. Assoc. Petrol. Geol. Bull., v. 56, p. 128–149.

Laubscher, H. P., 1965, Ein kinematisches Modell der Jurafaltung: Eclog. Geol. Helv., v. 58, p. 231–318.

———, 1971, The large-scale kinematics of the Western Alps and the Northern Apennines and its palinspastic implications: Am. J. Sci., v. 271, p. 193–226.

LeBorgne, E., LeMouël, J. L., and LePichon, X., 1971, Aeromagnetic survey of southwestern Europe: Earth Plan. Sci. Lett., v. 12, p. 287–299.

Maxwell, J. C., 1970, The Mediterranean, ophiolites, and continental drift, in Johnson, H., and Smith, B. L., eds., The megatectonics of continents and oceans: New Brunswick, N.J., Rutgers University Press, p. 167–193.

Nairn, A. E. M., and Westphal, M., 1971, A second virtual pole from Corsica, the Ota gabbrodiorite: Paleogeog., Paleoclim., Paleoecol., v. 3, p. 277–286.

Niggli, E., 1970, Alpine Metamorphose and alpine Gebirgsbildung: Fortschr. Miner., v. 47, p. 16–26.

Ogniben, L., 1970, Paleotectonic history of Sicily, in Alvarez, W., and Gohrbrandt, K. H. A., eds., Geology and history of Sicily: Tripoli, Petroleum Expl. Soc. Libya, p. 133–143.

Peters, Tj., 1969, Rocks of the alpine ophiolitic suite: Discussion on the paper "The origin of ultramafic and ultrabasic rocks" by P. J. Wyllie: Tectonophysics, v. 7, p. 507–509.

Reinhardt, B. M., 1969, On the genesis and emplacement of ophiolites in the Oman Mountains geosyncline: Schweiz. Mineral. Petrol. Mitt., v. 49, p. 1–30.

Sander, N. J., 1970, Structural evolution of the Mediterranean region during the Mesozoic era, in Alvarez, W., and Gohrbrandt, K. H. A., eds., Tripoli, Petroleum Expl. Soc. Libya, p. 43–132.

Smith, A. G., 1971, Alpine deformation and the oceanic areas of the Tethys, Mediterranean and Atlantic: Geol. Soc. Am. Bull., v. 82, p. 2039–2070.

Staub, R., 1924, Der Bau der Alpen: Beitr. Geol. Karte der Schweiz, N. F., v. 52, 272 p.

———, 1949, Betrachtungen über den Bau der Südalpen: Eclog. Geol. Helv., v. 42, p. 215–409.

Steinmann, G., 1905, Geologische Beobachtungen in den Alpen. II: Die Schardt'sche Überfaltungstheorie und die geologische Bedeutung der Tiefseeabsätze und der ophiolithischen Massengesteine: *Ber. Naturforsch. Ges. Freiberg, Breisgau*, v. 16, p. 44–65.

Stille, H., 1940, *Einführung in den Bau Nordamerikas*: Berlin, Borntraeger, 717 p.

Temple, P. G., 1968, Mechanics of large-scale gravity sliding in the Greek Peloponnesos: *Geol. Soc. Am. Bull.*, v. 79, p. 687–700.

Trümpy, R., 1965, Zur geosynklinalen Vorgeschichte der Schweizer Alpen: *Umschau*, v. 65, p. 573–577.

———, 1972, Notes accompanying lectures in Jerusalem (Israel) and Bethlehem (USA).

Van Bemmelen, R. W., 1933, Die Anwendung der Undationstheorie auf das Alpine System in Europa: *Proc. Kon. Ned. Akad. Wetensch. Amsterdam*, v. 36, p. 740–749.

———, 1960, New views on East-Alpine orogenesis: *Rep. 21st. Int. Geol. Congr. Copenhagen*, v. 18, p. 99–116.

Van der Voo, R., 1969, Paleomagnetic evidence for the rotation of the Iberian peninsula: *Tectonophysics*, v. 7, p. 5–56.

Zijderveld, J. D. A., Hazeu, G. J. A., Nardin, M., and Van der Voo, R., 1970, Shear in the Tethys and the paleomagnetism in the Southern Alps, including new results: *Tectonophysics*, v. 10, p. 639–661.

P. CHOUKROUNE

M. SEGURET

Laboratoire de Géologie Structurale
Université des Sciences et Techniques du Languedoc
Montpellier, France

Tectonics of the Pyrenees: Role of Compression and Gravity[1]

The structural geometry and tectonic evolution of the Pyrenees are becoming well understood (Mattauer and Henry, in press; Séguret, 1972; Choukroune, 1969, 1970). However, the problems of dynamics introduced by the study of this chain have not yet been attacked in the light of recent data. In this paper we intend to assess the mechanisms that caused the deformations produced during Pyrenean tectogenesis. We shall examine two types of models that are *a priori* possible. The first model calls basically upon a compression at the level of the crust and explains the tectogenesis by means of a relative approachment of the Iberian and European plates. In such a case the general resultant of the forces applied to the chain is close to the horizontal. The second model, by contrast, calls upon an essentially vertical force and explains the formation of the Pyrenean structures by gravitative phenomena.

To arrive at a choice between these two solutions, or at a compromise between them, we shall analyze (1) the geometry of the structures for which the models must account and (2) the manner of deformation and its evolution. The second consideration constitutes, in our opinion, an effective means of testing the hypotheses.

General Geometry of the Chain: Structural Zones

The Pyrenees are part of an almost linear east–west chain, extending for more than

---

[1] We thank R. Scholten for helpful remarks and for the translation of the text. The basic contribution by F. Arthaud, M. Mattauer, and Ph. Matte, who made available the main results of their own studies and provided critical advice in the writing of the present paper, is gratefully acknowledged.

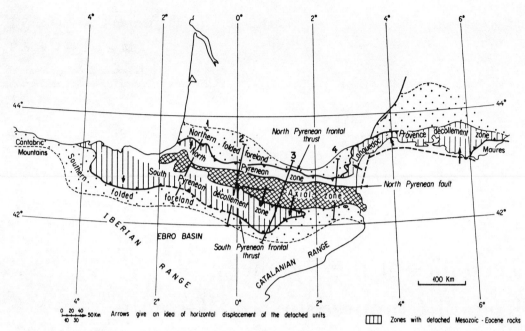

FIG. 1 Main structural subdivisions of the Pyrenean orogenic belt (slightly modified after Mattauer and Henry, in press).

1000 km from Asturia in the west to Provence in the east. Figures 1, 2, and 3 show its major structural subdivisions and the general style of the chain. An axial zone of Paleozoic rocks affected by the Hercynian orogeny is flanked by the North Pyrenean and South Pyrenean zones of deformed Mesozoic, represented largely by Cretaceous strata. The cross sections of Fig. 2 show the relative symmetry on both sides of the axial zone, with northward vergence on the north side and southward vergence on the south. Thus the Pyrenees may be viewed in gross aspect as a chain with double vergence.

The North Pyrenean Zone

This zone is essentially characterized by the presence of great vertical fractures and by folds that commonly have steep axial planes. The fractures, the most important of which is the North Pyrenean fault, correspond to ancient late Hercynian faults in the basement (De Sitter, 1953; Mattauer, 1968). During Mesozoic extension the most evident component of movement was vertical along these faults (whose throw is locally as great as 5000 m), but it has been recognized that important horizontal movements occurred during this period (LePichon and others, 1970).

Ultramafic rocks (lherzolites) and other undersaturated rocks (episyenites and picrites) were emplaced at least in part during this Cretaceous phase of extension. The age of emplacement of the lherzolites is post-Aptian and pre-Cenomanian in the western Pyrenees, as shown by veins penetrating no higher than Aptian strata and by lherzolite pebbles in the Cenomanian. In the eastern Pyrenees the age is known only to be pre-tectonic, hence pre-Eocene (Avé Lallemant, 1967). The episyenites and picrites are intrusive into the Upper Cretaceous.

With certain exceptions (in Languedoc, and at Orthez and Corbières) the Mesozoic rocks were not significantly displaced with respect to the Hercynian basement during the compressional phases. Preexisting faults were reactivated in complex fashion, either as reverse faults or as wrench faults. Locally,

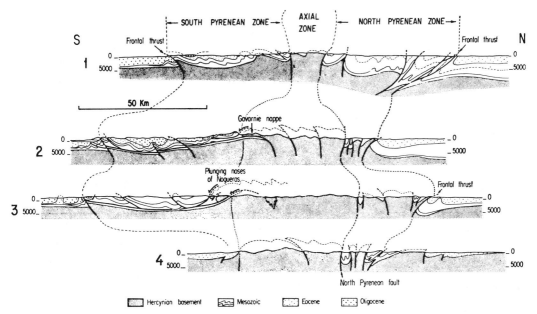

FIG. 2 Generalized cross sections of the Pyrenees. Section lines are shown in Fig. 1.

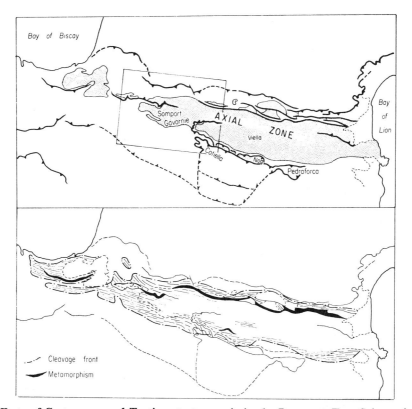

FIG. 3 Effects of Cretaceous and Tertiary tectogenesis in the Pyrenees. Top: Schematic structural map; rectangle indicates area of Fig. 4. Bottom: Map showing main strike directions of cleavage (dashed lines), cleavage front, and belts of metamorphism.

the structures are abruptly curved (Fig. 4). Deformation is very commonly inhomogeneous, with frequent variations in plunge of the extensional strain axis in the planes of cleavage.

A syntectonic, low-pressure, high-temperature metamorphism is developed along the North Pyrenean fault, doubtless in connection with the movement along this fundamental feature of the chain (Choukroune, 1970). In the eastern Pyrenees it has been demonstrated that this metamorphism took place at the end of the Cretaceous, along with the major tectogenesis in this region (Mattauer and Proust, 1965; Choukroune, 1970). In more westerly areas there is nothing to indicate that a tectonic phase occurred at this time, and the entire deformation is attributed to the Pyrenean phase of middle to late Eocene age.

The South Pyrenean Zone

In contrast to the North Pyrenean zone, this zone is characterized by a general *décollement* of the post-Triassic strata and by their feeble internal deformation. The main element is a structure composed of nappes and detached masses whose southward displacement attains 50 km in some places (Séguret, 1972).

Along the entire south flank the development and deformation of the nappes is related to the middle to late Eocene phase. No evidence for a phase at the end of the Cretaceous has been found. However, post-Oligocene deformation occurred in the central part of the South Pyrenean zone, and post-Miocene tectonics locally affected the western part, in the Basque country.

Selection of a Transversal Reference Section

In considering the degree to which gravitative phenomena have contributed to Pyrenean tectogenesis it is necessary to reason at the scale of the chain itself. In other words, the geometry of the structures and the state of deformation must be examined on the basis of a cross section which takes in the North and South Pyrenean zones as well as the axial zone. In addition, the following factors also determine the choice of our reference section:

1. The section must permit us to compare structures of the same age. Since the deformation in the southern zone was essentially late Eocene in age, the eastern part of the chain is eliminated from consideration, for in this part the North Pyrenean zone was deformed at the end of the Cretaceous.
2. The section must also permit us to evaluate the effect of the Pyrenean deformation on the axial zone of Hercynian rocks, and should therefore cross a maximum number of outcrops of the post-Hercynian cover in this zone.

Following these guidelines we have chosen the western end of the Paleozoic axial zone (Fig. 4). Here all of the South Pyrenean units are represented, it can be shown that the structures on the north and south slopes are of the same age, and numerous remains of the deformed post-Hercynian cover are present in the axial zone. Figure 4 shows the structural style along this section. We will discuss in sequence the South Pyrenean, North Pyrenean, and axial zones.

The South Pyrenean Zone

Ignoring the Oligo-Miocene molasse of essentially post-Pyrenean age, the south flank of the central and western Pyrenees may be divided on a structural basis into an autochthonous terrain on the one hand, and on the other hand a number of units that have been displaced southward over greater or lesser distances (Figs. 5 and 6) (Séguret, 1972).

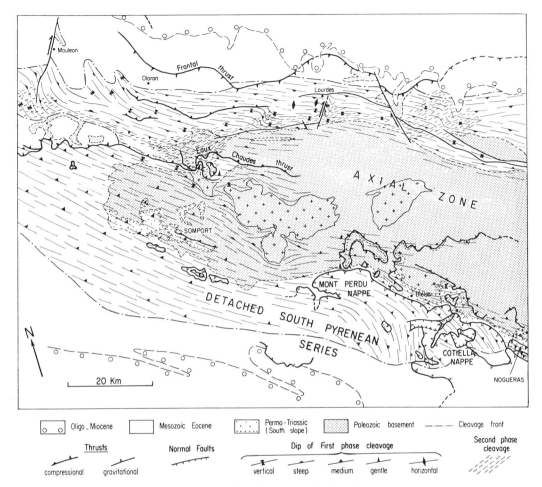

FIG. 4 Structural map of the study area (location shown in Fig. 3).

Autochthonous Terrain

The autochthonous terrain is represented in part by undeformed or little deformed sequences of strata in front of detached masses. In addition, it includes the southern portion of the axial zone.

Units with Southward Displacement

These units include both nappes with abnormal contacts and simple detached sequences without abnormal contacts. A given unit may generally be observed in both forms, with one form passing into the other both laterally and in a foreword sense. From bottom to top three main structural subdivisions are recognized: the plunging noses of Nogueras, the Gavarnie-Mont Perdu nappe and detached mass, and the Central South Pyrenean unit.

The Plunging Noses of Nogueras. The plunging noses or *têtes plongeantes* of Nogueras are synformal anticlines composed of a Hercynian core of Devonian to Carboniferous rocks and its Permo-Triassic cover. They rest on the Triassic cover of the axial zone. In the more westerly parts an axial plane cleavage invariably dips south. The autochthonous

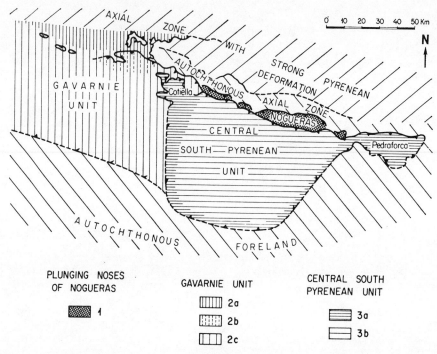

FIG. 5   Main structural units in the central part of the southern flank of the Pyrenees (from Séguret, 1972). (1) Plunging noses of Nogueras; (2) Gavarnie unit; (2a) Paleozoic thrust on Cretaceous; (2b) Cretaceous thrust on Cretaceous; (2c) detached series; (3) Central South Pyrenean unit; (3a) Cretaceous thrust on Eocene; (3b) detached series.

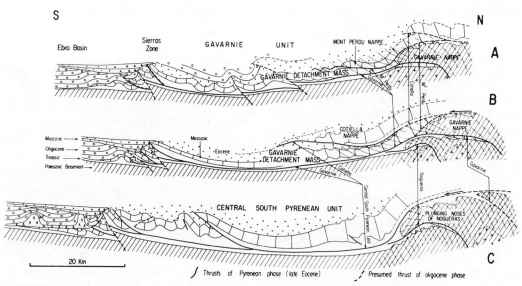

FIG. 6   Schematic sections showing the relations between the different units of the southern flank of the central Pyrenees (after Séguret, 1972).

146

substratum of the plunging noses is affected by folding and by a north-dipping fracture cleavage. Locally, the north-dipping cleavage in the autochthon passes gradually into the south-dipping cleavage of the allochthonous units, suggesting that the emplacement and refolding of the Nogueras units correspond to a single deformation of the Hercynian rocks (Séguret, 1972).

The Gavarnie–Mont Perdu Unit. This unit is composed of the Gavarnie nappe and the Gavarnie–Mont Perdu detachment mass. The nappe is composed of upper Paleozoic rocks, resting on the Cretaceous. In the Gavarnie area itself the minimum displacement of this nappe is 9 km, but it attains 15 km in the Bielsa area farther east. In front of the Paleozoic rocks of the nappe lies its former cover of Cretaceous to Eocene strata, detached at the level of the Triassic. It constitutes the Gavarnie–Mont Perdu detachment mass.

In the northern part of the detached mass are numerous slices, each of which constitutes a small, higher unit similar to the Mont Perdu mass (Fig. 6, section A). The basal contacts of these small units are often refolded on a scale of several meters to several hundreds of meters by folds of southward vergence, formed during cleavage development. Since these folds are contemporaneous with the overthrusting of the Paleozoic rocks, it is apparent that the small slices themselves were emplaced before the nappe, but we believe that this superposition of two apparently successive phases in the frontal zones is merely the result of a single evolution of deformation in the root zone (Séguret, 1972).

The Central South Pyrenean Unit. This unit comprises two zones of abnormal superposition: to the west the Cretaceous of the *Cotiella nappe* rests on the Gavarnie–Mont Perdu detachment mass, and to the east the Cretaceous of the *Pedraforca nappe* rests on autochthonous Eocene of the Sierra del Cadí. Between these two nappes the Mesozoic and Tertiary are detached at the level of the Triassic. The central South Pyrenean unit has been displaced the farthest: 50 km in the case of the Cotiella nappe (25 km for the underlying Gavarnie–Mont Perdu unit, plus 25 km for the Cotiella nappe across this unit), and 30 km in the case of the Pedraforca nappe.

In a general way it is seen that the intensity of deformation decreases as one goes upward through the various structural units. The autochthon is very commonly affected by folding with flow cleavage and thickening of the hinges. This changes into a fracture cleavage in the Nogueras masses and into a solution cleavage in the cover of the Gavarnie nappe. Cleavage is very rare in the central South Pyrenean unit, where mild folding is accomplished by flexuring and shear jointing.

The North Pyrenean Zone

In contrast to the south flank, it is not possible to identify separate structural units in the north flank of the chain. The western part of this zone is bounded to the north by the frontal North Pyrenean thrust and to the south by vertical faults or north-dipping thrusts, such as the Eaux Chaudes overthrust. The frontal thrust constitutes a significant structural boundary, along which the fold belt was displaced 2–5 km to the north over the mildly deformed foreland platform strata.

The Mesozoic rocks affected by the tectogenesis may be grossly divided into a Jurassic to Aptian sequence of essentially calcareous rocks and an Albian to uppermost Cretaceous sequence of shaly and sandy strata. The mappable structures and the accompanying microstructures are due to two superposed phases of folding (Choukroune, 1969), as discussed below.

Major Phase of Folding

At the scale of the western North Pyrenean zone, this phase of deformation is responsible for east–west folds, with axial plane cleavage

S_1 commonly dipping steeply to the north or south. The intensity of the deformation caused during this phase decreases from south to north. In the southern metamorphic part of the zone, where the limestones are deformed by flowage, the folding is synmetamorphic and isoclinal. Northward, there is a rapid gradation from flow cleavage to predominant fracture cleavage as flattening becomes less important; the folds are of the type produced by a combination of flattening and flexuring. Beyond the cleavage front the dominant mechanism of deformation is a combination of flexuring and shear-jointing (Arthaud and Mattauer, 1969a, 1969b).

However, a detailed study of the structures reveals certain local complications: (1) the strike of the cleavage changes abruptly in some places, as at Lourdes (Choukroune, 1969) and Mauléon (Richert, 1968); (2) the cleavage, although generally upright, may locally flatten and then steepen again, as at Lourdes (Choukroune, 1969); and (3) the plunge of fold axes measured in S_1 may vary, as at Oloron. These detailed complications show the inhomogeneous nature of the deformation. This inhomogeneity is due in part to normal variations in mechanical properties within the Jurassic–Cretaceous sequence. In part it is also due to the presence of precompressional structures, specifically the faults that were active during Cretaceous extension, which strike along two principal directions, one close to north–south, the other close to east–west. By juxtaposing rocks with different mechanical properties, and by constituting terrain discontinuities of their own, these faults contributed greatly to the heterogeneity of deformation in response to compression. For the most part, the south-striking faults became wrench faults; significant curvatures in their trend can often be observed at both ends, as at Mauléon (Richert, 1968) and Lourdes (Choukroune, 1969). The east–west faults developed into reverse faults with southward or northward directions of thrusting, and also into wrench faults.

Second Phase of Folding

The second phase is manifest particularly in the southern part of the western North Pyrenean zone, where it is represented by folds with axes striking between N50E and N90E and with northward vergence. They are accompanied by an S_2 fracture cleavage which generally has a low southward dip. This second cleavage likewise undergoes abrupt local changes in strike. It may appear in the form of two conjugate cleavage sets that together imply a subvertical direction of shortening. This second phase is essentially postmetamorphic.

Pyrenean Structures in the Hercynian Axial Zone

In the Hercynian rocks of the axial zone it is not possible to find direct evidence of Pyrenean deformation. To evaluate the effect of Pyrenean compression on the axial zone it is necessary to analyze, on the one hand, data from the post-Hercynian cover (such as the Permian at Somport and the Cretaceous of the border zone), and on the other hand the deformation of post-Hercynian veins found in the Paleozoic rocks.

Structures in the Cover

The structure of Eaux Chaudes (Figs. 4 and 9) is composed of a southward overthrust of recumbently folded Paleozoic and Cretaceous rocks on the autochthonous Upper Cretaceous, which rests discordantly on the Paleozoic (Ternet, 1965). The displacement is as much as 7 km. In the root zone of the thrust the axial planes of the folds and the cleavage strike east–west and are subvertical; away from the root the cleavage flattens and finally becomes horizontal.

At the western end of the axial zone the Cretaceous strata on top of this zone are characterized by the presence of cleavage folds and southward displacement on a minor thrust. In the same zone, near the Somport pass, the red pelites of the Permian show a strong slaty cleavage with low-grade chlorite metamorphism and large-scale overturned folds on a scale of the order of 1 km or more, with minor folds on a scale of meters.

Thus all post-Hercynian rocks in the southern, western, and northern border of the axial zone have been intensely deformed. The folds with amplitudes of hundreds to thousands of meters have east–west axial plane cleavage, which generally dips between 45 and 90° but may locally become horizontal.

The Hercynian rocks, on the other hand, are deformed by folds of approximately southward orientation whose pre-Permian age is shown by the fact that they are unconformably overlain by Permian or Cretaceous strata. These folds are cut by a steep, approximately east–west cleavage which passes upward into that of the Permian and Cretaceous rocks. This cleavage must therefore be considered Pyrenean in age, as is confirmed by the study of post-Hercynian veins.

Structures of Post-Hercynian Veins

Intrusive masses and numerous veins testify to acid Permian volcanism in the entire western part of the axial zone. Studies of the deformation of these veins by Mattauer (1964) and Mattauer and Matte (personal communication) show that (1) competent veins in brittle country rock are little deformed except for offsets along microfaults spaced on the order of 10 cm apart and (2) competent veins in a plastic country rock are cleaved and folded or, if the country rock is very plastic (as in the case of limestone deformed by flowage), they have suffered stretching and boudinage. These data demonstrate that the entire Hercynian terrain of the axial zone in the region studied has been deformed by internal yielding during Pyrenean tectogenesis. It is noteworthy that the cleavage generally dips steeply to the north.

The Roles of Compression and Gravity

Models

Several theoretical models may be proposed for the formation of the Pyrenean structures (Fig. 7). We shall discuss these in light of the nature of the deformation implied by each model. Thus for each of the models we shall indicate the attitude of the plane of flattening and the orientation of the principal stress directions, as well as the character of the deformation, whether continuous or discontinuous.

First Model: Compression. Here the reverse faults and overthrusts in the Paleozoic are supposedly the result of a fundamental compression of the entire crust, related to a collision between the Iberian and European plates. The resulting overall shortening must then be oriented close to the horizontal and at right angles to the chain. In the lower structural levels (Arthaud and Mattauer, 1969a, 1969b) this shortening will show up in the development of cleavage. If the direction of extension is vertical, this cleavage will of necessity be close to vertical and parallel to the chain. However, it is noteworthy that the reverse and thrust faults and the cleavage of the overthrust structures have a tendency to decrease in dip and "lie down" in the higher structural levels owing to the increased gravitational potential (Ramberg, 1967).

Second Model: Gravity Gliding at Shallow Levels. Essentially, this form of tectonics corresponds to simple translations along gliding surfaces. In this case, normal faults result from discontinuous stretching in the rear of transported units, which may proceed as far as actual tectonic denudation. In the front of these units there may be shortening. This model cannot account for the structures of the north slope.

Third Model: Gravity Gliding at Deep Level. Van Bemmelen (1955) has proposed that all of the Pyrenean structures together are basically the result of the rise of a deep,

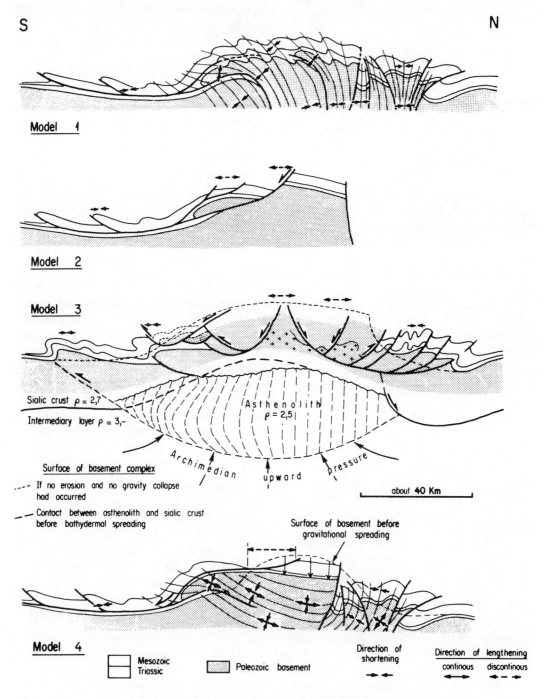

FIG. 7 Four geodynamic models of the formation of the Pyrenean chain (model 3 after Van Bemmelen, 1955; model 4 according to Price's hypothesis for the Canadian Rocky Mountains).

low-density astenolith blister. Gravitative compensation for this uplift is thought to occur in two ways: (1) by superficial gliding above the Triassic décollement horizon and (2) by deep gliding involving the entire sialic crust. This model implies a transversal shortening in the border zones of the chain (i.e., in the North and South Pyrenean zones) and, by contrast, an extension by normal faulting (discontinuous stretching) in the axial zone.

Fourth Model: Gravity Flow in Deep Levels. Price and Mountjoy (1970; Price, 1971, 1973) envisage an original model for the Canadian Rockies. We have tried to adapt and test this model in the Pyrenees. They call upon an initial rising of the central part of the chain, owing either to initial compression or to deep gravitative phenomena. The increase in energy potential caused by this uplift is thought to create a flow by gravity which affects the entire uplifted mass. This flow, which occurs in latest Precambrian and younger rocks, is at right angles to the axis of the chain and is thought to cause thrusting against the structural slope.

From the viewpoint of deformation, this model implies the development of a subhorizontal cleavage in the core of the chain with continuous extension at right angles to the axis of the chain. Thus, in Price's view, there would be no tectonic denudation because the stretching is here a continuous and not a discontinuous phenomenon. Nevertheless, even though such a stretching mechanism may be visualized in the deep zones, it does not apply to the more superficial levels, where the rocks behave in a brittle way. Thus a discontinuous stretching, leading to tectonic denudation, should take place in the upper structural levels (Arthaud and Mattauer, 1969a, 1969b).

Analysis of Deformation in the Study Area

Methods. In order to resolve the problem posed here, the essential element to be considered is the position in space of the cleavage plane. The different models presented carry in effect very definite implications with regard to the planes of flattening in the root areas of the chain. Furthermore, the orientations of the principal axes of internal strain and their relation to such geometric structural elements as the fold axes may also be of importance. At first approximation the cleavage planes may be considered as planes of flattening at right angles to the direction of maximum shortening and containing the axis of maximum extension and the intermediate strain axis.

The two axes in the cleavage plane may be given by the deformation of objects of known pretectonic shape and by syntectonic zones of crystallization (Fig. 8). Deformed objects include primarily (1) boudins of sedimentary strata and (2) stretched detrital pebbles. Syntectonic crystallization zones include (1) tension joints and (2) pressure shadows as revealed by mineral lineations in metamorphic rocks. Finally, the amount of shortening (represented by the strain ellipsoids in the sections of Fig. 9) may be estimated by taking into account the mechanisms of internal deformation (flowing, flattening, flexuring, or shear failure) (Arthaud and Mattauer, 1969a, 1969b).

Nature of Deformation along the Studied Section. The map of Fig. 4 and the composite cross sections of Fig. 9 show the general aspect of the cleavage along the section line. The axis of maximum extension is generally oriented along the dip of the cleavage. With reference to the fold geometry this is the *a* direction.

Table 1 summarizes the various modes of internal deformation in different stratigraphic and structural units as interpreted from the nature (or absence) of cleavage (flow, fracture, or solution cleavage) and from such features as tension gashes, striated joints, deformed minerals, disjointed or stretched pebbles or other entities, boudins, pressure shadows, stylolites, and dissolved objects. The same criteria can be employed to determine the strain axes. Figures 4 and 9 show that the

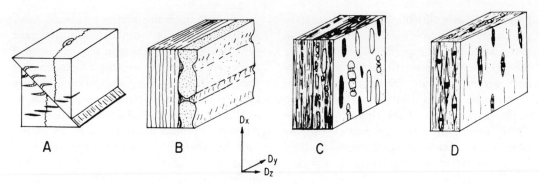

FIG. 8 Diagram showing methods for determination of principal strain axes. Dx = axis of maximum extension; Dy = intermediary strain axis; Dz = axis of maximum shortening. (A) By use of shear joints, tension fractures, and solution surfaces; (B) by use of boudinage in strata; (C) by use of deformation of pretectonic objects (pebbles); (D) by use of pressure shadows at the extremities of nondeformed, pretectonic objects.

planes of flattening are arranged in a fan around the vertical, which is compatible only with an overall horizontal shortening perpendicular to the chain.

However, the local presence of horizontal cleavage in overthrust structures needs to be explained. In the Gavarnie nappe it can be shown that this is not incompatible with a general subhorizontal shortening. The autochthon of the Gavarnie area is composed of Upper Cretaceous limestones that transgress directly onto Hercynian migmatites. Pyrenean deformation of these migmatites is effected by numerous reverse microfaults with south or northward dips (Debat, 1969), which indicate a horizontal north–south shortening and a vertical extension. By contrast, the Cretaceous limestones are deformed by laminar flow along subhorizontal planes. A well-marked a lineation in these planes shows that the extensional axis was oriented north–south and horizontally. In places where the shear surfaces in the migmatites cut the Cretaceous unconformity it is seen that the normally horizontal cleavage in the Cretaceous steepens progressively toward a subvertical position (Fig. 10). Hence in this area the great variability in the behavior

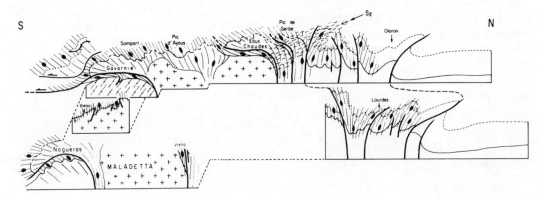

FIG. 9 Composite sections showing the aspect of planes of flattening in the central and western Pyrenees.

of the rocks under stress causes an extremely inhomogeneous strain, and the horizontal cleavage in the limestones is associated with horizontal shortening in the basement.

Specifically, the cleavage aspect in the limestones is related to two opposing couples: the southward displacement of the basement nappe gives rise to a low-angle cleavage, whereas the northward moving inbrications in the migmatites causes a steep cleavage. Since the two deformations are contemporaneous, the result is that the cleavage passes progressively from horizontal to vertical.

Discussion of the Models

Role of Compression. The study of the deformation in the Pyrenees shows that (1) the flattening planes are arranged in a fan around the vertical with maximum extensional strain in the dip direction, and (2) no

TABLE 1 Types of Internal Deformation in Different Stratigraphic and Structural Units

Structural Zones or Units	Lithologies and Ages	Mechanism of Deformation
North slope above cleavage front	Limestones (Jurassic and L. Cretaceous)	Shear-jointing
	Flysch (M. and U. Cretaceous)	Flexuring; shear-jointing
North slope—nonmetamorphic	Dolomites (Jurassic)	Shear-jointing
	Limestones (Jurassic and L. Cretaceous)	Flexuring; shear-jointing
	Flysch (M. and U. Cretaceous)	Flexuring; flattening
North slope—metamorphic	Dolomites (Jurassic)	Flexuring; shear-jointing
	Limestones (Jurassic and L. Cretaceous)	"Flowing"
	Flysch (M. and U. Cretaceous)	Flattening
Cretaceous of Eaux Chaudes	Limestones (U. Cretaceous)	Flattening
Cretaceous-terminus axial zone	Limestones } (U. Cretaceous)	Shear-jointing; flexuring
	Marls	Flattening
Permian of Somport (crest)	Pelites	Flattening
	Sandstones	
Gavarnie area (crest)		
Allochthonous	Flysch marls and sandstone (Eocene)	Flattening
	Marly limestones (Eocene)	Flexuring; flattening; shear-jointing
	Limestones and dolomites (Danian-Paleocene)	Flexuring; shear-jointing
	Sandy limestones (Campanian-Maestrichtian)	Flexuring; flattening; shear-jointing
	Dolomites (Coniacian-Santonian)	Shear-jointing
	Limestones (Cenomanian-Turonian)	"Flowing"
	Paleozoic (flysch, limestones)	?
Autochthonous	Limestones (Cretaceous)	Flattening
	Pelites and sandstones (Permian)	Flexuring; flattening
	Migmatites } Hercynian basement	Shear-jointing
	Granites	
Nogueras area (crest)		
Allochthonous	Sandstones and pelites (Permian) }	Flattening; flexuring; shear-jointing
	Pelites (Stephanian)	
	Paleozoic (flysch, limestones)	?
Autochthonous	Conglomerates, sandstones, pelites (Permian)	Flexuring; flattening
	Flysch (Visean)	Flattening
	Limestones (Devonian)	"Flowing"
South-Pyrenean unit (detached)	Limestones, marls, flysch (Mesozoic, Cenozoic)	Flexuring; shear-jointing

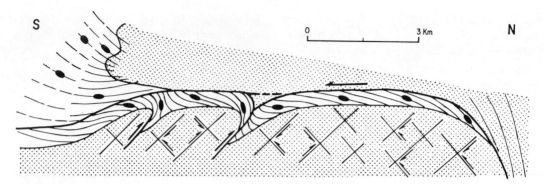

FIG. 10 Nature of deformation in a zone of horizontal cleavage (autochthon of the Gavarnie nappe).

tectonic denudation occurs at the western end of the axial zone.

Ignoring local shear couples related to overthrusts, the general shortening in the Pyrenees is close to the horizontal and perpendicular to the chain. When comparing the data of Fig. 9 with the theoretical models (Fig. 7) it clearly appears that model 1 is the only one consistent with the deformation in the Paleozoic basement and its attached cover. We see that in models 2, 3, and 4 some tectonic denudation of the axial zone ought to take place, and that models 3 and 4 do not account for the presence (model 3) or the orientation (model 4) of the flattening planes in the core of the axial zone.

Thus it must be recognized that horizontal compression is the fundamental motor of Pyrenean tectogenesis.

Role of Gravity. Superficial gravitational gliding (model 2) has played a far from negligible role in the detached units of the south slope. This is supported by the two following groups of arguments.

1. Extension features. Numerous indicators of extension are found in the northern part of the detached South Pyrenean units. Specifically, these include (1) normal faults that are contemporaneous with the emplacement of the units, as proved by the fact that several are truncated by the base of the Cotiella nappe, and (2) extension joints injected by Triassic material in the Pedraforca nappe, in the Gavarnie detachment mass north of the Cotiella nappe, in short, in the entire northern part of the central South Pyrenean unit, where their coalescence gives a "snakeskin" appearance to the rocks. Inasmuch as the total displacement of the northern part of the Central South Pyrenean unit reaches 50 km (as shown by the Cotiella nappe), it is possible, if not probable, that the emplacement of this unit was accompanied by a tectonic denudation of the central part of the axial zone.

2. Synsedimentary phenomena. The emplacement of the nappes and detached masses of the south slope took place during middle to late Eocene sedimentation (Séguret, 1972). The detached masses carried the base level of sedimentation along as they moved southward, and as a result their folding gave rise to numerous successively overlapping unconformities. By contrast the Pedraforca nappe slid into the Eocene basin of sedimentation and in the process olistolites broke away and moved forward separately, while an olistostrome was formed in front of the nappe. The detachment of the Mesozoic rocks of the south slope along the entire 20-km width of the Mesozoic basin is due to their gravitational sliding across the Triassic strata. Thanks to the syntectonic sedimentation it has been possible to demonstrate (Séguret, 1972) that this gliding occurred on a subhorizontal detachment surface, and that the

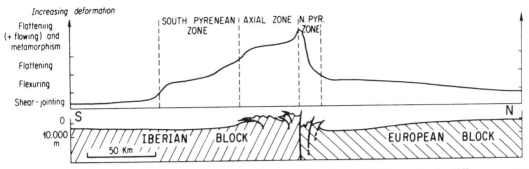

FIG. 11 Schematic diagram showing the predominant mechanisms of deformation in different parts of the Mesozoic cover.

average velocity of displacement of the units was of the order of $\frac{1}{2}$ cm/yr.

Conclusions

The analysis of the state of deformation in the rocks affected by Pyrenean tectogenesis shows that this deformation represents an overall horizontal north–south shortening at right angles to the chain, accompanied by maximum extension in a vertical direction. This could only have been caused by major crustal compression related to the collision between the Iberian and European plates. The total degree of shortening in different cross sections does not vary much (Mattauer, 1969, in press). It may be estimated at approximately 50 km, which gives a general idea of the amount of shift between the two approaching plates. The fact that extension is on the average vertical, rather than horizontal and parallel to the chain, is compatible with the linear aspect of this mountain belt.

This linearity is the result of the geodynamic evolution of the Pyrenean domain, which is directly related to the presence of a major ancient wrench fault (De Sitter, 1953; Mattauer, 1968). In effect, the faults of the North Pyrenean zone limit the Iberian and European plates, and it is in the vicinity of this discontinuity that deformation is most intense (Fig. 11). From this point of view the narrow zone at the edge of the two blocks constitutes the essential structural element of the Pyrenean chain.

In summary, the geologic history of the Pyrenees has been controlled by relative movements of the Iberian and European plates on either side of the faults of the North Pyrenean zone. These movements were composed of two simple types: (1) perpendicular to the edges of the plates (compression or extension), and (2) parallel to them (wrench faulting). During the Mesozoic the most evident phenomenon was extension, although LePichon and others (1970, 1971) believe that considerable left-lateral slip accompanied this extension. During the tectogenic phase at the end of the Cretaceous (confined to the eastern North Pyrenean zone) there occurred a relative approachment of the blocks (compression) and a strike-slip movement along the fault zone (Choukroune, 1970). During the middle to late Eocene tectogenic phase the process was essentially one of compression.

A result of this fundamental compression was the gliding of the nappes and detached masses of the south slope.

References

Arthaud, F., and Mattauer, M., 1969a, Présentation d'un nouveau mode de description tectonique: la notion de sous faciès tectonique: *Comptes Rendus Acad. Sci. Paris*, v. 268, p. 1019–1022.

———, 1969b, Niveau structural, faciès tectonique,

profil tectonique: *Comptes Rendus Acad. Sci. Paris,* v. 268, p. 1161–1164.

Avé Lallemant, H. G., 1967, Structural and petrofabric analysis of an alpine type peridotite: The lherzolite of the French Pyrenees: *Leidse Geol. Med.,* v. 42, p. 1–58.

Choukroune, P., 1969, Sur la présence, le style et l'âge des tectoniques superposées dans le Crétacé nord-pyrénéen de la région de Lourdes (Hautes Pyrénées): *Bull. Bur. Rech. Géol. Min.,* 2nd ser., sec. 1, no. 2, p. 11–20.

——, 1970, Contribution à l'étude structurale de la zone métamorphique nord-pyrénéenne: tectonique et métamorphisme des formations secondaires de la forêt de Boucheville (Pyrénées Orientales): *Bull. Bur. Rech. Géol. Min.,* 2nd ser., sec. 1, no. 4, p. 49–63.

Debat, P., 1969, Tectonique tertiaire dans les formations métamorphiques des vallées de Gavarnie et d'Heas (Hautes Pyrénées): *Comptes Rendus Som. Soc. Géol. France,* v. 2, p. 31–32.

De Sitter, L. U., 1953, La faille nord-pyrénéenne dans l'Ariège et la haute Garonne: *Leidse Geol. Med.,* v. 18, p. 287–290.

LePichon, X., Bonnin, J., and Sibuet, J. C., 1970, La faille nord-pyrénéenne: faille transformante liée à l'ouverture du Golfe de Gascogne: *Comptes Rendus Acad. Sci. Paris,* v. 271, p. 1941–1944.

LePichon, X., Bonnin, J., Francheteau, J., and Sibuet, J. C., 1971, Une hypothèse d'évolution tectonique du Golfe de Gascogne, *in Histoire structurale du Golfe de Gascogne,* Pub. Inst. Français Pétrole, Colloques et Séminaires no. 22, v. 2, p. 1–44.

Mattauer, M., 1964, Sur les schistosités d'âge tertiaire dans la zone axiale des Pyrénées: *Comptes Rendus Acad. Sci. Paris,* v. 259, p. 2891–2894.

——, 1968, Les traits structuraux essentiels de la chaîne pyrénéenne: *Rev. Géogr. Phys. Géol. Dyn.,* v. 10, p. 3–12.

——, 1969, Sur la rotation de l'Espagne: *Earth Planet. Sci. Lett.* v. 7, p. 87–88.

Mattauer, M., and Proust, F., 1965, Sur la présence et la nature de deux importantes phases tectoniques dans les terrains secondaires des Pyrénées: *Comptes Rendus Som. Soc. Géol. France,* no. 5, p. 132–133.

Mattauer, M., and Henry, J. (in press), The Pyrenees, *in Data for orogenic studies*: Geol. Soc. London, Spec. Publ.

Price, R. A., 1971, Gravitational sliding and the foreland thrust and fold belt of the North American Cordillera: Discussion: *Geol. Soc. Am. Bull.,* v. 82, p. 1133–1138.

——, 1973, Large-scale gravitational flow of supracrustal rocks, southern Canadian Rockies (this volume).

Price, R. A., and Mountjoy, E. W., 1970, Geologic structure of the Canadian Rocky Mountains between Bow and Athabasca rivers. A Progress report: *Geol. Assoc. Can., Spec. Paper no. 6,* p. 7–24.

Ramberg, H., 1967, The Scandinavian Caledonides as studied by centrifugated dynamic models: *Bull. Geol. Inst. Univ. Uppsala,* v. 43, p. 1–72.

Richert, J. P., 1968, Analyse structurale de la zone charnière entre le Béarn et le Pays Basque: *Bull. Serv. Carte Géol. Als. Lorr.,* v. 21, p. 137–164.

Séguret, M., 1972, Etude tectonique des nappes et séries décollées de la partie centrale du versant sud des Pyrénées; caractère synsédimentaire, role de la compression et de la gravité: Thèse Fac. Sci., Montpellier, Publ. U.S.T.E.L.A., Série Géol. Struct., v. 2, 155 p.

Ternet, Y. 1965, Etude du synclinal complexe des Eaux Chaudes: Thèse 3ème cycle, Fac. Sci., Toulouse.

Van Bemmelen, R. W., 1965, Tectogénèse par gravité: *Ann. Soc. Géol. Belg.,* Bull. 64, p. 95–123.

A. CAIRE
Laboratoire de Geologie Structurale
Faculté des Sciences de Paris
Quai St. Bernard, Paris, France

The Calabro-Sicilian Arc

Strictly defined, the Calabro-Sicilian Arc consists of the mountain belt around the Tyrrhenian Sea in Calabria and Sicily, but the Eastern Atlas and the Southern Apennines may also be considered part of it (Fig. 1). This paper outlines the stratigraphic and structural evolution of this large arc in the broader sense of the term. Details may be found in Caire (1957–1970).

The Calabro-Sicilian arc is a relatively recent feature; an arc shape was imprinted on tectonic units with an earlier evolution that did not point to a later arc. This can also be seen from the obliquity of the Tyrrhenian coastline cutting through numerous Alpine structures. Consequently, the arc might be considered a post-Alpine phenomenon, since its most recent and characteristic property, its shape, is post-Alpine.

The Calabro-Sicilian arc is of the same size as the Alpine arcs of Gibraltar, the Western Alps, and the Carpathians. All of these areas are similar in that their central parts show strong tectonic divergence and have recently subsided.

Figure 2 shows the constancy of the main structural theme of the arc. The major tectonic units overlie each other consistently in the same succession. The highest and presumably most internal tectonic unit consists of flysch-type rocks in North Africa, Sicily, and Calabria, and ranges in age from Tithonian (Late Jurassic) to earliest Miocene. Associated with these rocks are the ophiolites in northern Calabria and the northern Apennines. The Sahara platform (i.e., the foreland of the Atlas) and its equivalents, the Ragusa platform in Sicily and the Apulia–Monte Gargano platform in the Apennines, are the most external, autochthonous units of the complex. Between the internal flysch and the foreland are several intermediate thrust units, derived from juxtaposed facies belts along the trend of the Calabro-Sicilian arc.

The Atlas

Main Structural Pattern

The Atlas in Algeria and Tunisia is separated from the Sahara platform by the South Atlas

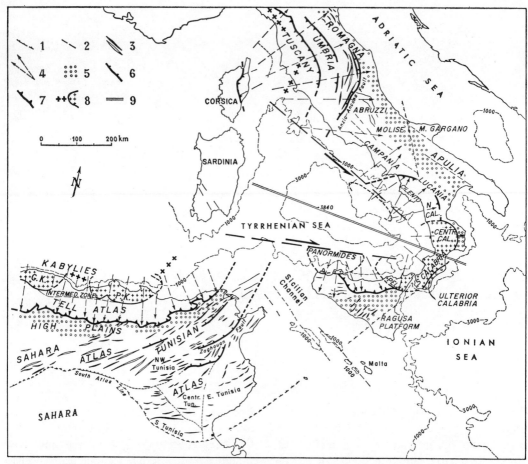

FIG. 1 Schematic structural map of the Calabro-Sicilian arc, the Southern Apennines, and the Eastern Atlas: (1) Structural boundary between the arc and its foreland (Sahara, Ragusa platform, Monte Gargano-Apulia); (2) structural directions; (3) major anticlines and synclines; (4) displacement directions of the nappes and their resedimented parts; (5) Neogene foredeeps; (6) Zaghouan and Anzio-Ancona lines; (7) erosional fronts of nappes; (8) core of the Kabylies and the Tuscan ridge; (9) axis of the Tyrrhenian Sea. Note the general structural symmetry with respect to the axis of the Tyrrhenian Sea.

FIG. 2 Schematic sections of the Calabro-Sicilian arc, the Apennines, and the Eastern Atlas, showing the generalized relations between the various tectonic units rather than their actual configurations along any specific cross section. The sediments, faults, and other deformations which occurred after the main tectonic phase have been omitted, except in the external zones. Cross section f is a synthetic section in which the characteristics of sections a–e have been combined: (1) Ligurian nappe, ophiolites, argille scagliose; (2) neo-autochthonous Miocene, transported to the external Apennines along with fragments of the Ligurian and Tuscan nappes; (3) Macigno (Oligocene flysch) with sedimentary klippen of Alberese flysch and olistostromes of Ligurian material; (4) detached parts of the Tuscan nappe; (5) and (6) Tuscan nappe and underlying Triassic rocks; (7) autochthonous Tuscan rocks (shown without the later uplift which produced the Apuan Mountains); (8) foredeep with resedimented parts of the Tuscan and Ligurian nappes; (9) Lucania nappe (internal flysch, Cilento flysch, and argille scagliose); (10) lower Miocene; (11) Lagonegro-Stigliano zone (= Sclafani zone), lengthwise relay of the external flysch 23 with external Numidian; (12) resedimentation basis with fragments of the Lucania nappe; (13) allochthonous and autochthonous of the Southern Limestone Apennines and Monte Gargano-Apulia foreland; (14) internal flysch (multicolored argille scagliose) with internal Numidian; (15) transgressive Oligo-Miocene; (16) limestone cover of 17; (17) metamorphic basement of the Calabria-Peloritani Mountains; (18) Mio-Pliocene resedimentation basin of Ulterior Calabria; (19) internal

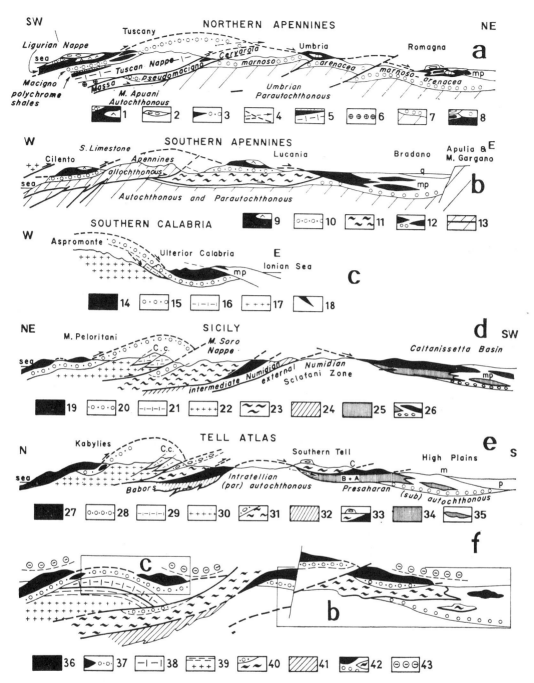

flysch, with internal Numidian, and Reitano zone; (20) Oligo-Miocene molasse; (21) limestone cover of 22; (22) metamorphic basement; (23) Monte Soro nappe (external flysch); (24) intermediate Numidian and Panormides; (25) external Numidian of the Sclafani zone; (26) resedimentation basin of Caltanissetta; (27) internal (Ultrakabylan) flysch; (28) transgressive Oligo-Miocene with sedimentary klippen; (29) limestone cover of the basement; (30) Kabylan basement; (31) external (Subkabylan) flysch and Neogene detrital formations; (32) intermediate zone (Babors); (33) dislocated flysch covered by lower Miocene; (34) A and B nappes; (35) South Tellian foredeep with sedimentary klippen of the A and B nappes; (36) internal rock sequences; (37) transgressive Oligo-Miocene with detritus of internal tectonic units; (38) limestone cover which slid from its base before the arrival of unit 36; (39) pre-Alpine basement; (40) external flysch units; (41) external Numidian and intermediate zone in North Africa; (42) resedimentation basins (Neogene foredeeps); (43) sediments deposited after nappe development (these sediments have been omitted from the internal parts of sections a–e).

line, a fault of fundamental importance along which both thrust and strike-slip movements have taken place (Fig. 1). The structures to the north of this line have an en echelen configuration with an average NE–SW direction in the Southern Atlas. In the Northern or Tell Atlas, east–west directions dominate, but they interfere locally with transverse deformations whose trends vary between N–S and NE–SW. The latter manifested themselves during the Mesozoic, before the Alpine orogeny made its definitive imprint on the structural grain. They reasserted themselves in Mio-Pliocene times as fractures, torsions, or axes of subsidence or uplift after the major Miocene phase of Alpine tectonism.

Figure 3 shows a particularly representative cross section across the Tell Atlas. The three nappes, A, B, and C, moved southward toward the Miocene foredeep and arrived in their final positions during the Miocene. Their frontal parts were "resedimented" in the foredeep as fragments ranging from a cubic millimeter to "sedimentary klippen" more than a cubic kilometer in size (Fig. 4), while at the same time clay and sand were fed into the foredeep from the south. Resedimentation was not in the form of chaotic sediment flows or submarine landslides (olistostromes), but as loose blocks, little deformed, and with the original stratification preserved. The lowest sedimentary klippen came from areas that were paleogeographically nearby, whereas the higher ones were derived from zones more to the north. This is similar to the emplacement of the allochthonous tectonic units, which did not arrive as a complex pile but successively, beginning with those from an area near the foredeep (see Caire, 1970b, Fig. 8). From the Miocene through the Pliocene the foredeep migrated to the south, and subsequently disintegrated into smaller basins. On the sills separating these basins erosion locally removed all sediments down to the basement. Elsewhere, the foredeep sediments are hidden by recent deposits, and for these reasons it is not always easy to recognize the foredeep as such.

Comparison between Atlas and Apennines

Structural affinity between the Atlas and the Apennines is evident from Figs. 2 and 5. Below the internal nappes (unit C in northern Tunisia and the Ligurian nappe of the Northern Apennines) the rocks are detached from their basement and faulted and displaced toward the exterior part of the mountain belt (Fig. 5). The two belts are both bordered by faults or flexures that probably belong to the same structural lineament. The lineament includes the Zaghouan fault in Tunisia and the Anzio-Ancona fault in Italy (Fig. 1). Strike-slip movements have been important along these faults, but the major movement, especially along the Zaghouan fault, consisted of overthrusting. In Tunisia the rocks east of the fault are folded and cut by oblique and strike-slip faults; in the Southern Apennines strike-slip faults are superposed on the detachment cover.

Mechanism of Overthrusting in the Atlas

Deep-reaching fractures developed in the Prealpine basement during the Miocene paroxysmal phase of orogeny, leading to uplifting of the Kabylie massif and its southward overthrusting across its sedimentary margin (Fig. 3). In a larger perspective this may be seen in terms of a deep-seated northward push and underthrusting of the African continent.

Detachment and gravitational sliding of the sedimentary cover characterize the supratectonics. Coherent sliding is dominant in the limestone and sandstone units, whereas in shales, marls, and pelitic flysch deposits turbulent sliding is more typical. The thickness of the layer of Triassic evaporites and shales at the base of the allochthonous units is uniform in some places but elsewhere is irregular and locally even zero, according to the character and position of the thrust sheets. Over this lubricant the movement of

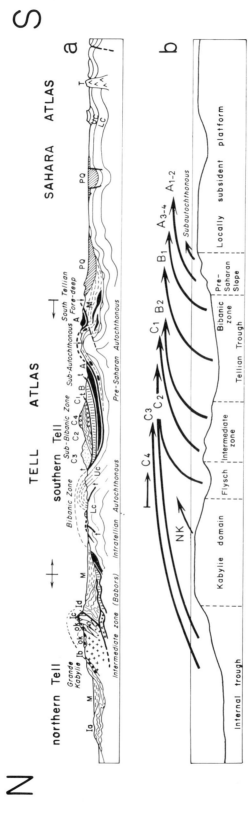

FIG. 3 (a) Schematic composite section across the Tell Atlas and the Sahara Atlas, along the meridian of the Grande Kabylie (GK in Fig. 1). Legend form south to north: (T) Triassic salt domes; (UC) Upper Cretaceous; (LC) Lower Cretaceous; (PQ) Pliocene to Quarternary fault trough (Tunisian Atlas) and furrow (northern edge of the Sahara Atlas); (M) Miocene; (A, B, C) allochthonous units; (t) Triassic below units B and C; (C_1) Senonian conglomerate; (C_2) Cretaceous flysch; (C_3) Numidian Oligocene flysch and Medjana molasse; (C_4) lower Miocene; (M) Miocene of the Soummam and Tizi-Ouzou basin, deposited after development of the nappes; (Id, Ic, Ib, Ia) South Kabylan flysch, limestone cover, Kabylan basement, North Kabylan flysch; ok: Kabylan Oligo-Miocene. (b) Senonian paleogeography and Miocene overthrusting.

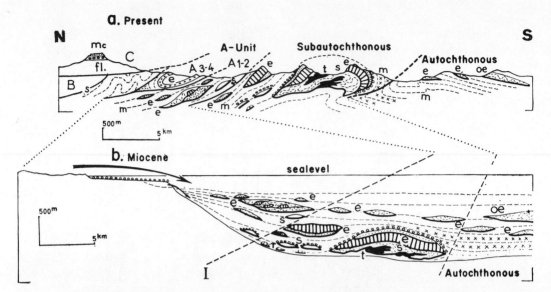

FIG. 4 The South Tellian foredeep. (a) Composite section of the Southern Tell and the Presaharan Autochthonous in the Hodna area: (t) Tell-Triassic; (s) Senonian; (e) Eocene; (o) Oligocene; (fl) flysch; (m) lower Miocene; (mc) lower Miocene conglomerates transgressive on nappe unit C. (b) Reconstructed section for the early Miocene, after the emplacement of sedimentary klippen but before thrusting: (t) Triassic; (T) Turonian; (s) Senonian; (e) Eocene; (o) Oligocene. The Eocene masses embedded in the Miocene are found in both normal and overturned positions. After the early Miocene, part of the foredeep was only weakly deformed (the Autochthonous), another part was uplifted and thrust towards the south (the Subautochthonous), and the northern part of the trough was thrust as unit A1-2 over the Subautochthonous. Thrust units A3-4, B, and C do not appear on section b, since they had not advanced far enough southward at this stage, but the position of A3-4 cannot have been far away, because sedimentary klippen of Eocene age and boulders above the Senonian were derived from these units.

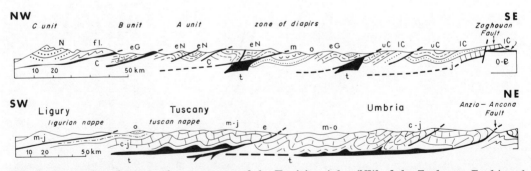

FIG. 5 Comparison between the structures of the Tunisian Atlas (NW of the Zaghouan Fault) and the Northern Apennines (W of the Anzio-Ancona fault). Letters t, j, c (or C),-e, o, and m indicate Triassic, Jurassic, Cretaceous, Eocene, Oligocene, and Miocene, respectively; fl. denotes flysch, and single letter N denotes Neogene.

the Tell allochthonous units was relatively easy and reached dozens of kilometers.

The large-scale gravity sliding of the sedimentary cover took place within an exceptionally brief period (early Miocene in the Tell of Algeria and Tunisia). French structural geologists (e.g., P. Termier and M. Lugeon) since the beginning of this century have compared this very special process with a *vent orogénique, tempête,* or *bourrasque* (orogenic storm, hurricane, or squall). It may also be compared with *balayage,* a sweeping or dragging of all superficial masses in various directions, whereby their frontal parts are bent, and torn-out debris is resedimented in contemporaneous troughs. Such mass movements stand in contrast to the pulsating vertical uplifts that preceded them, and they may have been caused by brief earthquakes that repeatedly affected the base of the sedimentary cover, as if this cover were lying on a tilted vibrating table. Local detachments and overthrusting of the sedimentary cover and local resedimentation phenomena may simply result from regional uplift alone. The small tangential displacements that did not occur during the major tectonic phases, in particular the repeated northward sliding toward the Mediterranean coast, may be explained in this fashion. The emplacement of sedimentary klippen in the Cretaceous of the Babors domain, in the upper Eocene of the Subkabylan domain, in the Oligocene of the Kabylies, and in the foredeep of the Southern Tell may be similar phenomena.

Several possible causes may be envisioned of the disequilibrium that led to the overthrusting of the large allochthonous units in the Atlas (see Caire, 1957, 1970b):

1. The entire Tell region could have been tilted southward, resulting in an important height difference between the internal domains and the foredeep. The first rocks that slid into the foredeep came chiefly from the two unstable Mesozoic zones flanking the Tellian trough on the south and north: the pre-Saharan slope and the Intermediate zone (Fig. 3). The trough itself subsided continuously from Triassic to Oligocene time and became only afterward involved in the overthrusting.
2. According to another hypothesis tilting was only a local and relatively brief phenomenon, propagating itself in space and time. It may have been localized at the flank of a southward-migrating ridge.
3. A third hypothesis ascribes the movements of the sedimentary cover to the same process that caused the thrust faults in the crystalline basement of the Kabylies. According to this hypothesis the basement of the Tell Mountains was progressively foreshortened, causing first the uplift of the unstable epicontinental slope, intermediate zone and Kabylie domain, and finally the compression of the contents of the basins and the spilling of the sedimentary fill across the neighboring areas.

The last two hypotheses may well be combined. In such a view the deep-seated compressive force would lead to thrusting in the Kabylies and to the progressive uplift first of the unstable zones and next of the basins. The migration of the ridge formed in this fashion and the associated seismic shocks would account for the horizontal sweeping movements during the paroxysmal tectonic phase.

Sicily

During Alpine time Sicily was connected with Africa and the Italian mainland and, together with Calabria, formed the transition between the Atlas and the Apennines. Certain North African facies and structural characteristics are in places almost identical to those found in northern Sicily and Calabria. This is especially true for the most northerly domains. Thus there is a close resemblance between the Kabylie domain and the Peloritani Mountains (Fig. 1), specifically with regard to their limestones, their Oligocene–Miocene cover, and the bordering flysch rocks. It is thus possible to consider Sicily and its geological extension, southern Calabria, as the eastern end of the Berber orogenic belt. At the same time, Sicily possesses some characteristic features of its own, which distinguish it from the Atlas. For example, it has no "Tellian" Triassic of germanic facies, the miogeosynclinal Tell basin does not continue into Sicily, and facies zones unknown in Africa make their appearance. Finally, Sicily and Calabria have some properties in common with the Apennines. This is especially true for the *argille scagliose* (scaly shales), the flysch of internal origin that covers large parts of Sicily. The Sicilian argille scagliose constitute a facies transition between the internal flysch of North Africa and the argille scagliose of the Apennines. These three formations all come from an internal domain common to the Berber Mountains, Sicily, and the Apennines, but whose various segments were diversified at different times during the Cretaceous, Eocene, and Oligo-Miocene.

The Ragusa Platform

The foreland of the Sicilian mountain chain is the Ragusa platform (Fig. 6), an external unit which, like Apulia-Monte Gargano on the Italian mainland (Fig. 1), remained sheltered from the tangential Alpine movements. The platform, characterized by a Mesozoic to Neogene epicontinental cover, may be considered as an extension or equivalent of the Sahara platform or of a stable unit similar to the Moroccan or Oran Meseta. A system of NNE–SSW normal faults separating the platform from the Caltanissetta basin provided access channels for the Monte Lauro (Iblei Mountains) basalts in the early Miocene and for the Etna volcanics in the Quaternary. In the extreme southeastern part of the Ragusa platform basaltic flows occur in the Upper Cretaceous, a counterpart of those in the Sfax area of Tunisia.

Central Sicily

Central Sicily comprises the western part of the island and its eastern part between the Ragusa platform and the Peloritani Mountains (Fig. 6). The upper structural level is represented by Mio-Pliocene basins that went through a tectonic-sedimentary evolution much comparable to that of the Southern Tell foredeep. Exposures of their substratum reveal the existence of three facies zones (labeled 6*a*, 6*b*, and 7 in Fig. 6). From late Miocene to early Pliocene times these zones were affected by an "orogenic wave" advancing from north to south. The rocks were detached from their Triassic base, moved to the south as small overthrust sheets, and dispersed as sedimentary klippen and debris in the marly sediments of the foredeep. The clearest example of this resedimentation phenomenon is displayed in the Caltanissetta basin (Beneo, 1956b; Rigo de Righi, 1956; Ogniben, 1953, 1954; Broquet, 1970; Broquet and others, 1966). It contains 3000 m of Miocene sediments, including olistostromes and olistolites (Flores, 1959; Jacobacci, 1965) (Fig. 2*d*).

Farther north the Sclafani zone (labeled 8a in Fig. 6) was covered by the Oligo-Miocene of the external Numidian zone (8b). Intercalated in the latter are sedimentary klippen of the Panormide zone (9), which range in size from boulders to mountains (see cross section, Fig. 6) and consist of Mesozoic to Eocene rocks (Caire, 1970b, Fig. 14).

The Flysch Nappes

In the northeastern part of Sicily the paleogeographic zones show much affinity with their corresponding zones (or extensions) in North Africa, and they also occur in a tectonic framework with the same architecture as in Algeria.

The flysch deposits of the Monte Soro nappe (Figs. 2*d* and 6) are of Late Jurassic (Tithonian) to Cretaceous age and its southern part is overlain by conglomeratic

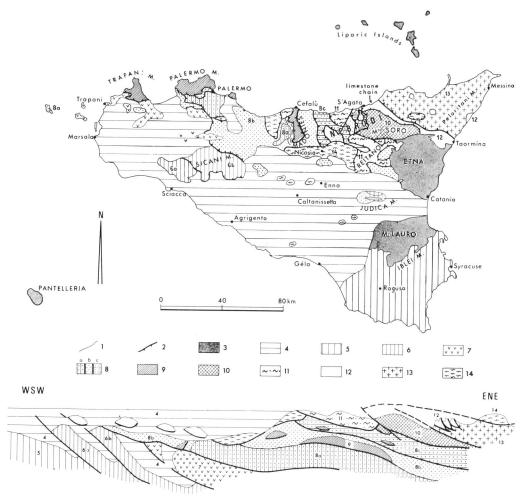

FIG. 6 Schematic map and composite section of Sicily. The postorogenic sediments in the northeastern part of the island are not represented in the section: (1) stratigraphic contacts and minor faults; (2) major faults; (3) recent volcanism; (4) Mio-Pliocene and Quaternary basins; (5) Ragusa-Monte Iblei platform; (6) Sciacca and Campofiorito-Cammarata zones; (7) Vicari zone; (8a) Mesozoic and Eocene of the Sclafani zone (Mascle, 1970); (8b) external Numidian; (8c) intermediate Numidian; (9) Panormides; (10) Monte Soro nappe (external flysch and discordant Oligo-Miocene molasse); (11) Reitano nappe (intermediate between the Peloritan domain and the internal flysch); (12) Limestone Chain (former cover of the Peloritan massif); (13) Peloritan massif; (14) argille scagliose nappe with internal Numidian (internal flysch) (Duée, 1970).

Senonian identical to the Senonian of the C_1 unit in Algeria (Fig. 3) and by detrital Eocene. The internal units of the flysch are unconformably covered by Oligo-Miocene of molasse facies, which is identical to the Oligo-Miocene of the Peloritani Mountains to the north.

In the argille scagliose nappe (labeled 14 in Fig. 6) the Numidian shales form the matrix, in which are dispersed various-sized fragments of other formations, including Lower Cretaceous flysch, Cenomanian radiolarites and "Fish-shales," and Eocene limestones. Rocks of the same facies are also found in the internal zones of the Atlas. The Reitano nappe (11) overlies the Numidian units and the argille scagliose nappe. Its rocks are equivalent to those of the Medjana series in Algeria.

The Peloritan Massif

The Peloritan massif (labeled 13 in Fig. 6) is the Sicilian Kabylie. The massif consists of a metamorphic basement intruded by granites, partially overturned and thrust toward the south. Its Mesozoic to Eocene cover forms the Limestone Chain (12), similar to the cover of the Kabylies in Algeria. The three external zones of the Peloritan cover contain Permo-Triassic rocks of Verrucano type (red sandstones and conglomerates) and a thin sequence of dolomites and limestones of Jurassic to Eocene age, replete with diastems and Fe-Mn precipitation zones and poor in detritus. Evidently, this sequence represents a platform in the geosyncline similar to the Briançonnais platform in the Alps (Debelmas and Kerckhove, 1973). The external Longi-Gallodoro zone and its phyllitic basement are traversed by hundreds of sedimentary dikes and sills (Truillet, 1970). Dating of their filling shows that these were formed from Liassic to early Eocene time. They bear witness to an almost continuous instability of the sea bottom, which culminated in the resedimentation of limestone slabs several hundred meters long during the early Eocene.

The Lutetian tectonic phase (Eocene) manifested itself by the thrusting and sliding of the Limestone Chain over its phyllitic basement. Some slabs were transported like ice floes and piled on top of each other; others became embedded in the strongly deformed phyllites. The next events were the deposition of conglomerates during the middle Eocene, the thrusting of the highest tectonic unit of the Limestone Chain (the Novara zone) over these conglomerates, and the discordant deposition of Oligo-Miocene molasse on all other formations. Finally, the Peloritan domain was thrust at least 15 km to the south, its frontal slices overriding the Monte Soro nappe.

The Tectonic Evolution of Sicily Compared to that of Algeria

The shearing in the basement and the detachment and sliding of the sedimentary cover in Sicily are phenomena entirely comparable to those in North Africa. In Sicily, however, the "orogenic wave" was much more prolonged. Starting from the Peloritani Mountains in the Eocene, the Panormides were affected in the Oligocene, and the most external zones in the Miocene. In some southern areas, the Sicani Mountains (Fig. 6), for example, the "wave" was still active during the Pliocene.

The resedimentation phenomena are also much more developed and spectacular in Sicily (see Broquet, 1970). This is caused in part by the fact that some stratigraphic sequences, such as the argille scagliose, were very thick and shale-rich and thus gave rise easily to olistostromes. More rigid formations produced sedimentary klippen or olistolites. We may conclude that the "orogenic storm" in Sicily was much stronger and more constant than in Atlas. The "storm" struck the crests of the nappes, removing swarms of debris from the zone that was raised by the migrating wave and carrying it into the adjacent foredeep. The foredeep likewise lasted much longer (from the Oligocene through the early Pliocene) than the foredeep in North Africa. As in Algeria, it moved slowly toward the south and finally fragmented and disorganized.

Calabria

The backbone of southern Calabria consists of schists and crystalline rocks, with a strip of sedimentary formations of "Alpine" age to the southeast (Fig. 7). A cross section through this strip of "Ulterior Calabria" shows that its architecture is very similar to that of the external part of the Peloritani Mountains. During the Eocene tectonic

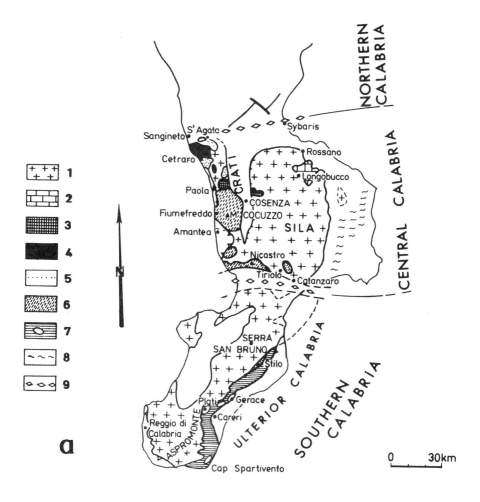

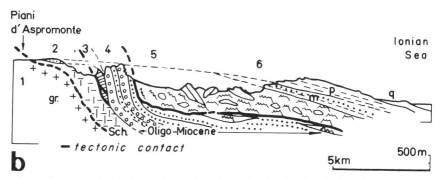

FIG. 7 Central and Southern Calabria, and a section through Ulterior Calabria: (a) (1) Granite, gneiss, mica schists, and phyllites; (2) autochthonous cover of the basement; (3) ophiolites; (4, 5, 6) epimetamorphic flysch; (7) internal flysch (argille scagliose with Numidian) and Oligo-Miocene molasse; (8) chaotic flysch east of Sila; (9) Sangineto and Catanzaro zones. Postorogenic deposits shown in white. (b) (1) Aspromonte-Serra San Bruno basement; (2) phyllites (parautochthonous); (3) slices of the Limestone Chain; (4) discordant Oligo-Miocene molasse; (5) chaotic flysch of internal origin (multi-colored argille scagliose) with internal Numidian; (6) postorogenic middle to upper Miocene (m), Pliocene (p), and Quaternary (q).

phase the limestone cover of the basement slid slabwise to the east, much as in the case of the Novara zone of the Peloritani Mountains. Oligo-Miocene of molasse facies constitutes the next higher tectonic unit, and is in turn overlain by the argille scagliose nappe containing many sedimentary klippen. The postorogenic Mio-Pliocene contains several olistostromes of argille scagliose material.

Central Calabria consists mainly of crystalline rocks which, together with those of southern Calabria and the Peloritani Mountains, constitute an asymmetric fan structure, overthrust to the south in Sicily and to the north in Calabria (Caire, 1970b, Figs. 19 and 20). The overthrusting is rather minor in Sicily but is strongly pronounced in Calabria, where it is accompanied by glaucophane metamorphism. The crystalline rocks, together with their sedimentary cover, have been thrust over the Southern Limestone Apennines, which appear in tectonic windows on the western side of central Calabria (Dubois, 1969, 1970). A "Ligurian" sequence of argille scagliose with ophiolites has been overridden as well.

In northern Calabria allochthonous flysch of internal origin covers the Southern Limestone Apennines (Fig. 2b). Pertinent references are Bousquet (1965), Grandjacquet (1969), and Ogniben (1969).

Southern Apennines

Southeastern Italy north of Calabria consists of the Monte Gargano-Apulia foreland (which corresponds to the Ragusa platform of southeastern Sicily), the Southern Limestone Apennines (labeled 3 in Fig. 8, and comparable to the Panormides of northwestern Sicily and the Babors of the Tell Atlas), with the resedimentation basin of Molise-Bradano between.

The foreland is characterized by a thick Mesozoic carbonate sequence. It represents the margin of the "Dinaric Promontory," a promontory of the African platform which was pushed over the Alps like a *traineau écraseur* (crushing sledge) according to P. Termier.

The Southern Limestone Apennines are in part thrust eastward across the more external Lagonegro-Stigliano zone (labeled 5 in Fig. 8b), which corresponds to the Sclafani zone of Sicily (8a in Fig. 6). According to Manfredini (1965) the Limestone Apennines are autochthonous in Campania, Latium, and Abruzzi; according to Accordi (1964, 1966) and Fancelli and others (1966) they have been thrust northeastward. The NW–SE faults and folds of the Southern Apennines are cut by transverse faults parallel to the Anzio-Ancona fault. D'Argenio (1966) interprets these as sinistral and dextral strike-slip faults with displacements ranging from a few kilometers to as much as 30 km. Some of the faults were already active during the Mesozoic.

Allochthonous units of internal flysch and ophiolites locally overlie the Southern Apennines (Fig. 8b) and occupy the highest structural position. They show many similarities with the ophiolite-bearing Ligurian nappe of the Northern Apennines and with the internal flysch complexes of Algeria and Sicily.

The Molise-Bradano basin constitutes the foredeep of the Southern Apennines and is a classical area for the study of resedimentation phenomena (Beneo, 1956a, 1956b; Boenzi and others, 1968; Crostella, 1967; Crostella and Vezzani, 1964; Jacobacci, 1962; Ogniben, 1955, 1962; Roda, 1967). Mio-Pliocene molasse deposited in this basin contains intercalations of large olistostromes, the major ones coming from the flysch complex on top of the Southern Limestone Apennines. Figure 9 shows how the foredeep migrated toward the exterior between the Miocene and the early Quaternary.

Tyrrhenian Sea

Seismic refraction measurements and a Bouguer gravity maximum both suggest that the mantle lies at a depth of approximately

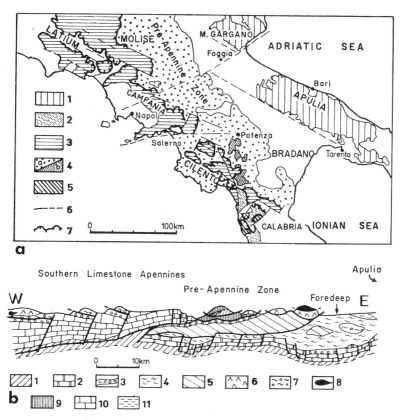

FIG. 8 (a) Schematic geologic map of the Southern Apennines after Grandjacquet (1963, Fig. 1, p. 786): (1) carbonate rocks of Apulia-Monte Gargano; (2) parautochthonous Calabrian rocks; (3) allochthonous carbonate rocks of the Southern Limestone Apennines; (4a) flysch complex (argille scagliose and various other flysch formations); (4b) Lagonegro unit; (5) schistes lustrés and ophiolites; (6) major faults; (7) boundaries of the allochthonous units. Recent sediments shown in white. (b) Composite schematic section without scale (after Grandjacquet, 1962, Fig. 1, p. 108): (1) epimetamorphic Triassic (gypsum, greenschists, marble, limestone and dolomite); (2) Mesozoic and Cenozoic neritic rocks of the Bradanian foredeep; (3) discordant Miocene molasse above the Southern Limestone Apennines; (4) upper Miocene and Pliocene with olistostromes; (5) Lagonegro unit (in fact tectonically doubled); (6) epimetamorphic flysch associated with unit 8; (7) internal flysch complex; (8) ophiolites of glaucophane-lawsonite metamorphic facies, radiolarites, and Calpionella limestones; (9) "Argille e Calcare" unit (Upper Cretaceous to Eocene flysch); (10) Jurassic to lower Miocene carbonate rocks of the Southern Limestone Apennines (thrust towards the east and very simplified in this figure); (11) external Quaternary.

10 km below the abyssal plain of the Tyrrhenian Sea. Earthquake foci occur to a maximum depth of 480 km and lie within an inclined Benioff zone which dips from the Ionian to the Tyrrhenian Sea (Fig. 1) (Peterschmitt, 1956). The intersection of this zone with the surface outlines the Pliocene to Quaternary arc. A progressively expanding rise within the mantle below the Tyrrhenian Sea may account for the southward-moving "orogenic wave" during the Miocene and Pliocene in Sicily and for the gliding of the sedimentary nappes and the tectonic divergence away from the uplift (Caire and others, 1960). The central part of the uplift collapsed to form the present Tyrrhenian sea bottom.

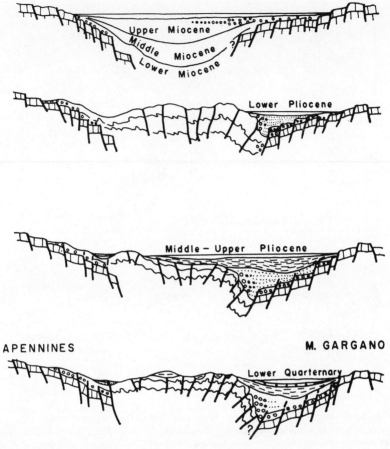

FIG. 9 Evolution of the Mio-Pliocene foredeep of the Southern Apennines (after Jacobacci, 1962). Note the migration of the foredeep toward the northeast from Miocene to Quaternary time. North of Monte Gargano the Adriatic Sea is now the foredeep of the Apennines.

According to Van Bemmelen (1969) the rise of the upper mantle and the accompanying oceanization occurred because the lower part of the sialic crust was absorbed into the mantle. The upper part was removed by the combined result of erosion and tectonic sliding. According to Ritsema (1969) the earthquakes of the Tyrrhenian Sea are probably related to the drifting of Calabria and Sicily toward the southeast over the stable basin of the Ionian Sea; the accompanying extension in the rear would explain the oceanization of the Tyrrhenian Sea and the present-day volcanism of the Liparic Islands.

Conclusion

The regional descriptions have shown (1) the unity of the structural framework of the Calabro-Sicilian arc, (2) the correspondence between the Eastern Atlas and the Apennines, and (3) the close proximity of the tectonic units of most internal origin to the foreland, the internal units being superimposed on the most external units.

The Alpine structures were formed in two phases, one during the Eocene and the other during the Neogene (Mio-Pliocene). The large overthrust sheets moved during the

Miocene, their frontal part sliding into a foredeep that moved toward the exterior of the arc. Study of these complex structures permits the reconstruction of a Mesozoic to Eocene paleography and reveals the existence of five major belts (see Figs. 1 and 2):

1. An internal belt, characterized by Jurassic to Miocene flysch.
2. A discontinuous belt occupying a relatively high position in the geosyncline (a geanticline or platform) and consisting of a condensed pelagic carbonate sequence with many hiati (Kabylies, Peloritan-Calabrian domain, and the Tuscan ridge).
3. A series of juxtaposed zones which succeed each other along the arc, for example, the belt stretching from the Babors via the Panormides to the Southern Limestone Apennines, and the External Flysch-Sclafani-Langonegro zones.
4. The "Pre-Saharan" belt, including the

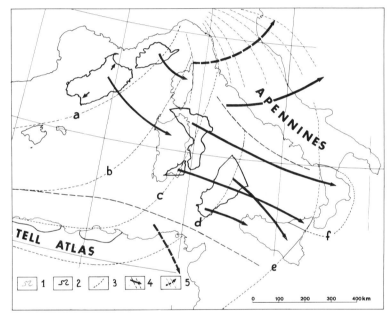

FIG. 10 Hypothetical restoration of the major paleogeographic divisions and their subsequent migration in the central Mediterranean region: (1) present coastlines; (2) presumed positions of Corsica, Sardinia, Calabria, the Peloritan massif of northern Sicily, and the central Sicilian domain in Mesozoic time (arrows in Sardinia show Alpine movements corresponding to those in the Pyrenees); the coastlines are only indicated as theoretical references; (3) Mesozoic paleogeographic trend lines deduced from the presumed initial positions shown by 2; (4) major migrations within the inner portion of the Tyrrhenian fan, with anticlockwise rotation of the Corso-Sardinian block as confirmed by paleomagnetic data (Hospers and Van Andel, 1969) (more localized movements of individual parts, oblique or transverse to arrows, are not shown); (5) major migration of the outer parts of the Tyrrhenian fan, including the Northern Apennines and the Eastern Atlas [(a) border of the European continental platform; (b) separation between the future Alpides and Dinarides; (c) axis of rise or platform within the geosyncline, represented by the Kabylie and Peloritan-Calabrian massifs and the Tuscan ridge; (d) arc represented by the Tell Atlas and Central Sicily, joining the Apennines in Umbria or the Southern Limestone Apennines; (e) border of the African continental platform and its promontories, the Ragusa platform and Apulia; (f) presumed external border of the Peloritan-Calabrian block] (see Caire 1970b, p. 338).

Sahara and Tunisian Atlas and the Vicari-Sciacca zones (Fig. 6) in Sicily.
5. The African platform and its promontories (Sahara, Ragusa platform, and Apulia).

A reconstruction of these belts has to remain qualitative and without dimensions, as it is hard to measure distances and original surface areas of Mesozoic zones which have been overthrust, considerably foreshortened, and partially eroded. Nevertheless, Fig. 10 presents a provisional attempt at a restoration of the major paleogeographic divisions during the Mesozoic.

Various types of orogenic mobility can be recognized: pulsatory movements, faulting of the basement (as in the Kabylies and the Peloritan-Calabrian massif), compression of troughs, expulsion of their sediments, gravitational sliding, and strike-slip faulting.

Gravity tectonics was active at different scales and in different tectonic environments. Induced by basement deformation, it was confined to orogenic zones in time and space. Induced by regional uplifts, such as the late epeirogenic arching, gravity tectonics may operate in various directions and have a chaotic aspect.

Rocks with "plastic" properties, including the Triassic of germanic facies in the Atlas and Northern Apennines and the phyllites in the Kabylies and the Peloritan-Calabrian massif, seem to have facilitated the sliding movements. However, movement also occurred in marls and flysch, and locally at the contact between "rigid" rocks. The importance of rocks with "plastic" properties should thus not be overemphasized, since gravity tectonics can operate without such rocks.

References

Accordi, B., 1964, Lineamenti strutturali del Lazio e dell'Abruzzo meridionali: *Mem. Soc. Geol. Ital.*, v. 4, no. 1, p. 595–633.

——, 1966, La componente traslativa nella tettonica dell'Appennino laziale-abruzzese: *Geol. Rom.*, v. 5, p. 355–406.

Beneo, E., 1956a, Accumuli terziari da risedimentazione (olistostroma) nell'Appennino centrale e frane sottomarine. Estensione tempo-spaziale del fenomeno: *Bull. Serv. Geol. Ital.*, v. 78, p. 291–321.

——, 1956b, Il problema delle "Argille Scagliose-Flysch" in Italia e sua probabile soluzione. Nuova nomenclatura: *Boll. Soc. Geol. Ital.*, v. 75, p. 53–68.

Boenzi, F., Ciaranfi, N., and Pieri, P., 1968, Osservazione geologiche nei dintorni di Accettura e di Oliveto Lucano: *Mem. Soc. Geol. Ital.*, v. 7, no. 3, p. 379–392.

Bousquet, J. -CL., 1965, Sur l'allure et la mise en place des formations allochtones de la bordure orientale des massifs calabro-lucaniens: *Bull. Soc. Géol. Fr.*, v. 7, p. 937–945.

Broquet, P., 1970, Observations on gravitational sliding: The concept of olistostrome and olistolite, *in Geology and history of Sicily*: Tripoli, Petrol. Expl. Soc. Libya, p. 255–259.

——, Caire, A., and Mascle, G., 1966, Structure et évolution de la Sicile occidentale (Madonies et Sicani): *Bull. Soc. Géol. Fr.*, v. 8, p. 994–1011.

Caire, A., 1957, Etude géologique de la région des Biban (Algérie) (Thèse Sc. Paris): Publ. Serv. Carte Géol. Algérie, n.s., Bull. 16.

——, 1961, Remarques sur l'évolution tectonique de la Sicile: *Bull. Soc. Géol. Fr.* v. 3, p. 545–558.

——, 1962, Les arcs calabro-siciliens et les relations entre Afrique du Nord et Apennin: *Bull. Soc. Géol. Fr.*, v. 4, p. 774–784.

——, 1963, Phénomènes tectoniques de biseautage et de rabotage dans le Tell algérien: *Rev. Géogr. Phys. Géol. Dyn.*, v. 5, p. 299–325.

——, 1964, Comparaison entre les orogènes berbère et apenninique: *Ann. Soc. Géol. Nord.*, v. 84, p. 163–176.

——, 1970a, Sicily in its Mediterranean setting, *in Geology and history of Sicily*: Tripoli, Petrol. Expl. Soc. Libya, p. 145–170.

——, 1970b, Tectonique de la Méditerranée centrale: *Ann. Soc. Géol. Nord.*, v. 90, p. 307–346.

——, Glangeaud, L., and Grandjacquet, Cl., 1960, Les grands traits structuraux et l'évolution du territoire calabro-sicilien (Italie méridionale): *Bull. Soc. Géol. Fr.*, v. 2, p. 915–938.

Crostella, A., 1967, Rapporti fra serie autoctone e serie allochtone nell'Alto Aventino (Abruzzi sud-orientali): *Mem. Soc. Geol. Ital.*, v. 6, no. 2, p. 121–136.

——, and Vezzani, L., 1964, La geologia dell'Appennino foggiano: *Boll. Soc. Geol. Ital.*, v. 83, p. 1–23.

D'Argenio, B., 1966, Zone isopiche e faglie trascorrenti nell' Appennino centro-meridionale: *Mem. Soc. Geol. Ital.*, v. 4, p. 279–299.

Debelmas, J., and Kerckhove, Cl., 1973, Large gravity nappes in the French–Italian and French–Swiss Alps (this volume).

Dubois, R., 1969, Le passage latéral des prasinites de Rose-Fuscaldo aux épanchements jurassiques de Malvita et ses conséquences sur l'interprétation de la suture calabro-apenninique: *Comptes Rendus Acad. Sci. Fr.*, v. 269, p. 1815–1818.

——, 1970, Phases de serrage, nappes de socle et métamorphisme alpin à la jonction Calabre-Apennin. *Rev. Geogr. Phys. Géol. Dyn.*, v. 12, p. 221–254.

Duée, G., 1970, The geology of the Nebrodi mountains of Sicily, *in Geology and history of Sicily*: Tripoli, Petrol. Expl. Soc. Libya, p. 187–200.

Fancelli, R., Ghelardoni, R., and Pavan, G., 1966, Considerazioni sull'assetto tettonico dell'Appennino calcareo centro-meridionale: *Mem. Soc. Geol. Ital.*, v. 5, no. 1, p. 67–90.

Flores, G., 1959, Evidence of slump phenomena (olistostromes) in areas of hydrocarbons exploration in Sicily: *Proc. 5th World Petrol. Congr., New York*, sec. 1, p. 259–272.

Grandjacquet, Cl., 1962, Importance de la tectonique tangentielle en Italie méridionale: *Rev. Géogr. Phys. Géol. Dyn.*, v. 4, p. 109–113.

——, 1963, Schema structural de l'Apennin campano-lucanien: *Rev. Géogr. Phys. Géol. Dyn.*, v. 5, p. 185–202.

——, 1969, Les phases tectoniques et le metamorphisme tertiaire de la Calabre du Nord et de la Campanie du Sud (Italie): *Comptes Rendus Acad. Sci. Fr.*, v. 269, p. 1819–1822.

Hospers, J., and Van Andel I., 1969, Paleomagnetism and tectonics, a review: *Earth Sci. Rev.*, v. 5, p. 5–44.

Jacobacci, A., 1962, Evolution de la fosse mio-pliocène de l'Apennin apulocampanien (Italie méridionale): *Bull. Soc. Géol. Fr.*, v. 4, p. 691–694.

——, 1965, Frane sottomarine nelle formazione geologiche: *Boll. Serv. Geol. Ital.*, v. 86, p. 65–85.

Manfredini, M., 1965, Sui rapporti fra facies abruzzese e facies umbra nell' Appennino centro meridionale: *Boll. Serv. Geol. Ital.*, v. 86, p. 87–112.

Mascle, G., 1970, Geological sketch of western Sicily, *in Geology and history of Sicily*: Tripoli, Petrol. Expl. Soc. Libya, p. 231–243.

Ogniben, L., 1953, "Argille Scagliose" ed "Argille Brecciate" in Sicilia: *Boll. Serv. Geol. Ital.*, v. 75, no. 1, p. 281–289.

——, 1954, Le "Argille Brecciate" Siciliane: *Mem. Ist. Geol. Min. Univ. Padova*, v. 18, p. 1–92.

——, 1955, Le argille scagliose del Crotonese: *Mem. Not. Ist. Geol. Applic. Napoli*, v. 6, p. 3–72.

——, 1962, Stratigraphie tectono-sédimentaire de la Sicile, *in Livre P. Fallot*: *Soc. Géol. Fr., Mém. hors-sér.*, v. 2, p. 203–216.

——, 1969, Schema introduttivo alla geologia del confine calabro-lucano: *Mem. Soc. Geol. Ital.*, v. 8, p. 453–763.

——, 1970, Paleotectonic history of Sicily, *in Geology and history of Sicily*: Tripoli, Petrol. Expl. Soc. Libya, p. 133–143.

Peterschmitt, E., 1956, Quelques données nouvelles sur les séismes profondes de la mer Thyrénienne: *Ann. Geoph.*, v. 9, p. 305–334.

Rigo de Righi, R., 1956, Olistostromi neogenici in Sicilia: *Boll. Soc. Geol. Ital.*, v. 75, no. 3, p. 185–215.

Ritsema, A. R., 1969, Seismic data of the West Mediterranean and the problem of oceanization: *Verh. Kon. Ned. Geol. Mijnb. Gen.*, v. 26, p. 105–120.

Roda, C., 1967, I sedimenti neogenici autoctoni ed alloctoni della zona di Cirò-Cariati (Catanzaro e Cosenza): *Mem. Soc. Geol. Ital.*, v. 6, no. 2, p. 137–149.

Truillet, R., 1970, The geology of the eastern Peloritani Mountains of Sicily, *in Geology and history of Sicily*: Tripoli, Petrol. Expl. Soc. Libya, p. 171–186.

Van Bemmelen, R. W., 1969, Origin of the Western Mediterranean sea: *Verh. Kon. Ned. Geol. Mijnb. Gen.*, v. 26, p. 13–52.

P. ELTER
L. TREVISAN

Istituto di Geologia e Paleontologia
University of Pisa
Pisa, Italy

Olistostromes in the Tectonic Evolution of the Northern Apennines

In this paper various aspects of diverse types of olistostromes in the Northern Apennines will be described, and the authors will attempt to interpret their structure and kinematics in the framework of the tectonic evolution of the Northern Apennines.

In 1955 Flores introduced the term "olistostrome" to indicate heterogeneous, more or less intimately mixed deposits that are interbedded with normal sedimentary rocks and interpreted them as the result of submarine sliding. The phenomenon for which this new term was created had been observed in Sicily in various manifestations and dimensions, but subsequently the term olistostrome was applied to comparable phenomena in the Apennines and other mountain chains.

Some authors have tried to give new descriptions and definitions of the terms olistostrome and olistolite, discussing the genesis and mode of occurrence of these phenomena (e.g., Abbate and others, 1970). We do not intend to enter into these discussions but prefer to illustrate some types of submarine slumping and their relation to the causative tectonic events.

In the Northern Apennines Elter and Schwab (1959) were the first to interpret as olistostromes some deposits of breccia intercalated between flysch beds. Previously, all deposits for which no explanation had been found, and which were simply called "chaotic" in view of their complex structure, were generally named *argille scagliose* (scaly clays or shales), because of the frequent occurrence of shales that break down in outcrop into splinters of several millimeters. Hence an old term with a purely lithological meaning was expanded to indicate allochthonous and commonly strongly tectonized units. Today the term "argille scagliose" should no longer be used in this generalized sense, because the stratigraphy and structure of the various shale-rich allochthonous complexes have been studied in sufficient detail in almost the entire Apennine chain.

Types of Slides in Relation to the Provenance of the Material

The gravitative phenomena we want to examine belong to a group which, in its simplest aspect, involves slumping. Other, more complex types deserve to be better known. Figure 1A shows a model of a slump. We include in this term intercalated deposits made up either of beds deformed by differential sliding or of mud flows with a brecciated character, providing that the material is derived from the same units with which they are interbedded. For this process to occur, a slight tilting of the sea floor during sedimentation suffices.

Figure 1B shows a model that differs from the preceding model in that the slide unit includes in addition materials derived from formations older than the sequence in which the slide occurs.

Figure 2 shows an actual example from Sicily. To obtain this type of structure a slight folding or some other kind of structural deformation appears to be required. The slide seems to occur in front of the fold. This kind of phenomenon was first named *argille brecciate* (brecciated shales) and later olistostromes.

In other instances the materials intercalated in a stratigraphic sequence with evidence of sliding movement belong at least partly to the sediments of a different depositional basin. This third type of slide is evidently related to the emplacement of nearby allochthonous masses (Fig. 1C), which follows the slide after a brief interval. For these slides we propose the term "precursory olistostromes."

The three types of slides are consequently distinguished from each other by the different provenances of the constituent materials. These different provenances may be attributed to different conditions of the "tectonic landscape" at the moment the slides occurred.

In the next section we shall examine several types of olistostromes in relation to the evolution in time and space of the tectonic landscape of the Northern Apennines. The distribution of olistostromes and of the major tectonic units in this region is shown in Fig. 3.

Evolution of the Tectonic Landscape of the Northern Apennines

Character of Sedimentary Basins

In the geosyncline of the Northern Apennines an internal domain (southwest of the present mountain chain) can be distinguished from an external domain by the considerable differences between their sediments (Fig. 4A).

The internal area, called the Ligurian domain, is characterized by the fact that the oldest sediments (radiolarites) belong to the Malm (Upper Jurassic). These pelagic sediments rest directly on a substratum of basic and ultrabasic igneous rocks: more or less serpentinized peridotites, gabbros, and discontinuous diabase flows. In view of this, one may imagine that during Malm time the sialic crust had been stretched and that an oceanic bottom appeared from below (Decandia and Elter, 1969). The radiolarites and the overlying *Calpionella* limestone, with their noteworthy variability in thickness and local absence, seem to indicate relatively rapid upward and downward movements of the basin floor. The oldest sediments often contain intercalations of detritic ophiolitic material. Until the start of the Cenomanian, sedimentation continues with shales interbedded with gray pelagic limestones (*palombini*). The subsequent flysch sedimentation extends until the end of the Paleocene in the innermost part of the Ligurian domain, and until the end of the Eocene toward the northeastern margin of the domain.

The external part of the northern Apennine

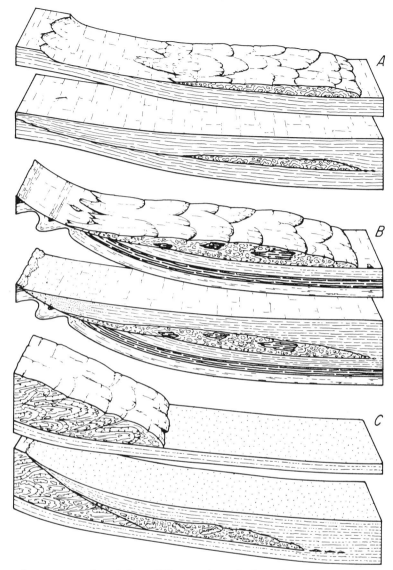

FIG. 1 Schematic representation of three different types of submarine slides. (A) Slumping (materials derived from same formation); (B) olistostromes (materials derived from other formations in the same sedimentary basin); (C) precursory olistostromes (materials derived from the front of an advancing allochthonous sheet).

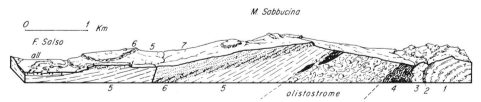

FIG. 2 Example of olistostromes near Caltanissetta (Sicily). (1) Shales (Tortonian: middle Miocene); (2) Tripoli; (3) gypsum (Messinian: upper Miocene); (4) white marls with foraminifera "trubi" (lower Pliocene); (5) blue shales of the Piacenza facies (Pliocene); (6, 7) yellow calcarenites and sands of the Asti facies (Pliocene). The olistostromes have the aspect of conglomerates with components of Tortonian shales and sequences of "trubi" strata.

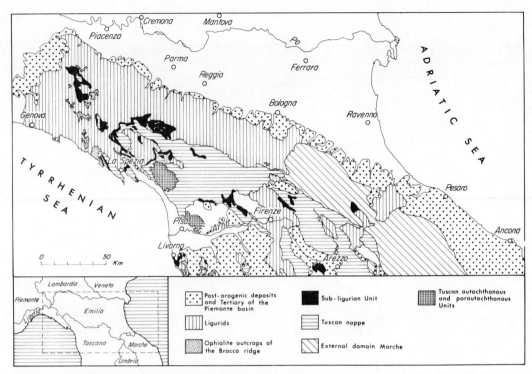

FIG. 3 Schematized map showing the distribution of the tectonic units of the Northern Apennines.

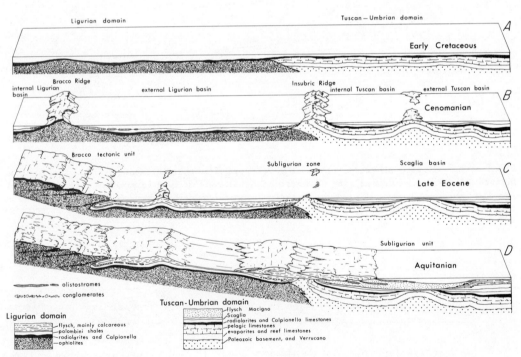

FIG. 4 Schematic representation of the tectonic evolution of the Northern Apennines from the Cretaceous to the Aquitanian.

geosyncline, called the Tuscan-Umbrian domain, is characterized by a sedimentary sequence in which the Middle Triassic rests discordantly on a metamorphosed continental-type basement involved in the Hercynian orogeny. The Upper Triassic is represented by evaporites and reef deposits, and the Jurassic by pelagic, calcareous, and siliceous rocks, locally interrupted by diastems. At the end of the Malm, formations of radiolarites and *Calpionella* limestone units similar to those of the Ligurian domain, were deposited in the Tuscan-Umbrian domain, suggesting that no barrier existed between the two areas at that time. Flysch sedimentation begins in the external domain in Oligocene time.

Cretaceous Tectonic Phase

The first tectogenic movements in the Northern Apennines manifest themselves toward the end of the Early Cretaceous (Fig. 4B). Several uplifts in the external domain are revealed by local discordances in the Cenomanian *scaglia rossa* (red scaly shale) and by an hiatus which corresponds to the Albian and Aptian stages or which may locally be even larger. Cenomanian conglomerates in the eastern part of the Ligurian domain, at the border between the internal and external domains, demonstrate that during the Cenomanian an emergent ridge (Insubric ridge of Elter and others, 1966) existed between the two areas. Crystalline pebbles in the conglomerate show that erosion had progressed to the metamorphic basement. The rise of this ridge is probably related to important tectonic movements which occurred in the Austroalpine region farther north.

A significant event in the Ligurian area, perhaps unrelated to the Insubric ridge, is the uplift of the Bracco ridge (Elter and Raggi, 1965), which consists of ophiolites and their sedimentary cover. This ridge divides the Ligurian domain into two parts, so that from then on there existed an internal and an external Ligurian basin. In both basins turbiditic flows came in both from raised portions of the Austroalpine domain and from the Hercynian massifs of Corsica and Sardinia. Thus began the flysch sedimentation of the Late Cretaceous.

The Bracco ridge was rapidly uplifted and attained its maximum development in the Cenomanian and Turonian. The elongate ridge with locally high and low parts was at some places above sea level, as shown by conglomerate lenses. In the two basins separated by the Bracco ridge sedimentation was different. The Val Lavagna shales were deposited in the internal basin; these marly-silty units tend to assume in their higher parts the aspects of flysch, with intercalations of sandstone beds. Away from the ridge a coarse sandy flysch, the Monte Gottero sandstone, is equivalent to the upper part of the Val Lavagna shales. In the external basin (NE of the ridge) flysch was deposited on shaly and sandy rocks of Cenomanian to Turonian age. This is the *Helminthoid* flysch of the Northern Apennines; Senonian to Paleocene in age, it is mainly calcareous to marly in composition and, because of common facies changes, is known under different formation names: Monte Caio, Monte Cassio, Serramazzoni, and others.

Olistostromes of the Cretaceous Phase

The uplift of the Bracco ridge is documented mainly by the appearance of detrital units in the basin sediments on the two sides of the ridge. These intercalated deposits have different aspects: in places they are turbidite sandstones with dominantly ophiolitic material, but more commonly they consist of layers or lenses of breccia with a pelitic matrix which are interpreted as olistostromes (Fig. 5). The breccias include clasts of ophiolitic rocks and Upper Jurassic to Cretaceous sediments that were originally on top of the ophiolites.

The locally abundant matrix is derived primarily from the source rock, but it appears in a few instances enriched with the sediments

FIG. 5 Aspect of an olistostrome of Cenomanian age composed of elements derived from ophiolites and their sedimentary cover, with abundant clay matrix.

over which sliding occurred. Breccias composed of ophiolites, *Calpionella* limestone, and radiolarites without a pelitic matrix are also known, and in some cases the matrix consists of sandstone or reworked and commonly chloritized hyaloclastites. The olistostromes with little clay matrix are localized in the vicinity of the ridge from which they originated, whereas those with a more abundant clay matrix may have moved over distances on the order of 50 km.

Associated with the latter type of olistostrome, but not necessarily enclosed in them, are large blocks, occasionally enormous in size (up to several cubic kilometers), and composed primarily of diabase and more or less serpentinized peridotite. In view of their relation to the surrounding sediments, these rocks are regarded as enormous clasts which slid over the muddy sea bottom.

The olistostromes are clearly more numerous in proximity to the ridge, and in many cases they substitute for the normal flysch sediments and the underlying rocks.

Olistostromes of this type are present on both sides of the Bracco ridge, but they are more frequent and extend farther out on the external side. This may indicate an asymmetry of the ridge, which is probably related to the future northeasterly vergence of the Apennines.

The olistostromes are generally intercalated in Cenomanian sediments, but in some areas as large as several square kilometers they rest directly on the ophiolites. Figure 6 provides an explanation of this phenomenon.

It seems certain that various generations

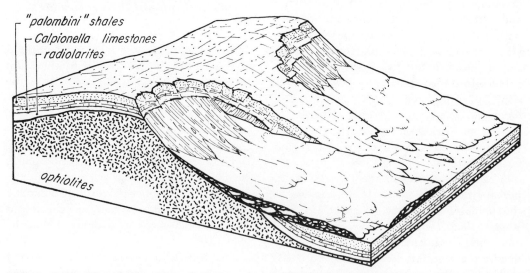

FIG. 6 Schematic representation of olistostromes on the NE flank of the Bracco ridge, showing the relationships to the substratum.

FIG. 7 Aspect of an olistostrome near the Bracco ridge. The arrows indicate a brecciated component derived from a preceding olistostrome.

of olistostromes occurred during the Cretaceous tectonic phase. It is, in fact, not uncommon to find fragments of older olistostromes within a younger one (Fig. 7). It can be concluded that the Bracco ridge has been the focus of increasing mobility during the Cenomanian, which may be considered as the precursor of the great Eocene paroxysmal phase in the Ligurian domain.

Eocene Tectonic Phase (Ligurian Phase)

Sedimentation in the Ligurian domain continued until the end of the Paleocene, predominantly in a flysch-type facies. During the Eocene an important phase of folding affected the internal Ligurian basin, the Bracco ridge, and part of the external basin (Fig. 4C). The upper part of the Bracco ridge moved toward the east in the form of a recumbent fold, as a result of which the sediments and the intercalated olistostromes that had been deposited on the exterior side of the ridge just prior to this movement were covered over an area of many square kilometers by the ophiolites that constitute the core of the fold. Other recumbent folds of a more internal origin, likewise with ophiolites in their cores, were piled on top of this fold. Part of this folded terrane was subsequently covered by sediments of late Eocene to middle Miocene (Helvetian) age. The most important formation of this unconformable sequence is the Oligocene Ranzano sandstone

Sedimentation was continuous in the extreme external part of the Ligurian basin, which did not participate in these structural movements. The dominantly calcareous flysch of this area grades laterally without apparent discontinuity into the Ranzano sandstone.

Olistostromes of the Eocene Phase

A part of the relief formed during the Ligurian phase was undoubtedly emergent and became the source area of the sandstones and associated conglomerates of the Ranzano formation. With respect to the Ligurian phase the Ranzano sandstone constitutes a molasse-type sediment. Since the marine domain had become more restricted, olistostromes are much less abundant. Nevertheless, several are intercalated in the Ranzano sandstone, and others occur sporadically in the external Eocene flysch. Some of these contain very large ophiolite blocks as well as portions of the basal complex, and they appear identical in composition to the Cenomanian ones. It is probable that they are "inherited" olistostromes, that is, due to reworking of the older ones.

Whatever the origin of the olistostromes of post-Ligurian age, they were successively formed in areas that are ever more external, and they are intercalated in ever younger sediments, thus indicating the relatively slow progress of the tectogenic movements toward the external side.

Aquitanian Tectonic Phase (Early Miocene)

The second or Ligurian tectonic phase was accompanied by large-scale overthrusting but remained confined to the internal part of the Ligurian domain. During the Cretaceous and the Eocene no important movements occurred in the Tuscan-Umbrian domain, apart from some vertical mobility as demonstrated by partial hiati in the formations of the *scaglia* (scaly shales). In the Oligocene a trough was formed in the external domain, which was filled by thick deposits of sandy flysch, the Macigno sandstone. The first overthrusting that involved the advancement of an allochthonous sheet of internal domain rocks over the external domain took place at the end of the Oligocene and the beginning of the Miocene (Fig. 4D).

The first tectonic unit to reach the Tuscan-Umbrian domain consists almost completely of Eocene rocks, composed subordinately of calcareous rocks and sandstones, but predominantly of a thick succession of shales with interbedded limestones. With some reservation the most external part of the Ligurian domain may be considered as the source area of this tectonic unit. Most probably its upper part advanced still farther, and it is now found in areas farther to the east as the calcareous flysch of "Alberese." The mechanism of this forward movement over the external domain is considered to be gravitational, if for no other reason than that a rock complex consisting mainly of shales cannot transmit a large pushing force from the rear.

This tectonic unit, previously called the shale-limestone unit, is here named the Subligurian unit. Its forward movement into the sedimentation trough of the Macigno was not instantaneous, but more or less gradual, as demonstrated by the presence of olistostromes in various horizons of the Macigno formation.

Olistostromes of the Aquitanian Phase

When the Subligurian unit advanced into the Macigno trough, small and large submarine slumps detached themselves from the frontal part, which advanced separately over many kilometers, and which may be observed today as brecciated deposits interbedded with the Macigno turbidites. All of the Eocene formations of the Subligurian unit can be found in the clastic components of the olistostromes, with a slight prevalence of limestones over sandstones. Angular fragments of shale are also encountered, indicating that diagenesis was already in progress prior to tectogenesis.

FIG. 8 Olistostromes of very small size included in Oligocene flysch (Macigno) of the Tuscan sedimentary basin. The calcareous components are slightly rounded.

FIG. 9 Same as Fig. 8.

The olistostromes have various aspects. The easiest recognizable is that of a lenticular mass composed of clastic elements predominantly in the range between 10 and 20 cm, and embedded in a disorganized manner within an abundant shale matrix. The corners of the clasts are slightly rounded, suggesting a rolling movement over a relatively long distance. The smaller olistostromes in particular have this aspect (Figs. 8 and 9).

The second type of olistostrome is composed of blocks most commonly between 0.2 and 1 m in size, with angular corners and to some degree fractured. The matrix consists of the underlying sediment with some traces of stratification. In some places, the blocks are separated from each other. It is thought that this type of olistostrome was transported over a short distance only, without rolling movement.

The olistostromes of the third type are characterized by very large blocks composed of many beds (Fig. 10). The lower boundary commonly appears tectonic in nature, and it is almost invariably marked by stretching of sedimentary beds as a result of traction. In many blocks the beds are overturned,

FIG. 10 Aspect of an olistostrome with preserved sequences of strata.

which, however, does not necessarily indicate rolling movements because the possibility that they were derived from the overturned flanks of recumbent folds cannot be excluded.

The various characteristics just described may be found together in a single olistostrome and, in some cases at least, may indicate the distance to the area of origin, whether nearby or far away.

Figure 11 shows a cross section between the Vara and Taro rivers in the Northern Apennines with two olistostromes of respectable dimensions. Mapping of the area made it possible to reconstruct the sequence of events (Fig. 12).

The truncation of the rear part of the two olistostromes is visible in outcrops a short distance south of the area represented in Fig. 11. Renewal of Macigno sedimentation after the second olistostrome was followed by the arrival of the allochthonous masses of the Subligurian unit. Figure 13, based on many field observations, shows schematically the relative positions of the olistostromes in the Macigno turbidite unit, which, including its highest members, exceeds 1500 m in thickness. It may be concluded that the advance of the Subligurian unit into the Macigno basin was gradual and extended over a relatively long period of time. These major olistostromes very likely correspond to renewals or accentuations of the forward movement of the allochthonous mass. As forerunners and indicators of overthrust movements they therefore have a special significance in the evolution of the tectonic landscape and deserve to be distinguished as "precursory" olistostromes.

The precursory olistostromes also provide testimony (and perhaps the only testimony) to the long time involved in the gravitative gliding of an allochthonous mass over an appreciable distance. Assuming that the Macigno trough had a width of 50 km and taking the length of the late Oligocene (i.e., the period of Macigno deposition) at 5 million years, the average velocity of the allochthonous mass would be in the order of 1 cm/yr.

The Tuscan Tectonic Phase (Tortonian: Middle Miocene)

A paroxysmal tectonic phase, characterized by large-scale overthrusting, occurred during the middle Miocene, particularly during the late part of this stage, the Tortonian. During this phase the tectonic units of the Northern Apennines reached their present positions (Fig. 3) and the large overthrust movements eventually ceased. The sediments of the internal Ligurian basin (sandstone and limestone flysch of Late Cretaceous age) were thrust across the ophiolite-bearing tectonic units derived from the Bracco ridge, which were stretched and laminated in the process. The tectonic pile of mutually overthrust Ligurian units moved as a whole toward the east and covered the external Tuscan-Umbrian domain. During this advance, overthrusting occurred in the external domain as well, resulting in structural repetition of

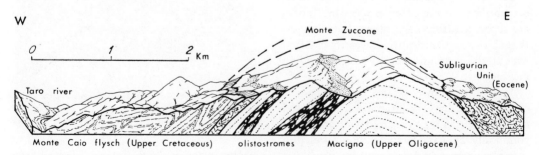

FIG. 11 Section showing two precursory olistostromes in the Oligocene flysch (Macigno) in relation to the tectonic window of the Taro south of Piacenza.

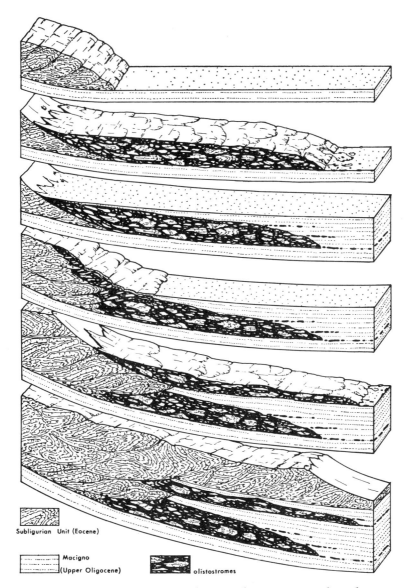

FIG. 12 The block diagrams show, from top to bottom, the reconstructed emplacement mechanism of the two olistostromes of Fig. 10. In the first two diagrams the front of the Subligurian Unit advances into the Macigno trough and gives rise to the first olistostrome. Subsequently a further advance of the allochthonous unit causes the detachment of a second olistostrome which includes reworked parts of the first. In the end, the ultimate advance of the Subligurian Unit interrupts completely the deposition of the Macigno flysch.

the Tuscan sequence over large areas (Fig. 14). The individuality of the Tuscan overthrust sheet is particularly manifest in the Apuane Alps, where the autochthonous terrane, exposed in a window, has the appearance of a metamorphic nucleus.

During this same period great horizontal translations occurred south of the Arno

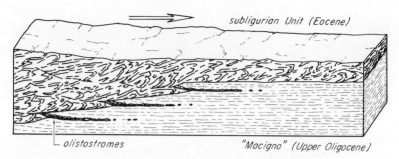

FIG. 13 Schematic representation of the relation between the gradual advance of the Subligurian Unit into the Macigno sedimentary basin and the distribution of the olistostromes.

River as well. Moreover, large areas of tectonic denudation occur here behind overthrust Tuscan units that were detached from their substratum and displaced eastward by sliding (Fig. 15). As a result, Tuscan rocks replaced the original cover of Miocene flysch in the Umbrian domain over a distance of several tens of kilometers ("cover substitution"). The tectonic denudation to the rear is demonstrated by extensive gaps in the Tuscan domain, where allochthonous Ligurian units rest directly on the Tuscan substratum of Upper Triassic evaporites, replacing the original Tuscan cover. These large gaps in the wake of advanced cover sheets are taken as proof that the movements were gravitative in origin.

Although olistostromes already reached into parts of the external domain during the advance of the Subligurian unit in the Aquitanian tectonic phase, it was not until the Tortonian tectonic phase that they arrived in the most external part of the Apennines, intercalating themselves in the autochthonous sediments of the Adriatic and Po basins. They had the character of large frontal slides, consisting of highly chaotic material derived from all Ligurian tectonic units, and mixed with scrambled slivers of the most recent autochthonous units of the external domain. In some cases they include extensive sheets of little-deformed sedimentary beds. They are large-dimensional gravitative forerunners of the front of the allochthonous Ligurian mass. Whether these phenomena qualify as olistostromes is open to discussion, for in this case it becomes difficult and arbitrary to define the boundary between olistostromes and gravitative overthrust sheets.

In any case, the allochthonous masses

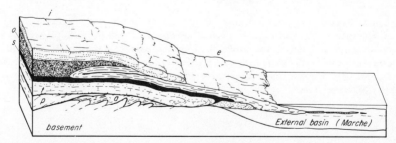

FIG. 14 Schematic representation of the relations between the various tectonic units at the end of the Tortonian phase. The olistostromes formed in the most external belt of the Apennines, corresponding to the Tuscan-Umbrian basin of sedimentation. (*i*) Internal Ligurian Unit; (*e*) External Ligurian Unit; (*o*) ophiolites; (*s*) Subligurian Unit; (*t*) Tuscan nappe; (*p*) Tuscan parautochthon (metamorphic); (*a*) Tuscan autochthon (metamorphic).

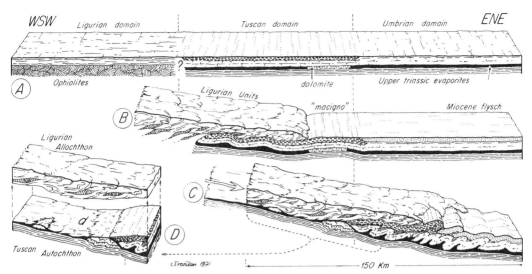

FIG. 15 Schematic representation of gravitative tectonics in a transversal section across the Apennines about 50 km south of Florence. (A) The three basins of deposition; (B) overthrusting of the Ligurian units onto the Tuscan domain; (C) gravitational sliding and folding along the Upper Triassic evaporites, whereby the Tuscan cover and the Ligurian units come to rest on the Umbrian area and, behind them, the cover of the substratum in the Tuscan domain is replaced by Ligurian units; (D) detail of C, showing the gap left by tectonic denudation (d) in the Tuscan domain.

arrived at the external Po-Adriatic margin of the Northern Apennines at the beginning of the Tortonian and during the Messinian (late Miocene) after a stepwise journey. These movements brought to a close the last episode of the gravitative horizontal movements in the Northern Apennines.

The Last Tectonic Pulsations and the Final Tensile Phase

After the Tortonian there were only compressive phenomena in the Marchesian part of the external domain. These take the form of folds which slowly die out in the direction of the Adriatic Sea. In Tuscany, on the other hand, extensional tectonism prevailed instead, characterized by normal faults with subsidence of graben in which accumulated neo-autochthonous sediments of late Miocene to Pliocene age, and uplift of horsts that carried Pliocene deposits more than 80 m above sea level. Late Miocene to Quaternary volcanic processes are related to these extensional movements. Meanwhile, the west coast of the Northern Apennines north of the mouth of the Arno subsided. These late movements of a clearly extensional and rigid style on the one hand, and of an attenuated compressive nature on the other, modeled the present-day topography of the Northern Apennines and are the cause of the asymmetry between the Tyrrhenian and Adriatic drainage patterns.

References

Abbate, E., Bortolotti, V., and Passerini, P., 1970, Introduction to the geology of the Northern Apennines: Sed. Geol., v. 4, no. 3/4, p. 207–250, 521–558.

Baldacci, F., Elter, P., Giannini, E., Giglia, G., Lazzarotto, A., Nardi, R., and Tongiorgi, M., 1967, Nuove osservazioni sul problema della falda toscana e sulla interpretazione dei flysch arenacei tipo macigno dell'Appennino settentrionale: Mem. Soc. Geol. Ital., v. 6, no. 6, p. 213–244.

Beneo, E., 1955, Les résultats des études pour la recherche petrolifère en Sicile: Proc. 4th World Petrol. Congr., Roma, p. 1–13.

———, 1956, Accumuli terziari da risedimentazione

(olistostroma) nell'Appennino centrale e frane sottomarine: Estensione tempospaziale del fenomeno: *Boll. Serv. Geol. d'Italia*, v. 78, no. 1–2, p. 291–319.

Decandia, F. A., and Elter, P., 1969, Riflessioni sul problema delle ofioliti nell'Appennino settentrionale (nota preliminare): *Atti Soc. Toscana Sc. Natur. (Pisa)*, serie A, v. 76, no. 1, p. 1–9.

Elter, G., Elter, P., Sturani, C., and Weidmann, M., 1966, Sur la prolongation du domaine ligure de l'Apennin dans le Monferrat et les Alpes et sur l'origine de la nappe de la Simme s. l. des préalpes romandes et chablaisiennes: *Arch. Sci. Genève*, v. 19, no. 3, p. 279–378.

Elter, P., and Raggi, G., 1965, Tentativo di interpretazione delle brecce ofiolitiche cretacee in relazione con movimenti orogenetici nell'Appennino Ligure: *Boll. Soc. Geol. Ital.*, v. 84, no. 5, p. 1–12.

Elter P., and Schwab, K., 1959, Nota illustrativa della carta geologica all'1:50.000 della regione Carro-Zeri-Pontremoli: *Boll. Soc. Geol. Ital.*, v. 78, no. 2, p. 157–187.

Flores, G., 1956, Lettera al presidente della Società geologica Italiana: *Boll. Soc. Geol. Ital.*, v. 75, no. 3, p. 221–222.

———, 1959, Evidence of slump phenomena (olistostromes) in areas of hydrocarbons exploration in Sicily: *Proc. 5th World Petrol. Congr., New York*, sec. I, p. 259–275.

Jacobacci, A., 1965, Frane sottomarine nelle formazioni geologiche: Interpretazione dei fenomeni olistostromici e degli olistoliti nell'Appennino e in Sicilia: *Boll. Serv. Geol. d'Italia*, v. 86, p. 65–85.

Merla, G., 1964, Centro di studio per la geologia dell'Appennino, I sez. Firenze: Attività svolte nel periodo 1951–1963: *La Ricerca scientifica*, suppl. v. 3, no. 3, ser. 2, Consiglio Nazionale delle Ricerche, Roma, p. 107–126.

Rigo de Righi, F., 1956, Olistostromi neogenici in Sicilia: *Boll. Soc. Geol. Ital.*, v. 75, no. 3, p. 185–215.

Tongiorgi, E., and Trevisan, L., 1953, Livret-guide de l'excursion AS (Sicile): *IV Congrès INQUA*, p. 1–36.

Trevisan, L., 1960–1963, La paléogéographie du Trias de l'Apennin septentrional et central et ses rapports avec la tectogénèse: *Livre a la mém. du Prof. Paul Fallot, Soc. Géol. Fr.*, v. 2, p. 217–225.

———, 1962, Considérations sur deux coupes à travers l'Apennin septentrional: *Bull. Soc. Géol. Fr.*, v. 4, p. 675–681.

JACQUES DEBELMAS
CLAUDE KERCKHOVE

Department of Geology
University of Grenoble
Grenoble, France

Large Gravity Nappes in the French-Italian and French-Swiss Alps

The Alps have been regarded as one of the classical mountain ranges for the study of large-scale tangential tectonics since 1900. Its overthrust structures are still among the most frequently quoted examples of nappes. However, until about 1938 little attention had been paid to the mechanism of transport of these nappes, which were described merely from a geometric point of view.

Only after World War II did Alpine geologists recognize much more clearly the structures and mechanisms operative at various structural levels. At the deepest level the gneissic nappes of the Italian and Swiss Pennine Alps were emplaced in realms of deep-seated compression. At the uppermost levels are the "skin-deep" nappes which moved mainly through gravity sliding.

Although only these skin-deep nappes are within the scope of the present paper, this does not imply that gravity-caused phenomena cannot occur at lower levels, but if they do, it must be in a less direct and complete way.

The three superimposed groups for which there is proof or strong evidence of emplacement wholly or predominantly through gravity are the Prealpine nappes, the Piemont Schistes Lustrés nappe and the Helminthoid Flysch nappe of the Embrunais nappe system.

Paleogeographic and Structural Setting[1]

Paleogeographically, the external Alpine Dauphinois or Helvetic zone (Fig. 1) is a continental shelf, which subsided in some places and notably toward its inner margin, the Ultradauphinois or Ultrahelvetic zone. It is bound to the east by a major structural break extending along the entire length of the Western Alpine arc. Beyond it we find the internal Alpine zone, whose paleogeographic

---

[1] For more details see Trümpy (1960) and Debelmas and Lemoine (1970).

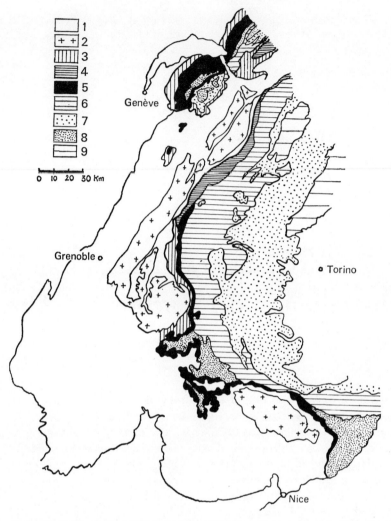

FIG. 1 Structural units of the French Western Alps. (1) External zone (cover); (2) external crystalline massifs; (3) Ultradauphinois zone; (4) Valais zone; (5) Subbriançonnais zone; (6) Briançonnais zone; (7) Piemont zone; (8) Helminthoid flysch s.l.; (9) Austroalpine zone s.l.

and structural evolution was much more varied. Its tectonics are also more complex, featuring large nappes that are dynamically miscellaneous and in part gravitational in origin. We shall describe them later but will first summarize the paleogeographic pattern of these internal zones.

Crossing them eastward, we first come upon a paleogeographically complex area which acted initially as a cordillera and later as a furrow. Its several paleogeographic units are discontinuous and succeed one another along the entire length of the range. This area includes the Valais zone (restricted to the northern part of the French–Italian Alps) and the Subbriançonnais zone. Farther east, we come upon a broad geanticline, the Briançonnais zone, and finally we reach the eugeosynclinal trough proper, the Piemont zone, with its monotonous marly and ophiolite-bearing sediments.

We are here at the core of the Alpine

geosyncline, whose contents deserve detailed description as it gave birth to two of the largest Alpine nappes. From bottom to top, the Piemont sediments consist of (1) thick Triassic dolomites, (2) the Jurassic to Cretaceous "Schistes Lustrés" with ophiolitic bodies, metamorphosed at the end of sedimentation and during the subsequent Alpine folding, and (3) the Helminthoid Flysch, an unmetamorphosed Upper Cretaceous sequence, which probably extended originally beyond the deposition area of the Piemont Schistes Lustrés, overlapping still more easterly but as yet poorly known units of the Austroalpine zone.

At the time of the Alpine paroxysm, both the flysch and the Piemont schists were detached from the bottom of their basin and glided gravitationally westward, parting from each other in the process. Thus came into being two of the previously mentioned large nappes, the Helminthoid Flysch nappe and the Piemont Schistes Lustrés nappe.

The Helminthoid Flysch nappe was the earlier one and included numerous slices of various origin, derived from the zones upon which it was thrust. These slices are of rather minor importance in the southern French Alps (Embrunais area). They are far more extensive in the Prealps where additional nappes make their appearance as a stack-up of several units, differing from one another both in origin and nature and quite distinct from the flysch nappe, although they are actually its lower "fellow-travelers."

The Pre-Alpine Nappes

On both sides of the Rhône valley above Lake Geneva, the Chablais Prealps and Romande Prealps (Figs. 2 and 3) are composed of a complex pile of cover nappes that originated from behind the Mont Blanc massif. At their base lie units derived from the Ultrahelvetic zone, followed by a unit probably derived from the Valais realm (nappe of the Niesen Flysch), and these in succession by others from the Subbriançonnais zone (nappe of the Préalpes Médianes Plastiques), the Briançonnais zone (nappe of the Préalpes Médianes Rigides), and probably the Piemont zone or still more internal realms (nappes of the Les Gets Flysch, Simme Flysch, and Helminthoid Flysch).

The thrust nature of the Prealps was already demonstrated at the end of the nineteenth century by Schardt (1893), who even then considered their motion to be of gravitational origin. Lugeon and Gagnebin (1941) also accepted this concept.

This hypothesis is supported by two kinds of data observable in the lowest units, the Ultrahelvetic and Préalpes Médianes nappes: (1) diverticulation phenomena and (2) traction phenomena.

Diverticulation Phenomena

This mechanism was first described and named by Lugeon (1943), for the Ultrahelvetic nappes at the base of the Prealpine nappes, chiefly in the Romande Prealps. The Ultrahelvetic nappes are made up of the sediments filling the deepest and more internal parts of the external Alpine basin. This material was separated into slices which slid successively off one another toward the exterior of the range. Thus every slice is made up of only part of the normal succession of the original sequence and the more recent this part, the lower the location of the nappe in the stack-up. With much simplification,[2] we see from top to bottom (Fig. 4):

4. Bex Laubhorn nappe (Triassic and Liassic).
3. Arveyres nappe (Dogger and flysch).
2. Anzeinde nappe (Malm and Lower to mid-Cretaceous).
1. Plaine Morte nappe (Upper Cretaceous and flysch).

---

[2] For more details, see Badoux (1963).

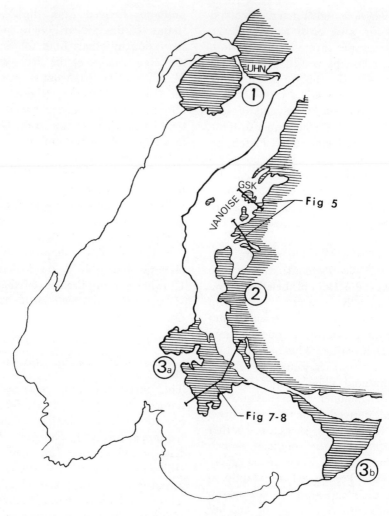

FIG. 2 Location of the studied area. (1) Prealps (UHN: Ultrahelvetic nappes); (2) Piemont Schistes Lustrés nappes (GSK: Grande Sassière klippe); (3) Helminthoid Flysch nappes; (3a) Embrunais area; (3b) Maritime French-Italian Alps.

In short, the highest slice glided down first and arrived first in its final position, maintaining its normal succession of strata in right-side-up position. Then a second slice glided down and covered the first in the same manner, and so on.

Only gravitational gliding can explain such a structure, all the more so since every slice, being only a few hundred meters thick and having an incompetent marly facies, would have been unable to transmit any sort of push from behind.

Traction Phenomena

According to their tectonic behavior the units overlying the Ultrahelvetic nappes, notably the nappes of the Préalpes Médianes, are classically subdivided into the Médianes Plastiques (derived from the Subbriançonnais zone) and Médianes Rigides (derived from the Briançonnais zone). The former are made up of a marly sequence, 1.5–2 km thick, which was easily folded above a base of gypsiferous Triassic. The Médianes Rigides

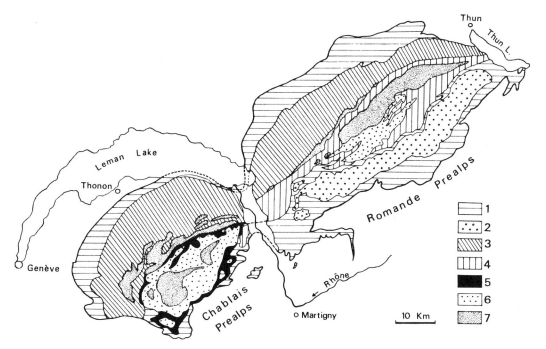

FIG. 3 Structural map of the Prealps. Units shown are, from bottom to top: (1) Ultrahelvetic units; (2) Niesen Flysch nappe; (3) nappe of the Préalpes Médianes Plastiques; (4) nappes of the Préalpes Médianes Rigides; (5) lens-bearing flysch nappe (probably Médianes); (6) Breccia nappe; (7) Simme Flysch nappe (and lateral equivalents).

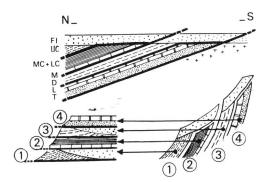

FIG. 4 The Ultrahelvetic sequence and its slicing off into nappes (diverticulation phenomenon). (1) Plaine Morte nappe (Upper Cretaceous and flysch); (2) Anzeinde nappe (Malm and Lower and Middle Cretaceous); (3) Arveyre nappe (Dogger and flysch); (4) Bex-Laubhorn nappe (Triassic and Liassic).

are made up of a slab of dolomitic Triassic and Upper Jurassic limestones and are endowed with a tectonic style commonly regarded as being much more rigid and brittle. This concept was largely developed and extended by Lugeon and Gagnebin (1941), who described the fracturing of the Médianes Rigides slab into little-folded or unfolded blocks, enveloped by a marly matrix of mainly Ultrahelvetic Flysch (e.g., see Lugeon and Gagnebin, 1941, Figs. 3, 5, 6, and 8). These authors concluded that such features were best explained by traction and that this traction was clearly due to gravity. However, things are not quite so clear-cut. First, the separation of the slab into blocks could also be ascribed simply to erosion acting upon a more or less jointed and cracked floor just before deformation took place; disjointing of such already fissured material could give rise to disharmonic structures such as those described by Lugeon and Gagnebin.

On the other hand, it may well be that gravity was not the only traction agent. Traction may have resulted also from stretching of the Médianes Rigides under an overlying unit, itself horizontally in motion toward the outer part of the Alps, and later eroded away (e.g., the Simme nappe).

Finally, recent studies performed in the Romande Prealps (e.g., Lonfat, 1965) do not appear to support everywhere the postulated existence of disjointed and floating blocks. In several places they invoke the existence of synclines and anticlines, differing little from those of the Médianes Plastiques. Should these conclusions be extended in the future to the whole Médianes Rigides unit, Lugeon and Gagnebin's argument would completely disappear.

However, another facet of these phenomena was recently brought to light by a study of the Préalpes Médianes Plastiques (Badoux and Mercanton, 1962). Aiming at a paleogeographic reconstruction, these authors tried to unfold the Chablais Médianes Plastiques folds so as to obtain the initial width of the basin along a selected cross section. Their first attempt was made by using a curvometer, but they also checked the result in another way. Knowing the average original thickness of a given stratigraphic unit (the Malm limestone) as observed in its less folded parts, and measuring its surface area on the cross section, a mere division gives the initial width. The results were quite different, the curvometer yielding a 40 percent shortening against 10 to 20 percent by the surface area method (see Fig. 10 in Lemoine, 1973). We may deduct from this that there was appreciable stretching during the gliding of the nappe. In the case of deeply buried nappes, such stretching might be ascribed to vertical compression, but we are here at a very shallow level, for these nappes glided into the brackish or lacustrine basin of the Oligocene Swiss molasse and do not exhibit regular or large-scale cleavage. In other words, the units moved superficially, and it is therefore perfectly conceivable that stretching occurred together with the folding of the strata under their own weight. Nevertheless, some doubt persists, since stretching might equally well have been caused by traction due to an overriding unit, now wholly eroded away.

In summary, it appears that only diverticulation of the Ultrahelvetic units cannot be explained without gravitational gliding. The other features observed in the Préalpes Médianes and interpreted as the result of traction phenomena, provide permissive evidence, but not proof, of gravity gliding.

The Piemont Schistes Lustrés Nappe

As previously stated, part of the contents of the eugeosynclinal Alpine trough was expelled outward, probably after undergoing a beginning of schistosity and metamorphism. Various considerations indicate that metamorphism was completed after the emplacement of this nappe over its substratum. Thus the recrystallization and schistosity implied by the term "Schistes Lustrés" do not prove emplacement of the nappe at a low structural level and with a "Penninic" style linked to compressive stress.[3] On the contrary, in the Vanoise area (Fig. 2) where the Schistes Lustrés nappe shows its most extensive and spectacular development, it exhibits features suggesting emplacement by gravity tectonics.

First, at the end of the Eocene or the beginning of the Oligocene, the nappe overrode an eroded terrain in the Briançonnais zone. The sedimentary cover of this zone emerged at the end of the Eocene and was then folded, eroded, and deeply dissected. When the incoming nappe flowed over it, it molded itself into the residual relief with the aid of a huge mass of gypsum which it

---

[3] This is not to say, of course, that there is necessarily a contradiction between low-level Penninic style and "gravitational tectonics" in Van Bemmelen's broad sense of that term. However, such a relationship is quite indirect, and in this paper we do not apply the term gravity tectonics to deformational mechanisms of that type.

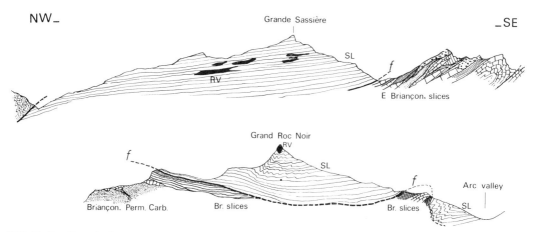

FIG. 5 La Grande Sassière massif and Grand Roc Noir massif; Piemont Schistes Lustrés klippen (s.l.) upon Briançonnais slices (Vanoise). RV = ophiolites. The unfolded disposition of the Piemont schists is probably due to the traction of the gliding mass by gravity. Note also some small disturbances due to the later backthrusting.

carried at its base and which filled the depressions.[4] But the nappe also tore up a few blocks of the Briançonnais cover, which are now enveloped in the gypsum as large, tectonically isolated masses ("blocs-klippes" of French authors). Where gypsum was absent, such as north of the Val d'Isère, the frictional effect was stronger and exerted instead a planing or truncating action quite different from the ductile stretching of the lower structural levels.

Thus the first conclusion is that the Schistes Lustrés nappe was emplaced at a very superficial level and was, in fact, epiglyptic.

Furthermore, from observations of some of the large masses of Schistes Lustrés overlying the Vanoise massif, such as the klippe of La Grande Sassière, north of Val d'Isère, or the Grand Roc Noir massif near Lanslebourg (Fig. 5), it appears that the layers are not folded at all, or only very gently so. This is an indication (though not a proof) of emplacement by traction and not by a push from the rear. Once more, at this ground level, one can hardly imagine any force, other than gravity, that could exert such a pulling action.

By the end of this first paroxysmal event, that is, by the end of the middle Oligocene, the masses of Schistes Lustrés underwent eastward "backthrusting" (*rétrocharriage* of French authors). This motion is not solely of gravitational origin: it most likely accompanies an underthrusting (*sous-charriage*) of the underlying Piemont basement of the Gran Paradiso massif below the Briançonnais basement of Vanoise-Ambin. Such a motion is indicated by microfolding and incipient fracture cleavage. But what is important to point out is that this motion, even though it is not a simple gravity sliding on a slope, is the response of an inert mass, subject only to gravity, to a more deeply seated orogenic phenomenon. Such an explanation is in reasonable agreement with the fact that the fold axes related to the backthrusting exhibit rather random orientations (Ellenberger, 1958).

---

[4] The paleogeographic origin of this gypsum is still not very well known. Probably it is of Carnian or Norian age and belongs to the inner Briançonnais realm, and it was carried away together with the Piemont Schists nappe when the latter began to override the Briançonnais zone. Discussion of this problem lies outside the scope of the present paper.

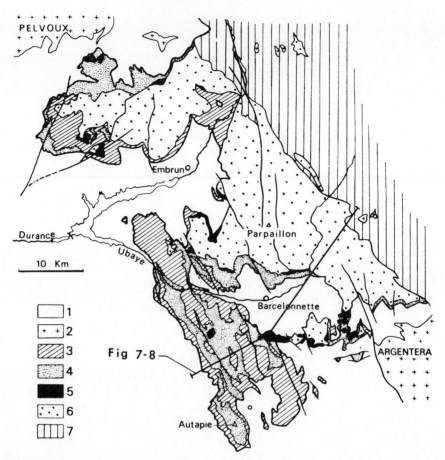

FIG. 6 Structural map of the Embrunais nappes. (1) External zone; (2) external crystalline massifs; (3) Subbriançonnais slices; (4) Autapie Flysch nappe; (5) Basal slices of the Parpaillon nappe; (6) Parpaillon Flysch nappe; (7) Briançonnais zone.

The Embrunais Nappes

In the Western Alps, the Embrunais-Ubaye area (Fig. 2) may be considered as a true showcase of nappe tectonics. It is there that the first large Alpine nappes were described by the end of the last century. More recently the Embrunais Flysch was used as a model for Gignoux' "gravity flow" theory (1948). Even though his flow concept is now out of date, there is no doubt that the kinematics of the Embrunais-Ubaye nappes results in its main features from a gravitational movement, and that this is in all cases true for the final emplacement of the nappes.

General Structure

Nappes are preserved in a wide gap between the external crystalline massifs: Pelvoux to the north and Argentera to the south (Fig. 6). The crystalline basement is here hidden below a Triassic to Eocene sedimentary cover 7 km thick despite the absence of lower and middle Eocene. The thrust is approximately 40 km in extent, as shown by the window of Barcelonnette in the Ubaye valley and by the half-window of Embrun in the Durance valley.

The allochthonous units may be separated into two groups. The lower group is made up of numerous Subbriançonnais slices arranged in an irregular pattern, with a series of varied facies ranging from Upper Triassic to upper Eocene. These slices are wrapped up in a complex of Upper Cretaceous flysch units of internal paleogeographic origin (Piemont to Austroalpine zones) which form the Autapie nappe (Kerckhove, 1969). One of these units exhibits a characteristic Helminthoid calcareous facies. Various kinds of evidence show that during an early phase in the late Eocene the Autapie nappe came to rest upon the Subbriançonnais area, where no thrusting had as yet occurred. The details of this nappe are described below. The Autapie nappe thus formed a pseudo-cover of the Subbriançonnais units. Together with them it drifted away and was folded in Oligocene time. The Autapie nappe is preserved today in nappe synclinoria.

The upper group of allochthonous units is made up of the large nappe of typical Embrunais Helminthoid Flysch, or Parpaillon nappe, separated from its original Piemont area at the level of a thick marly sequence associated with variegated and manganiferous slates, the so-called basal complex of probably Cenomanian to Turonian age. The overlying Upper Cretaceous Helminthoid Flysch may locally grade into a sandstone facies, the Embrunais sandstone.

The Parpaillon nappe exhibits large structures one to several kilometers in size and comprised of recumbent folds and thrusts; in addition, there are numerous folds on a scale of hundreds of meters with variously oriented axes, whose interpretation is still not clear. The basal thrust of the nappe contains numerous slices of various sizes, types, and origins which in some places form small mountains, such as south of Barcelonnette.

The emplacement of the Parpaillon nappe and its basal slices probably occurred in the early Miocene, after a folding phase involving the autochthonous formations and the overlying Subbriançonnais units with the Autapie nappe as a pseudo-cover.

To summarize, the structure of the Embrunais-Ubaye area is the result of both imbrication and superposition of three nappes, including, in sequence of development, the Autapie, Subbriançonnais, and Parpaillon nappes. These are described below.

Autapie Nappe

The Upper Cretaceous to Eocene flysch of this nappe conformably overlies the black, pelitic Upper Eocene flysch (*Flysch Noir*) of the underlying Subbriançonnais sequence. This is the reason why the Autapie Flysch was long confused with the Subbriançonnais flysch. Actually, a chaotic horizon of shales containing blocks of sandstone and limestone tens of centimeters in diameter can locally be observed very clearly at the contact. These *schistes à blocs* are up to 100 m thick. Their matrix yields a redeposited microfauna of late Eocene age, but the blocks are rubble from an Upper Cretaceous to Paleocene calcareous flysch, doubtless the Helminthoid Flysch of the Autapie nappe or its lateral equivalents. Thus the *schistes à blocs* probably formed as an "olistostrome," that is, a layer deposited by submarine avalanches or rockfalls. These originated from the front of the Autapie nappe and glided into a basin where nummulitic shales were being deposited, pushing ahead of them piles of blocks that covered progressively the previously formed olistostrome (Fig. 7).

The Autapie nappe glided as far as the external zone, where the upper Eocene is also the last-deposited member. The age of the *schistes à blocs* in the area where the Subbriançonnais structural units are the substratum of the Autapie nappe is apparently the same as in the external zone, where this substratum is autochthonous. Since these two areas were about 100 km apart, the gliding must have occurred rapidly, almost instantaneously on a geologic time scale. This phenomenon implies the existence of a surface sloping toward the center of the residual marine basin, which was probably also migrating in the same direction.

Considering these features, as well as the fact that the Autapie nappe is thin as

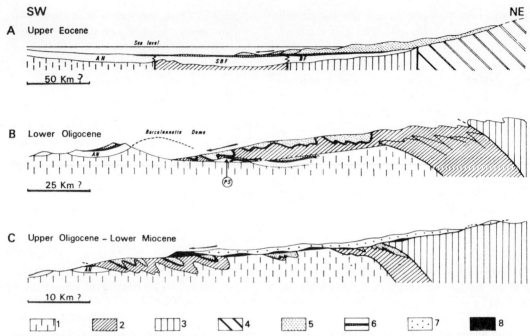

FIG. 7 Scheme of nappe emplacement mechanism in Embrunais-Ubaye area. (1) Autochthonous realm = external alpine zone; (2) Subbriançonnais nappe and slices; (3) Briançonnais realm and nappes; (4) Piemont and further realms; (5) Autapie nappe; (6) olistostrome; (7) Parpaillon nappe; (8) Basal slices of Parpaillon nappe; AN = autochthonous nummulitic cover; SBF = Subbriançonnais Flysch; BF = Briançonnais Flysch; PS = parautochthonous slice (morphotectonic).

compared to its width (approximately 500 m and 100 km, respectively), we can only conclude that the emplacement of this nappe was gravitational in nature.

Subbriançonnais Nappes

These nappes vary greatly in their stratigraphic sequences and are now cut up into numerous imbrications of all sizes. After an early phase of folding in the beginning of the Eocene, the Subbriançonnais Flysch was deposited: it was followed by the arrival of the submarine Autapie nappe (Fig. 7). Then, Subbriançonnais material, unstuck by thrusting, moved as far as the external zone in the Embrunais depression. This part of the external zone had previously been deeply eroded as far down as the Dogger (Middle Jurassic) and was surrounded by a topography composed of autochthonous Upper Cretaceous and nummulite-bearing strata. Hence the area exhibited a relief inversion phenomenon, for it was a dissected dome subsequently turned into a depression into which the nappes moved.

Just as in the case of the Schistes Lustrés nappe, this setup provides a clear example of epiglyptic thrusting, with its morphotectonic phenomena (e.g., the formation of parautochthonous slices from residual outliers), its lubricating basal level (Triassic gypsum), and its emplacement into a morphologic basin, created in Oligocene time between the old Pelvoux and Argentera massifs. The cleavage that occurs in these Subbriançonnais units was caused by subsequent deformation, which also affected the autochthonous rocks and the thrust surface.

All these facts suggest a gravitational emplacement of these units, at least for the Embrunais-Ubaye area.

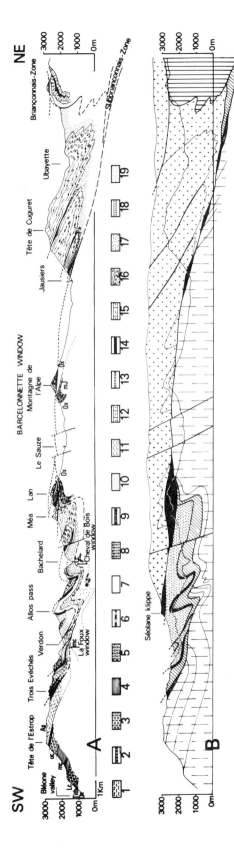

FIG. 8. (A) Present Embrunais-Ubaye cross section, including the Barcelonnette window. (1–8) "Autochthonous basement": (1) Grès d'Annot (upper Eocene); (2) middle to upper Eocene shales and limestone; (3) Upper Cretaceous limestones and limestone; (4) Middle Cretaceous shales; (5) Lower Cretaceous limestones and shales; (6) Tithonian limestones; (7) "Terres Noires" (Upper Jurassic shales); (8) Middle Jurassic limestones; (9) olistostromes (upper Eocene in autochthonous and Subbriançonnais realms). (10–15) "Subbriançonnais and Briançonnais nappes, basal slices of the Parpaillon nappe": (10) upper Eocene flysch or similar formations; (11) Upper Cretaceous planktonic limestones; (12) Upper Jurassic reef limestone (Seolane Subbriançonnais realm); (13) Upper Triassic and Lower Jurassic basal formations (Subbriançonnais units); (15) Middle Triassic Briançonnais limestones. (16–19) "Helminthoid Flysch nappes": (16) Helminthoid Flysch (Autapie and Parpaillon nappes); (17) "Dissociated flysch" of the Autapie nappe; (18) Embrunais Sandstone of the Parpaillon nappe; (19) Shaly basal formation of Parpaillon nappe. (B) The same section, before Quaternary erosion. More schematic, same symbols as in Fig. 8A (the present relief is shown by a line as in A).

Parpaillon Nappe

This nappe exhibits the features of both the Autapie and the Subbriançonnais units (Fig. 7). It is a very broad but relatively thin tectonic unit (approximately 1000 m) resting unconformably on all of the preceding units, which had been folded and eroded at the end of Oligocene time. The basal imbrications range in size from hundreds to thousands of meters. Their remarkable discontinuity indicates probable pulling motions, as previously described in the case of the Prealps.

Furthermore, these slices, which include Briançonnais material, are detached outliers indicative of the highly superficial gravity gliding of the overlying nappe from the Briançonnais zone toward the Embrunais-Ubaye depression. We have here a clear example of tectonic "scraping" of the Briançonnais zone where only scarce klippen of Helminthoid Flysch (with its "Embrunais sandstone" facies) remain.

On the other hand, basal truncation (*troncatures basales* of French authors) of the folds of the Parpaillon nappe, clearly observable on both sides of the Durance valley, indicates that gliding occurred after the rocks had been folded and cut off at the base. Folds were accentuated or flattened during gliding, and cleavage developed parallel to the horizontal axial planes, but apparently no new important folds were formed at that time. Hence, although we accept gliding by gravity as a probable process, we can no longer accept the concept of "gravity flow" of Gignoux (1948) in this area.

To summarize the problem of the Embrunais nappes (Fig. 8), we may conclude that at least part of the movement of all three units was by gravitational gliding, even if the beginning of the motion involves other causes, including in particular contraction of the basement. The gravitational gliding itself is controlled by the existence of depressions within the boundaries of the tectogenetic area. In the case of the Autapie nappe, the depression was a marine basin, probably a narrow arm of sea between the Subalpine continent to the west and the Piemont relief to the east and connected with the southern Tethys and the northern seas. Within this marine basin we may hope to find the same submarine thrusts as below the Autapie nappe. In fact, Autapie-like flysch has been observed in the Maritime French-Italian Alps to the south and the Prealps to the north.

In the case of the epiglyptic Subbriançonnais and Parpaillon nappes the depression is a continental one, caused by subsidence of the autochthonous basement in a limited area. This subsidence is also evident in the Argentera massif area of the Maritime French-Italian Alps and in the Helvetic molasse basin, which was rapidly sinking at the time the Prealpine nappes were emplaced by gravitational gliding.

References

Badoux, H., 1963, Les unités ultrahelvétiques de la zone des cols: *Eclog. Geol. Helv.*, v. 56, p. 113.

———, and Mercanton, Ch., 1962, Essai sur l'évolution tectonique des Préalpes médianes du Chablais: *Eclog. Geol. Helv.*, v. 55, p. 135–188.

Debelmas, J., and Lemoine, M., 1970, The Western Alps: Paleogeography and structure: *Earth Sci. Rev.*, v. 6, p. 221–256.

Ellenberger, F., 1958, Etude géologique du Pays de Vanoise: *Mém. Expl. Carte Géol. Fr.*, 561 p.

Gignoux, M., 1948, La tectonique d'écoulement par gravité et la structure des Alpes: *Bull. Soc. Géol. Fr.*, v. 18, p. 739–761.

Kerckhove, Cl., 1969, La "zone du Flysch" dans les nappes de l'Embrunais Ubaye: *Géol. Alpine (Grenoble)*, v. 45, p. 5–204.

Lemoine, M., 1973, About gravity gliding tectonics in the Western Alps (this volume).

Lonfat, F., 1965, Géologie de la partie centrale des Rochers de Château d'Oex: *Matér. Carte Géol. Suisse*, n.s., 120 p.

Lugeon, M., 1943, Une nouvelle hypothèse tectonique: la diverticulation: *Bull. Soc. Vaud. Sc. Nat.*, v. 62, no. 260, p. 301–303.

———, and E. Gagnebin, 1941, Observations et vues nouvelles sur la géologie des Préalpes Romandes: *Mém. Soc. Vaud. Sc. Nat.*, v. 47, p. 1–90.

Schardt, H., 1893, Sur l'origine des Préalpes Romandes; *Eclog. Geol. Helv.*, v. 4, p. 129–142.

Trümpy, R., 1960, Paleotectonic evolution of the Central and Western Alps: *Geol. Soc. Am. Bull.*, v. 71, p. 843–908.

M. LEMOINE

Department of Geology
Ecole Nationale Supérieure des Mines de Paris
Paris, France

About Gravity Gliding Tectonics in the Western Alps[1]

The Western Alps may perhaps be called the "fatherland" of Alpine gravity tectonics (for Prealpine nappes, see Schardt, 1898). This theory, which grants to gravity forces a predominant or exclusive part in the mechanism of folding and overthrusting, is a genetic one; it may therefore be accepted or rejected, but in any case requires discussion. In this paper, only a few characteristic Alpine examples will be discussed (see also Debelmas and Kerckhove, 1973) in order to set forth the reasons for adopting the theory and to see how it fits with both geological and rock mechanics data. Since the main examples in the Western Alps deal with detached cover units, we shall restrict our subject to a limited part of the theory, the one that Van Bemmelen (1955) called "epidermal" gravity tectonics.

The structure of the Western Alps arc will not be described here (see Trümpy, 1960; Debelmas and Lemoine, 1970). Let us merely recall that three main palaeogeographic and structural realms may be distinguished (Fig. 1a): (1) the External or Helvetic (and Ultrahelvetic) realm; (2) the Penninic realm, divided into Valais eugeosyncline (external), Subbriançonnais zone, Briançonnais zone (ridge), and Piemont eugeosyncline (internal); and (3) the Austroalpine and South Alpine realm.

In nearly all these subdivisions the rock sequence can be separated into (1) a basement (pre-Triassic, mainly crystalline rocks)

---

[1] I am greatly indebted to Professor R. Scholten of The Pennsylvania State University and to Dr. P. Ch. de Graciansky of the Ecole des Mines de Paris, for language corrections, for critical reading of the manuscript, and especially for helpful suggestions.

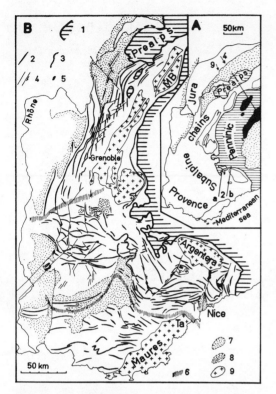

FIG. 1 (A) Simplified structural map of the Western Alps, showing the external zone (Subalpine Chains: see Fig. 1B), the Penninic nappes, and the Austroalpine nappes (shown in black: Dent-Blanche nappe and "root-zone"). Dotted: main molassic basins. Location of sections of Figs. 2, 4, and 9 is indicated. (B) Structural sketch-map of the Subalpine Chains. (1) Penninic nappes; (2) axes of main anticlines; (3) thrust faults; (4) faults, wrench-faults; (5) diapiric structures (Upper Triassic evaporites); (6) approximate boundary between (a) the Northern Subalpine Chains (to the north), (b) the Southern Subalpine Chains (middle), and (c) the Provence Chains (to the south); (7) Oligocene-Miocene-Pliocene molasse; (8) unconformable Upper Cretaceous of Dévoluy; (9) external crystalline massifs (pre-Triassic basement) (MB = Mont Blanc; Ta = Tanneron).

and (2) a sedimentary cover (Mesozoic and Cenozoic, mainly marine). These two groups may show very different tectonic styles[2]; moreover, they are nearly always separated by a *décollement* or "lubricating" horizon of evaporitic and argillitic Triassic rocks. This therefore gives rise to *cover folds* (as those of the Subalpine chains in the external domain) and *cover nappes* (as those of the Prealps). Cover folds or nappes may also occur by disharmonic décollement or shearing within the cover along incompetent argillitic or serpentinite layers.

First Example—The Subalpine Chains

In the external part of the French Alps the Subalpine Chains are low mountains, built up of folded Mesozoic and Cenozoic sedimentary rocks. Their pre-Triassic basement lies at depths ranging from -1000 to -4000 m, but it rises to the east and crops out in the so-called External Crystalline Massifs (Mont Blanc, Belledonne, Pelvoux, Argentera), reaching an altitude of $+2000$ to $+5000$ m. One or two Triassic evaporitic–argillitic horizons permit the general décollement of the cover, which is folded and in places thrust independently of its basement.

According to the age of folding, two main subdivisions may be distinguished (Fig. 1b): (1) the Northern Subalpine Chains, which continue northward into the Jura folds on their external side and into the Helvetic nappes on their internal side, and which show one single set of nearly N–S folds of Neogene age; (2) the Southern Subalpine Chains, which show the intersection of two fold systems: (*a*) a first group of nearly E–W folds, Late Cretaceous to middle Eocene in age ("Pyrenean–Provençal" folds)[3] that were refolded during the Oligocene–Neogene Al-

---

[2] At least in the upper tectonic levels of the chain. In deeper parts, the case is rather different: here metamorphism and migmatization are associated with deformation.

[3] These folds also occur farther to the south, in the Provence chains (Fig. 1b), but this region does not belong to the Alps proper, because no further folding occurred.

pine deformations; and (b) a second group of N–S to NW–SE "Alpine" folds and thrusts that interfere with the first group. The latter are mainly of Neogene age, similar to those of the Northern Subalpine Chains of which they are more or less the extension.

It is not possible to describe the whole area in this paper. We will discuss only two sections across the Southern Subalpine Chains in the Maritime Alps, between the Argentera Massif in the north and the Provençal domain in the south.

These sections (Fig. 2) cut across nearly E–W folds and thrusts known to be for the most part as young as Neogene (post-Oligocene, and post-Miocene in the case of some thrusts); nevertheless, they belong to the E–W system of earlier Provençal folds. In fact, some of them at least can be proved to have started during the Provençal folding on the basis of unconformities between Eocene beds and already eroded E–W folds (e.g., Fig. 2b).

The basement of the folded cover appears here in the Argentera Massif to the north, in the Permian Dôme de Barrôt in the middle, and in the Provençal Massif of Maures–Esterel–Tanneron to the south (Fig. 2b). This basement appears to be at least slightly deformed; in section b its mean altitude grows from a few hundred meters in the south to 3000–4000 m in the north, but in section a it increases by only a few hundred meters.

The décollement of the post-Triassic cover, accompanied by shearing or "truncation" at the base of the thrusts, can be deduced from direct field observation, especially around the Dôme de Barrôt (Fig. 2b), but also from the shape of the folds, which excludes the participation of pre-Triassic formations in the cores of anticlines and at the base of the thrusts.

The southern vergence of folds and thrusts is almost general, a feature that suggests the direction of movement of the gliding cover. Moreover, a disharmonic thrust, the so-called Roya "intercutaneous thrust" (Fallot and Faure-Muret, 1949b), involving Upper Jurassic and Lower Cretaceous rocks but overlain by nearly flat Upper Cretaceous and Tertiary (Fig. 2b), leads to the same conclusion.

All these observations are emphasized by some authors (e.g., Fallot and Faure-Muret, 1949a), who propose gravity gliding of the cover upon an inclined basement. Others (e.g., Goguel, 1949) emphasize that alternative explanations may be put forward, especially pushing at the rear of the cover (Fig. 3), either by the Argentera Crystalline Massif or by the Penninic nappes which crop out to the north and probably existed also farther south before erosion.

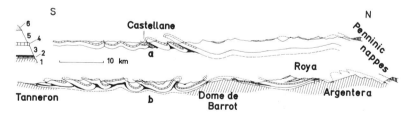

FIG. 2 Two sections across the Southern Subalpine Chains in the Maritime Alps (after Bordet, Fallot, Goguel) (see location on Fig. 1A). (1) Basement (mainly pre-Carboniferous crystalline rocks in Tanneron and Argentera massifs, Permian sedimentary rocks in the Dôme de Barrôt); (2) Triassic (evaporite-bearing Upper Triassic in black); (3) Jurassic limestones and marls; (4) Upper Jurassic (Kimmeridgian-Portlandian) limestones; (5) Cretaceous; (6) Tertiary. These sections show the general décollement of the post-Triassic cover, especially section b, which illustrates the complete tectonic independence between the Permo-Werfenian basement (Dôme de Barrôt), the Middle Triassic limestones (recumbent folds between two evaporite horizons), and the post-Triassic cover.

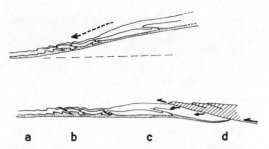

FIG. 3 Theoretical diagrams showing possible explanations of cover folding in the Southern Subalpine Chains (Maritime Alps: see Fig. 2). Upper schematic section shows free gravity gliding. Lower section shows pushing at the rear by Penninic nappes (a third possibility would be a basement shortening below the folded cover). (a) Southern Provençal foreland; (b) frontal folds (with "frontal damp down"); (c) rear part, much thicker series, almost unfolded; (d) area now covered by Penninic nappes. Dotted: Triassic at the base of the folded cover. Hatched: Penninic nappes.

One of the arguments advanced in favor of gravity is based on the fact that folding is more intense and the folds are more compressed in the south, whereas only slight and broad undulations can be seen in the northern area. This is a kind of "frontal damp-down" (*amortissement frontal* of Fallot and Faure-Muret, 1949a) or "tamping" (*refoulement* of Gèze, 1962), which may at first sight correspond to the case of a free gliding cover blocked in front, so that it suffers frontal shortening only. This avoids the difficulty of imagining the transmission of a rear compressive force without folding of the rear part of the cover. But one may also agree with Goguel (1949), who points out that the southern, strongly folded cover is only about 2 km thick, whereas the slightly folded northern cover is twice as thick and would therefore require much greater energy to fold it than to cause mere slipping along the basal incompetent evaporite-bearing Triassic. In that case, compressive forces applied at the rear may be transmitted by the cover and induce frontal folding only, in addition to the general gliding.

Finally, we must point out that there is also a slope problem (see also below). The higher the angle required for free gliding, the higher the top of the slope. In the case of a mere 3° slope, the top has to reach an altitude of 4–5 km, which happens to be the present altitude of the Argentera Massif. The folding, however, began very early (Late Cretaceous to middle Eocene: Provençal folding), whereas the uplift of the Argentera Massif very likely occurred only in Oligocene–Miocene–Pliocene time. Moreover, the folds occur also in transverse sections where the crystalline massifs do not occur (see Figs. 1b and 2a).

In conclusion, we may say that these Subalpine folds were possibly in part gravitational in origin, especially at the end of their formation, but that there is no need, let alone evidence, to explain them exclusively by gravity gliding.

Second Example—The Prealps

The Prealps, where gravity tectonics was proposed as early as 1898 by H. Schardt, are large tectonic outliers resting on the Helvetic and Molasse zones (Figs. 1 and 4). They build two main mountainous regions, the Chablais in France and the Préalpes Romandes in Switzerland, where three main groups of cover nappes can be distinguished (Figs. 8 and 9):

1. *A basal group*, the Ultrahelvetic nappes.
2. *A middle or main group*, of Penninic origin, comprising: (a) the Médianes Plastiques or simply Plastiques nappe, a 1.5–2-km-thick limestone-marl series derived from the Subbriançonnais zone and affected by folds of the concentric type; (b) the Médianes Rigides or Rigides nappe, derived from the Briançonnais middle-Penninic platform, and consisting of 1–1.5-km-thick, rigid (unfolded) limestone and dolomite plates; (c) the Brèche (Breccia) nappe, derived from the bound-

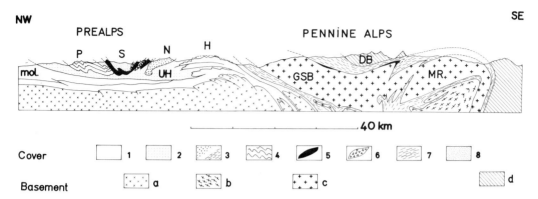

FIG. 4 Schematic section of the Western Alps across the Pennine Alps and the Préalpes Romandes (after Bearth, Gagnebin, Lombard, Lugeon, Tercier, and others). Cover units: (1) autochthonous and Helvetic (H, Helvetic nappes; mol, molasse); (2) Ultrahelvetic (UH); (3) Valais (Niesen nappe, N, in the Prealps; Valais-Simplon Schistes Lustrés below GSB basement nappe); (4) Subbriançonnais (P, Plastiques nappe); (5) Briançonnais Rigides nappe in the Prealps, and cover remnants on the GSB basement nappe); (6) Brèche nappe; (7) Piemont Schistes Lustrés: overthrust on the GSB nappe, they are tectonically substituted for the previous Subbriançonnais-Briançonnais cover, which lies now in the Prealps ("cover substitution" of Ellenberger); (8) Simme nappe (S). Basement units: (a) autochthonous and external crystalline massifs; (b) lower Penninic (Simplon-Ticino nappes); (c) upper Penninic (GSB, Grand Saint Bernard nappe; MR, Monte Rosa nappe); (d) Austroalpine (DB, Dent Blanche nappe).

ary between the Briançonnais and Piemont zones and consisting of a mildly folded limestone-shale series, 1–1.5 km thick, with numerous intercalations of calcareous breccias.

3. *An upper group* (Simme nappe *sensu lato*), resting on the Plastiques, Rigides, and Brèche nappes, and comprising mainly Upper Cretaceous flysch, in addition to ophiolite-bearing shale, the whole being of composite origin, partly Penninic (Piemont), partly South Alpine.

The Niesen nappe, of Valais origin, only represented in the Préalpes Romandes, is somewhat of a special case, and will not be dealt with here.

Considering only the overthrusting of the preceding three main groups and that of the Helvetic nappes, several successive stages can be recognized. In a somewhat simplified manner, we may distinguish (Fig. 5):

1. Overthrusting of the Simme nappe upon the not yet deformed realms of the future Plastiques, Rigides, and Brèche nappes (middle–late Eocene).
2. Overthrusting of the Ultrahelvetic nappes upon the Helvetic realm prior to thrusting in that realm (late Eocene–early Oligocene).
3. Main overthrusting of the Plastiques, Rigides, and Brèche nappes, carrying on their backs the previously emplaced Simme nappe, and dragging beneath them the Ultrahelvetic material they encountered during their displacement. The whole passed over the as yet unfolded Helvetic realm and arrived at the external border of the chain (Oligocene).
4. Folding and overthrusting of the Helvetic nappes, together with the previously diverticulated Ultrahelvetic nappes (Miocene).

The early submarine overthrusts of (1) and (2) above occurred in flysch basins and interrupted the flysch sedimentation. In part, at least, they correspond to the so-called diverticulation mechanism (see below).

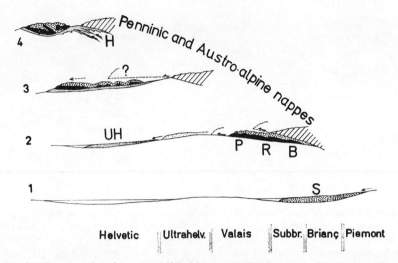

FIG. 5 Schematic diagram showing a possible kinematic development for the overthrusting of the Prealpine and Helvetic nappes. These theoretical sections are greatly simplified in order to show only the main aspects of a possible kinematic development. The present stage (following that of top diagram) is that of Fig. 4. (1) Upper Eocene: Overthrusting of the Simme nappe (S) upon the Subbriançonnais, Briançonnais and outer Piemont realms (mainly submarine, in flysch basins), very probably by gravity gliding. (2) Beginning of Oligocene: diverticulation of the Ultrahelvetic nappes (UH) upon the Helvetic realm (mainly submarine, in a flysch basin) by free gravity gliding (but perhaps induced by a push from the rear). Beginning of the décollement of the Plastiques (P), Rigides (R), and Brèche (B) nappes, carrying on their backs the Simme nappe. This décollement was very probably induced by the more internal Piemont Schistes Lustrés and Austroalpine Dent Blanche nappes (see Fig. 4). At this time began probably the "cover substitution" (Piemont Schistes Lustrés taking the place of the detached Subbriançonnais and Briançonnais covers: see Fig. 4). Cleavage of the rocks of some Prealpine nappes (mainly Simme and Brèche) may have been generated at this time. (3) Oligocene: The Prealpine nappes passed across the Helvetic realm. Their emplacement may have been at least in part by free gravity gliding, especially at the very end. (4) Miocene: Folding and overthrusting of the Helvetic nappes (H), together with the previously diverticulated Ultrahelvetic nappes, and probably under a certain overburden, accompanied by cleavage generation. The overthrusting of the Niesen nappe occurred probably at this time, but is not shown on this simplified diagram. Length of section 1 is about 200–300 km.

The Diverticulations

This term was first applied by Lugeon (1943) to the Ultrahelvetic nappes, which are at present folded together with the Helvetic nappes (Fig. 6). Simplifying briefly (for more details, see Badoux, 1963), we may distinguish three main Ultrahelvetic nappes. Each comprises a part of the stratigraphic sequence with normal polarity, but the lowest unit is made up of the younger part of the series, and the uppermost unit of the older. These small nappes are each only a few hundred meters thick and include mainly limestones and marls that could easily be folded. How-

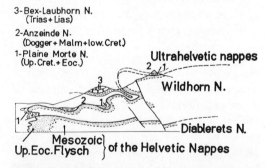

FIG. 6 Simplified section of the Ultrahelvetic nappes now resting upon the upper Helvetic Wildhorn nappe (after Lugeon). In Fig. 4 this section is located between letters H and N, with the same orientation.

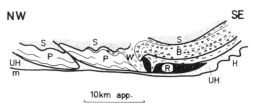

FIG. 7 Theoretical and simplified diagram of the diverticulation mechanism, as imagined by Lugeon (1943) to explain the overthrusting of the Ultrahelvetic nappes. (T) Upper Triassic (with evaporites); (L) Liassic; (D) Dogger; (M) Malm; (C_1) Lower Cretaceous; (C_2) Upper Cretaceous; (E) Eocene (flysch); (P) Plaine Morte nappe (C_2 + E); (A) Anzeinde and Arveyes nappes (D + M + C_1 + E); (B) Bex-Laubhorn nappe (T + L) (see also Fig. 6). 1, 2, 3 = successive stages of décollement and downslope gliding.

FIG. 9 Schematic section showing the geometric relations between the Prealpine nappes of Chablais (after Lugeon and Gagnebin, Badoux, and Caron). Autochthonous: (H) Helvetic folds; (m) molasse; (UH) Ultrahelvetic nappes; (P) Plastiques nappe; (R) Rigides nappe (disjointed blocks); (B) Brèche nappe; (S) flysch of the Simme nappe *sensu lato*; (W) Wildflysch (schistes à blocs et lentilles): olistostromes with large olistolites resting upon the Plastiques, Rigides, and Brèche nappes, below the Simme nappe.

ever, they were only folded in Miocene time, after their emplacement, and together with the underlying Helvetic units (Fig. 6). Hence their preliminary thrusting without folding despite their plasticity and thickness can hardly have been achieved by a push at the rear. Lugeon (1943) suggested that three slabs (*diverticules*) became successively detached along incompetent marly or evaporite horizons and slid to the bottom of the Helvetic flysch basin (Fig. 7).

A similar mechanism can be proposed for the different parts of the Simme nappe (*sensu lato*), which arrived almost unfolded upon the latest deposits (Eocene flysch) of the future Plastiques, Rigides, and Brèche nappes. Moreover, the Simme flysch deposits (Fig. 8) are separated from the underlying nappes by a certain thickness of olistostrome-like material (*schistes à blocs et lentilles* or "Wildflysch" of Caron, 1966), as shown in Fig. 9.

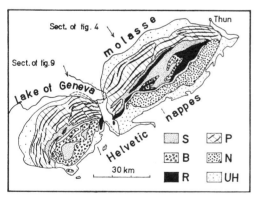

FIG. 8 Structural map of the Prealpine nappes of Chablais (south of Lake of Geneva) and Préalpes Romandes (between Geneva and Thun Lakes). (S) Simme nappe *sensu lato*; (B) Brèche nappe; (R) Rigides nappe; (P) Plastiques nappe (with main anticlines, thrusts, and faults); (N) Niesen nappe; (UH) Ultrahelvetic nappes. The Niesen and Plastiques nappes are continuous bodies, but the Rigides and Brèche nappes are represented by isolated blocks, which probably were disjointed before or during their overthrusting (see also Figs. 4 and 9).

Overthrusting of the Main Prealpine Nappes

This overthrusting started either by décollement (e.g., along the evaporite-bearing Triassic at the base of the Plastiques nappe) or by décollement and truncation (e.g., basal truncation of the isolated Rigides and Brèche plates). According to Schardt (1898) and Lugeon and Gagnebin (1941), the transport of these nappes occurred mainly or exclusively by the force of gravity acting on a suitable slope.[4] These authors apparently thought that these relatively thin nappes seem incapable of transmitting a compressive force, and also that one could not easily find a tectonic body responsible for applying this pushing force. In fact, for a long time, the Prealpine nappes were considered to be Austroalpine, that is, to belong to the uppermost Alpine units. We now know, however, that the Plastiques, Rigides, and Brèche nappes are Penninic in origin, and it is therefore possible that they were, at one time, overlain by Austroalpine crystalline nappes, such as the Dent Blanche nappe (see Fig. 5). Moreover, at least some of the Prealpine nappes, especially the Brèche and Simme nappes, show regional cleavage, and we may therefore suppose that they were deformed under a fairly thick load during some stage of their overthrusting. It is therefore very likely that the first stages of overthrusting occurred by compressive movements, only the end of it being possibly gravitational (Fig. 5).

Lugeon and Gagnebin offer also some presumed evidence for such gravitational gliding, emphasizing what we may call *traction tectonics*. For example, the Rigides and Brèche nappes are not continuous bodies throughout the entire Prealps, but are separated into large, unfolded blocks (Figs. 4, 8, and 9), which could have been disjointed during their overthrusting. In the opinion of these authors, this disjunction into blocks can be explained only by traction, not by compression, and traction can result only from the action of gravity acting on a free gliding slab.

Another observation, also leading to "traction tectonics," was described by Badoux and Mercanton (1962), in the Plastiques nappe of the Chablais massif. In an attempt to unfold the rocks for the purpose of palaeogeographic reconstruction, these authors used different methods. Employing the Malm limestone in a transverse section, they first used a curvometer. In a second method they divided the cross-sectional surface area of the Malm by its mean thickness in areas of little disturbance. The results were highly different because there are some parts where the Malm limestones are highly stretched, not only in overturned flanks, but also in both flanks of some recumbent folds. Figure 10 provides an attempt to visualize their results: 40–45 percent shortening by the curvometer method, but only 20–25 percent by the surface/thickness method. Similar results can be obtained in employing the total pile of sediments of the nappe. These authors interpret this to mean that the nappe has been first stretched (+45 percent), and thereafter folded (−45 percent), the final result being only a 20 percent shortening (approximate figures).

These conclusions seem very probable, but the question remains whether traction will, in effect, cause plastic stretching rather than rupture. Furthermore, one may raise the question whether stretching, and perhaps

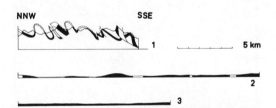

FIG. 10 Schematic representation of the result of unfolding for the folds of the Plastiques nappe in Chablais, as explained by Badoux and Mercanton (1962). The section is south of Lake of Geneva (see Figs. 8 and 9). The Malm limestones are shown in black. Section 1 shows the present stage (11 km), section 2 shows the result of unfolding with a curvometer (about 20 km), and section 3 shows the result of unfolding using the Malm surface thickness method (about 13–14 km). From stage 2 to stage 1 shortening by folding would be about 40–45 percent, but from 3 via 2 to 1 it is only 20–25 percent.

---

[4] A slope of 100–150-km length, or more. This figure will lead us to the problem of the altitude at the top of the slope (see below).

also disjunction into blocks, cannot be obtained as well under a thick overburden, as in a rolling-mill. Moreover, we are not sure that stretching did not occur also in a direction perpendicular to the section, that is, parallel to the axes of the folds. These folds are indeed arcuate within the nappe (Fig. 8), a feature suggesting longitudinal stretching.

Thus, in the final analysis, the Prealps, the "fatherland" of Alpine gravity gliding tectonics, may lead us or constrain us in certain cases to accept the theory, but they also offer us some difficulties. More difficulties will be encountered when dealing with rockmechanics.

The Reasons for Adoption of the Theory of Gravity Tectonics

The Alpine examples we have just discussed provide a fairly good selection of the reasons why some tectonicians choose gravity tectonics to explain various Alpine folds or overthrusts.

In some areas, the theory is adopted chiefly because any other suitable explanation cannot be found. Of course, the case of the Subalpine Chains is somewhat different, because alternative solutions do exist. In many other areas, however, both in the Alps and elsewhere, we feel "driven to the wall," and forced to explain the motion of nappes by gravity as the sole possibility. First, there are cases where a geological body capable of applying the compressive force, that is, of pushing at the rear edge of the nappe, cannot be found. In other cases, we face the problem of the transmission of this force: the thin overthrust slices (*diverticules*), made up of rather plastic material, were not folded, as they would of necessity be if pushed at the rear.

In addition, several other lines of evidence have been tentatively offered, especially the supposed results of "traction tectonics," that is, the disjunction of "rigid" nappes into separate blocks and the plastic stretching of the Plastiques nappe before its folding. Still other presumptions have at times been proposed. For example, one may point to the presence of olistolites of nappe material, now lying below the nappe, inside olistostromelike formations. Block-clays, olistostromes, and olistolites certainly moved by gravity, and this leads us to infer that the nappe that followed and covered them similarly might have moved by gravity (see, e.g., Sandulescu, 1967, who states that there is a genetic connection between the two processes).

Nevertheless, we have seen that such lines of evidence, even though they may be quite strong, are still open to discussion, and that they are *presumptions rather than evidence*.

Test of the Theory of Gravity Tectonics

In adopting gravity tectonics without further discussion, even if we feel that there is no other explanation, we indeed remain dissatisfied. Therefore we feel it necessary to examine all implications and consequences of the theory, and to test it according to two different but inseparable points of view:

1. The mechanical point of view: Are the known mechanical properties of the material involved consistent with the kinematic development we imagine?
2. The palaeogeographical and palaeomorphological point of view: Is the slope we are constrained to imagine geologically likely?

In fact, the behavior of geologists facing these problems has varied. Although some have experimented with scaled-down models, such models cannot be accepted as true evidence. Small slides, often catastrophic in nature, such as the Vaiont slide of 1963 in Italy (Selli and others, 1964) and the Chilean slides of 1960 (Tazieff, 1961) have been observed, but they are much smaller and possibly more rapid (see below) than true gliding of large nappes, with which they can perhaps not be compared.

In many cases, positive or negative criticism of the theory calls for rock mechanics data. Geologists often (unconsciously) feel that mathematical equations, curves, and figures given by physicists are an image of the truth (a kind of "fetishism"). Still we must remember that the mathematical tool is only applied to a *model*, not to the genuine geological body or phenomenon. The chosen model has to be as simple as possible to permit calculations, and this simplicity is fascinating. But even though a large modern computer may be used, the figures obtained should never be taken as an image of the truth. *The use of "exact sciences" can only help us to choose between "likely" or "unlikely," not between "true" and "false."*

The Slope Problem

Gravity gliding requires a slope. In most cases the initial slope can no longer be observed, having been deformed by subsequent crustal movement or destroyed by erosion. For example, crystalline massifs such as that of Mont Blanc, and, more generally, the Helvetic zone, were uplifted after the Prealps traveled across them.

We are therefore forced to imagine this slope, which should be characterized by its regularity and continuity, and consider not only the likelihood of its existence in accordance with the general palaeogeographical evolution, but more specifically its angle with the horizontal, its length, and therefore the altitude of its summit. All these aspects are to be discussed in the context of both palaeogeography and rock mechanics.

For example, in the case of the Prealps, the travel-distance of the Plastiques, Rigides, and Brèche nappes from their original Penninic basement to their present position was about 100–150 km. These nappes arrived at a very low altitude, because they interrupted the deposition of continental or brackish Oligocene sediments resting on marine beds. Therefore, if we imagine a single regular and continuous slope, their altitude of departure depends on the angle of the slope: 3–5 km for a 2° slope, 8–15 km for 5°, 17–25 km for 10°, more than 55 km for 30°, and so on. The minimum angle permitting gliding depends on rock mechanics considerations.

The Minimum Angle of Slope Permitting Gliding and the Choice of a Mechanical Model

If we wish to know the minimum angle consistent with gliding, we must consider the mechanical properties of the rocks and also know something about the tectonic mechanism involved. In other words, we have to choose a model, which must be a simple one, much simpler than the very complex natural phenomenon that is under discussion.

Several mechanical models can be taken into consideration.[5] They may be classified into two main categories: gliding may occur (1) by true friction of rock on rock or (2) by plastic deformation of a basal "lubricating" horizon (e.g., the Triassic evaporites).

1. Plastic Deformation of a Basal Lubricating Horizon. For a simple model of this kind, Goguel (1948) showed that the limit angle depends on the weight of the nappe, that is, on its thickness: the thicker the nappe, the lower the limit-angle of the slope. Choosing an elastic limit or yield point of 40 bars for evaporite-bearing Triassic, Goguel found that a 2-km-thick layer may glide on a 4–5° slope, a 4-km-thick layer on a 2° slope, and so forth. These figures, of course, are to be taken only as orders of magnitude, and are linked with the choice of a certain model. Hence we find that the Prealpine nappes ought to have departed from an altitude of 5–15 km. Therefore we have to imagine the rising of a great subaerial mountain range, which partly disappeared after the over-

---

[5] We shall see that different models may lead us to very different results!

thrusting and which must consist essentially of a rather permanent and regular slope, despite the fact that this slope will be immediately attacked by erosion (but see below). Finally, we may ask ourselves if this whole reconstruction coming out of our brain is reasonable! And the difficulty seems at first sight greater in the case of marly or argillaceous décollement horizons, as in the case of the Ultrahelvetic nappes, for if, in calculating our model, we use results from laboratory experiments, the value of the elastic limit of marls is five to ten times greater than for evaporite rocks. The limit angle therefore has to reach much higher values: 20° to 30° instead of 4° to 5°, for instance. We are inevitably drawn into an impossibility.

2. *The Friction Model*. In this case, as shown, for example, by Hubbert and Rubey (1959), the minimum angle of the slope will be the friction angle, a physical constant of materials and rocks. For most common rocks this angle has an approximate value of 30°, which would mean that the Prealps departed from an altitude of 55 km or more! We must also note here that this simple friction model leads us to a limit-angle that does not depend on the thickness of the nappe, contrary to the preceding model. We see here that the choice of a model already introduces a part of the conclusions we are expecting from its use. Hsü's (1969) pure friction model is somewhat different: the limit angle depends on the weight of the nappe, and the figures obtained are higher.

Concluding this preliminary review of the two main mechanical models, we see that models that are too simple lead us into impossibilities, contradictions, and paradoxes. We shall see further that a number of additions or improvements could perhaps bring us to a more satisfying conclusion.

Length, Regularity, and Permanence of the Slope

Our reasoning concerning the altitude of departure of gliding nappes, that is, the top of the slope, was much too simple, partly because we imagined a single permanent, regular, and very long slope[6] for the entire journey of the gliding nappes (100–150 km in the case of the Prealps). Consequently, we face impossibilities of two kinds: (1) according to the chosen model, the top of the slope is much too high; and (2) it is very difficult to imagine a truly continuous and regular subaerial slope which, in spite of erosion, has to remain regular during the entire time required for gliding (perhaps a few million years; see below).

In fact, the problem of the regularity and permanence of the slope changes according to different cases. The basement slope required for the gliding of the Subalpine folds was, of course, not attacked by erosion. Submarine slopes in flysch basins on which glided the Ultrahelvetic or Simme nappes may have been more or less regular. But the extensive *subaerial* slope required for the main overthrusting of the Prealps leads us into difficulties.

To get around these difficulties, two different routes may be followed. The first one is a kind of guile—some geologists imagine a forward-moving slope, a kind of "wave" pushing on its front the simultaneously gliding nappe, as does a sea-wave with a surf-rider (Fig. 11). With this hypothesis, the slope might be much shorter. But this is of course artificial, even if we link this idea with a supposed general orogenic polarity of the chain. One must take into account that such a model needs to fit with both the length of the nappes and the speed at which they traveled (Fig. 11). In fact, as far as length is concerned, the Plastiques, Rigides, and

---

[6] And we also unconsciously imagine a planar slope, such as the one drawn on the schematic representation of the model!

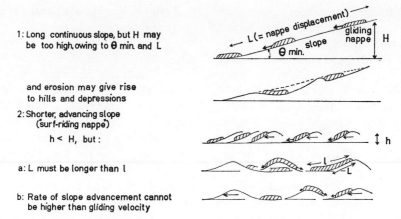

1: Long continuous slope, but H may be too high, owing to θ min. and L

and erosion may give rise to hills and depressions

2: Shorter, advancing slope ("surf-riding nappe")

h < H, but:

a: L must be longer than l

b: Rate of slope advancement cannot be higher than gliding velocity

FIG. 11 Theoretical diagrams showing the main aspects of the slope problem.

Brèche nappes together were at least 50 km long at the time of their emplacement, so that they had to travel on a slope at least 50 km long, and the difficulty is not much eliminated.

We must therefore try to find a second way to get around the contradiction, and search for an improved mechanical model, leading to a much lower limit angle, as we shall see further.

In the case of subaerial gliding, as in the second stage of the Prealps motion, the slope, attacked by erosion, will not remain regular for long, with the birth of depressions acting as "nappe-traps," or hills acting as obstacles (Fig. 11) against the motion of the nappes.[7] Could these obstacles have been pushed, detached, and dragged along by a "bulldozer-nappe"? This leads us to the problem of kinetic energy, and therefore the velocity of the gliding nappes. The velocity problem is also important because the lower the velocity of the nappes, the greater the time during which erosion may destroy the slope (and the nappes as well!).

Velocity, Kinetic Energy, Obstacles, and the Principle of Uniformitarism

We know very little about the velocity of a moving nappe, whether gravitational or not. Stratigraphic data give us only a minimum figure: for example, the Plastiques, Rigides, and Brèche nappes were overthrust after the middle Eocene and arrived at their present position during the Oligocene, which allows for few million years for a 100–150-km displacement,[8] implying a minimum mean velocity of 1–10 cm/yr. Nevertheless, the velocity could have been much higher, and in reality we know nothing about it. Let us remember that the catastrophic Vaiont slide of 1963 has been proved to have had a maximum velocity of 60 km/hr (Selli and others, 1964). However, we are not sure that such a catastrophic event, involving a small volume of rocks (0.3 km³) and a somewhat steeper slope (45° to 5–10°), may be compared to the gliding of large cover nappes (800 km³ for the Brèche nappe of Chablais only!).

Here enters the problem of uniformitarian-

---

[7] For the Tell nappes of North Africa, Caire (1965) believes that the nappes glided upon a "pediment-slope," continuously being created in front of the moving nappes. But this does not eliminate the problem of the altitude of departure and implies a very slow motion of the gliding nappes.

[8] And also for preliminary uplift, because during middle Eocene the nappes still were below sea level in the flysch basin.

ism. First, we know that for most mechanical models, such as those by Goguel (1948) and Hsü (1969), the thicker the nappe, the lower the limit angle of the slope consistent with gliding. This leads us to a kind of paradox: extensive and thick nappes should be more probable than smaller ones, and all the more than small slides such as the Vaiont slide, because they could glide on a lesser slope. But the Vaiont slide has been observed before, during, and after its motion, whereas true large gliding nappes in motion are unknown to date. If nappe gliding is a very slow phenomenon, it is possible that nappes are still gliding somewhere on the earth, even if we can neither measure nor observe their movement. If the normal velocity is higher, however, we can assert that large *subaerial* nappes are not moving today, either by gravity or otherwise. In the same way, recent *submarine* slides have been described (see, e.g., Heezen and Drake, 1964; Moore, 1964; Rona and Clay, 1967; Glangeaud and others, 1968; Mary and Dangeard, 1970), but they were merely inferred from seismic or topographic data and have not actually been observed to be in fact slides, either during displacement or in displaced position.

If the velocity is really low, the kinetic energy of the nappe will be very small. For example, the kinetic energy of the whole assemblage of Prealpine nappes of the Chablais and the Préalpes Romandes (6600 km^3), moving at a rate of 1 m/yr, is 1000 times less than the kinetic energy of a single automobile moving at a speed of 100 km/hr. Would, in that case, large obstacles such as hills be pushed and dragged along? A depression, bottoming out to a lesser slope, will perhaps stop the nappe (Fig. 11), and a simple calculation shows that this will occur after only a very short time (a fraction of a second). The very slowly gliding geological body will certainly not continue by its own momentum.

In conclusion, we must confess that we cannot say much about this problem of velocity and kinetic energy, except that it cannot be neglected. Perhaps the tectonic slides were catastrophic, but in that case uniformitarianism does not hold, for we do not observe them at present, at least not true large ones. If, on the contrary, the nappes moved slowly, erosion has time to act and irregularities of the slope may perhaps stop the movement. In the final analysis it becomes clear that we first need a more accurate mathematical study, if this is possible (however, see Hsü, 1969).

Toward Some Possible Mechanical Explanations: Using Improved Models

One of the greatest difficulties we encountered was that of the altitude at the top of the slope. If the angle of slope were very small, say, 1° or less, we would not face serious problems. Our first review of the possible models brought us to prohibitive figures. However, these models were much too simple; the genuine tectonic mechanism is certainly much more complicated, for several factors have been neglected, including first the *role of fluids*, especially that of water.

Water is not a lubricant, but it could take a prominent part in facilitating deformation. For example, laboratory experiments show that the presence of water, especially at high confining pressure, lowers the elastic limit of limestones. In fact, deformation of limestone, gypsum, and halite is known to require lower stress limits if solution and recrystallization occur (a well-known mechanism in the deformation of glacial ice).

The main role of water, however, is in its pressure, because abnormally *high internal fluid pressures* modify the effective normal stress. In a general way, both laboratory experiments and model calculations show that the higher the ratio of interstitial fluid pressure to lithostatic pressure, the lower the effective strength of rocks. Hubbert and Rubey (1959) have shown that a 90 percent pressure ratio lowers the limit angle at which gliding of a nappe can be initiated from 30 to 3°. According to Hsü (1969), who used a somewhat different model,[9] the angle depends on the weight (and therefore the thickness) of the nappe, and the figures obtained are higher. But one may expect from such models a better explanation of gravity gliding, if it is a reality.[10]

Abnormally high fluid pressures, as high as 90 percent of the lithostatic pressure, are known to exist at depth in many regions (Hubbert and Rubey, 1959; Goguel, 1969). They may occur when a submarine gliding nappe arrives on water-filled and incompletely compacted sediments, such as flysch (Goguel, 1969), and also, in a more general way, through differential compaction. Byramjee (1966) showed that some cases of known overpressures may be linked with the presence of evaporite layers. Gypsum dehydration (leading to anhydrite plus water) is another process creating a lowering of the effective strength of evaporites; Heard and Rubey (1966) showed that such a decrease of strength may occur at temperatures in the range of 100–150° C under a 5-kb confining pressure, corresponding to conditions at depths of 100–2000 m. This could perhaps have occurred below some Prealpine nappes and possibly also in some parts of the Subalpine Chains.

Another possibly important process may be *water vaporization due to friction heat* at the base of the gliding mass. In the case of the Vaiont slide, for example, Habib (1967) has shown that after only a small displacement (a few meters) friction may lead to sufficient heating in the friction plane to vaporize the interstitial water. This induces therefore a sudden rising of fluid pressure, causing the slide to become catastrophic. Such a mechanism may occur during nappe transport, but only at relatively shallow depths, because the critical pressure of water corresponds to a depth of 840 m, beyond which fluid dilatation will be less high (Goguel, 1969). The same heating may also induce dehydration of gypsum (Goguel, 1969). In all these cases, model calculations show that a very short gliding motion, if it occurs in a sufficiently brief time, may induce a sudden release or vaporization of water. Thus, if the movement is initiated by *earthquakes*, which cause the base of the gliding mass to act as a "shaking-table," it may suddenly become a catastrophic slide (Goguel, 1969). This mechanism may perhaps explain, for example, the catastrophic slides that occurred, on very gentle slopes, during the 1960 Chile earthquake (Tazieff, 1961), and (why not?) the gravitational movement at shallow depths of some nappes in the geologic past.

Conclusions

The last part of the discussion leaves open a door toward a solution to the contradictions we encountered. Various improved models may indeed explain gliding on a much gentler slope, a possibility that partly eliminates the problem of the altitude and erosion of the slope.

With respect to the Western Alps, as discussed here, we may say the following:

1. In the case of the Prealps, we cannot escape gravity gliding as the mechanism for diverticulation, that is, for early submarine overthrusting of unfolded, thin, incompetent slabs, often preceded by submarine "avalanches." This conclusion applies to the first stage of the Ultrahelvetic and Simme nappes. Considering now the second stage of the Prealps motion, gravity gliding occurred perhaps *at the end* of the movement of Plastiques, Rigides, and Brèche nappes, carrying outliers of the Simme nappe on their backs (Fig. 5). However, in view of the problem of altitude and erosion of the slope, and

---

[9] See also the discussion of Hubbert and Rubey's model by various authors (references in Hsü, 1969).
[10] And, more generally, of nappe overthrusting, whether gravitational or not.

because rocks of some of these nappes show regional cleavage, gliding cannot be accepted as the sole mechanism, and it is therefore very likely that overthrusting of these nappes began as a result of strong compressive deformation under a relatively thick overburden.[11]

2. In the Subalpine Chains, gravity gliding may have played a role during their folding, but it is very likely that this role was secondary and occurred mainly *at the end* of the folding.

In conclusion, we see that due consideration may constrain us to adopt gravity tectonics in certain instances, even though reluctantly. In that case, we must try to find a mechanical explanation, in order to escape the paradoxes. In other cases, however, the theory can only be rejected or accepted according to one's conviction about its universality.

A tectonic theory tries to reconstruct the course of past events of which we can observe only the final result. In genetic reconstructions the geologist has fewer possibilities than, for example, the physicist. A physical theory also deals with models, and it follows concrete observation of facts. But it may be followed by experiments capable of testing the theory, which may repeat again and again the whole course of the event or phenomenon under inquiry. As geologists we cannot produce true experiments, that is, the repetition of past tectonic events.[12] We can only observe those events that occur at the present time on the earth's surface, but it is possible that some events of the geologic past do not occur today, or else are too slow to be observed, least of all from beginning to end. Even if the uniformitarian principle is correct, we remain disarmed in this case. Let us repeat, finally, that consideration of both geological observations and rock mechanics data allows us only to choose between "likely" and "unlikely," not between "true" and "false."

This paper will certainly not appear as a plea for gravity tectonics, but no more is it a charge against it. All that has been tried here is to emphasize that such tectogenic theories can never be proved—we can only attempt to know if they are plausible. The rest is a matter of subjective choice.

References

Badoux, H., 1963, Les unités ultrahelvétiques de la zone des cols: *Eclog. Geol. Helv.*, v. 56, p. 1–13.

———, and Mercanton, C., 1962, Essai sur l'évolution tectonique des Préalpes Médianes du Chablais: *Eclog. Geol. Helv.*, v. 55, p. 135–188.

Byramjee, R., 1966, Argiles non compactées et pressions anormales—Cas du Nord-Sahara: *Rev. Inst. Fr. Pétrole*, v. 21, p. 1067–1077.

Caire, A., 1965, Morphotectonique de l'autochtone présaharien et de l'allochtone tellien: *Rev. Géogr. Phys. Géol. Dyn.*, v. 7, p. 267–276.

Caron, C., 1966, Sédimentation et tectonique dans les Préalpes: "flysch à lentilles" et autres complexes chaotiques: *Eclog. Geol. Helv.*, v. 59, p. 950–957.

Debelmas, J., and Kerckhove, C., 1973, Large gravity nappes in the French–Italian and French–Swiss Alps (this volume).

Debelmas, and Lemoine, M., 1970, The Western Alps: Palaeogeography and structure: *Earth Sci. Rev.*, v. 6, p. 221–256.

Fallot, P., and Faure-Muret, A., 1949a, Sur l'extension du décollement de la série de couverture subalpine: *Comptes Rendus Acad. Sci. Paris*, v. 228, p. 616–619.

———, 1949b, Sur un mode particulier de charriage: *Comptes Rendus Acad. Sci. Paris*, v. 228, p. 789–792.

Gèze, B., 1962, Distinction d'un type de nappes à enracinement frontal: les refoulements: *Comptes Rendus Somm. Soc. Géol. Fr.*, v. 2, p. 38–39.

Glangeaud, L., Bellaiche, G., Genesseaux, M., and Pautot, G., 1968, Phénomènes pelliculaires et épidermiques du Rech Bourcart (Golfe du Lion) et

---

[11] The same conclusion may be proposed for the Schistes Lustrés nappe (see Debelmas and Kerckhove, 1973), which underwent first a strong compressive polyphased deformation with metamorphism, and therefore could have glided by gravity at the very end only of its displacement.

[12] "Experiments" on small-scale models can never be taken as evidence. They are imagined and constructed by us to put into concrete form a more or less intuitive idea, so that we find in the result only what we put into the construction of the experiment.

de la mer Hespérienne: *Comptes Rendus Acad. Sci. Paris*, v. 267 (D), p. 1079–1083.

Goguel, J., 1948, Introduction à l'étude mécanique des déformations de l'écorce terrestre (2nd. ed.): *Mém. Expl. Carte Géol. Fr.*, 530 p.

———, 1949, A propos du glissement de la couverture au Sud-Ouest du Massif de l'Argentera: *Comptes Rendus Acad. Sci. Paris*, v. 228, p. 698–699.

———, 1969, Le rôle de l'eau et de la chaleur dans les phénomènes tectoniques: *Rev. Géogr. Phys. Géol. Dyn.*, v. 11, p. 153–164.

Habib, P., 1967, Sur un mode de glissement des massifs rocheux: *Comptes Rendus Acad. Sci. Paris*, v. 264, p. 151–153.

Heard, H. C., and Rubey, W. W., 1966, Tectonic implications of gypsum dehydration: *Geol. Soc. Am. Bull.*, v. 77, p. 741–760.

Heezen, B. C., and Drake, C. L., 1964, Grand Banks slump: *Am. Assoc. Petrol. Geol. Bull.*, v. 48, p. 221–225.

Hsü, K. J., 1969, Role of cohesive strength in the mechanics of overthrust faulting and of landsliding: *Geol. Soc. Am. Bull.*, v. 80, p. 927–952.

Hubbert, M. K., and Rubey, W. W., 1959, Role of fluid pressure in mechanics of overthrust faulting: *Geol. Soc. Am. Bull.*, v. 70, p. 115–166.

Lugeon, M., 1943, Une nouvelle hypothèse tectonique: la diverticulation: *Bull. Soc. Vaud. Sci. Nat.*, v. 62, p. 260–261.

———, and Gagnebin, E., 1941, Observations et vues nouvelles sur la géologie des Préalpes Romandes: *Mém. Soc. Vaud. Sci. Nat.*, v. 47, p. 1–90.

Mary, G., and Dangeard, L., 1970, Les phénomènes de glissement dans le domaine marin: *Rev. Géogr. Phys. Géol. Dyn.*, v. 12, p. 313–324.

Moore, J. G., 1964, Giant submarine landslides on the Hawaiian Ridge: *U.S. Geol. Survey Prof. Paper 510-D*, p. D95–D98.

Rona, P. A., and Clay, C. S., 1967, Stratigraphy and structure along a continuous seismic reflection profile from Cape Hatteras, North Carolina, to the Bermuda Rise: *Jour. Geophys. Res.*, v. 72, p. 2108–2130.

Sandulescu, M., 1967, La nappe de Haghimas—Une nouvelle nappe de glissement dans les Carpates orientales: *Assoc. Geol. Carpato-Balkan., VIIIe Congr. (Beograd)*, p. 179–185.

Schardt, H., 1898, Les régions exotiques du versant nord des Alpes Suisses: *Bull. Soc. Vaud. Sci. Nat.*, v. 34, p. 114–219.

Selli, R., Trevisan, L., Carloni, G. C., Mazzanti, R., and Ciabatti, M., 1964, La frana del Vaiont: *Giorn. di Geol., Ann. Mus. Geol. Bologna*, (2), v. 32(1), p. 1–154.

Tazieff, H., 1961, Signification tectonique des glissements de terrain accompagnant le grand séisme du Chile: *Comptes Rendus Acad. Sci. Paris*, v. 251, p. 2204–2206.

Trümpy, R., 1960, Palaeotectonic evolution of the Central and Western Alps: *Geol. Soc. Am. Bull.*, v. 71, p. 843–908.

Van Bemmelen, R. W., 1955, Tectogénèse par gravité, *Bull. Soc. Belge Géol.*, v. 64, p. 95–123.

H. LAUBSCHER

*Geologisches-Paläontologisches Institut
Universität Basel
Basel, Switzerland*

Jura Mountains

Gravity sliding as an easy answer to the difficult problems in dynamics posed by Jura tectonics has been advocated repeatedly and on different scales. Thrusts in individual folds (Aubert, 1945; Glangeaud, 1949) have been ascribed to local erosion and sliding into the erosional depressions. A systematic geometric-kinematic analysis reveals, however, that except for occasional large landslides these thrusts are all regional tectonic features that owe their existence not to local but to regional movements (Laubscher, 1961, 1965).

Sliding of the whole Jura system was proposed as early as 1892 by Reyer, and the notion has been resuscitated, though as a possibility rather than a necessity, by Pierce (1966, p. 1268); however, the stratigraphic situation unambiguously demands an uphill push at the time of the folding. If gravity sliding indeed played an important role it will have to be sought in the rotation of the entire sedimentary sequence away from the Alps because of overloading on the Alpine side of the Molasse basin (Laubscher, 1961).

Thus the decisive factor is the tectonic configuration on the northern side of the Alps at the time of the Jura folding (late Pontian). The stratigraphic record, though very fragmentary, suggests a rather subdued relief for that time, and chances for gravity sliding as the principal motor do not look promising.

Evidently, a first step toward an evaluation of the nature of the driving motor for Jura tectonics is the construction of a kinematic model depicting successive movements in the Autochthonous and Helvetic realms during the Neogene orogenic phase. This has been attempted schematically in Fig. 6 and the following text is essentially a discussion of this figure with some additional illustrations of especially important points. Details are omitted since only the large-scale aspects bear on the problem here investigated.

The Situation before the Onset of Neogene Movements

Figure 1 is a simplified sketch of the main tectonic elements here discussed, in their

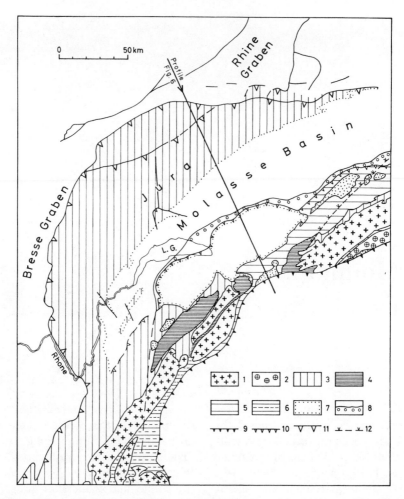

FIG. 1 Location of the main tectonic elements discussed in the text. (1) External massifs; (2) Gotthard massif; (3) Jura Mountains, Chaînes Subalpines, and "Autochthonous"; (4) Morcles-Doldenhorn nappe; (5) higher Helvetic nappes (Diablerets-Wildhorn); (6) Helvetic to Ultradauphinois; (7) nappes of the Prealps and Ultrahelvetic (Paleo-Alps); (8) Subalpine Molasse; (9) thrust front of Penninic nappes south of massifs; (10) thrust front of Prealps and Helvetic border chain; (11) front of Jura Mountains and Chaînes Subalpines; (12) eastern limit of Middle Triassic evaporites; (L.G.) Lake Geneva.

present position. Figure 2 is an attempt at reconstruction of their position before the onset of Neogene folding. In map view this problem is made particularly difficult because both the Jura and the source area (see Fig. 5) for its movements in the Alps are located on the external side of the sharp northwestern corner of the arc of the western Alps. The regional three-dimensional kinematics are rather involved in this area (Laubscher, 1970b, 1971a, 1971b). I shall not discuss this problem here but point out only the following salient features (compare also Fig. 6a).

1. The Helvetic realm, except perhaps for its southernmost part, was not yet directly affected by orogenic movements. It was a foreland, covered by the sedimentary décollement nappes of the Ultrahelvetic and Penninic belts, much as the southern

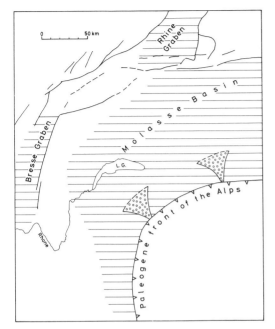

FIG. 2 Paleotectonic situation before the beginning of Neogene orogeny. Reconstruction is based on arguments similar to those developed for the construction of Fig. 6, except that the three-dimensional nature of the kinematics is here hinted at. The abrupt bend in the northwestern corner of the western Alps which characterizes the area here discussed is far from kinematically homogeneous. Discordantly intersecting fronts of movement are evident (for example, the Paleogene front of the Alps delineated by the relics of Paleogene foreland nappes, and the NE-SW alignment of the external Aiguilles Rouges-Belle donne massifs south of Lake Geneva as shown in Fig. 1). This is only a particularly obvious example of the superposition of different kinematic tendencies which have probably been in play throughout the development of the western Alps. In the Alps movement of nappes is mainly to the north, onto the northern foreland, with a smaller component to the west. In the foreland the Rhine-Bresse graben system delineates an embryonic lithospheric plate boundary, with an EW (or ESE-WNW) separation not exceeding 3 km. From plate kinematics it follows that movement in the connecting zone between the two graben must be essentially sinistral strike-slip (transform fault zone).

part of the Molasse basin now is covered by décollement nappes (see Lugeon and Gagnebin, 1941). The décollement masses extended northwest as far as the southern part of what is now called "autochthonous." From their margin outward jutted the Oligocene river deltas with their huge piles of conglomerates (see Trümpy and Bersier, 1954).

2. South of the Helvetic realm Penninic and Ultrahelvetic basement units formed the structural backbone of the Alps, much as they do now, except that they were apparently more deeply buried and that the external massifs belonging to the Helvetic realm did not yet exist; perhaps the northernmost part of the Gotthard massif (Ultrahelvetic) or its unexposed western equivalent formed a series of Paleogene external massifs.

3. North of the Helvetic realm the Oligo-Miocene Molasse basin sloped gradually up toward the northwest with a dip slightly smaller than its present one (see Lemcke and others, 1968). The culmination was defined by the southern and eastern lip of the Rhine–Bresse graben system. This system was one of east–west dilatation, with a sinistral transform fault zone establishing the connection between the two graben (Laubscher, 1970a).

4. The potential foreland instabilities for an expanding orogeny were characterized by this faulted culmination in the north and west and by a sheet of Triassic evaporites at the base, extending east as far as a line connecting the eastern ends of Lake Brienz and Lake Zürich (Fig. 1; see Büchi and others, 1965).

Some Notions, Arguments, and Postulates Concerning the Rule that Governed the Neogene Movements

The principal fact is that the Neogene Helvetic décollement nappes originally covered a depositional zone some tens of kilometers

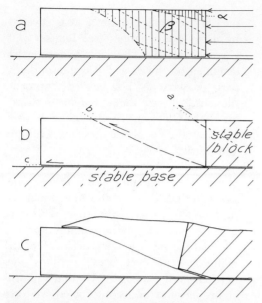

FIG. 3 The problem of shear zones emerging from deep crustal layers. Models of mountain building based on present views of large-scale movements of the lithosphere demand detachment of the light upper crust from the lower crust and the lithospheric upper mantle. Shear zones are required to connect two surfaces that bound a wedge-shaped body of movement: the surface of the foreland and the zone of crustal detachment. (a) Problems of stability which arise in this wedge are best, if schematically demonstrated by means of the simple décollement model of Hafner (1951), simplified in Fig. 3a. Stability within the block depends on (1) material properties (distribution of strength, anisotropy), and (2) frictional resistance at the base of the block and consequently the level of stress at the onset of décollement. As the boundary force increases, the region of instability expands from the corner at the upper right into the block (e.g., from stability boundary α to β). Potential shearing surfaces within the unstable wedge are shown as dashed lines. (b) Shearing surfaces a, b, c are three typical cases. a is located within the initial domain of instability α; it is kinematically ineffective as it does not continue into the basal zone of detachment (the pushing block on the right is stable by assumption). Shearing surface b is the first to achieve a kinematic connection between the zone of detachment and the surface, whereas activation of c means décollement of the whole block. (c) The combined shearing surface has two opposite curvatures, one concave upward and one convex upward. Actual displacement along such a surface entails folding with internal deformation of the wedge.

(~30 km) wide whose crystalline basement has somehow disappeared, evidently as the result of basement compression; autochthonous and Ultrahelvetic basement, originally 30 km apart, are now in close juxtaposition (see Trümpy, 1969). Consequently, purely vertical movement of the basement, with gravity sliding of the cover resulting in the piling up of the Helvetic nappes, is out of the question.

Basement nappes in the Alps, with few exceptions, involve only the uppermost few kilometers of the continental crust. For geometric reasons, the same seems to be true for the external massifs (Laubscher, 1970b). The combination of intracrustal décollement zones—which somewhere must penetrate to the surface—and décollement in the sediments leads to the notion of the *décollement wedge* (Fig. 3). In the conceptual context of plate tectonics these wedges are the external terminations of light, shallow rock masses which are peeled off the drowning lithosphere in geotectonic sinks and pushed toward the continent to form the Alpine-type mountain system (see Laubscher, 1970b).

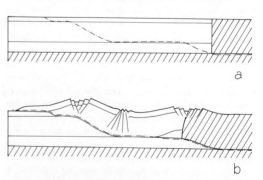

FIG. 4 The effect of layering. (a) Incompetent layers provide zones of weakness which modify the shape of the shearing surface; this now consists of a series of horizontal planes connected by oblique shears. (b) Movement of the wedge is now accompanied by folding at several points. The rigid base calls for stretching in the anticlines and compression (with cleavage at depth) in the synclines. No attempt is made to go beyond a qualitative demonstration, because dynamic conditions are hard to assess quantitatively.

As the orogenic movements engulf more and more of the foreland, or in the parlance of plate tectonics, as the foreland lithosphere continues to move into the sink (the orogenic zone), ever more external portions of the shallow crust are peeled off. New décollement wedges develop in the former foreland. Although this is a well-known tendency sometimes enjoying the status of a "law" in geology, exceptions do occur (see, e.g., Trümpy, 1969).

The décollement wedge has an irregular lower boundary. In a first approximation this is composed of subhorizontal zones of weakness connected in shallow portions by oblique shears of the Mohr type (Figs. 3, 4).

In deeper portions wide zones of cataclastic flow and "plastic" deformation are found, particularly recumbent folds where the main mechanism seems to be a rolling of the comparatively competent upper limb over a series of frontal hinges (Fig. 5), though folding and cleavage are by no means absent from the upper limb. Such recumbent folds are found in basement nappes, such as the basement lobes of the Aar Massif, particularly the Jungfrau lobe (Scabell, 1926; Collet and Paréjas, 1931), and the Truns lobe (Käch, 1969), as well as in sedimentary cover nappes, such as the Morcles-Doldenhorn (Lugeon, 1940). No discussion of the detailed kinematics within such recumbent folds is here

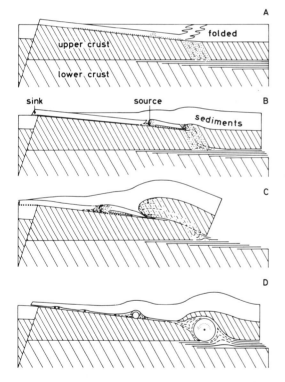

FIG. 5 Simplified scheme of Fig. 6. The shearing zone comprises both laminar and rotational movements on different scales. Observation shows that particularly in lower levels recumbent folds are ubiquitous. Their essential role is transport of the overlying wedge by rolling it over the hinge or system of hinges (D). Even in the shallow portions (left), representing the Jura décollement, transport of this type occurs if the material properties are right. This is evident in the Triassic anhydrites of the Bölchentunnel, where flow folding (distinguishable from synsedimentary deformations) on a meter to centimeter scale could be observed (Laubscher, 1967; Wohnlich, 1967). No attempt is made here to discuss the detailed kinematics involved in the large-scale phenomenon which is commonly called a "recumbent fold"; for careful field studies see Labhart (1968) and particularly Steck (1968). In a tectonic system, as is customary in transport phenomena generally, mass may be said to be displaced from a source area into a sink area. The source area for a décollement sheet may be reasonably defined as that area in the rear where the forward movement of the sheet has left a gap, either directly exposed as in the case of gravity sliding, or covered by those masses that have transmitted the push ("substitution de la couverture"). The sink area is the frontal part of the sheet where masses not previously present are piled up as folds or thrusts. For a general discussion of the concepts see Laubscher (1965).

(A) A continuous zone of instabilities develops at the base of a potential wedge. It emerges from the base of the upper crust (right, horizontally ruled) obliquely through the upper crust (dots) into the base of the sediments (dots, left). Clouds of dots denote places of pronounced rotational instabilities. The folded sediments on the right are nappes of the Paleo-Alps. (B) Initial movements of the wedge. (C) Final stage. (D) Abstraction of (B) showing the hinges of recumbent folds as rolling cylinders. The base of the wedge is characterized by an irregular distribution of sliding and rolling friction.

attempted (see explanation for Fig. 5).

Somewhere in the shallower frontal part this zone of the rolling hinges will somehow pass into décollement and Mohr-type shear zones which connect with the surface. Of particular interest is the case where a rolling hinge pushes away the sedimentary foreland series above a basal décollement zone. I have implicitly postulated such a mechanism for the front of the Morcles-Doldenhorn (Laubscher, 1961) by assuming that the gap in the sedimentary cover of the Aiguilles Rouges and the western Gastern Massif is the source for the Jura folds: either the sedimentary cover has slid away, leaving a gap into which the Morcles nappe subsequently moved, or the front of the Morcles nappe pushed away the sedimentary cover, thus creating the gap by its own efforts. Recently, Trümpy (1969) has advocated a similar mechanism for the Cavestrau–Griesstock complex in the eastern Helvetic domain (see also Käch, 1969). Now, a plausible paleotectonic reconstruction demands that at the time of these movements the area of the Morcles nappe was a low, and gravity gliding to account for a gap at that time is a rather hopeless proposition (compare Fig. 6E, F).

Movement of the wedge over its irregular lower boundary causes folding of the moving masses (Fig. 4). Nappe anticlines with external stretching may be postulated above convex-upward segments of the base, nappe synclines with compression above concave-upward segments. The detailed kinematics will be exceedingly complex, but I think it possible that such phenomena as extensional faults in the Helvetic nappes (e.g., *Untervorschiebungen*: Schaub, 1936) commonly attributed to gravity gliding of the frontal parts of the nappe away from the rootward parts are—at least partly—due to such distortions of the moving wedge. Other extensional features in nappes may be attributed to relative displacements of superjacent masses, as in the case of the stretched middle limbs of recumbent folds.

Field evidence for the sequence of events is complex and often conflicting. For a discussion about the relative age of movements in the western Helvetic Alps see, for example, Lugeon (1940). That author arrives at the conclusion that the Morcles nappe is the oldest, contrary to the general rule. The reason is that the normal limb of the Morcles nappe is folded into a series of anticlines, whereas the overlying Diablerets nappe has a slightly curved but originally smooth base. I do not think, however, that this argument is decisive. In fact, there are no decapitated folds whose apexes have been carried away by the Diablerets nappe, and even if there were, it would mean only that there have been folds, but not that these folds have been part of an already existing nappe. Indeed, it is easy to visualize kinematic schemes that produce the present geometry with the Diablerets nappe preceding the Morcles nappe. For example, a smooth overthrust of the Diablerets nappe may first produce by friction (or, alternatively, by diffusion of movement downward) a series of localized décollements with folding (rotation), and as the base of the moving wedge is displaced to a deeper level, the whole sedimentary cover is sheared off and rolled over a series of frontal hinges. In the overall picture this view is more satisfactory and is made use of in the model of kinematic development (Fig. 6). Since Jura folding is dated as post-early Pontian it is probably one of the latest movements of the Neogene phase, and if it is causally connected with the movement of the Morcles nappe the latter ought to be one of the latest structures also.

A Kinematic Model of the Neogene Orogenic Phase in the Western Helvetic Alps and the Jura

The basic principles briefly put forth in the foregoing section may be used for an attempt at reconstruction of the Neogene sequence of movements in a general way (Fig. 6). From the present geometry, adapted from profiles

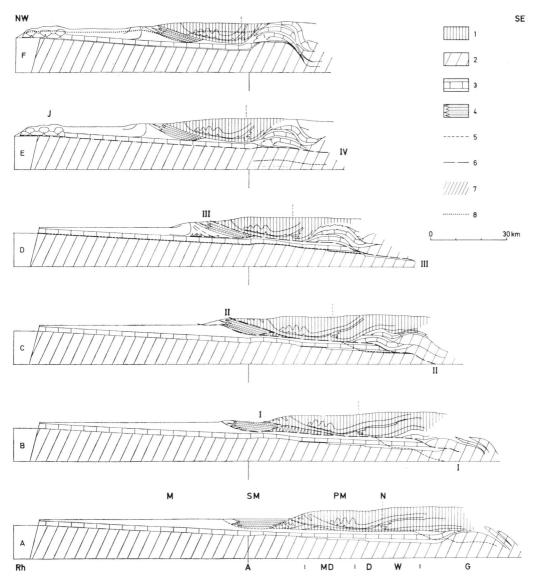

FIG. 6 A kinematic model of the system Jura-Helvetic nappes. (1) Paleo-Alps; (2) basement; (3) Mesozoic of the Paleogene foreland (only Malm-Cretaceous of Helvetic realm); (4) Subalpine Molasse; (5) location of thrusts activated in the subsequent stage; (6) main thrusts (I-IV); (7) zone of stretching in the wake of the Subalpine Molasse; (8) approximate trend of present surface; (Rh) Rhine graben-Bresse graben low; (M) Molasse basin; (SM) Subalpine Molasse; (PM) Préalpes Médianes; (N) front of Niesen nappe; (A) northern foot of Aiguilles Rouges Massif; (MD) Morcles-Doldenhorn; (D) Diablerets; (W) Wildhorn; (G) Paleogene Gotthard Massif.

The Niesen nappe, assumed here to have reached its position relative to the other units of the Prealps in the Paleogene, has been suspected by Badoux (1945, p. 69) to have been the latest arrival in the area, postdating the Wildhorn nappe. The argument is similar to that for a younger age of the Diablerets nappe with respect to the Morcles nappe, and similar objections may be raised. Indeed, displacement of the base of the moving wedge after stage C (development of Wildhorn nappe) back again to the base of the Niesen nappe involves such difficulties as to require quite incontrovertible evidence. However, for the general illustration of movements another marker in the Paleo-Prealps would have served equally well.

3 and 4 of the *Generalkarte* 1:200,000 *der Schweiz*, the approximate palinspastic position before the onset of Neogene movements (Fig. 2) is obtained as explained in general terms above and more specifically below. Then, successive wedges are developed as characterized on Figs. 4 and 5.

Visualization is facilitated by fixing attention on a particular marker unit, which is then followed through the different stages, by comparing its successive relative positions with a control point in the basement, for example, the north foot of the Aiguilles Rouges Massif (A in Fig. 6). Such a marker is, for instance, the frontal part of the Niesen nappe (N in Fig. 6). Its present position is in front of all the Helvetic nappes and the Aiguilles Rouges Massif. Its pre-Neogene position is found by (1) smoothing out Jura folds and thrusts, thereby moving the whole edifice of the Prealpine and Helvetic nappes back by about 20 km with respect to the basement control point (Laubscher, 1965), (2) pushing back the imbricated slices and thrust sheets of the Subalpine Molasse by at least 10 km (e.g., see Habicht, 1945), and (3) pushing back the Prealps, which now cover the Subalpine Molasse, by another 20 km, the amount being fixed by compression measured in the Helvetic nappes that carried them forward (see below).

The smoothing out of the Helvetic nappes is accomplished with the aid of the following notions:

1. The Morcles nappe (compression about 20 km) has pushed the Jura décollement sheet away frontally (see Fig. 5). Smoothing out of the Jura (20 km) then automatically relocates the Morcles nappe without calling for any additional corrections inside the corresponding wedge.
2. The Diablerets and Wildhorn nappes, with roughly an additional 30 km of original depositional width in this area, require an outlet of their respective wedges of movement at the base and on top of the Subalpine Molasse. As the amount of thrusting at the base of the Subalpine Molasse may be estimated at about 10 km, the movement on top and over the original internal border of the Molasse is found to be 20 km.

These operations replace the front of the Paleogene Prealps and their contact with the discordantly onlapping Oligocene conglomerate fan close to the middle of the Aiguilles Rouges Massif. If no internal Neogene compression of importance is assumed, the Niesen marker then is located as shown in Fig. 6*A* (see legend) on top of the Wildhorn depositional zone.

As this marker now lies in front of the pile of Helvetic nappes, the first wedge bases must have emerged from the basement south of, or in the southernmost part of the original Wildhorn domain, and thence followed a décollement zone in the Helvetic and particularly Ultrahelvetic flysch at the base of the Prealps, until its final emergence to the surface at the front of the Prealps (Fig. 6*B*, thrust I). The frontal part of the early wedges involving the internal margin of the molasse poses some special problems, which will be discussed separately below.

Basal thrust I (Fig. 6*B*) displaces the Niesen marker to the front of the Diablerets domain (Fig. 6*B*); basal thrust II (Fig. 6*C*) creates the Wildhorn nappe and displaces the marker onto the Morcles domain. Thrust III (Fig. 6*D*), the Diablerets thrust, pushes it farther onto the Aiguilles Rouges domain, and thrust IV, the Morcles-Jura thrust, carries it to its approximate present position just north of the Aiguilles Rouges. After having formed a rough idea of the sequence of movements by following the Niesen marker, a few additional points will be briefly elaborated on.

The first concerns the relation between the profile chosen for Fig. 6 and another profile situated east of Lake Thun (2 in Fig. 1), where the Prealps have disappeared and in their place the Helvetic border chain ("Randkette," Ra in Fig. 1) emerges. Relics of the Prealpine nappes are present also in this area, but instead of lying in front of the Helvetic nappes they are perched on top of them. Their position with respect to the underlying Helvetic sediments is still approximately the same as it had been at the end of the Paleogene phase, according to the reconstruction shown in Fig. 6A and discussed previously. In the Neogene phase they were carried passively on top of the Helvetic nappes. This evidently requires that all the thrusts passed essentially at the base of and within the Helvetic sedimentary sequence in this area, whereas in profile I thrusts I to III followed the base of the Prealps for considerable distances. Consequently, there must have been a transverse oblique shear at about the position of Lake Thun, where the thrusts at the base of the Prealps in the southwestern domain dipped laterally into those at the base of and within the Helvetic sequence (particularly the base of the Cretaceous) in the northeastern domain.

There are other important changes connected with this transverse zone. For instance, it marks approximately the eastern end of the Morcles–Doldenhorn and Diablerets nappes (Fig. 1). At the same time a normal autochthonous sequence appears on top of the continuation of the Aiguilles Rouges–Gastern massif, that is, the source area of the Jura movement (see Laubscher, 1961). This means that the basal thrust IV (the Jura thrust) in this area must connect the base of the Wildhorn nappe with the basal décollement of the eastern Jura. The Wildhorn nappe thus assumes here the role of the Morcles nappe. Why this happened is a question of three-dimensional kinematics and dynamics outside the scope of this paper. I shall only point out briefly that primary facies differences evidently played a role: the Triassic evaporites, predestinating the northern part of thrust IV, disappear along a line striking NE from the eastern Gastern massif and crossing obliquely the Helvetic realm (Fig. 1). Similarly, the Liassic trough, whose sediments form the core of the Morcles–Doldenhorn recumbent fold, also disappears in this area.

A special problem is posed by the Subalpine Molasse (compare Fig. 6a, b, c, d, and the profiles of the *Geologische Generalkarte der Schweiz*). Its thrust slices dip invariably more steeply than the basement below the Molasse basin; the profiles show a tectonic truncation of the thick foreland molasse by the generally thinner slices of the Subalpine Molasse. This creates difficulties for the palinspastic reconstruction (see Fig. 6A). The general situation has been discussed previously: somewhere in the central part of the Aiguilles Rouges domain the Oligocene delta issued from and discordantly spread across the front of the Paleoalps. The base of the molasse imbrications generally consists of marine Rupelian, which apparently was the décollement horizon. The deeper horizons, including the lower Rupelian to upper Eocene "autochthonous" flysch and its passage to the molasse were left behind (see Lugeon and Gagnebin, 1941; Trümpy, 1969, for eastern Switzerland).

Space requirements suggest that frontal movements may have persisted through the early Oligocene, though the main part of the Paleogene phase ended before the Oligocene. The result would be as indicated in Fig. 6A: the upper Oligocene conglomerates, and perhaps also older ones (Holliger, 1955), rest discordantly on the frontal Prealps and, at their base, on a cushion of crumpled Ultrahelvetic to Helvetic flysch, containing imbrications of basal Oligocene Taveyanne sandstone derived from the cover of the Diablerets nappe. This picture, if faulty in some respects, should be sufficient for the level of kinematic sophistication here aimed at.

The present dip of the Oligocene conglomerate imbrications indicates that they have not simply been thrust but rotated in a clockwise direction (on profiles having north on the left). This implies stretching of sediments in the wake of the rotating molasse, and gouging out of basal Oligocene strata and their squeezing into the base of the foreland molasse. The shape of some of the anticlines of the "folded molasse" in front of the Oligocene imbrications suggests a core of tectonically accumulated middle Oligocene, and Habicht (1945) has mapped a marginal underthrust to the north of the Subalpine Molasse in eastern Switzerland. It follows that the leading edge of those wedges emerging in the Subalpine Molasse cannot even approximately be treated as simple thrusts, and a mechanism involving thrusting, rotation, stretching, gouging, and underthrusting has been incorporated into Fig. 6.

Another point is folding of the nappes. The model of Fig. 6 produces it as a consequence of the irregularity of the base of the wedge. Generally it is the result of movements involving masses below the nappes, and there are two principal cases of such movements—one with the horizontal component dominating and the other an essentially vertical one. The first is a thrust, and folding of the overlying wedge is either the result of an irregular base or of compression of at least the lower part of the wedge. The second is described by the "Sanford model" (Sanford, 1959), which consists of a segment of crust vertically pushed up from below and bounded by faulted flexures. In Fig. 6F the Aiguilles Rouges Massif is represented as a compressional fold above the crustal décollement layer, but vertical uplift may have played a role too (see Laubscher, 1970b).

Conclusions

The role of gravity in Jura folding depends critically on the geometry of the northern border of the Alps during Neogene orogeny. Palinspastic reconstruction reveals a situation that was not propitious for gravity sliding on a regional scale. Development of a simplified kinematic model tying the Jura into the Neogene system of the Alps must take into account the obvious role of basement compression. The result is a moving wedge with its sharp edge in the foreland. Its base consists of segments of décollement zones linked by oblique shears or hinges of recumbent folds. These wedges are pushed up against gravity, which only locally enters the picture as an independent motor. The Jura is the crumpled leading edge of one of the latest of these wedges. Its location is fixed by inherited geometry with its distribution of strong and weak domains. The system of inherited weaknesses consists of two principal components: (1) the Triassic evaporites with their eastern termination against a regional Triassic basement high; and (2) the Tertiary basement high bordering the Rhine–Bresse graben system on the south and east.

This picture, developed purely on local evidence, happens to fit perfectly into a plate tectonics concept of orogeny: the wedges are the shallow domains of low density peeled off the dense remainder of the lithosphere, which, below the central part of the orogen, sinks into the mantle by its own weight. This may have been the main role of gravity in Jura tectonics.

References

Aubert, D., 1945, Le Jura et la tectonique d'écoulement: *Bull. Lab. Géol., Minéral., Géophys., Musée Géol. Univ. Lausanne*, v. 83, p. 1–20.

Badoux, H., 1945, La géologie de la zone des cols entre la Sarine et le Hahnenmoos: *Beitr. Geol. Karte Schweiz*, N.S., no. 84, p. 1–70.

Büchi, U. P., Lemcke, K., Wiener, G., and Zimdars, J., 1965, Geologische Ergebnisse der Erdölexploration auf das Mesozoikum im Untergrund des schweizerischen Molassebeckens: *Bull. Ver. Schweiz. Petrol. Geol. Ing.*, v. 32, no. 82, p. 7–38.

Collet, L. W., and Paréjas, E., 1931, Géologie de la chaîne de la Jungfrau: *Mat. Carte Géol. Suisse*, N.S., no. 63.

Glangeaud, L., 1949, Evolution morphotectonique du Jura septentrional pendant le Miocène supérieur et le Pliocène: *Comptes Rendus Acad. Sci. Paris*, v. 229, p. 720–722.

Goguel, J., 1948, *Introduction à l'étude mécanique des déformations de l'écorce terrestre*, 2nd ed.: Mém. Serv. à l'Expl. Carte Géol. Dét. France, Paris. Imprimerie Nationale.

Habicht, K., 1945, Geologische Untersuchungen im südlichen sanktgallisch-appenzellischen Molassegebiet: *Beitr. Geol. Karte Schweiz*, N.F., no. 83.

Holliger, A., 1955, Geologische Untersuchungen der subalpinen Molasse und des Alpenrandes in der Gegend von Flühli (Entlebuch, Kt. Luzern): *Eclog. Geol. Helv.*, v. 48, p. 79–97.

Käch, P., 1969, Zur Tektonik der Brigelserhörner: *Eclog. Geol. Helv.*, v. 62, p. 173–183.

Labhart, T. P., 1968, Der Bau des nördlichen Aarmassivs und seine Bedeutung für die alpine Formungsgeschichte des Massivraumes: *Schweiz. Mineral. Petrogr. Mitt.*, v. 48, p. 525–537.

Laubscher, H. P., 1961, Die Fernschubhypothese der Jurafaltung: *Eclog. Geol. Helv.*, v. 54, p. 221–282.

———, 1965, Ein kinematisches Modell der Jurafaltung: *Eclog. Geol. Helv.*, v. 58, p. 231–318.

———, 1967, Geologie und Paläontologie: Tektonik: *Verh. Natf. Ges.*, Basel, v. 78, p. 24–34.

———, 1970a, *Grundsätzliches zur Tektonik des Rheingrabens*: Sci. Rep. no. 27, Int. Upper Mantle Proj., Stuttgart, E. Schweizerbart'sche, p. 79–87.

———, 1970b, Bewegung und Wärme in der alpinen Orogenese: *Schweiz. Mineral. Petrogr. Mitt.*, v. 50, p. 565–596.

———, 1971a, Das Alpen-Dinariden-Problem und die Palinspastik der südlichen Tethys: *Geol. Rundschau*, v. 60, p.819–833.

———, 1971b, The large-scale kinematics of the western Alps and the northern Apennines and its palinspastic implications.: *Am. J. Sci.*, v. 271, p. 193–226.

Lemcke, K., Büchi, U. P., and Wiener, G., 1968, Einige Ergebnisse der Erdölexploration auf die mittelländische Mollasse der Zentralschweiz: *Bull. Ver. Schweiz. Petrol. Geol. Ing.*, v. 35, p. 15–34.

Lugeon, M., 1940, Erläuterungen zu Blatt 19, *Diablerets*, Geol. Atlas d. Schweiz: Bern, A. Francke SA, 1/25,000.

———, and Gagnebin, E., 1941, Observations et vues nouvelles sur la géologie des Préalpes romandes: *Bull. Lab. Géol., Minéral., Géophys., Musée Géol. Univ. Lausanne*, v. 72, p. 1–90.

Pierce, W. G., 1966, Jura tectonics as a décollement: *Geol. Soc. Am. Bull.*, v. 77, p. 1265–1276.

Reyer, E., 1892, *Ursachen der Deformationen und der Gebirgsbildung*: Leipzig, Wilhelm Englemann, 40 p.

Sanford, A. R., 1959, Analytical and experimental study of simple geologic structures: *Geol. Soc. Am. Bull.*, v. 70, p. 19–52.

Scabell, W., 1926, Beiträge zur Geologie der Wetterhorn-Schreckhorn-Gruppe (Berner Oberland): *Beitr. Geol. Karte d. Schweiz*, N.F., no. 57, pt. III, p. 1–62.

Schaub, H. P., 1936, Geologie des Rawilgebietes: *Eclog. Geol. Helv.*, v. 29, p. 337–407.

Steck, A., 1968, Die alpidischen Strukturen in den zentralen Aaregraniten des westlichen Aarmassivs: *Eclog. Geol. Helv.*, v. 61, p. 19–48.

Trümpy, R., and Bersier. A. 1954, Les éléments des conglomérats oligocènes du Mont-Pélerin: *Eclog. Geol. Helv.*, v. 47, p. 119–166.

———, 1969, Die helvetischen Decken der Ostschweiz. Versuch einer palinspastischen Korrelation und Ansätze zu einer kinematischen Analyse: *Eclog. Geol. Helv.*, v. 62, p. 105–142.

Wohnlich, H. M., 1967, Kleintektonische Bruch- und Fliess-deformationen im Faltenjura: Unpublished dissertation, Basel.

Geological Maps

Geologische Generalkarte der Schweiz, 1 : 200,000:
Bl. 1, Neuchâtel, 1944;
Bl. 2, Basel–Bern, 1942;
Bl. 3, Zürich–Glarus, 1950;
Bl. 5, Genf–Lausanne, 1946;
Bl. 6, Sion, 1942; Bern, Kümmerly & Frey.

RUDOLF TRÜMPY

Swiss Federal Institute of Technology
Zürich, Switzerland

The Timing of Orogenic Events in the Central Alps[1]

The Problem

Episodicity or Continuity of Orogenic Movements

The question of continuity or episodicity of orogenic movements received considerable attention by geologists a few decades ago and is gaining fresh interest today, due to efforts to explain the genesis of mountain belts in light of the plate tectonics model. The controversy can be related to the names of two great geologists. In 1924, Hans Stille put forth his theory on the episodicity and contemporaneity of orogenic movements. He held that these movements had been of relatively short duration, interrupted by much longer times of orogenic quiescence, and that these orogenic phases had been contemporaneous all over the world. The opposite view was taken in 1949 by James Gilluly, in a celebrated presidential address to the Geological Society of America. He demonstrated conclusively that, in some well-documented instances at least, folding movement had been continuous over a long time span, and that it was impossible to distinguish discrete phases.

In the appraisal of these two conflicting ideas, there is or was a certain significant geographic control: the Stille theory was taken very seriously in eastern Europe, it was considered obsolete in America, and in western Europe most geologists used Stille's

---

[1] I have benefited from many fruitful discussions with colleagues in Switzerland and abroad; Emilie Jaeger's help for understanding the significance of radiometric data was especially appreciated. P. Beck-Mannagetta, A. Gansser, K. J. Hsü, P. Hunziker, A. G. Milnes, R. Oberhauser, S. Prey, and E. Wenk read the manuscript or parts of it at various stages and offered most useful criticism. The manuscript was completed in March, 1971.

phase terms without attaching too much importance to them.

The case for continuity of tectonic deformation is an excellent one. It has been conclusively verified for vertical movements and for gravity adjustments to such movements, and also for some types of folds, which seem to be subordinate either to vertical movements or to large strike-slip movements. On the other hand, in many mountain chains we are finally left with very little time available for a complex sequence of processes, such as stripping of cover, folding, thrusting, metamorphism, emplacement of granitoid bodies, erosion, and transgressive overstep of postorogenic deposits. For mechanical reasons as well, it often seems difficult to envisage repeated movement along one thrust plane over a very long period of time, especially if certain segments of this thrust plane had become inactive by subsequent folding, by metamorphism, or by the emplacement of granitoid bodies.

The Case of the Alps

The problem of the timing of orogenic events is here discussed on the basis of Alpine geology. It is needless to repeat that the Alps do not represent *the* model for all or even most other orogenic belts, but rather one possible example. They have the advantage of being fairly familiar, for historical reasons, to geologists from abroad.

For the study of the timing of orogenesis, the Alps are neither better nor worse than many other mountain systems. One major shortcoming is the strong uplift, especially in the Central Alps, which has destroyed most of the synorogenic detrital formations inside the chain. Furthermore, the paleontological dating of some of the Tertiary formations surrounding the Alps is poor. On the other hand, this very uplift affords excellent opportunities for examining the deeper structures of the Alpine edifice; we have at our disposal a great number of geological observations, accumulated during more than a century of investigations, and a large and rapidly growing body of radiometric data is becoming available.

For the Western and Eastern Alps, two outstanding summaries have been presented recently, by Debelmas (1963a, 1963b) and by Oberhauser (1968). The purpose of this paper is to fill the gap between these two syntheses, which is the central segment or the Swiss Alps *sensu lato*.

Little is to be said of the methods used, except that we do not feel competent to speculate on the age of the microtectonic structures. Owing to the scarcity of synorogenic sediments within the chain, much of the evidence is circumstantial, and the sources of error of course increase with each step of reasoning.

The Alps as a Multiple Orogen

Because of the extreme crustal shortening, the Alps are a geometrically and paleogeographically complex mountain chain, in spite of their relatively small size.[2] It appears that they are also complex with respect to the timing of deformation (Fig. 1). The folding and consolidation of different parts of the orogenic belt proceeded in several steps, each in itself complex and extending over a certain time span. These episodes of compressional tectonics (at least two during the Late Cretaceous, a most important one at the end of the Eocene, and two during the Neogene) were separated by longer periods of relative quiescence, characterized by uplifting, erosion, and high-angle faulting.

To this concept of the Alps as a multiple mountain chain we have been led especially by a comparison with the Carpathians (for a summary, see Andrusov, 1964). The Central Carpathians have definitely been folded and consolidated in Late Cretaceous time; later on, they were subjected only to block faulting

---

[2] A schematic cross section through the western part of the Swiss Alps may be found in the contribution to this volume by Lemoine (Fig. 4, p. 205) (Editors' note).

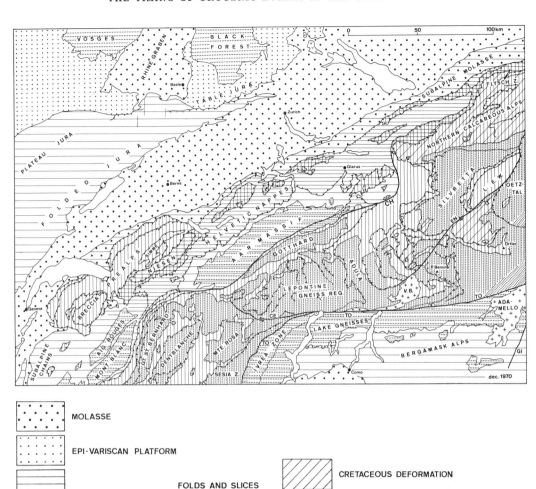

FIG. 1 Map of Switzerland, showing times of deformation (based on a draft by A. Spicher and R. Trümpy for the International Tectonic Map of Europe). (CE) Centovalli-Simplon fault; (CH) Brig-Chur fault zone; (EN) Engadine fault; (GI) Giudicara fault zone; (TO) Tonale (or Insubric) fault; (L.E.W.) Lower Engadine Window; (VB) Val Bregaglia granitoids.

and to vertical movements. The Neogene Flysch Carpathians, or Neo-Carpathians, were formed much later, when a new root zone or subduction zone [3] developed along the northern margin of the consolidated block of the Paleo-Carpathians. The Pieniny

---

[3] We are here using "subduction" more or less in its original sense (Ampferer's *Verschluckung*), i.e., for the down-drawing of *crustal* elements. The existence of crustal subduction zones in the Alps must also involve subduction of lithospheric plates at depth, but this cannot be observed directly.

Klippen belt may represent the superficial expression of such an external or secondary subduction zone; from the paleogeographic analysis of the Pieniny rocks it becomes evident that a very broad and complex area is represented by these isolated slices, and that they have been deformed both by the Paleo-Carpathian and Neo-Carpathian orogenies (see, e.g., Birkenmajer, 1970).

The Cretaceous folding of the Paleo-Carpathians can be followed into the southeastern part of the Alps, where we have ample evidence of pre-Coniacian folding and thrusting as well as of Late Cretaceous metamorphism. This Cretaceous (or "Paleo-Alpine") deformation also affected parts of the future Central and Western Alps.

The main body of the Central Alps, comprising the Penninic zone, the western part of the Austroalpine nappes, and probably also the Gotthard massif, owes its deformation and metamorphism essentially to the Paleogene (or "Meso-Alpine") orogenic phase. It took place during a surprisingly short time span, between early upper Eocene and late lower Oligocene.

The Helvetic nappes, on the northern margin of the chain, were only formed during a later, Mio-Pliocene (or "Neo-Alpine") phase. A new subduction zone developed north of the consolidated Meso-Alpine block, along the present valleys of the Upper Rhône and the Vorderrhein (Anterior Rhine). The dating of the Neo-Alpine deformation in the Central Alps is not very accurate, so that it is difficult to decide whether the early Pliocene folding of the Jura Mountains was a separate act or not.

The Cretaceous Orogenies

Eastern Alps

Evidence of Cretaceous folding and thrusting has been known for a long time and from various regions of the Eastern Alps, although the discussion on the relative importance of Cretaceous and Tertiary movements is still going on.

For a summary of the Cretaceous tectonic phases, we refer to Tollmann (1963, 1964a, 1964b, 1966) and especially to Oberhauser (1964, 1968). Polygenic breccias with components of Hallstatt nappe rocks and of phyllites in the Lower Cretaceous of the Tyrolean nappes south of Salzburg furnish a first indication of tectonic movements. A "pre-Cenomanian" (more probably pre-upper Albian) phase is indicated by an angular unconformity over gentle folds in lower units of the western Calcareous Alps. Important Albian or even late Aptian movements also occurred in the Salzburg area. At the same time, flysch sedimentation begins—probably quite a good criterion for compressive orogenic events. The abundance of chrome spinel grains in the Flysch and Gosau groups of the northern Austroalpine units, from the Cenomanian up to the Campanian (Woletz, 1962), implies the emergence of ultrabasic rocks, presumably in the southernmost part of the Penninic eugeosynclinal realm.

The most important Cretaceous phase of folding and thrusting in the Eastern Alps predates the deposition of the Gosau group (Coniacian to lower Paleocene; in some basins, marine sedimentation continued into the Eocene). In the higher units of the Northern Calcareous Alps, especially east of the Inn valley, the Gosau beds overstep on folds and décollement thrusts, especially on those of the highest, "Juvavic" nappes.

The pre-Gosau phase is well bracketed upward (pre-Coniacian) but not as well downward, since middle and early Upper Cretaceous formations are rare in the area in which the Gosau formations occur. However, beds of at least early Aptian age have been involved in the pre-Gosau folding. Intra-Gosau movements (especially Campanian) are also documented and must have wrought a major change in Alpine paleogeography, as the detritus from ultrabasic rocks (chrome spinel) disappears rather suddenly.

The Cretaceous phases in the Eastern Alps were accompanied by metamorphism of greenschist to albite-epidote-amphibolite grade. The Rb-Sr ages of micas from the southeastern part of the Oetztal thrust mass and from the Triassic sediments overlying it furnish values of 90 to 77 million years; similar ages have been found east of the Tauern window (Oxburgh and others, 1966; Schmidt and others, 1967; Lambert, 1970). The radiometric ages would correspond to Turonian–Santonian; the metamorphism itself should be somewhat older. We suspect that the belt of Cretaceous metamorphism extends west into the southeastern part of Graubünden (Val Müstair) and into the upper reaches of Valtellina (?), but there is yet no proof of this.

The Cretaceous folding was thus very pronounced in the more southerly part of the Eastern Alps. Tollmann (1963) even proposes that the Penninic nappes now exposed in the Tauern window were already overthrust by Austroalpine sheets in the Early Cretaceous. This, however, is highly improbable (Oberhauser, 1964, 1968), since the formations of the Flysch belt on the northern margin of the Alps are certainly of Penninic origin and must represent the sheared-off cover of the Tauern Schistes Lustrés; these formations include beds of latest Cretaceous age, and at their eastern extremity, in the Wienerwald, they even comprise Eocene sediments. In Graubünden, upper Paleocene rocks underlie the Austroalpine thrustsheets as far south as Oberhalbstein, and in the Lower Engadine window the presence of Maestrichtian is proved by fossils (Torricelli, 1956), while that of Paleocene may at least be suspected.

Central and Western Alps

The problem of Cretaceous folding in the Western Alps is a difficult one. In the Devoluy mountains of Dauphiné (France), strong Turonian folds trend E–W, at right angles to the Mio-Pliocene folds (see Glangeaud and d'Albissin, 1958). Eocene (Pyrenean) folds occur in the same area and farther west (Flandrin, 1966). It is probable that this Dévoluy orogeny extended eastward, into the internal zones of the Alps, and it is not even excluded that it represents a western extension of the pre-Gosau folding. In the southernmost part of the Helvetic realm, the Maestrichtian Wang formation (shales and siliceous limestones) lies unconformably on Campanian to Barremian, in the northern Ultrahelvetic nappes even on Upper Jurassic formations. This Campanian phase, apparently coeval with the main intra-Gosau folding of the Eastern Alps, produced strong warping but not necessarily true folds.

All other indications of Cretaceous folding in the Penninic and Helvetic nappes are very vague. Lugeon's idea (1938) of a transgression of Maestrichtian Niesen Flysch on superimposed slices of Triassic and Jurassic rocks has been criticized by Badoux (1970). Staub (1937) suggested that Upper Cretaceous flysch overstepped on the Schams nappes of Central Graubünden. This view receives some support from recent investigations (W. Nabholz, personal communication).

Some Late Cretaceous folding is probable for parts of the two Penninic eugeosynclines. Flysch sedimentation begins in the Cenomanian, both in the Piemont geosyncline (including its Ligurian offshoot) and in the more external Valais eugeosyncline. The Cretaceous Simme and Helminthoid Flysch, among others, belong to the former, the Niesen, Wägital, Prätigau, and Vorarlberg Flysch to the latter facies belt. Toward the end of the Cretaceous Period (Campanian–Maestrichtian) flysch sedimentation encroached northward on the southeastern part of the former Ultrahelvetic continental slope (Cretaceous part of the Schlieren and Sardona Flysch).

It seems highly probable that typical flysch sedimentation occurs while compressional deformation is going on in an adjoining area. Cretaceous folding is probable for part of the Piemont belt—especially for its southernmost (e.g., Arosa) zone, which must have undergone a Cenomanian phase (see below)—and for part of the Valais belt;

but we have only vague indications as to the age and nature of this disturbance. A Cretaceous metamorphism in the Piemont belt of the Western Alps is also suspected by Vialon (1967), although on equivocal evidence.[4]

On the other hand, there are broad zones in the Central and Western Alps for which we have good evidence that no major Cretaceous disturbance took place, sedimentation being more or less continuous and conformable all through the period. This is the case for the Helvetic–Dauphiné realm (except for the small areas affected by the Turonian Devoluy and the Campanian pre-Wang folding) and for the complex Briançonnais platform, between the Valais and Piemont eugeosynclines (represented also in the Klippen and Breccia nappes of the Prealps).

The Significance of 180 to 110 Million Year Radiometric Data

Radiometric ages between 180 ($\pm$ Pliensbachian) and 110 million years ($\pm$ Aptian–Albian) have been signaled from different areas of the Central Alps, including the Middle Penninic nappes of Graubünden (Hanson and others, 1969) and Valais (Hunziker, 1970), and the Ivrea zone of the Southern Alps (McDowell and Schmid, 1968). Some of these "Jurassic" and "Cretaceous" ages are clearly mixed ages, due to Tertiary heating and crushing of Hercynian granitoids or metamorphic rocks (e.g., Mittagfluh granite in the Aar massif; see Jäger and Faul, 1959). For others, this easy way out does not seem practicable. The "Mesozoic" ages have been measured especially in the frontal part of gneiss nappes. Hanson and others (1969) have suggested that this was due to a special tectonization of these nappe fronts. It seems more probable that the particular situation of the nappe fronts is due to the circumstance that they were deformed under lower temperature than the proximal parts of the same nappe bodies. For the time being we must accept the indication that some event—crystallization, heating, cooling, or shattering—affected some Penninic rocks during the Jurassic and the Early Cretaceous.

All that the field geologist can contribute to this question is essentially negative: there are no folding phases known in the Western Alps that might fit into this time span. The existence of Pliensbachian folding in the Helvetic nappes (Collet and Paréjas, 1946) has been disproved by Baer (1959); Nabholz' (1951) assumption of Liassic thrusting in the Lower Penninic nappes of Graubünden rests on uncertain evidence. Of course, this is no reason to exclude the possibility of folding; but there is no positive indication for it and, in those units where the stratigraphy is well established, no folding can be detected.

The following events might have had some bearing on the radiometric evolution of Alpine rocks:

1. In the Pliensbachian (ca. 180 million years according to the 1964 Phanerozoic time scale), a first deepening of geosynclinal troughs, beginning of oceanic conditions (weakening of the crust?).
2. In the late Bajocian or Bathonian (ca. 170–165 million years), a still poorly understood event effecting major changes in Penninic paleogeography and giving rise to flyschlike intercalations in several sequences.
3. Between Barremian and Albian (ca. 115–100 million years), the outflow of most basaltic ophiolites and possibly also the emplacement of serpentinite bodies.

The "Paleocene Restoration" and Some Remarks on the Migration of Flysch Troughs

Sedimentation in the northernmost part of the Alpine geosynclines, on the Helvetic

---

[4] Recent radiometric work (E. Jäger and others) seems to indicate that Late Cretaceous deformation of the lower Austroalpine, Piemont, and possibly also Valais belts was stronger than here suspected (footnote added during impression).

miogeosynclinal shelf and the complex Ultrahelvetic slope, shows a curious repetition of tectonic cycles.

In the Helvetic area proper, the Lower Cretaceous has shallow-water limestone and shale facies. The glauconitic, condensed "Gault" formations (upper Aptian to lower Cenomanian) mark a change in general conditions. Upper Cenomanian to lower Campanian are represented by pelagic deposits, limestones (Seewen formation) overlain by shales (Amden shales). A flysch-like sedimentation appears with the Maestrichtian Wang formation. Farther south, in the Ultrahelvetic Schlieren and Sardona sheets, there is typical Maestrichtian flysch.

This Upper Cretaceous sequence (neritic → condensed → pelagic → flysch) represents the normal evolution of sediments preceding an orogenic deformation. But the Paleocene, instead of being a phase of folding, is a time of emersion without major tectonic events in the Helvetic nappes. Shallow seas again transgressed from the south during the later part of the early Eocene, the middle Eocene, and the early upper Eocene. They deposited Nummulite limestones, quartz sandstones, and Assilina glauconites, followed by pelagic Globigerina shales and finally by flysch again, this time for good. In the Ultrahelvetic nappes, thick-bedded Paleocene quartz sandstones are intercalated between typical Maestrichtian and typical lower to upper Eocene flysch (Sardona quartzite of the Sardona sheet and Guber sandstone of the Schlieren sheet).

This "Paleocene restoration" marks a break in the evolution of the geosyncline. The author takes this as an indication that the Cretaceous phases of folding did not simply continue into the Paleogene phases, but that they were followed by an episode of relative tectonic quiescence. The next act of the Alpine drama does not begin until Eocene time.

Evolution of flysch sedimentation apparently does not follow a simple and straightforward pattern in the course of time. The same is true for its evolution in space; there is not a single wave of migrating flysch troughs. Flysch sedimentation begins in the Cenomanian, both in the southern part of the Piemont belt and in the Valais belt. In the Eocene it reaches the intervening Briançonnais belt and also the Helvetic belt. Wunderlich's (1967) method of determining the origin of flysch nappes simply by the age of the flysch formations is fallacious.

The Paleogene Orogeny

Date: Western Alps

The Paleogene (Eocene and/or early Oligocene) phase is certainly the main orogeny in the whole of the Penninic realm, in part of the Ultrahelvetic belt, in the western and northern part of the Austroalpine nappes, and in part of the Southern Alps. The classical Alpine metamorphism can also be ascribed to this phase, although Late Cretaceous crystallization is proved for part of the Austroalpine nappes and may be suspected in part of the Penninic eugeosynclinal belts.

For the Western Alps, we may refer to the excellent summaries by Debelmas (1963a, 1963b). Modern outlines of structural and paleogeographic evolution will be found in Barbier and others (1963) and in Debelmas and Lemoine (1970).

Debelmas stresses the existence of a preliminary phase of late Paleocene or early Eocene age. In many units, this disturbance is merely attested by a break in sedimentation and may correspond to nothing more than the general uplift of the "Paleocene restoration." In others, the date of the (gravitational?) advance of flysch nappes, especially of the Helminthoid Flysch slipsheets, is established by the age of the youngest underlying formations, which generally also show flysch facies, on the assumption that flysch sedimentation was terminated by the arrival of the slipsheets in the basin. This assumption is in itself quite plausible, but it can be proved only exceptionally, in cases where the underlying flysch passes upward into wildflysch with boulders and olistolites from the overriding nappe. In other cases, the arrival of higher units was apparently preceded by some erosion of the lower units, which implies that it may have taken place considerably later than the deposition of the latest underlying formations preserved. In spite of these reservations, the superposition of the exotic slipsheets of Piemontese or more internal (easterly) origin upon formations ranging from lower Eocene in the east to upper Eocene in the west may well be accepted as an indication of gravitational movements following a generalized uplift of the easternmost parts of the Alpine edifice. We refer especially to the important work of Kerckhove (1969).

In a more external part of the Western Alps, there is one area at least for which strong compressional deformations of pre-late Eocene age are definitely proved. This is the *cordillère tarine* (Barbier, 1948, 1956), on the margin of the Ultradauphinois zone. The Aiguilles d'Arves Flysch, which is dated as upper Eocene in its upper part and may reach into middle Eocene in its lower part, transgresses onto sharp folds and slices of older rocks. The downward bracket of this folding is not well established, but pelagic limestones of Upper Cretaceous age occur nearby, so that it is presumably Paleocene or early to middle Eocene. The widespread occurrence of coarse Eocene flysch in the Ultradauphinois and Ultrahelvetic belts of the Western Alps implies the existence of contemporary uplifts, presumably due to folding; the cordillère tarine may just be an exceptional outcrop area where the overstep of flysch on fold structures can actually be observed.

At any rate, the main deformation of the Western Alps occurred somewhat later, at the end of the Eocene and/or the beginning of the Oligocene. The time of this Mesoalpine paroxysm is very well established in the Western Alps. It affected the whole Penninic belt, the Ultrapenninic units to the east and probably also part of the Ultradauphinois belt to the west.

The lower bracket is determined by the age of the youngest formations preserved in this belt. The flysch of the Briançonnais belt is certainly in part of middle Eocene (Lutetian) age and may comprise the early upper Eocene; a similar age is assumed for the nonfossiliferous flysch of the Vanoise mountains in Savoy. In the Median Prealps, most of the flysch is upper Paleocene and lower Eocene, but in at least one place upper Lutetian foraminifera have been found (Frutiger, in Badoux, 1962, p. 37). Farther west, there are large masses of upper Eocene (Priabonian) flysch, but it is not certain whether they were overridden during the main thrusting or only during the subsequent gravity sliding. At any rate, the Meso-Alpine orogeny is post-Lutetian, very probably post-early Priabonian.

The best upper bracket is furnished by the angular unconformity of molasse beds in Liguria, on the inner margin of the southwestern Alps (Lorenz, 1961, 1962). Conglomerates, deposited over a hilly landscape, truncate folded, thrust, and metamorphosed Penninic rocks. The conglomerates contain no diagnostic fossils, but they grade upward into marine sands and shales of middle Oligocene ("Rupelian"[5]) age; they are considered to belong to the lower Oligocene (Lattorfian or "Sannoisian").

This leaves a very brief time—certainly no greater than the interval between late Lutetian and early middle Oligocene, more probably between early Priabonian and late Lattorfian—for the "Meso-Alpine" orogeny. Debelmas, by somewhat more circumstantial evidence drawn from the small Tertiary

---

[5] Roth (1970) has shown that the type Rupelian is of the same age or probably even younger than the type Chattian. Use of Oligocene stage names (and of Tertiary stages in general) is hazardous. Unfortunately, the Tertiary formations surrounding the Alps contain few pelagic micro- and nannofossils or even none at all, so that modern biostratigraphic zonation cannot be applied yet.

basins in the Basses-Alpes (see also Chauveau and Lemoine, 1961), manages to encompass the deformation even more narrowly, in the first half of the early Oligocene. Taking into account the status of Tertiary biostratigraphy based on benthonic foraminifera and megafossils, one should probably avoid overprecision.

A complex sequence of events must be crowded into this very short time span, corresponding only to a few foraminiferal zones and at most to 6 million years. It must have witnessed the initial premetamorphic stripping of sedimentary covers, folding and thrusting of the whole belt (including the later phase of backthrusting at the eastern margin of the Briançonnais belt), metamorphism, uplift, and erosion.

Date: Central Alps

In the Penninic zone of central and eastern Switzerland, the youngest formations known are of late Paleocene or early Eocene ("Ilerdian") age. Eocene flysch is abundant in the Ultrahelvetic-Helvetic realm, and some intra-Eocene faulting and folding is present in the Helvetic realm (Jeannet, 1941; Brückner, 1946; Frei, 1963). The main deformation must at any rate be later than the youngest intra-Alpine sediments of the Penninic realm (upper Lutetian[6]; see above).

The evidence for the end of the paroxysm is less straightforward than in the Western Alps; but by somewhat more devious reasoning we arrive at the same result. Cornelius (1928) and Milnes (1969) have shown that the evidence rests mainly on the Bregaglia (Bergell) granite and quartz diorite ("tonalite") complex. In its southwestern (lower) part, this granitoid body is more or less conformable to the Penninic country rock gneisses and forms a nappe-like structure (Moticska, 1970; H. R. Wenk, 1970), whereas the contact on the northeastern (upper) margin is clearly intrusive and cuts across nappe boundaries (Gyr, 1967). The emplacement of at least the "magmatic" part of the massif is thus definitely later than the main thrusting. Zircons from moraine boulders gave lead-α ages of 25 and 30 million years, with a very wide margin of error (Grünenfelder and Stern, 1960).[7] The Rb-Sr ages on biotites (Jäger and others, 1967; Jäger, personal communication) increase from the southwest part of the granitoid body (18–22 million years) to its northeast part (25–27 million years). These ages are definitely younger than the emplacement of the granite body and in accordance with the biotite ages in the surrounding country rock. They must certainly be considered cooling ages.

Pebbles and large boulders of Bregaglia granitoids occur in the molasse of the Como–Chiasso area, on the southern foot of the Alps (Longo, 1968). The lowest member of the Como Molasse contains a poor fauna of arenaceous foraminifera and is assigned an early Oligocene age by Cita (1957, 1958), also by comparison with more diagnostic faunas in the vicinity. It is followed by a large mass of unfossiliferous conglomerates, which are in turn overlain by more sandy and shaly layers containing a benthonic microfauna of the *Uvigerina mexicana* zone (middle to late Oligocene). Pebbles of quartz diorite (tonalite) and of leucocratic granite are found at the base of the conglomerate mass, while large boulders of absolutely typical Bregaglia granite appear somewhat higher up.[8] This would imply that the Bregaglia body—late

---

[6] Middle Eocene has meanwhile also been found in the Wägital Flysch, from the Valais belt (H. J. Kuhn, thesis Univ. Zürich, in press) (footnote added during impression).

[7] This age has since been confirmed by more detailed studies (Jäger and others, in press). It casts some doubt on the paleontological dating of the Como Molasse; a revision of its fauna is under way (footnote added during impression).

[8] Staub (1934), who was aware of the importance of the transgressive "Tongrian" in Liguria, thought that the Como conglomerates were much younger.

tectonic to post-tectonic with respect to the main thrusting—was emplaced, cooled, and unroofed by erosion of several (about 5?) kilometers of overlying rock before the middle Oligocene.[9]

Complexity of Orogenic Movements

This very short available time span—a few million years at most—witnessed a very complex sequence of tectonic events and concomitant pressure and temperature changes in the part of the Alpine edifice folded during the Meso-Alpine orogeny, including those areas that had already undergone some deformation during the Cretaceous (Paleo-Alpine) phases or during the Eocene preliminary movements.

Microtectonic studies in various parts of the Alps seldom fail to produce evidence of at least three and often more sets of s-planes and l-linears (see, e.g., Chadwick, 1968; Chatterjee, 1961; Milnes, 1968). Correlation with large-scale structures and still more with stratigraphically dated orogenic movements is in most cases still in its beginning, and much more work must be done before a synthesis can be attempted.[10]

Megascopic structures also cannot be interpreted without assuming complex changes of both mechanism and apparent direction of movement. The Schams nappes of central Graubünden furnish a particularly intricate and not yet sufficiently understood example (see Streiff, 1939; Kruysse, 1967).

A little farther to the east, the relationship of Austroalpine nappes (Fig. 2) is much easier to understand (see Arbenz, 1920; Cadisch and others, 1919).

Five movements can be distinguished:

A. Development of folds trending SW–NE in the basement rocks and Triassic cover of the Silvretta nappes.
B. Main movement of the Austroalpine nappes (Staub's Grisonide phase).
C. Thrusting of the Silvretta nappe, with an apparent component to the west (Staub's Tyrolide phase). The thrust plane truncates the Hercynian basement structures as well as the folds (A).
D. Folding of the thrust plane (C) along axes trending WSW–ENE.
E. Strike-slip movement along the Engadine fault.

Phase (A) may be Paleo-Alpine or Meso-Alpine, but (B) and (C) are definitely Meso-Alpine, and (D) and (E) are probably late Meso-Alpine.

Metamorphic events also show a complex development. In many cases, an initial phase of high-pressure, low-temperature metamorphism can be seen, followed by crystallization under more "normal" $P - T$ conditions. The first phase may be related to an older deformation; but it may also have been produced in the same act: rapid loading will cause pressures to rise almost instantaneously, whereas the rise in temperature will take much longer.

Subsequent Faulting: The Tonale Line and Related Faults

The great Tonale fault, or Insubric line (see Fig. 1), delimits the area of Alpine northward thrusting and metamorphism very sharply toward the south (Cornelius and Furlani, 1930; for a recent summary, see Gansser, 1968). The fault is clearly postmetamorphic, since it truncates all isograds of Alpine metamorphism. The dating of the movement along

---

[9] E. Wenk (personal communication) has pointed out that the boulders in the Como conglomerates might also be Hercynian granites from the southern part of the Tambo nappe, near Chiavenna, which are very similar to the Tertiary Bregaglia granites. This, however, would make our point still stronger, since the Tambo nappe occupies a much deeper structural zone than the upper ("magmatic") part of the Bregaglia mass.

[10] Voll (in Nabholz and Voll, 1963, p. 790) concludes from similar structures in the Glarus nappes and in the Lower Engadine window that the two areas were deformed during the same act or sequence of acts. This is contrary to the well-established time difference of deformation in the two regions and may serve as an example for unwarranted extrapolation of isolated observations. The same criticism applies to Chatterjee's work (Milnes, 1968).

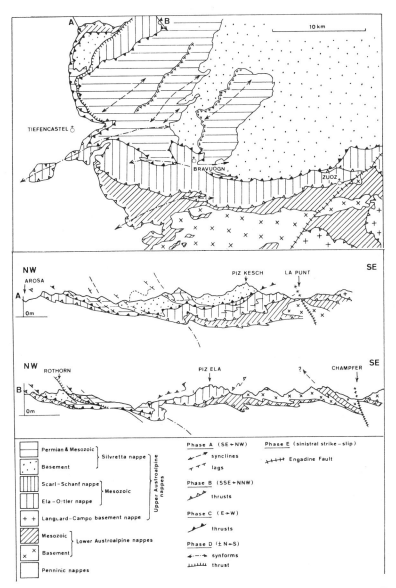

FIG. 2 Map and cross section through the Austroalpine nappes of central Graubünden, eastern Switzerland.

this fault zone is poor; it must have been active during the (middle and?) late Oligocene and the Miocene, but probably no longer during the Pliocene, when the Alps as a whole were leveled down to a chain of fairly low hills. Today, the Tonale fault shows no movement, and it is seismically inactive.

Geometrically, the Tonale fault is clearly related to the main or Meso-Alpine root zone, or rather subduction zone, of the Alps. This belt of steep-standing gneissic rocks has received some attention during the last decades. Several observations have been made which did not seem to fit into the simple classical picture of a root zone. The existence of closed folds with subvertical axial planes was especially troubling to some geologists; even the great Monte Rosa nappe seems to degenerate eastward into a fold of this kind (Blumenthal, 1952; Reinhardt, 1966). However, these folds are not in contradiction to the concept of a subduction zone; in some instances they may also be due to a late deformation within the Meso-Alpine cycle, posterior to the development of the nappes.

The parallelism of the late Eocene to early Oligocene subduction zone and of the late Oligocene to Miocene Tonale fault is not perfect, and the amount of throw varies considerably along the strike. It is modest in the east, in the Gailtal area, where the Southern Alps are offset against Upper Austroalpine (or Middle Austroalpine?) units and grows progressively stronger westward. South of Brenner Pass, the Austroalpine roots are almost completely cut out. The fault is then displaced to the SSE, by the Giudicaria line, where the offset diminishes again. From Dimaro, north of the Tertiary Adamello batholith, the Tonale line resumes its E-W direction and truncates successively the proximal bodies of the Upper and Lower Austroalpine nappes. In the Mera and Ticino valleys the Tonale line reaches its maximum throw, with Alpine sillimanite to the north against a little sericite to the south, 16-million-year-old basement rocks against 300-million-year-old basement rocks, and Lepontinic (Lower Penninic) gneiss nappes against the autochthonous basement of the Southern Alps. The vertical displacement must be of the order of 15 km at least.

Quite possibly, there is also some strike-slip movement, but its importance and even its direction are still uncertain. The roots of all the Austroalpine nappes and of the higher Penninic nappes are cut out completely.[11]

Westward, the throw of the Tonale or Insubric disturbance diminishes again, as a considerable part of the movement is taken over by the Simplon-Centovalli fault (see below). In Piemont, the Tonale fault seems to run into the subduction belt itself, as rocks with strong Alpine deformation occur also southeast of it.

We are thus led to distinguish two separate but certainly related phenomena: an older subduction zone formed about 39 million years B.P., and, more or less on the same site but in places discordantly, a later, Oligocene and Miocene fault belt. The same sequence can be recognized, as we shall see later, along the external, Miocene subduction zone. The lag in time is remarkable. One is of course tempted to ascribe the Insubric faulting to the crustal disturbance caused by the Meso-Alpine subduction.

Two other faults or fault zones appear to be subordinate to the Tonale line. One is the Engadine line, which runs more or less SW-NE from the upper Val Bregaglia along the Engadine valley. It is a complex, sinistral strike-slip fault with a lateral displacement of up to 15 km (first suggested orally by Ch.Ch. Biq; see Trümpy, 1971), which displaces the thrust planes of Penninic and Austroalpine nappes as well as zones of Alpine metamorphism. To the NE, it passes itself into the steep, late, and discordant thrust of the Oetztal over the Silvretta block. Its SW continuation is not yet elucidated. There seems to be some relationship between the Engadine fault and the emplacement of the Bregaglia granitoid body, which occupies the sharp angle between the Engadine and Tonale fault; both strike-slip faulting and

---

[11] There is no reason to assume that these higher units did not overlie the Ticino culmination. Outliers of Penninic and Austroalpine cover nappes lie just north of the culmination, along Lake Lucerne, and the Molasse conglomerates of the Rigi contain abundant Ultrapenninic (Simme) and Austroalpine pebbles. The isograde surfaces are remarkably parallel to the inferred structural surface of late Oligocene times; the crosscutting of nappe boundaries (Wenk, 1956) is not in contradiction to this fact (see especially Niggli and Niggli, 1965). It does not seem absolutely necessary to postulate a special "heat-dome" in the Ticino area. A definite thermic bulge, penetrating into zones of relatively low pressure, only surrounds the Bregaglia granitoid body. [The author had already written this footnote when he received the very important paper by Niggli (1970), and he was greatly relieved to find the guesswork of a naive stratigrapher more or less confirmed by the views of a leading petrologist.]

granite intrusion are late Meso-Alpine events.

The other fault zone is the Simplon–Centovalli line (Bearth, 1956a, 1956b), which branches off from the Tonale line in the Ticino valley and runs first W, then NW. It is also an essentially postmetamorphic fault, with a component of dextral strike-slip. Hunziker (1970) has recently analyzed the complex movements along the postcrystalline disturbance on the basis of the offset of belts of apparent mineral ages—a most interesting attempt, even though not all his conclusions are yet sufficiently corroborated by detailed field studies.

Radiometric Ages, Subsequent Uplift, and Foreland Sedimentation

The Rb/Sr ages for biotites from the central Swiss Alps generally furnish dates of 10 to 25 million years, with a broad belt of 16 to 18 million years in the most highly metamorphic zone of the Lepontinic gneisses and the youngest values in the Simplon area to the NW. Muscovite ages from the same rocks are of the order of 8 million years older than the corresponding biotite ages (Jäger and others, 1967). It is now universally accepted that these essentially "Miocene" ages do not represent a major episode of deformation or crystallization, but rather the time of the closing of the mineral systems, through cooling down to temperatures of about 300°C (biotite) and 500°C (muscovite). Assuming an average temperature gradient, this would correspond to a velocity of uplift and corresponding erosion of the order of 0.6 mm/yr. By considerably more sophisticated reasoning, based on heat-flow measurements in tunnels, Clarke and Jäger (1969) arrive at 0.4–1 mm/yr. Jäckli (1957) gives 0.58 mm/yr for the denudation in the present Rhine basin of Graubünden, so that the order of magnitude is quite plausible. Füchtbauer's (1964, p. 221) estimate of denudation based on molasse sedimentation (1300 m during the "Chattian" and Aquitanian) would lead to a somewhat slower rate, of the order of 0.2 mm/yr. This figure, however, does not account for the material transported southward or washed out of the Molasse basin.

Radiometric ages agreeing with the stratigraphically deduced timing of the Paleogene deformation and metamorphism (latest Eocene to early Oligocene, ca. 39 ± 3 million years) might be expected in areas of lower regional metamorphism, where the mineral systems closed with or shortly after crystallization. In N–S aligned amphiboles from the Gotthard massif, Steiger (1964) found K-Ar ages of 46 million years, whereas randomly oriented amphiboles gave the usual, regional 23- to 30-million-year values. Recently, Hunziker (1970) determined 38 ± 2 million years as an average of 50 phengite and biotite samples from the Monte Rosa nappe.

Deposition of conglomerates in the foreland Molasse basin began in the late Oligocene ("Chattian") and continued until late Miocene (Tortonian, possibly Messinian). Several kilometers of detritus (up to 6 km close to the Alpine margin) were laid down. The pebbles are derived from Paleogene nappes, in western and central Switzerland mainly from Ultrapenninic units (related to the Simme and Dentblanche-Sesia nappes), in eastern Switzerland from Austroalpine nappes. Pebbles from the Neogene Helvetic nappes do not appear before the Helvetian or Tortonian and remain quite subordinate. Rare ophiolite pebbles are found about at the Oligocene/Miocene boundary; they become frequent in the Miocene and show successively higher grades of Alpine metamorphism in the younger Molasse formations (Dietrich, 1969).[12]

---

[12] Dietrich's remarkable work contains unacceptable conclusions insofar as he tries to derive all slightly metamorphic ophiolite pebbles from their present outcrop area in the Platta nappe of the Oberhalbstein valley. This is in flat contradiction to all we know about rates of erosion, past and present, in the Alps. The ophiolite detritus in the older Molasse must come from the westward prolongation of the higher Penninic nappes, above the present Lower Penninic nappes of western Graubünden.

It is remarkable that the bulk of radiometric ages (30–11 million years) and the bulk of Molasse sedimentation (late Oligocene to late Miocene) correspond so nicely. The two phenomena are related insofar as they are both determined by the postorogenic uplift of the mountain chain. This uplift only took place after a long delay, 5–10 million years, following the (39 ± 3 million years) compressional deformation. The uplift seems to have been more or less continuous, whereas the folding took place in a relatively short time.

Formerly, most geologists—with the notable exception of Cornelius (1928; Cornelius and Furlani, 1930)—had a tendency to consider coarse molasse deposition as coeval with paroxysmal phases in the Alps (e.g., Staub, 1934; Trümpy, 1960; Gasser, 1968). This concept is wrong; molasse sedimentation is not related to the paroxysm itself but to the postparoxysmal uplift.

The lag in time reflects the anisostatic or anti-isostatic nature of orogenic processes, especially if we compare it with the almost instantaneous reaction of normal crust to load changes, such as the recent rise of Fennoscandia or Greenland after the melting of the Würm-Wisconsin glacier.

Gravitational Tectonics after the Paleogene Orogeny

The Prealpine nappes of western Switzerland furnish one of the best examples of gravitational structures in the Alps (Lugeon and Gagnebin, 1941).

Their initial individualization and their departure (*mise en marche*) is apparently due to a compressional phase; the basal thrust planes cross-cut stratigraphic units. At any rate, this departure from the central part of the Alps[13] must have taken place before the onset of metamorphism, which means before the end of the early Oligocene. The formations of the Prealpine nappes (Ultrahelvetic nappes, Median Prealps, Helminthoid Flysch, and Simme nappe) are not metamorphic; only two latecomers, the Niesen nappe and the Breccia nappe (including the exotic sheets on its back) show some slight metamorphism. It is therefore logical to conclude that the individualization of the Prealpine nappes, the youngest formations of which are upper Lutetian, date from the main Meso-Alpine orogeny, between early upper Eocene and late lower Oligocene (see also Fallot, 1956).

Their subsequent emplacement (*mise en place*) seems to be largely due to gravitational sliding. Most of the units (excluding the Niesen nappe, which was emplaced only during or after the Neogene orogeny) show pull-apart structures along their rear (SE) margins. The timing of this gravitational movement is very nicely established. The Prealpine nappes override the "Red Molasse," which appears in the Val d'Illiez window; its age is given as "lower Chattian" by the geologists, as "middle Stampian" by the vertebrate paleontologists. On the other hand, the Pèlerin conglomerates, north of Lac Léman (Lake Geneva) and in front of the Prealps, contain abundant pebbles from the Simme nappe and also, to a lesser degree, from Ultrahelvetic units which apparently cropped out at the very margin of the late Oligocene Alps (Trümpy and Bersier, 1954; Elter and others, 1966). The size of the pebbles proves that, by that time—in the "late Chattian," toward the end of the Oligocene—Prealpine nappes were in a nearby position and had already overridden the rocks exposed in the Val d'Illiez window.

We can thus fix the gravitational emplacement of the Prealps in the late Oligocene, at about 30 million years. This is considerably later than the preceding compressional deformation (ca. 39 million years). Like foreland detrital sedimentation and radiometric cooling ages, the emplacement of gravity

---

[13] It is now generally accepted that the Préalpes Médianes nappe has been stripped off from the present Middle Penninic Great St. Bernard nappe, which shows metamorphism of greenschist to albite-epidote-amphibolite grade.

nappes indicates uplift, presumably its rather early stages. The long delay between folding and uplift is again noteworthy.

In eastern Switzerland, the Oligocene episode of gravity sliding is exemplified by the overthrust of South-Helvetic, Ultrahelvetic, and Penninic flysch nappes upon the still unfolded future Helvetic nappes and Autochthonous belt. This movement is later than early Oligocene, since beds of this age, such as the famous Glarus fish slates,[14] are preserved in the underlying North-Helvetic flysch; but the upper bracket cannot be established as well as in more westerly regions.

The late Meso-Alpine advance of Prealpine nappes in the Rhone valley, between middle and late Stampian, caused Stille (1924) to define his "Helvetic phase." Some authors (e.g., Tollmann, 1964a, 1964b) have been misled by this term, which they believed to indicate the date of deformation of the Helvetic zone. The folding of the Helvetic belt is definitely later (late Oligocene to early Pliocene) than the "Helvetic phase," which only affected the "spillover" of the Meso-Alpine edifice.

The Neogene Orogeny

The Helvetic Nappes and the External Subduction Zone

The third segment of the Alps comprises essentially the Helvetic nappes and the autochthonous basement massifs (Fig. 1). Except for the Gotthard massif and its Ultrahelvetic cover, this belt had been hardly affected by the Paleogene orogeny; there was only some local faulting and subordinate folding in the late Eocene. The Oligocene gravitational slip-sheets of Ultrahelvetic and more southerly origin arrived on a fairly level surface, which had previously been only slightly tilted and had probably undergone some erosion. The folding of the Helvetic nappes cannot have begun before late Oligocene; it is essentially of Miocene date.[15] During this Miocene deformation, the Ultrahelvetic sheets behaved as if they were the youngest, incompetent formations of the Helvetic units (see, e.g., Mercanton, 1963).

The younger age limit of the Helvetic folding is not very satisfactory. The Helvetic nappes butt against and override beds of the subalpine Molasse, of late Oligocene and early Miocene ("Chattian," Aquitanian, and doubtful Burdigalian) age. The contact can be interpreted as an epiglyptic thrust (preceded by some erosion of the Molasse). The first, rare pebbles of Helvetic origin are known from the Helvetian (Leupold, Tanner, and Speck, 1942). These two weak indications might suggest placing the main Helvetic phase somewhere around the Burdigalian, with considerable margin of error. On the other hand, the entire Molasse group has been affected by some compression up to formations of Tortonian or maybe Messinian age. In fact, it is difficult to get a better bracket than simply "Miocene" for this Neo-Alpine phase, at least in Switzerland. More information is available farther east.

The Helvetic folding was in itself complex. The author has recently (1969) tried to analyze the sequence of events in the Glarus Alps, where he has been led to distinguish three events after the post-Meso-Alpine sliding of Ultrahelvetic cover sheets: (1) initial overfolding; (2) main thrusting and individualization of digitations; and (3) late uplift with concomitant gravitational readjustments. Phases 1 and 2 represent the

---

[14] The rocks of the Helvetic zone, especially in the Glarus Alps, are affected by low-grade metamorphism (Frey, 1968). It is not yet clear whether this metamorphism is due to the loading by Meso-Alpine nappes or whether it took place only later, during the Neo-Alpine orogeny.

[15] New radiometric age determinations (Frey and Schindler, in press) seem to indicate that at least the first phases still fall within the Oligocene (footnote added during impression).

main "Neo-Alpine" orogeny, whereas phase 3 took place only after a considerable interval during which part of the overburden was removed by erosion. In other parts of the Helvetic belt, Ayrton (1969), Labhart (1966), and Spörli (1966) have also shown that the Helvetic deformation was not a simple process, though we do not find sequences of events as complicated as those in the "Meso-Alpine" part of the Alps.

The Helvetic nappes are derived from the sheared-off cover of the Tavetsch massif, which has since been drawn down along the external zone of crustal subduction, between the Aar–Montblanc and Gotthard basement wedges (see Trümpy, 1963). During the Oligocene, the central, Meso-Alpine chain had become consolidated, and the internal, southern subduction zone ceased to be active. As the European continental plate reassumed its southward and downward movement in the Miocene, it was drawn beneath the northern edge of this consolidated block. The continued subsidence of the Molasse basin may well be related to this same movement.

Subsequent Uplift, Gravitational Tectonics, and Faulting

It is well known that the uplift of the external massifs (Aiguilles Rouges/Montblanc, Aar) occurred only after the folding and thrusting of the Helvetic nappes (see, e.g., Arbenz, 1934). In detail, this needs some specification, since some basement wedges already formed during the folding itself (see, e.g., Käch, 1969); but, in a general way, there is no doubt that the originally subhorizontal thrust planes of the Helvetic nappes were later arched by the uplift of the crystalline wedges.

We do not know exactly when this uplift took place. The great present elevation of the Montblanc and Finsteraarhorn mountain groups might indicate that this uplift continued well into the Pleistocene.

This uplift of the massifs remobilized older thrusts and created new ones. Effects of these posthumous, post-nappe deformations are apparent in the entire belt of the Helvetic nappes. In eastern Switzerland, they have the character of gravity slides and do not necessarily indicate crustal shortening (Schindler, 1959; Spörli, 1966; Trümpy, 1969). Farther to the west, in the Bernese Oberland, where the late uplift of the Aar massif reaches its maximum intensity, the Helvetic thrust surface has itself been folded, and between Grindelwald and Meiringen it is even inverted.

The Penninic Niesen nappe (Fig. 1) was also emplaced during this late Neo-Alpine sliding phase. Its last movement is later than that of the Helvetic nappes (Badoux, 1945). At the time of its gliding, the culminations of the Aar and Montblanc massifs were already in existence; the Niesen nappe is found only in front of the axial (Wildstrubel) depression between the two, and it traveled farthest at its SW and NE ends, where the available slope was greatest. The other Prealpine nappes show quite a different disposition; they lie in front of the Montblanc culmination as well as in front of the Wildstrubel depression.

We have seen that the Tonale–Insubric fault developed after the Paleogene (Meso-Alpine) orogeny, more or less on the site of the internal subduction zone. A similar phenomenon is observed along the northern Miocene (Neo-Alpine) subduction zone. Young longitudinal faults, with strong and late uplift of the northern block and possibly also with a strike-slip component, determine the Rhone, Urseren, and Vorderrhein valleys between Martigny and Chur. At Chur they swing north, around the plunging nose of the Aar massif. It is significant that these faults are not yet wholly dead: there is some seismic activity, and Jäckli (1951) as well as Eckardt (1957) have described late Pleistocene or early Holocene faults that offset local moraines.

The Jura Mountains

The age of the folding of the Jura Mountains is well established. Faulting, in connection with the Rhine graben structures, began in the late or possibly middle Eocene; it continued through the Oligocene and into early Miocene. Some gentle folds are subordinate to these Oligocene faults. Minor local folding went on in the Miocene; it may be interpreted as a reaction to the main deformation of the Helvetic belt. All these disturbances were not yet linked to the general décollement of the sedimentary cover characteristic of the main phase of folding. The folding of the Jura is definitely later than its youngest Molasse formations, that is, post-Messinian. Formerly, it was believed that two phases could be distinguished, one before and one after the "Pontian." Laubscher (1962), however, has shown that the Jura folding is entirely of post-Pontian age. By painstaking analysis of badly exposed and not well-dated sediments on the NW border of the Jura, Liniger (1964) has established a first slight folding after the Pontian followed by the main phase after the middle Pliocene and before the early Pleistocene (Villafranchian). His demonstration, although it involves a fair amount of circumstantial evidence, is convincing.

The relationship between the time of the Helvetic folding and that of the Jura Mountains is not yet clear, because of the poor dating of the former. In the Helvetic belt, late (post-nappe) folds occur only west of the Reuss valley (Spörli, 1966); since this valley lies just opposite the eastern end of the Jura fold belt, one might assume that these posthumous folds were coeval with the Jura folding. East of the Reuss valley, the uplift of the Aar massif would be almost entirely of isostatic nature, whereas a compressional component seems to exist west of this same valley. In the French Subalpine Chains, at any rate, the existence of two phases, one before and one after the "Pontian," is obvious (Goguel, 1936).

By the effect of the Jura folding, the Mesozoic cover rocks underwent a shortening of 15–20 km, while the underlying basement remained undisturbed. The state of seismic exploration of Switzerland being most unsatisfactory, we do not know where the corresponding shortening of the basement top must be sought. The most likely place to expect this small and late ("Epi-Alpine") crustal subduction zone would lie north and below the external basement massifs, which would then represent the margin of the consolidated Alpine block by Pliocene time. Incidentally, the Alps had been worn down to a low chain of hills at the moment of the Jura folding; their present morphology is essentially due to a late Pliocene to early Pleistocene uplift, which was strongest in the segment facing the Jura Mountains and the northern half of the external zone of the French Alps.

Conclusions

If we recapitulate the sequence of orogenic events in the Central Alps, we arrive at the following time table (compare also Fig. 3):

1. Main Hercynian orogeny before the late Namurian ("Sudetic phase"), ca. 310 million years B.P.
2. Late Hercynian disturbances in the Permian, locally leading to the emplacement of large granite bodies; Permian volcanism.
3. Initiation of geosynclinal subsidence during the Triassic (possibly late Permian in the southeastern Alps).
4. Acceleration of subsidence from the Early Jurassic (Pliensbachian) onward, leading to deep-sea conditions. Tensional faulting along the margins between troughs and platforms.
5. Oceanic stage during late Middle Jurassic, Late Jurassic, and Early Cretaceous; emplacement of ophiolites between Late Jurassic and Albian.
6. "Paleo-Alpine" folding and thrusting, in several episodes, between Albian and Maestrichtian. Strong folding and thrusting, with metamorphism, in the SE. Deformation and flysch sedimentation in the two eugeosynclines, starting on the southern margin of each and migrating northward (Fig. 3).

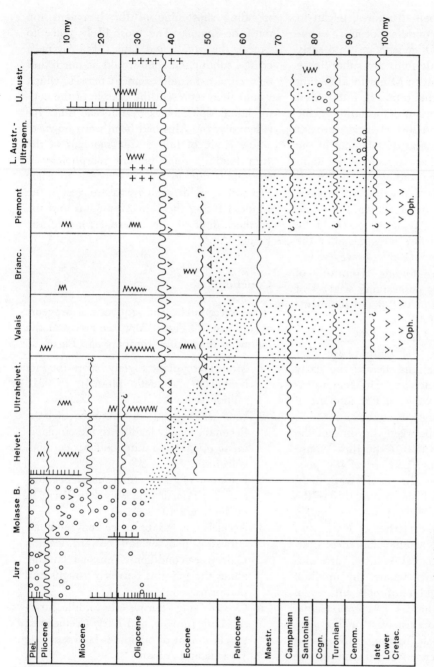

FIG. 3 Progress of Alpine orogenic events. An attempt at diagrammatic representation of orogenic events in the Central Alps.—*Warning*: Readers should not be misled, by the rather apodictic form of this preliminary sketch, to overlook the very slender evidence on which parts of it are based!

7. "Paleocene restoration," stage of relative quiescence and general uplift in the early Paleocene. Possibly followed by some gravity sliding.
8. Some folding and gravity sliding between late Paleocene and late Eocene.
9. Main, "Meso-Alpine" deformation between late upper Eocene and late lower Oligocene. End of flysch sedimentation. Essential part of Alpine metamorphism, together with and followed by the emplacement of Periadriatic granitoid plutons; initiation of Tonale fault. Crustal shortening at least 300 km.
10. Middle and late Oligocene: postparoxysmal uplift of the Meso-Alps. Gravity sheets (especially Prealpine nappes and Penninic flysch nappes) along northern margin. Movement along Tonale, Engadine, and Centovalli faults (continuing into Miocene). Cooling of Middle and Upper Penninic nappes and Gotthard massif (p.p.); beginning of Molasse sedimentation.
11. Late Oligocene and Miocene: "Neo-Alpine" orogeny (Helvetic belt); external subduction zone. Possibly deformation in the northern part of the Meso-Alpine block. Continued uplift in Central Alps (cooling of Lepontinic gneiss region) and continued Molasse sedimentation.
12. Late Miocene and Pontian: postparoxysmal uplift of external basement massifs, with some gravity movements. Rhone–Rhine fault zone (continuing into Holocene). Latest Molasse sediments to N of Alps; latest cooling ages (Simplon area).
13. Pliocene: Alps leveled down; strong sedimentation in the Italian plain. Folding of Jura Mountains.
14. Late Pliocene and Pleistocene: renewed uplift of Central and Western Alps.

This makes for a rather complicated picture, with some parts clear and others considerably blurred. When the author set out to gather information on the timing of orogenic events, he started as a convinced Gillulyan; to his own surprise, he has ended up as a moderate Stillean. Apparently, compressional deformations took place during distinct, relatively short periods; uplift and erosion followed only after a significant delay. Compressional movements appear to be episodic, whereas vertical, strike-slip, and gravity movements are more or less continuous. This also applies to the folding generated by such movements.

In the Alps, at least, the movements involving crustal shortening seem to be spasmodic rather than continuous. For the major, Paleogene ("Meso-Alpine") deformation only a very short time span is available; it can hardly have started before early upper Eocene and it was over before the end of the early Oligocene. This leaves 8 million years at most, during which a crustal shortening of at least 300 km was accomplished; the minimum rate of horizontal movement would be 4 cm/yr. This estimate is certainly on the low side; the actual duration of the phase may have been shorter (but certainly not longer) and the crustal shortening greater (certainly not less). A rate of 6 cm/yr is perhaps more probable. Fallot (1960) gave speeds of only 5–6 mm/yr; but he allowed for less shortening and for a much longer time—the entire Oligocene epoch—than we may allow now, after Lorenz' and Cita's paleontological results on the postorogenic formation in Liguria and Lombardy have become known.

For the Neogene deformation, with a minimum crustal shortening in central Switzerland of about 60 km, it is not possible to give such figures, since the timing is much less well bracketed. Taking into account the mechanics of overthrusting, Hsü (1969) has recently computed 2 cm/yr for the horizontal movement along the Glarus thrust. His calculations rest on a number of (plausible) assumptions. Of course 2 cm/yr is a magic number, since it coincides with the calculated rate of present North Atlantic ocean floor spreading. For the Meso-Alpine orogeny, and also for the well-dated folding of the Jura

Mountains and of the French Subalpine Chains, the rates are significantly higher. Rates of this order concern compressional nappe movements; in gravitational sliding, the velocities may be lower. But even for these phenomena, Wunderlich's (1965, p. 456) estimate of "some mm per century" seems to be far too low.

Vertical uplift and erosion of the chain, on the other hand, run at rates of the order of 0.5–1 mm/yr. Similar figures are obtained whether one considers the present rate of denudation, molasse sedimentation, heat flow data, or the apparent age differences of metamorphic minerals.

The essential deformation of the Alpine edifice took place at depth. It was not immediately preceded or accompanied by the development of morphological bulges; stratigraphic analysis of the surrounding areas during the times just before compressional deformation fails to produce space available for geotumors from which the nappes might be derived. It is noteworthy that we find very few detrital formations corresponding in age to the late upper Eocene and/or early lower Oligocene main folding of the Alps; and those we have—the North-Helvetic flysch with Taveyannaz sandstones—are derived from a chain of andesitic and basaltic volcanoes rather than from developing nappes. Only some wildflysch deposits may be strictly synorogenic.

Uplift follows only after the compressional deformation, and often with an appreciable delay of several million years. This uplift is accompanied by vertical and strike-slip faulting, mainly in the area of former subduction zones. Radiometric ages and detrital molasse sedimentation reflect this uplift, not the folding itself.

This subsequent uplift will allow the gravitational sliding of superficial décollement sheets. The Ultrahelvetic and Prealpine nappes were emplaced, through gravity movements, in the middle and late Oligocene, considerably after the late Eocene/early Oligocene main orogeny. The late sliding phase of the Helvetic nappes took place only during and after the post-nappe uplift of the Aar massif. Glangeaud (1956) was probably the first author to make the sharp distinction between the synchronous, rapid compressional deformations and the subsequent gravitational readjustments, which are stretched out over a longer time span.

The recognition of gravity as a motor of nappe movements, to which R. W. Van Bremmelen has made outstanding contributions, has had an essential impact on our understanding of Alpine structures. It did away with the simplistic and monocausalistic interpretations of classical Alpine geology. In chains where the superficial part of the structure is still preserved, as in the northern Apennines or in the Flysch Carpathians, which the mid-Oligocene Alps may have resembled, its importance is obvious. But the "real" tectonics, involving crustal shortening of many hundreds of kilometers, went on in deeper levels and cannot be ascribed to vertical tectonics and concomitant gravitational reactions alone. It is impossible to understand the Alpine structures and especially the kinematic development of these structures without assuming plate movements of some kind, although these movements need not conform to our present, still crude model. The fact that a major disturbance affected the northern Atlantic 39 million years ago (Laughton and others, 1970), at the very moment of the main crustal shortening in the Alps, is a remarkable coincidence indeed.

References

Andrusov, D., 1964, *Geologie der tschechoslowakischen Karpaten*: Berlin, Akad.; Bratislava, Slowak. Akad. Wiss., 263 p.

Arbenz, P., 1920, Über die Faltenrichtungen in der Silvrettadecke Mittelbündens: *Eclog. Geol. Helv.*, v. 16. p. 116–119.

——, 1934, *Die helvetische Region: Geol. Führer Schweiz*: Basel, Wepf, 2a, p. 96–120.

Ayrton, S., 1969, Déformations des séries autochtone

et helvétique au SE du massif du Mont Blanc: *Eclog. Geol. Helv.*, v. 62, p. 95–104.

Badoux, H., 1945, La géologie de la Zone des Cols entre la Sarine et le Hahnenmoos: *Mat. Carte Géol. Suisse*, N.S., v. 84, 70 p., 4 pls.

——, 1962, Géologie des Préalpes valaisannes (Rive gauche du Rhône): *Mat. Carte Géol. Suisse*, N.F., 113, 78 p.

——, 1970, Les Klippes Niesen du Chamossaire (Alpes vaudoises): *Bull. Soc. Vaud. Sci. Nat.*, v. 70, no. 332, 6 p.

Baer, A., 1959, L'extrémité occidentale du Massif de l'Aar (Relations du socle avec la couverture): *Bull. Soc. Neuchâteloise Sci. Nat.*, v. 82, p. 5–160.

Barbier, R., 1948, Les zones ultradauphinoise et subbriançonnaise entre l'Arc et l'Isère: *Mém. Carte Géol. Fr.*, 291 p., 7 pls.

——, 1956, L'importance de la tectonique "anténummulitique" dans la zone ultradauphinoise au N du Pelvoux: la Chaîne arvinche: *Bull. Soc. Géol. Fr.*, v. 6, p. 355–370.

——, and others, 1963, Problèmes paléogéographiques et structuraux dans les zones internes des Alpes Occidentales entre Savoie et Méditerranée, in Livre à la Mémoire du Professeur Paul Fallot: Paris, Soc. Géol. France, v. 2, p. 331–375.

Bearth, P., 1956a, Zur Geologie der Wurzelzone östlich des Ossolatales: *Eclog. Geol. Helv.*, v. 49, p. 267–278.

——, 1956b, Geologische Beobachtungen im Grenzgebiet der lepontinischen und penninischen Alpen: *Eclog. Geol. Helv.*, v. 49, p. 279–289.

Birkenmajer, K., 1970, Pre-Eocene fold structures in the Pieniny Klippen Belt (Carpathians) of Poland: *Stud. Geol. Polonica*, v. 31, 77 p.

Blumenthal, M. M., 1952, Beobachtungen über Bau und Verlauf der Muldenzone von Antrona zwischen der Walliser Grenze und dem Locarnese: *Eclog. Geol. Helv.*, v. 45, p. 220–262.

Brückner, W., 1946, Neue Konglomeratfunde in den Schiefermergeln des jüngeren helvetischen Eocaens der Zentral- und Ostschweiz: *Eclog. Geol. Helv.*, v. 38, p. 315–328 (1945).

Cadisch, J., Leupold, W., Eugster, H., and Brauchli, R., 1919, Geologische Untersuchungen in Mittelbünden: *Vierteljschr. Natf. Ges. Zürich*, v. 64 (Heim–Festschrift), p. 359–417.

Chadwick, B., 1968, Deformation and metamorphism in the Lukmanier region, Central Switzerland: *Geol. Soc. Am. Bull.*, v. 79, p. 1123–1150.

Chatterjee, N. D., 1961, The alpine metamorphism in the Simplon area, Switzerland and Italy: *Geol. Rundschau*, v. 51, p. 1–72.

Chauveau, J. C., and Lemoine, M., 1961, Contribution à l'étude géologique du synclinal tertiaire de Barrême (moitié nord): *Bull. Carte Géol. Fr.*, v. 264, p. 79–157.

Cita, Maria B., 1957, Studi stratigrafici sul terziario subalpino lombardo: Sintesi stratigrafica della gonfolite. *Riv. Ital. Pal. Strat.*, v. 63, p. 79–157.

——, 1958, Litofacies e biofacies della Gonfolite Lombarda. *Boll. Soc. Geol. Ital.*, v. 77, p. 39–48.

Clark, S. P., and Jäger, E., 1969, Denudation rate in the Alps from geochronologic and heat flow data: *Am. J. Sci.*, v. 267, p. 1143–1160.

Cornelius, H. P., 1928, Zur Altersbestimmung der Adamello- und Bergeller Intrusion: *Sitz.-Ber. Akad. Wiss. Wien, math.-naturw. Kl.*, Abt. 1, v. 137, p. 541–562.

——, and Furlani-Cornelius, M., 1930, Die insubrische Linie vom Tessin bis zum Tonalepass: *Denkschr. Akad. Wiss. Wien, math.-naturw. Kl.*, v. 102, p. 207–301.

Debelmas, J., 1963a, Plissement paroxysmal et surrection des Alpes franco-italiennes: *Trav. Lab. Géol., Grenoble*, p. 125–171.

——, 1963b, Essai sur le déroulement du paroxysme alpin dans les Alpes franco-italiennes: *Geol. Rundschau*, v. 53, p. 133–153.

——, and Lemoine, M., 1970, The Western Alps: Palaeogeography and structure: *Earth-Sci. Rev.*, v. 6, p. 221–256.

Dietrich, V., 1969, *Die Ophiolithe des Oberhalbsteins (Graubünden) und das Ophiolithmaterial der ostschweizerischen Molasse, ein petrographischer Vergleich* v. 1: Berne, Europ. Hochschulschriften, Reihe 17, Erdwiss., 179 p.

Eckardt, P. M., 1957, *Zur Talgeschichte des Tavetsch, seine Bruchsysteme und jungquartären Verwerfungen*: Diss. Univ. Zürich, 96 p.

Elter, G., Elter, P., Sturani, C., and Weidmann, M., 1966, Sur la prolongation du domaine ligure de l'Apennin dans le Monferrat et les Alpes et sur l'origine de la nappe de la Simme s.l. des Préalpes romandes et chablaisiennes: *Arch. Sci., Genève*, v. 19, p. 279–378.

Fallot, P., 1960, Le problème de l'espace en tectonique: *Festschrift Ernst Kraus*: Berlin, Akademieverlag, Kl. III, v. 1, p. 48–58.

Flandrin, J., 1966, Sur l'âge des principaux traits structuraux du Diois et des Baronnies: *Bull. Soc. Géol. Fr.*, v. 7, p. 376–386.

Frei, R., 1963, *Die Flyschbildungen in der Unterlage von Iberger Klippen und Mythen*: Zürich, Schmidberger und Müller, 175 p.

Frey, M., 1969, Die Metamorphose des Keupers vom Tafeljura bis zum Lukmanier-Gebiet: *Beitr. Geol. Karte Schweiz*, N.F., v. 137, p. 11–298.

Gansser, A., 1968, The Insubric Line, a major geotectonic problem: *Schweiz. Min.-Petr. Mitt.*, v. 48, p. 123-143.

Gasser, U., 1968, Die innere Zone der subalpinen Molasse des Entlebuchs (Kt. Luzern): Geologie und Sedimentologie. *Eclog. Geol. Helv.*, v. 61, p. 230-319.

Gilluly, J., 1949, The distribution of mountain building in geologic time: *Geol. Soc. Am. Bull.*, v. 60, p. 561-590.

Glangeaud, L., 1956, Corrélation chronologique des phénomènes géodynamiques dans les Alpes, l'Apennin et l'Atlas nord-africain: *Bull. Soc. Géol. Fr.*, v. 6, p. 876-891.

———, and d'Albissin, M., 1958, Les phases tectoniques du NE du Dévoluy et leur influence structurologique: *Bull. Soc. Géol. Fr.*, v. 8, p. 675-688.

Goguel, J., 1936, Description tectonique de la bordure des Alpes de la Bléone au Var: *Mém. Carte Géol. France*, v. 35, 360 p.

Grünenfelder, M., and Stern, T. W., 1960, Das Zirkon-Alter des Bergeller Massivs: *Schweiz. Min.-Petr. Mitt.*, v. 40, p. 253-259.

Gyr, T., 1967, *Geologische und petrographische Untersuchungen am Ostrande des Bergeller Massivs*: Diss. Eidgen. Techn. Hochschule, Zürich, 125 p.

Hanson, N. G., Grünenfelder, M., and Soptrayanova, G., The geochronology of a recrystallized tectonite in Switzerland—The Roffna-gneiss: *Earth Plan. Sci. Lett.*, v. 5, p. 413-422.

Harland, W. B., Gilbert Smith, A., and Wilcock, B., eds., 1964, *The Phanerozoic time-scale (Symposium A. Holmes)*: London, Geol. Soc., 458 p.

Hsü, K. J., 1969, A preliminary analysis of the statics and kinetics of the Glarus Overthrust: *Eclog. Geol. Helv.*, v. 62, p. 143-154.

Hunziker, J. C., 1970, Polymetamorphism in the Monte Rosa, Western Alps: *Eclog. Geol. Helv.*, v. 63, p. 151-161.

Jäckli, H., 1951, Verwerfungen jungquartären Alters im südlichen Aarmassiv bei Somvix-Rabius (Graubünden): *Eclog. Geol. Helv.*, v. 44, p. 332-337.

———, 1957, Gegenwartsgeologie des bündnerischen Rheingebietes. Ein Beitrag zur exogenen Dynamik alpiner Gebirgslandschaften: *Beitr. Geol. Schweiz*, Geotech. Ser., v. 36, 131 p.

Jäger, E., and Faul, H., 1959, Age measurements on some granites and gneisses from the Alps: *Geol. Soc. Am. Bull.*, v. 70, p. 1553-1558.

Jäger, E., Niggli, E., and Wenk, E., 1967, Rb-Sr-Altersbestimmungen an Glimmern der Zentralalpen: *Beitr. Geol. Karte Schweiz*, N.F., v. 134, 67 p.

Jeannet, A., 1941, Geologie der oberen Sihltaler-Alpen (Kanton Schwyz): *Ber. Schwyz. Natf. Ges.*, v. 3, (1938/41), p. 3-24.

Käch, P., 1969, Zur Tektonik der Brigelserhörner: *Eclog. Geol. Helv.*, v. 62, p. 173-183.

Kerckhove, C., 1969, La "zone du Flysch" dans les nappes de l'Embrunais-Ubaye (Alpes occidentales): *Géol. Alpine*, v. 45, 202 p.

Kruysse, H., 1967, Geologie der Schamser Decken zwischen Avers und Oberhalbstein (Graubünden): *Eclog. Geol. Helv.*, v. 60, p. 157-235.

Labhart, T. P., 1966, Mehrphasige alpine Tektonik am Nordrand des Aarmassivs. Beobachtungen im Druckstollen Trift-Speicherberg (Gadmental) der Kraftwerke Oberhasli AG: *Eclog. Geol. Helv.*, v. 59, p. 803-830.

Lambert, R. St.-J., 1970, A Potassium-Argon Study of the margin of the Tauernfenster at Döllach, Austria: *Eclog. Geol. Helv.*, v. 63, p. 197-205.

Laubscher, H. P., 1962, Die Zweiphasenhypothese der Jurafaltung: *Eclog. Geol. Helv.*, v. 55, p. 1-22.

Laughton, A. S., and Berggren, W. A., 1970, Deep sea drilling project leg 12: *Geotimes*, v. 15, p. 10-14.

Leupold, W., Tanner, H., and Speck, J., 1942, Neue Geröllstudien in der Molasse: *Eclog. Geol. Helv.*, v. 35, p. 235-246.

Liniger, H., 1964, Beziehungen zwischen Pliozän und Jurafaltung in der Ajoie: *Eclog. Geol. Helv.*, v. 57, p. 57-90.

Longo, V., 1968, *Geologie und Stratigraphie des Gebietes zwischen Chiasso und Varese*: Graz, Müller (Diss. Univ. Zürich), 181 p.

Lorenz, C., 1961, Le bassin oligocène de Bagnasco (Italie, prov. de Cuneo): *Bull. Soc. Géol. Fr.*, v. 3, p. 50-58.

———, 1962, Le Stampien et l'Aquitanien ligures: *Bull. Soc. Géol. Fr.*, v. 4, p. 625-784.

Lugeon, M., 1938, Quelques faits nouveaux des Préalpes internes vaudoises (Pillon, Aigremont, Chamossaire): *Eclog. Geol. Helv.*, v. 31, p. 1-20.

———, and Gagnebin, E., 1941, Observations et vues nouvelles sur la géologie des Préalpes romandes: *Bull. Soc. Vaud. Sci. Nat.*, v. 47, 90 p.

McDowell, F. W., and Schmid, R., 1968, Potassium-Argon ages from the Val d'Ossola section of the Ivrea-Verbano zone (Northern Italy): *Schweiz. Min.-Petr. Mitt.*, v. 48, p. 205-210.

Mercanton, C. H., 1963, La bordure ultra-helvétique du massif des Diablerets: *Mat. Carte Géol. Suisse*, N.F., v. 116, 75 p.

Milnes, A. G., 1968, Strain analysis of the basement nappes in the Simplon region, northern Italy: *23rd Int. Geol. Cong. (Prague)*, v. 3, p. 61-76.

———, 1969, On the orogenic history of the Central Alps: *Jour. Geol.*, v. 77, p. 108-112.

Moticska, P., 1970, Petrographie und Strukturanalyse des westlichen Bergeller Massivs und seines Rahmens: *Schweiz. Min.-Petr. Mitt.*, v. 50, p. 355-443.

Nabholz, W. K., 1951, Beziehungen zwischen Fazies und Zeit: *Eclog. Geol. Helv.*, v. 44, p. 131–158.

———, and Voll, G., 1963, Bau und Bewegung im gotthardmassivischen Mesozoikum, bei Ilanz (Graubünden): *Eclog. Geol. Helv.*, v. 56, p. 755–808.

Niggli, E., 1970, Alpine Metamorphose und alpine Gebirgsbildung: *Fortschr. Miner.*, v. 47, p. 16–26.

———, and Niggli, K. R., 1965, Karten der Verbreitung einiger Mineralien der alpidischen Metamorphose in den Schweizer Alpen (Stilpnomelan, Alkali-Amphibol, Chloritoid, Staurolith, Disthen, Sillimanit): *Eclog. Geol. Helv.*, v. 58, p. 335–368.

Oberhauser, R., 1964, Zur Frage des vollständigen Zuschubes des Tauernfensters während der Kreidezeit: *Verh. Geol. Bundesanstalt Wien*, v. 1, p. 47–52.

———, 1968, Beiträge zur Kenntnis der Tektonik und der Paläogeographie während der Oberkreide und dem Paläogen im Ostalpenraum: *Jahrb. Geol. Bundesanstalt Wien*, v. 111, p. 115–145.

Oxburgh, E. R., 1966, Potassium-Argon age studies across the southeast margin of the Tauern window, the Eastern Alps: *Verh. Geol. Bundesanstalt Wien*, p. 17–33.

Paréjas, E., 1946, Indices d'une orogénèse dans le lias moyen du Ferdenrothorn et autres observations: *Comptes Rendus Soc. Phys. Hist. Nat. Genève*, v. 63, p. 53–54.

Reinhardt, B., 1966, Geologie und Petrographie der Monte Rosa-Zone, der Sesia-Zone und des Canavese im Gebiet zwischen Valle d'Ossola und Valle Loana (Prov. di Novara, Italien): *Schweiz. Min.-Petr. Mitt.*, v. 46, p. 553–678.

Roth, P. H., 1970, Oligocene calcareous nannoplankton biostratigraphy: *Eclog. Geol. Helv.*, v. 63, p. 799–881.

Schindler, C. M., 1959, Zur Geologie des Glärnisch: *Beitr. Geol. Karte Schweiz*, N.F., v. 107, 136 p.

Schmidt, K., Jäger, E., Grünenfelder, M., and Grögler, N., 1967, Rb-Sr- und U-Pb-Altersbestimmungen an Proben des Oetztalkristallins und des Schneeberger-Zuges: *Eclog. Geol. Helv.*, v. 60, p. 529–536.

Spörli, B. K., 1966, *Geologie der östlichen und südlichen Urirotstock-Gruppe*: Diss. Eidg. Techn. Hochschule Zürich, 160 p.

Staub, R., 1934, Grundzüge und Probleme alpiner Morphologie: *Denkschr. Schweiz. Natf. Ges.*, v. 69, 183 p.

———, 1937, Gedanken zum Bau der Westalpen zwischen Bernina und Mittelmeer, I. Teil: *Vierteljschr. Natf. Ges. Zürich*, v. 82, 140 p.

Steiger, R., 1964, Dating of orogenic phases in the central Alps by K-Ar ages of hornblende: *Jour. Geophys. Res.*, v. 69, p. 5407–5421.

Stille, H., 1924, *Grundfragen der vergleichenden Tektonik*: Berlin, Borntraeger, 443 p.

Streiff, V., 1939, *Geologische Untersuchungen im Ostschams (Graubünden)*: Diss. Univ. Zürich, 236 p.

Tollmann, A., 1963, *Ostalpensynthese*: Wien, Deuticke, 256 p.

———, 1964a, Zur alpidischen Phasengliederung in den Ostalpen: *Anzeiger math.-naturw. Kl. Österr. Akad. Wiss.*, v. 10, p. 237–246.

———, 1964b, Uebersicht über die alpidischen Gebirgsbildungsphasen in den Ostalpen und Westkarpaten: *Mitt. Ges. Geol. Bergbaustud.*, v. 14 (1963), p. 81–88.

———, 1966, Die alpidischen Gebirgsbildungsphasen in den Ostalpen und Westkarpaten: *Geotekt. Forsch.*, v. 21, 156 p.

Torricelli, G., 1956, Geologie der Piz Lad-Piz Ajüz-Gruppe (Unterengadin): *Jahrb. Natf. Ges. Graubünden*, v. 85, 83 p.

Trümpy, R., 1960, Paleotectonic evolution of the Central and Western Alps: *Geol. Soc. Am. Bull.*, v. 71, p. 843–908.

———, 1963, Sur les racines des nappes helvétiques, *in Livre à la mémoire du Professeur Paul Fallot*, v. 2: Paris, Soc. Géol. France, p. 419–428.

———, 1969, Die helvetischen Decken der Ostschweiz: Versuch einer palinspastischen Korrelation und Ansätze zu einer kinematischen Analyse: *Eclog. Geol. Helv.*, v. 62, p. 105–138.

———, 1973, Zur Geologie des Unterengadins: *Wiss. Mitt. Schweiz. Nat.-Park* (in press).

———, and Bersier, A., 1954, Les éléments des conglomérats oligocènes du Mont-Pèlerin: *Eclog. Geol. Helv.*, v. 47, p. 120–164.

Vialon, P., 1967, Quelques remarques sur l'étude géologique du massif cristallin Dora-Maira (Alpes cottiennes internes, Italie) et de ses abords: *Trav. Lab. Géol. Grenoble*, v. 43, p. 245–258.

Wenk, E., 1956, Die lepontinische Gneissregion und die jungen Granite der Valle della Mera: *Eclog. Geol. Helv.*, v. 49, p. 251–265.

Wenk, H. R., 1969, Geologische Beobachtungen im Bergell. I. Gedanken zur Genese des Bergeller Granits: Rückblick und Ausblick: *Schweiz. Min.-Petr. Mitt.*, v. 50, p. 321–348.

Woletz, G., 1962, Zur schwermineralogischen Charakterisierung der Oberkreide- und Tertiärsedimente des Wienerwaldes: *Verh. Geol. Bundesanst.*, v. 2, p. 268–272.

Wunderlich, H. G., 1965, Zyklischer Bewegungsablauf beim Vorrücken orogener Fronten und der Mechanismus des Deckschollentransports nach dem Surf Riding- Prinzip: *Geol. Mijn.*, v. 44, p. 440–457.

———, 1967, Orogenfront-Verlagerung in Alpen, Apennin und Dinariden und die Einwurzelung strittiger Deckenkomplexe: *Geol. Mijn.*, v. 46, p. 40–60.

E. CLAR

Institut für Geologie
Universität Wien
Vienna, Austria

Review of the Structure of the Eastern Alps[1]

The Eastern Alps are structurally characterized by the extensive superposition of parts of the Austroalpine (*Ostalpin*) nappe system on the eastern continuation of the Helvetic and Penninic systems of Switzerland. The main objective of this paper is to help clarify the mechanism of this large-scale tectonic transport, which locally exceeded 100 km. I will therefore deal mainly with the central and northern parts of the Eastern Alps and will touch only briefly on the problems of the Southern Alps. In reviews of East Alpine tectonics questions of nappe-classification and nomenclature often play a large role but tend to cloud understanding. I shall try to keep them to a minimum.

The best introduction to the geology of the Eastern Alps is by means of the geological map of Austria and accompanying explanations (*Geologische Übersichtskarte der Republik Österreich*, 1:1,000,000, 1964: *Erläuterungen*, 1966). Other recommended references are those of Oxburgh (1968) and Gwinner (1971). Figure 1 is a generalized tectonic map, and three simplified cross sections are shown in Fig. 2.

The Eastern Alps, like the Western Alps, combine east-west-trending orogenic zones that had different evolutions with regard to sedimentation as well as manner and time of deformation. Counter to general usage we will start our short description in the south in order to contrast from the beginning the Austroalpine system and the "Infra-Austroalpine" units (Helvetic, Klippen, and Penninic zones) overridden by it.

---

[1] I would like to thank Dr. R. Scholten and Dr. K. A. De Jong for their painstaking translation of the German text and for valuable comments and suggestions that were incorporated in the final version of the manuscript. I am gratefully indebted to Dr. W. Schlager for aid and numerous discussions.

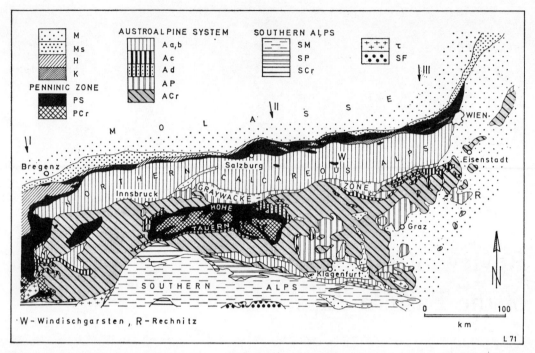

FIG. 1 Simplified tectonic map of the Eastern Alps. (M) Molasse; (Ms) Subalpine molasse; (H) Helvetic zone; (K) Klippen zones; (PS) sediments; (PCr) crystalline basement of the Penninic belt; (Aa, Ab, Ac, Ad) different tectonic positions (a-d) of the Austroalpine Mesozoic cover; (AP) Paleozoic cover of the Austroalpine crystalline basement; (ACr) crystalline basement of the Austroalpine zone; (SM, SP) Mesozoic and Paleozoic covers of the crystalline basement of the Southern Alps; (τ) "Peri-adriatic" intrusives; (SF) southern flysch zone. Arrows I-III indicate cross sections of Fig. 2.

Southern Alps

The Southern Alps are separated from the central part of the Eastern Alps (Fig. 1) by the "Peri-Adriatic fault system," which consists here of the Insubric and Pusteria-Gail faults, with the Judicaria fault between. This system is known to be a zone of strong compression and mylonitization, and the discordant intrusions of Alpine age close to the faults (the "periadriatica": Bregaglia, Adamello, Eisenkappel, and others) underline its importance and deep-reaching influence. However, the Mesozoic covers on either side certainly belong to the same geosynclinal belt and are not fortuitously combined parts of different orogenic systems ("Alpides" and "Dinarides"). It now seems established that the main movement along this fault system was vertical (Cornelius and Cornelius-Furlani, 1930; Van Hilten, 1960; Agterberg, 1961) and that megatectonic longitudinal translation did not occur.

In the context of Austroalpine tectonics the Southern Alps, with its Hercynian basement, acted above all as a block, whose northward movement (or rotation; see below) shortened the central crystalline zone near Innsbruck by about 50 percent. This block was itself affected by steep Alpine faults, descending stepwise toward the Po plain, where Tertiary sediments have been partially overriden along thrusts with a relative southward movement. Many of the deformations in the Mesozoic cover of the Southern Alps have been explained as the result of the action of gravity. As early as 1929 Ampferer described overthrusts, whose mechanism and localization were determined by deep erosional carving (*Kerbwirkung* and *Reliefüber-*

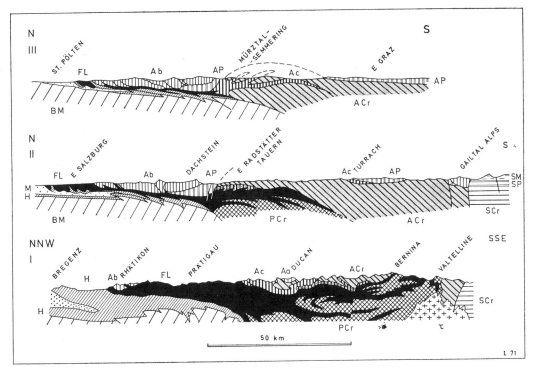

FIG. 2 Simplified north-south cross sections through the Eastern Alps. Locations shown in Fig. 1. (BM) Bohemian massif; (FL) Rheno-Danubian flysch; (H) Helvetic zone and Klippen zones. (Section I after Trümpy, 1970).

schiebung). Regional descriptions and interpretations partly or wholly based on the concept of gravity tectonics are given by De Sitter (1949) and De Sitter and De Sitter-Koomans (1949), Fallot (1950, 1955), Signorini (1951, 1952), and Van Bemmelen (1966, 1970) and his school (e.g., Engelen, 1963; De Jong, 1967; Vink, 1969). In addition, the well-known problem of the summit-overthrusts (*Gipfel-Faltungen*) is discussed by Accordi (1955; in Leonardi, 1967) and Colacicchi (1960).

Austroalpine System

Basement

Although the area of basement exposure north of the Peri-Adriatic fault system is largely devoid of younger sediments (Fig. 1), some remnants of the Mesozoic cover show that its structure and metamorphism are mostly Prealpine. In the southeast the kato- to epimetamorphic basement complex grades into limestones, shales, and volcanics of Ordovician to Devonian age via a metamorphic transition zone and is overlain disconformably by the post-Hercynian sequence. In the west the basement includes also rock masses with still older, pre-Hercynian radiometric ages. Furthermore, a weak superposed Alpine metamorphic phase produced biotite with a Late Cretaceous (80 million year) K/Ar age in the southeastern part of the Triassic cover south of Innsbruck; comparable K/Ar ages have been determined at some places farther east (Oxburgh and others, 1966). This metamorphism doubtless correlates with the Late Cretaceous (early Alpine) tectonic deformation of the Austroalpine sedimentary cover (see below).

The gigantic basement sheet of the Austroalpine system is confined at its base by a low-angle thrust which rises to the NNW, so that the overthrust mass becomes thicker and wider toward the east and south. On the geologic map (Fig. 1) the central zone of the Eastern Alps thus seems to be connected eastward with the so-called "Pannonian massif" (Beck-Mannagetta, 1967), consisting of several northeast-trending zones below the Hungarian basin. Here one problem has to be mentioned: according to Schmidt and Pahr (in Tollman, 1968) the Rechnitz Mountains (R in Fig. 1, east) represent a tectonic window of the Pennides within the Austroalpine basement nappe. The present author does not agree with this interpretation, but if it can be proved, large overthrust movements have occurred in the "Pannonian massif" as well.

During the formation of the Alpine nappes the metamorphic basement sheet was divided into blocks whose marginal parts were affected in different ways by the deformational processes. Close to the basal fault plane exposed along the Penninic windows (especially in the Hohe Tauern, Fig. 1) strong retrograde metamorphism occurred, whereas along other margins simple mylonitization took place, probably characterizing thrusting during later phases. At the northern front of these overthrust sheets at least parts of large masses of quartz phyllite and phyllitic gneiss exposed from west to east near Arlberg-Landeck, Innsbruck, Ennstal, and Semmering have been shown to be a product of phyllonitization and retrograde metamorphism (diaphthoresis).

Sedimentary Cover

When the nappes are placed back in their approximate original Prealpine positions it is seen that the Austroalpine Permo-Mesozoic rocks were generally deposited directly on the pre-Hercynian crystalline basement in the northern and western areas (except for an isolated occurrence of Ordovician volcanics and shales preserved in the western part of the Hohe Tauern window), but on Paleozoic with Hercynian nappe structures in the southeast.

The Mesozoic sequence continues upward into the Lower Cretaceous without major interruptions, but it shows quite remarkable variations in thickness and facies. Generally the rocks are referred to as miogeosynclinal and are up to 3 km thick. These dominantly carbonate rocks are most extensive in the Northern and Southern Calcareous Alps. Coarse-grained breccia deposits, indicating tectonic movements, originated in the northwest as early as the Triassic; they became a facies characteristic at the margin of the more rapidly sinking Pennine trough in the Early Jurassic, and they spread to the south and east during the Late Jurassic. Important Late Mesozoic orogenic movements, including major overthrusting, are demonstrated by angular unconformities and coarse mid-Cretaceous breccias along the northern and northwestern margin of the Calcareous Alps and by the Late Cretaceous, post-Turonian to lower Eocene Gosau formation farther south.

In the eastern part of the Northern Calcareous Alps the numerous outcrops of the transgressive Gosau formation show that subdivision into the main thrust plates (major nappe units) of the Mesozoic cover was accomplished during the Cretaceous, preceding their further deformation during the Tertiary. In the eastern Central Alps the Gosau formation is transgressive on metamorphic Paleozoic rocks, the Mesozoic strata having been removed by erosion or gravity sliding. Finally, in the extreme northeastern part of the Central Alps upper Eocene strata transgress across tectonically deep crystalline units of the Alpine framework.

Rocks of the formerly coherent Mesozoic cover of the Austroalpine system are now found in four tectonically different positions. The subdivisions are indicated by letters Aa–Ad in Fig. 1 and letters a–d in Fig. 3. Their tectonic positions are characterized

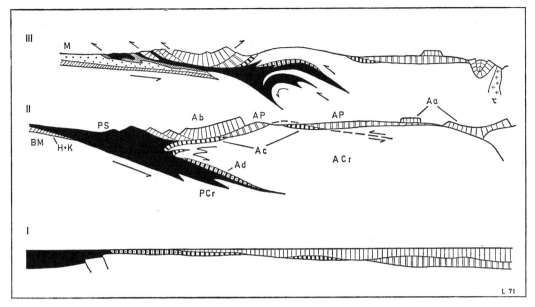

FIG. 3 Schematic representation of the tectonic evolution of the Eastern Alps in a cross section east of Salzburg (section II in Figs. 1 and 2). For key to symbols see Fig. 1.

below. It is likely that this subdivision was already achieved during the Cretaceous. These positions should not be confused with the tectonic units recognized by various authors. Commonly, positions a and b have been assigned to Upper Austroalpine units, c comprises all occurrences attributed either by Staub (1924) or by Tollmann (1963) to Middle Austroalpine (*Mittelostalpin*), whereas position d is typical for Lower Austroalpine (*Unterostalpin*) units.

Subdivision a. This "autochthonous" subdivision consists of Mesozoic rocks without major displacement and still in contact with their original Paleozoic and crystalline basement. Examples (Figs. 1 and 2) include the Ducan and Engadin Dolomites p.p. southwest of Innsbruck, Kalkkögel south of Innsbruck, Krappfeld northeast of Klagenfurt, and the Gailtal-northern Karawanken Mountains west to southeast of Klagenfurt.

Subdivision b. This "allochthonous" Austroalpine subdivision is composed of thrust plates that are completely detached from their original basement and now constitute the Northern Calcareous Alps. For details of the internal structure the reader is referred to the new tectonic maps by Tollmann (1966–1970). Depending on the different lithologic properties of the rocks, the Northern Calcareous Alps have been reduced to approximately half their original width by folding and by thrusting along predominantly south-dipping faults (Figs. 2 and 4). That they are allochthonous can clearly be seen in the Raetikon region at their western end (Fig. 2); between Salzburg and Vienna the existence of a flat-lying overthrust is proved by tectonic windows (W in Figs. 1 and 2) and deep oil exploration wells (Fig. 4). The equivalents of the Northern Calcareous Alps in the western Carpathians are also in an allochthonous position.

In two areas at the southern margin the Lower Paleozoic rocks of the "Graywacke zone," with which the Mesozoic cover is in transgressive relation, form part of the thrust mass and are separated from the central crystalline zone by major dislocations or (in

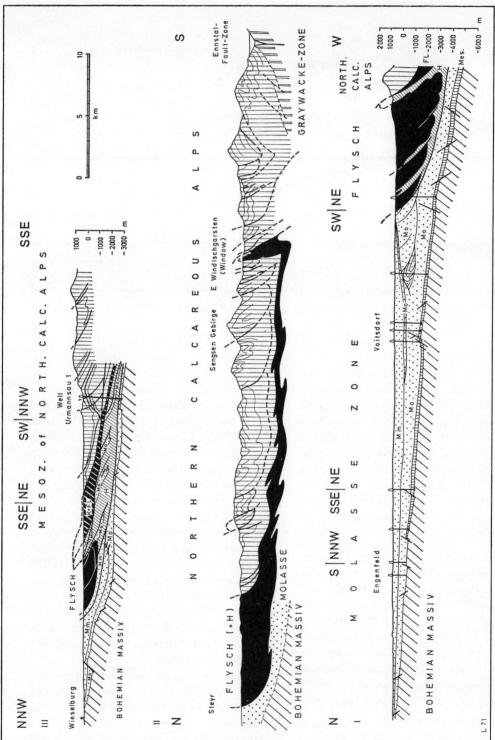

FIG. 4 Three cross sections through the northern front zones of the Eastern Alps. (I) Molasse zone between Salzburg and Linz (east of Gmunden) (after Janoschek and Götzinger, 1969); (II) section approximately south of Linz (after Plöchinger and Prey, 1968); (III) section through the well Urmannsau 1, approximately halfway between Linz and Vienna (after Kröll and Wessely, 1967); (H) Helvetic sequence and "Buntmergel" series; (K, FL) Klippen zones and flysch; (Mo) Oligocene molasse; (Mm) Miocene molasse.

the east) by tectonic intercalations of Mesozoic rocks of position c (see Figs. 1 and 2, and section II, Fig. 4).

The basal detachment plane rises northward at a low angle to the bedding and follows incompetent formations, such as the Permo-Triassic saltbeds. The most southerly tectonic units contain thrust slices down to the crystalline basement, whereas the northernmost units consist only of Upper Triassic and younger sediments. Both this aspect of the internal structure and the southward dip of the basal fault plane demonstrate, in the author's opinion, that the shearing off (décollement) of the sedimentary cover from its basement and the first overthrusting was caused primarily by an external compressive force. After a first detachment phase, however, further transport along the previously formed detachment surface occurred in all probability by gravitational sliding without a push from the south, for in this rearward area the Gosau formation is transgressive on the basement, so that compressive forces could no longer be transmitted by the sedimentary cover. During these later phases complex Upper Cretaceous structures have been transported over great distances across Cretaceous to Eocene flysch. The most impressive example, in the author's view, is that of the Weyer arc, which was very likely modeled after a similar structure in the crystalline basement before it slid 80 km (50 miles) to the north-northwest (Fig. 5).

Subdivision c. Mesozoic rocks in this third type of position were squeezed between the pre-Mesozoic Austroalpine basement blocks. They occur along overthrusts of varying displacement. The most important examples are (1) the Mesozoic rocks of the "Rannach series," which form the base of the Graywacke zone north and northwest of Graz (Fig. 1) and are recognized as Permo-Triassic by Metz (1953); (2) the Mesozoic intercalations in the basal part of phyllites northwest of Klagenfurt; and (3) the Meso-

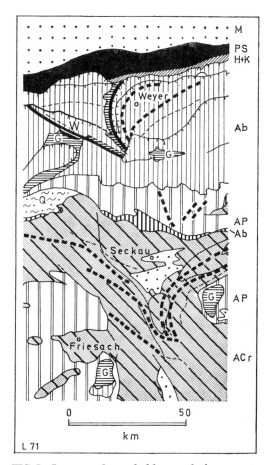

FIG. 5 Large-scale, probably correlative arcuate structures in the Austroalpine basement (south of Seckau) and the gravitationally displaced Mesozoic (Weyer arc). Folding started in pre-Gosau time (Late Cretaceous). Displacement is to the NNW across younger flysch in the Windisch-Garsten window, and probably amounts to about 80 km. (M) Molasse; (PS) flysch; (H + K) Klippen zones; (Aa) "autochthonous" Mesozoic south of Friesach; (Ab) rocks of the Northern Calcareous Alps; (AP) Graywacke zone and Paleozoic rocks near Graz; (Ac) Permo-Triassic "Rannach series"; (ACr) crystalline Austroalpine basement; (Q) quartz phyllite; (G) Gosau formation; (W) Windisch-Garsten window.

zoic of the Brenner area south of Innsbruck. According to Tollmann (1963) intercalations of this type indicate the tectonic limit between his Middle and Upper Austroalpine nappe systems. As shown in the scheme of Fig. 3 this continuation of the basal detachment fault of the Northern Calcareous Alps can be traced to the south downward between phyllites and crystalline rocks, where it disappears into a set of zones of mylonitization and internal deformation.

The Mesozoic rocks of the Semmering–Wechsel area (Fig. 1, east of Eisenstadt), which are involved in overturned, north-driven folds together with their crystalline and phyllitic base, are here comprised under the same position-type c. Far to the west, other examples belonging to this subdivision are the Mesozoic rocks of the Arosa Dolomites, Ela Örtler, Mauls, Vilgraten, and others.

Subdivision d. The fourth type of tectonic position, commonly classified as "Lower Austroalpine," is represented by Mesozoic rocks originally in the frontal part of the enormous Austroalpine overthrust mass and overriden by it as it moved into the subsiding Penninic trough. Probably they never formed a coherent tectonic unit. In the southern part of Graubünden large masses of crystalline rock accompany these elements, but at the margins of the Penninic Tauern window (Fig. 1) they are connected only with thin basement slices.

Several tectonic elements that are attributed to the Lower Austroalpine subdivision in most previous reviews are assigned to different positions in Fig. 1. Thus, as indicated, the Mesozoic of the Semmering nappe system is attributed to our subdivision c, and in westernmost Austria the Falknis and Sulzfluh nappes and the imbricate slices of the "Arosa Schuppenzone" (including the Tasna nappe of the Engadin window) are assigned to the Penninic system of tectonic units, following Trümpy (1970) and Streiff (1939). As thus defined, the Lower Austroalpine elements were nowhere transported farther north than the overturned front of the Austroalpine basement nappes (see Fig. 3). This is of importance when considering the mechanism of overthrusting, frontal folding, and gliding.

Infra-Austroalpine Tectonic Units

Penninic Tectonic Windows

At the western margin of the Eastern Alps the tectonic superposition of the Austroalpine system on the eastward-plunging units of the West-Alpine Penninic system is impressively exposed (Fig. 1; Fig. 2, section I). The highest Penninic units, derived from the south, make up the ophiolite-bearing "Arosa Schuppen zone," which has been dragged to the northern margin of the Calcareous Alps of Bavaria.

The minute tectonic window of Gargellen and the Engadin window reveal only relatively high Infra-Austroalpine tectonic units, including Mesozoic schists (*Bündnerschiefer*) and lower Eocene flysch. In the Tauern window (Fig. 1) the lowest granites and migmatites of Hercynian age may be considered as autochthonous. The Mesozoic cover consists of epicontinental Triassic and eugeosynclinal *Bündnerschiefer* of Jurassic to Early Cretaceous (?) age. Since the youngest formations have been sheared off, it cannot be determined whether the sedimentation continued until the Late Cretaceous and early Eocene, as it did farther west in the form of flysch deposition. The southern parts of the geosynclinal sediments were transported northward by folding and plastic flowage of nappes (*Obere Schieferhülle*). Large, partly submarine masses of basic and ultrabasic igneous rocks allow a comparison of this part of the geosyncline with the Piemont belt (Trümpy, 1960, Fig. 1) in the Western Alps.

The main thrust fault below the Austroalpine system crops out around the Tauern window and is characterized by an imbricate zone composed of upper Penninic units, partly composed of ultrabasites similar to those of the Arosa zone, and Austroalpine elements in a Lower Austroalpine position. All rocks of the Tauern window suffered postkinematic epi- to mesozonal metamorphism. The few datings performed have yielded values in the order of 20 million years (Oxburgh and others, 1966). A retrograde metamorphic zone at the base of the overlying Austroalpine crystalline rocks is apparently related to the same metamorphic phase. A comparable metamorphism with regard to mineral facies in the extreme eastern part of the Eastern Alps (Semmering–Wechsel region) is probably somewhat older.

The final deformation of the rocks of the Tauern window is postmetamorphic and is linked to the doming of the window and the steepening of the northern flank (Frasl and Frank, 1966; Exner, 1964). This young dome is situated where the central zone of the Eastern Alps has been compressed to its smallest width by the northward push of the Southern Alps (see Fig. 1). The axis of the dome of the Engadin tectonic window lies at an angle to the fold axis of the window rocks but is parallel to the Judicaria fault of the Periadriatic fault system.

The considerable shortening of the deep Penninic units is apparently reflected in the high negative Bouguer anomalies near the Tauern (Fig. 6); according to Bott (1954) the cause of these anomalies cannot lie deeper than 10 km. However, the extension of the anomalies into the foreland must have a different explanation; it can perhaps be attributed to the southward slope of the overridden foreland block (see sections, Figs. 2 and 4).

Rheno-Danubian Flysch

This name was suggested by Oberhauser (1968) to distinguish the flysch zone of the Eastern Alps from other flysch zones. The rocks of this zone range from the upper part

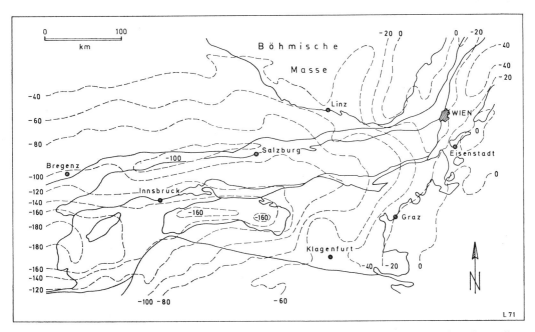

FIG. 6 Map of Bouguer anomalies in the northern part of the Eastern Alps, superposed on the outlines of the major tectonic zones shown in Fig. 1.

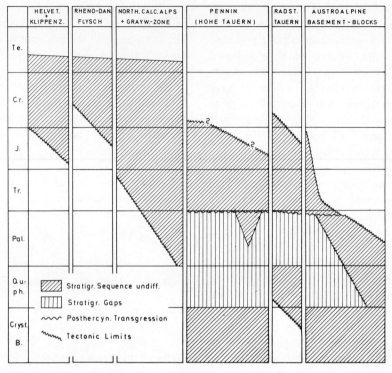

FIG. 7 Schematic representation of the stratigraphic content of the individual tectonic zones. The sheared off and tectonically transported sequences of the Rheno-Danubian flysch and the Northern Calcareous Alps apparently represent continuations of the cover of the Penninic zone and the Austroalpine basement blocks, respectively.

of the Middle Cretaceous into the middle Eocene. They are completely detached from their original basement, intensely folded, and dragged northward in front of and below the Austroalpine nappes of the Northern Calcareous Alps (see Fig. 4, sections I–III). The flysch was derived from a zone south of the Helvetic and north of the Austroalpine facies belt and is therefore considered Penninic in the sense of Trümpy (1960, 1970), Oberhauser (1968), and Prey (1968). Its age suggests that it was once the top part of the sediments of the Penninic trough, sheared off by the northward-moving Austroalpine nappes (Fig. 7). Both with regard to the problem of its derivation and that of the mechanism of its displacement it appears to be quite similar to the Helminthoid Flysch in the French Alps (see Lemoine, 1973). A link with the Carpathian flysch as the visible eastern continuation of the Rheno-Danubian flysch is once more suggested by the negative Bouguer anomalies, which extend ENE into the Carpathian flysch zone, obliquely crossing the structures of the Austroalpine nappes (Fig. 6). Their intensity decreases continuously and significantly. This may justify considering the flysch and Pienninian Klippen zones of the Carpathians as the Cretaceous to Tertiary continuation of the Penninic zone of the Eastern Alps, but with apparent loss of such properties as eugeosynclinal facies characteristics, deep-reaching compression, and metamorphism, and the tendency toward uplift in the final orogenic phase (Prey, 1965, 1968; Metz, 1965).

Helvetic and Klippen Zones

The folds of the Swiss Helvetic zone cross the Rhine and plunge eastward below the overthrust of the Rheno-Danubian flysch. Farther east the zone is found as small imbrications in front of the flysch or in small diapiric windows surrounded by flysch. East of Salzburg the same tectonic position is taken over by the "Gresten Klippen zone." Both zones comprising Jurassic to Eocene rocks have been sheared from their basement and transported northward similar to the Rheno-Danubian flysch and the Northern Calcareous Alps. It is probably important for the reconstruction of the original arrangement of the facies belts that the transition between the two zones takes place at about the same longitude where, farther south, the Penninic units of the Tauern window finally disappear eastward below the crystalline basement of the central zone.

From the extreme west eastward to the vicinity of the Vienna Forest the sequence of the main zones (Molasse, Helvetic or Klippen belt, Rheno-Danubian flysch, and Austroalpine Calcareous Alps) is a purely tectonic superposition. In the Vienna Forest region itself, however, it has not been definitely established whether flysch covers certain Klippen elements only structurally or also stratigraphically. As mentioned previously, it can be argued that the Klippen zones together with the flysch represent the continuation of the Penninic zone of the Eastern Alps with a gradual change in facies and tectonic character.

Molasse Zone

This zone is the middle to late Tertiary foreland trough of the Eastern Alps. It is an asymmetric basin (Fig. 4, section I). South of the Bohemian Massif the Prealpine metamorphic basement, with local remnants of the pre-Tertiary sedimentary cover, sinks slowly and stepwise southward beneath the Alps, where its position is known from geophysical data and drill holes. Sediments in the trough show onlap toward the foreland; southward they thicken to 3000 or 4000 m in front of the Austroalpine nappes, where a marginal zone of "Subalpine molasse" suffered folding and thrusting. Most of the molasse consists of detrital sediments of late Eocene to middle Miocene age. Steep faults in the lower part of the trough are truncated by a disconformity at the base of the lower Miocene Hall Formation. Different deformation of Oligocene as compared to younger strata indicate a low angle forward push of the Alps during molasse sedimentation. Near Perwang and Voitsdorf (Fig. 4, section I) an extensive allochthonous sheet more than 1000 m thick composed of Oligocene strata slid several kilometers to the north in late Oligocene time, in front of the encroaching Alpine nappes and across unfolded molasse of the same age (Janoschek and Goetzinger, 1969).

Tectonic Evolution

In the Eastern Alps part of the Hercynian orogenic belt was remobilized in Alpine time, but with the exception of a few cases, such as probably the Weyer Arc (Fig. 5), it is not yet possible to outline the influence of older structures on Alpine tectonics. As to the time of deformation, some attempts have been made to establish a rigid scheme of many orogenic "phases" in the sense of Stille (Tollman, 1966; Oberhauser, 1968). The present author prefers Trümpy's (1960) classification into five epochs of Alpine history.

First Epoch

In the first epoch the block that would later constitute the Southern and Austroalpine Alps became distinct from the germano-type foreland during the Middle Triassic. The block underwent relatively strong subsidence with deposition of sediments with distinctive facies characteristics and commonly considerable thickness.

Second Epoch

During the Jurassic and Early Cretaceous second epoch ("Schistes Lustrés time" of Trümpy) the area of greatest geosynclinal subsidence shifted toward the Penninic zone, where the shaly and calcareous beds of the Schistes Lustrés (*Bündnerschiefer*) accumulated. Breccias were strewn from the southern margin of the Penninic trough and from uplifts within the trough. One of the uplifts is overlapped by Malm. However, a Penninic geosynclinal trough can be shown to have existed only in the western part of the Eastern Alps (Hohe Tauern); farther east there is no direct evidence apart from the problematic Rechnitz region (see above) and it seems that the facies and tectonic character of this belt changed in this direction.

To the south, in the Austroalpine realm, smaller basins and swells developed as well, in part parallel to the later folding. In the future Lower Austroalpine zone at the northern margin of this realm, adjacent to the Penninic trough, coarse breccias were formed during the Late Jurassic and Early Cretaceous (?), with components of the crystalline basement and large blocks of Triassic dolomite that may have slid off the front of an advancing overthrust mass. Sedimentation in the Northern Calcareous Alps was brought to an end with the deposition of partly clastic Lower Cretaceous rocks.

Third Epoch

Although during the third epoch ("Flysch time" of Trümpy) the Rheno-Danubian flysch was deposited uninterruptedly from late Lower Cretaceous to early Eocene, this epoch can be divided into two periods in the Eastern Alps.

During the first period the main structures in the sedimentary cover of the Austroalpine units were shaped as a result of compressional deformation. The rocks arrived at the four position-types described above (Fig. 3), and the detachment nappes started sliding to the north. The northward push of the Austroalpine units was also felt at the southern part of the Penninic zone, which was upwarped so that crystalline basement rocks and the ophiolite-bearing zone of the southern part of the trough were subjected to erosion. This sequence of events has been reconstructed by Oberhauser (1968). Figure 8 shows how the forward march of the Calcareous Alps was accompanied by the termination of sedimentation in the northern zones. Three "phases" of this transport are distinctly marked in the sedimentary sequence: (1) in the Middle Cretaceous marine clastic sediments were deposited along the northern margin (*Randcenoman*); (2) after profound Turonian erosion, unconformable deposition of the Coniacian to lower Eocene Gosau formation was restricted to the area south of this zone; and (3) in the middle Campanian the source area of the Gosau clastics underwent a change: chromite as the characteristic heavy mineral disappeared in favor of garnet, probably because the southern Penninic ophiolites were covered by the overthrust of the Austroalpine nappes. The previously mentioned remobilization of the Austroalpine basement, or at least a transformation of mica into a closed system by cooling at this time, is indicated by radiogenic age determinations.

These considerations suggest a northward slope, which caused submarine gravity sliding of the Calcareous nappe system into the continuously subsiding Penninic trough. Nevertheless, the raising of the ophiolites to the level of erosion would seem unthinkable without compressional deformation of the southern flank of the Penninic zone.

Not before the second part of this epoch, from middle Eocene time on and approximately contemporaneous with the oldest nappe movements in the Western Alps, did the folding of the Penninic geosyncline and the continuing transport of the Austroalpine

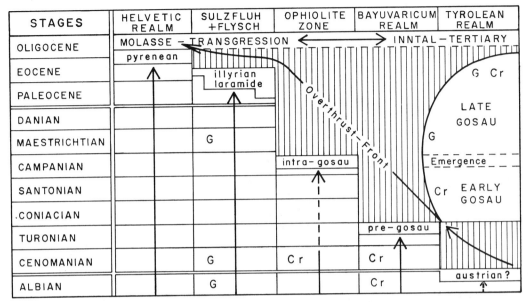

FIG. 8 Scheme showing the duration of sedimentation and successive times of tectonic closure of the main sedimentary zones by the advance of the Austroalpine thrust-front (after Oberhauser, 1968).

nappe pile close off sedimentation of the Rheno-Danubian flysch in what was presumably the northern part of the Penninic trough. Most probably the end of sedimentation was closely followed by the tectonic detachment of the flysch mass.

Thus there exists a remarkable contrast between two tectonic provinces in the Eastern Alps. Their dominant structures have different ages: the enormous Austroalpine nappes with their internal structures of Cretaceous age are superposed on more external zones which did not obtain their Alpine structure until the Tertiary (Figs. 5 and 9). Similarly, Cretaceous folding in the Western Alps appears to be restricted to the narrowed terminus of the Austroalpine realm (Simme).

Regarding the eastern continuation of the Eastern Alps, Andrusov (1968, 1969) has emphasized the juxtaposition in the western Carpathians of an external part (the Flysch Carpathians) with Tertiary nappes next to an internal part with Cretaceous nappes, and a mixture of both in the strongly compressed Pienninian Klippen zone between. Thus the special feature of the Eastern Alps within the Alpine–Carpathian belt is that the realm with the oldest structure is thrust in the form of large nappes over more external zones of younger structures. The transition into the Carpathian-type juxtaposition probably occurs in the easternmost part of the Eastern Alps (Figs. 4, 7, and 9); there the Penninic trough becomes less deep, as indicated by the Bouguer anomalies, and the Austroalpine basement must be connected with the Pannonian massif. The Tertiary thrusting of the Austroalpine plate would therefore amount to a clockwise rotational movement, as thrusting increases toward the west.

In the Penninic belt compression and subsequent uplift similarly increase westward, culminating in the enormous dimensions of the Swiss Alps. This rotation over about 30° of the Austroalpine block is not in contradiction with the counterclockwise movement of the South Alpine block postulated by Zijderveld and others (1970), which explains

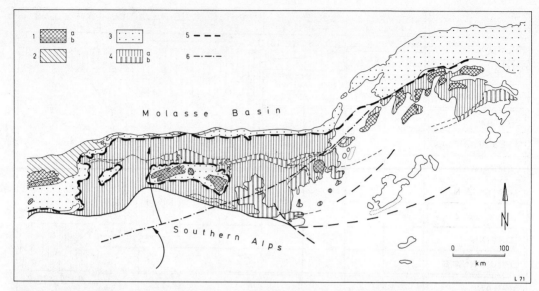

FIG. 9 Tertiary (second) part of Austroalpine overthrusting, seen as a rotational movement of an older block or plate across the adjacent "infra-Austroalpine," still active sediment trough. (1) Assumed autochthonous massifs [(a) Prealpine mineral facies; (b) Alpine mineral facies]; (2, 3) Helvetic and Penninic zones (including Klippen) with mostly Tertiary (middle Eocene and younger) structures; (4) Austroalpine zone with dominant Late Cretaceous structures [(a) crystalline basement with Mesozoic cover; (b) Paleozoic rocks]; (5) present erosional border of the Austroalpine nappe system (in the eastern, Carpathian section this line separates Tertiary nappe structures to the north from Late Cretaceous ones to the south) (after Andrusov, 1960); (6) approximate position of the Austroalpine front at the end of the Cretaceous.

better than previous hypotheses the west-trending fold bundles at the western margin of the Eastern Alps and the most recent thrust movements of the Oetztal and Silvretta nappes.

Fourth Epoch

This epoch is characterized as "Molasse time." The foreland trough sedimentation of the molasse started with the late Eocene and is locally transgressive on the Austroalpine units (Papp, 1958). From then on the deformed belt of the Eastern Alps became a mountain chain, and younger Tertiary rocks were deposited in local basins only and were caught along faults that remained active for a little while longer (including the faults along the longitudinal valleys north of the Karawank Mountains, along the Enns valley, and along the northern flank of the Kaisergebirge).

The northward displacement of sedimentation from the Rheno-Danubian flysch to the molasse represents a probably continuous shifting of the axis of subsidence toward the external side of the orogenic belt. However, the early Tertiary zone of continuous sedimentation (probably the autochthonous Helvetic belt and the southern molasse) is now covered by nappes, and the only more complete exposure of this interval is in the Waschberg zone near Vienna, because the amount of overthrusting in this marginal zone seems to decrease toward the east.

Fifth Epoch

During the upheaval of the mountain chain (in the fourth and fifth epochs) the major structural change is the domal uplift of the Pennine Hohe Tauern window and the broader central zone to the east, leading to steepening or overturning of the northern flank of the window (see the classic cross section by Cornelius, 1940). By mid-Oligocene time the Penninic nappes were exposed in this window and their detritus was carried northward over peneplains of the subsequent Calcareous Alps (*Augenstein Landschaft*) into the molasse trough (Tollmann, 1968).

Purely geometric reasons demand compression at the level of the Penninic nappes as the cause of the aforementioned arching of the window. As a secondary effect it caused the further, now purely gravitational displacement of the Austroalpine overthrust plate, composed of the rocks of the Calcareous Alps and the Graywacke zone, its front dragging the Rheno-Danubian flysch over the marginal molasse. A push against the rear of this plate is excluded as a possible mechanism at this time, because the central part of the Eastern Alps was already attacked by erosion down to the basement in the Late Cretaceous, as mentioned previously in connection with the Gosau formation.

The combined action of compressive arching and gravity sliding amounts to three contemporaneous tectonic movements (see Fig. 3, top section): (1) continuation of gravitational sliding of thrust masses to the north, perhaps more or less by the "surf-riding" mechanism (Wunderlich, 1965); (2) southward overthrusting of the same masses against the central arching zone; and (3) uplift of the whole into a coherent mountain chain. The final overthrust movements at the margin of the molasse become younger toward the east, and lasted until the middle Miocene near Vienna (Prey, 1968).

Tectonic Concepts and Evaluation of the Role of Gravity

Many of the tectonic concepts concerning the Eastern Alps deal only with the detailed description and classification of the tectonic units and treat the problem of the mechanism of folding and thrusting as a question of secondary importance (Kober, 1923; Staub, 1924; Cornelius, 1940; Tollmann, 1963). On the other hand, the Eastern Alps have also been the place where the pioneer of gravity tectonics, Edward Reyer of Vienna, applied his concepts (1892). As early as 1931 the great East Alpine geologist Otto Ampferer concluded on the basis of extensive field experience that: "in Querprofilen der Alpen wechseln Strecken der Pressung mit solchen der Freigleitung" (in cross sections across the Alps zones of compression alternate with zones of free sliding). In 1906 Ampferer deduced, for purely geometric reasons, the notion of down-sucking (*Verschluckung*) of the deep basement, and as far back as 1920 Schwinner presented an explanation for this phenomenon in the form of the general hypothesis of thermal convection currents, a concept subsequently applied in specific modifications to the Eastern Alps by Kraus (1951).

It was particularly Fallot (1955) who stressed the relative importance of gravity tectonics in the formation of East Alpine structures, and his concept is closest to the one presented here. There is also a basic similarity to the genetic interpretation of the Eastern Alps by Van Bemmelen (1960a, Fig. 5, sections II and III; 1960b) in that he makes a distinction between a deep pre-Tertiary evolution of the Austroalpine nappe system and a Tertiary period of up-arching of the central zone followed by "epidermal" gravity sliding. The present author, however, cannot agree with his assumption of the existence of cylindrical gliding surfaces that go down into the metamorphic zone, or with the assumption of an Adriatic tumor, neither of which lie in the field of observation in the Gailtal Alps or elsewhere. Structural features at the level of observation, schematically drawn by Van Bemmelen (1960a) in his Fig. 5, section II (on the flank of the tumor) and his Fig. 4 (on the flank of the asthenolith) give the impression of typical compressive phenomena at the "epidermal" and "bathydermal" levels and indicate shortening. In the author's view the structures that can be observed and described clearly show, on the scale of the Eastern Alps, that compression is the *primary* structural agent.

If, nevertheless, we include Van Bemmelen's concepts as one of the working hypotheses, we have at our disposal three different models to account for the tectonic driving forces in the Eastern Alps. The first model postulates thermal convection cells, which form orogenic structures by downsucking of the crust, resulting in a sialic mountain root (Exner, 1951; Clar, 1953). In the second model, apparently supported by the interpretation of new seismic data (Giese, 1968), it is assumed that the East Alpine structures are all part of a fundamental overthrust zone, dipping south at a low angle, along which frontal parts of the Gondwana block moved northward over the margin of the European foreland (Staub, 1924). This model is satisfactorily consistent with the new concept of plate tectonics and may in the final analysis again be reduced to the action of frictional drifting by thermal convection currents. As a third alternative, kinematics of the type of model 2 may be considered an effect of primary, fundamental gravitational tectonics in the Eastern Alps as advocated by Van Bemmelen (1960a, 1960b), if there are sufficient arguments for the existence of a gravitational gradient steep enough to cause not only transport but compressional deformation and large overthrusting of the sliding plate. Thus the choice between such models lies mainly outside the realm of East Alpine geology.

Summarizing the foregoing structural interpretation of the Eastern Alps, primary compression in the earth's crust, with the resulting changes in vertical mass distribution, gave rise to secondary phenomena of gravity tectonics, mainly by sliding. In the nappe framework of the Eastern Alps, these secondary effects lead to the largest observable tangential displacements.

References

Accordi, B., 1955, Le dislocazioni delle cime (Gipfelfaltungen) delle Dolomiti: *Ann. Univ. Ferrara*, Sez. IX, v. II, no. 2, p. 65–186.

Agterberg, F. P., 1961, Tectonics of the crystalline basement of the Dolomites in North Italy: *Geol. Ultraiectina, Utrecht*, no. 8, 232 p.

Ampferer, O., 1929, Einige Beispiele von Kerbwirkung und Reliefüberschiebung aus den Südtiroler Dolomiten: *Jahrb. Geol. Bundesanst. Wien*, v. 79, p. 241–256.

———, 1931, Beiträge zur Auflösung der Mechanik der Alpen: V. Forts.: *Jahrb. Geol. Bundesanst. Wien*, v. 81, p. 637–659.

Andrusov, D., 1960, Gedanken über das alpinkarpatische Falten-Decken-System: *Geolog. Sbornik Bratislava*, v. 11, p. 171–178.

———, 1968, *Grundriss der Tektonik der Nördlichen Karpaten*: Bratislava, Slov. Akad. Vied, 188 p.

Beck-Mannagetta, P., 1967, Über das Westende der Pannonischen Masse: *Mitt. Geol. Ges. Wien*, v. 59, p. 139–150.

Bott, M. H. P., 1954, The gravity field of the Eastern Alps: *Geol. Mag.*, v. 91, p. 377–383.

Clar, E., 1953, Zur Einfügung der Hohen Tauern in den Ostalpenbau: *Verh. Geol. Bundesanst. Wien*, p. 93–104.

———, 1965, Zum Bewegungsbild des Gebirgsbaues der Ostalpen: *Verh. Geol. Bundesanst. Wien*, Sonderheft G, p. 11–35.

Colacicchi, R., Le dislocazioni delle cime (Gipfelfaltungen) delle Dolomiti: Gruppo della Civetta: *Mem. Ist. Geol. Miner. Univ. Padova*, v. 22, 49 p.

Cornelius, H. P., 1940, Zur Auffassung der Ostalpen im Sinne der Deckenlehre: *Zeitschr. Deutsche Geol. Ges.*, v. 92, p. 271–312.

———, and Furlani-Cornelius, M., 1930, Die Insubrische Linie vom Tessin bis zum Tonalepass: *Denkschr. Akad. Wiss. Wien, math.-naturwiss. Kl.*, v. 102, p. 207.

De Jong, K. A., 1967, Tettonica gravitativa e raccorciamento crostale: *Boll. Soc. Geol. Ital.*, v. 86, p. 749–776.

De Sitter, L. U., 1949, Le style structural Nord-Pyrénéen dans les Alpes Bergamasques: *Bull. Soc. Géol. Fr.*, v. 19, p. 617–621.

———, and De Sitter-Koomans, C. M., 1949, The geology of the Bergamasc Alps, Lombardy, Italy: *Leidsche Geol. Mededel.*, v. 14 B, p. 1–257.

Engelen, G. B., 1963, Gravity tectonics in the northwestern Dolomites (N. Italy): *Geol. Ultraiectina, Utrecht*, no. 13, 92 p.

Exner, Ch., 1951, Der rezente Sial-Tiefenwulst unter

den östlichen Hohen Tauern: *Mitt. Geol. Ges. Wien*, v. 39–41, p. 75–81.

———, 1964, Erläuterungen zur Geologischen Karte der Sonnblickgruppe: *Geol. Bundesanst. Wien*, 170 p.

Fallot, P., 1955, Les dilemmes tectoniques des Alpes Orientales: *Ann. Soc. Géol. Belgique*, v. 78, p. 147–170.

Faupl, P., 1969, Geologische Studien an den kristallinen Schiefern des südlichen Wechselgebietes im Raum um Bruck an der Lafnitz, Steiermark: *Anz. Akad. Wiss. Wien, math.-naturwiss. Kl.*, p. 101–104.

Flügel, H., 1960, Die tektonische Stellung des "Alt-Kristallin" östlich der Hohen Tauern: *Neues Jahrb. Geol. Pal., Monatsh.*, p. 202–220.

Frasl, G., and Frank, W., 1966, Einführung in die Geologie und Petrographie des Penninikums im Tauernfenster: *Publ. Vereinigung Freunde Miner. Geol. Heidelberg, Der Aufschlusz*, Sonderheft 15, p. 30–58.

Geologische Bundesanstalt Wien, 1964, *Übersichtskarte der Republik Österreich*: map 1/1,000,000, explanations, 1966, 94 p.

Giese, P., 1968, Versuch einer Gliederung der Erdkruste im nördlichen Alpenvorland, in den Ostalpen und in Teilen der Westalpen mit Hilfe charakteristischer Refraktions-Laufzeitkurven, sowie eine geologische Deutung: *Geophys. Abhdlg., Freie Univ. Berlin*, v. 1, no. 2, 214 p.

Gwinner, M. P., 1971, *Geologie der Alpen*: Stuttgart, E. Schweitzerbart'sche, 477 p.

Janoschek, R., 1964, Das Tertiär in Österreich: *Mitt. Geol. Ges. Wien*, v. 56, p. 319–360.

———, and Göttzinger, K. G. H., 1969, Exploration for oil and gas in Austria, in: *The Institute of Petroleum*, p. 161–180.

Kober, L., 1923, *Bau und Entstehung der Alpen*: Berlin, Borntraeger; 2nd ed., 1955, Wien, Deuticke, 379 p.

Kraus, E., 1951, *Die Baugeschichte der Alpen*: Berlin, Akademie Verlag, v. I, 533 p., v. II, 482 p.

Kröll, A., and Wessely, G., 1967, Neue Erkenntnisse über Molasse, Flysch und Kalkalpen auf Grund der Ergebnisse der Bohrung Urmannsau 1: *Erdoel-Erdgas Zeitschr.*, v. 83, p. 342–353.

Lemoine, M., 1973, About gravity gliding tectonics in the Western Alps (this volume).

Leonardi, P., 1967, *Le Dolomiti: Geologica dei monti tra Isarco e Piave*: Trento, Consiglio Nazionale Ricerche, 2 vols., 1019 p.

Medwenitsch, W., and Schlager, W., 1964, Ostalpenübersichtsexkursion: *Mitt. Geol. Ges. Wien*, v. 57, no. 1, p. 57–106.

Metz, K., 1953, Die stratigraphische und tektonische Baugeschichte der steirischen Grauwackenzone: *Mitt. Geol. Ges. Wien*, v. 44, p. 1–84.

———, 1965, Das ostalpine Kristallin im Bauplan der östlichen Zentralpen: *Sitzungsberichte Öst. Akad. Wiss., math.-naturwiss. Kl.*, Abt. I, v. 174, p. 229–278.

Oberhauser, R., 1968, Beiträge zur Kenntnis der Tektonik und der Paläogeographie während der Oberkreide und dem Paläogen im Ostalpenraum: *Jahrb. Geol. Bundesanst. Wien*, v. 111, p. 115–145.

Oxburgh, E. R., 1968, An outline of the geology of the central Eastern Alps: *Proc. Geol. Assoc. London*, v. 79, no. 1, 46 p.

———, Lambert, R. St. J., Baadsgaard, H., and Simons, J. G., 1966, Potassium-Argon age studies across the southeast margin of the Tauern window: *Verh. Geol. Bundesanst. Wien*, p. 17–33.

Papp, A., 1958, Vorkommen und Verbreitung des Obereozäns in Österreich: *Mitt. Geol. Ges. Wien*, v. 50, p. 251–270.

Plöchinger, B., and Prey, S., 1968, Profile durch die Windischgarstner Störungszone im Raum Windischgarsten-St. Gallen: *Jahrb. Geol. Bundesanst. Wien*, p. 68–107.

Prey, S., 1965, Vergleichende Betrachtungen über Westkarpaten und Ostalpen im Anschlusz an Exkursionen in die Westkarpaten: *Verh. Geol. Bundesanst. Wien*, p. 68–107.

———, 1968, Probleme im Flysch der Ostalpen: *Jahrb. Geol. Bundesanst. Wien*, v. 111, p. 147–174.

Reyer, E., 1892, *Ursachen der Deformation und der Gebirgsbildung*: Leipzig, W. Engelmann, 40 p.

Schwinner, R., 1920, Vulkanismus und Gebirgsbildung: ein Versuch: *Zeitschr. Vulkanologie*, v. 5, p. 175–230.

Staub, R., 1924, Der Bau der Alpen: *Beitr. Geol. Karte Schweiz*, N.F., v. 52, 272 p.

———, 1950, Betrachtungen über den Bau der Südalpen: *Eclog. Geol. Helv.*, v. 42, no. 2 (1949), p. 215–408.

Streiff, V., 1939, Geologische Untersuchungen in Ostschams (Graubünden): Diss. Univ. Zürich, 236 p.

Tollmann, A., 1963, *Ostalpensynthese*: Wien, Deutike, 256 p.

———, 1966, Die alpidischen Gebirgsbildungs-Phasen in den Ostalpen und Westkarpaten: *Geotekton. Forsch.*, v. 21, p. 1–156.

———, 1966–1970, Tektonische Karte der Nördlichen Kalkalpen, parts 1–3: *Mitt. Geol. Ges. Wien*, v. 59, 61, 62.

———, 1967, Ein Querprofil durch den Ostrand der Alpen: *Eclog. Geol. Helv.*, v. 60, no. 1, p. 109–135.

———, 1968, Die paläogeographische, paläomorphologische und morphologische Entwicklung der Ostalpen: *Mitt. Österr. Geogr. Ges.*, v. 110, p. 224–244.

———, 1970, Für und wider die Allochthonie der Kalkalpen sowie ein neuer Beweis für ihren Fernschub: *Verh. Geol. Bundesanst. Wien*, p. 324–345.

Trümpy, R., 1960, Paleotectonic evolution of the Central and Western Alps: *Bull. Geol. Soc. Am.*, v. 71, p. 843–908.

———, 1970, Apercu général sur la géologie des Grisons: *Comptes Rendus Soc. Géol. Fr.*, v. 9, p. 330–364, 391–394.

Van Bemmelen, R. W., 1960a, New views on East-Alpine orogenesis: *21st Int. Geol. Congr. Copenhagen*, part 18, p. 99–116.

———, 1960b, Zur Mechanik der Ostalpinen Deckenbildung: *Geol. Rundschau*, v. 50, p. 474–499.

———, 1966, The structural evolution of the Southern Alps: *Geol. Mijnb.*, v. 45, p. 405–444.

———, 1970, Tektonische Probleme der östlichen Südalpen: *Geologija, Ljublijana*, v. 13, p. 133–158.

Van Hilten, D., 1960, Geology and Permian paleomagnetism of the Val-Di-Non area: *Geol. Ultraiectina, Utrecht*, no. 5, 95 p.

Vink, B. W., 1969, Gravity tectonics in Eastern Cadore and Western Carnia, NE Italy: *Geol. Ultraiectina, Utrecht*, no. 15, 64 p.

Wunderlich, H. G., 1965, Zyklischer Bewegungsablauf beim Vorrücken orogener Fronten und der Mechanismus des Deckschollentransportes nach dem Surf Riding-Prinzip: *Geol. Mijn.*, v. 44, p. 440–457.

———, 1969, Aufgaben und Ziele aktuogeologischer Forschung: Schwereverteilung und rezente Orogenese im Mediterrangebiet: *Zeitschr. Deutsch. Geol. Ges.*, v. 118, p. 266–284.

Zijderveld, J. D. A., Hazeu, G. J. A., Nardin, M., and Van der Voo, R., 1970, Shear in the Tethys and the Permian paleomagnetism in the Southern Alps, including new results: *Tectonophysics*, v, 10, p. 639–661.

HANS G. WUNDERLICH

Institute of Geology and Paleontology
University of Stuttgart
Stuttgart, West Germany

Gravity Anomalies, Shifting Foredeeps, and the Role of Gravity in Nappe Transport as Shown by the Minoides (Eastern Mediterranean)

In every branch of science knowledge evolves through distinctive periods, each of which carries its own fundamental problems, and the field of tectonics is no exception. The neptunism-plutonism controversy raged during the early nineteenth century, and the fixism-nappism discussion dominated the first half of the twentieth century. But there are also scientific problems of general interest that remain unsolved during more than one of those comparatively short periods of scientific history. Today, one of the most important geotectonic problems discussed is the question whether tectonic activity within young orogenic belts operates rapidly, much like such catastrophic events as earthquakes and volcanic eruptions observable within mobile tectonic arcs all around the world or, on the other hand, shows a more continuous evolution.

Mediterranean geology has played a role in these discussions ever since, in 1669, Nicolas Steno, called Nils Stensen in his native Denmark, wrote the very first book on tectonics, curiously entitled *De solido intra solidum naturaliter contento*. Near Florence, Tuscany, he compared faulting and folding phenomena of the Northern Apennines with those of karst holes known to him from Slovenia and Croatia. Karst holes often break down abruptly during only a few days or even hours, and therefore Steno thought orogenic movements generally to be also of a catastrophic nature. Since those early days of this pioneer of earth sciences, discontinuity in mountain building is widely accepted among geologists as shown by Suess, who wrote in 1885: "It is the breaking down of the globe we are confronted with," or by Stille, who, about 50 years ago, started his well-known theory of short but world-wide orogenic phases (Stille, 1924).

Since Gilbert's (1890) fundamental work on terraces of the former Lake Bonneville, geologists usually try to distinguish between such short-dated orogenic movements and, on the other hand, epeirogenic events free of faulting or folding activity during the comparatively long periods between the individual orogenic phases. In 1922, in his presidential address at the University of Göttingen, Stille attributed all present-day tectonic activity to epeirogeny. But since that time modern seismic research has clearly demonstrated that even today there are mobile zones in full orogenic activity as, for instance, central Japan where, at the Matsushima earthquake observatory, shocks may be registered every two or three minutes during the entire year.

The previously mentioned problem of periodicity or continuity of orogenic events cannot be solved only by the study of mountain chains whose orogenic activity became dormant long ago. To clarify how mountain building is really taking place we must look for orogenic belts that are active today. In doing so we have to bridge the gap between the specialized fields of geophysics and geotectonics as well as modern methods in geology.

Gravity Distribution and Mountain-Building Activity of the Present Time

How can we recognize orogenic belts active today? Unfortunately, actual orogeny is not visible by direct observation, for mountain building is one of the slowest of all events of dynamic geology. There are, however, some indications that may lead us to decide whether an orogenic belt is still active.

As pointed out repeatedly, young orogens often show strong negative gravity anomalies owing to an unbalanced state of mass equilibrium within the earth's crust and upper mantle. Orogenic disturbance of the earth's interior and a balanced gravity distribution do not harmonize with each other; even at first sight gravity distribution maps allow us to recognize, by the position and amount of anomalies, whether there is a disturbance of crustal masses. However, gravity anomalies as a sign of crustal disturbance may be due to orogenic movements of the past. Therefore, we have to look for additional signs marking orogens that are active today. In fact, one can distinguish orogens in a terminal stage of mountain building with negative gravity anomalies coinciding with the contour lines of the mountain chains, and others, in a less advanced or even initial stage of development, showing negative anomalies shifted outside toward the foreland, that is, to the convex side of mountain or island arcs. A geological and geophysical comparison of the Northern Apennines and the Alps has led the author to the conclusion that there is a remarkable difference in geodynamic evolution between these two young Alpine-type mountain chains.

The Northern Apennines represent an earlier, less complete stage of development, as shown by the fact that a strong negative gravity anomaly exists only over the northern part of the mountain chain, continuing northward across the southern part of the foredeep valley of the river Po. By contrast, the negative gravity anomaly in the Alps circumscribes only the mountain ranges themselves, with nearly normal gravity values in the northern and southern forelands, characterizing a mountain chain in the final stage of orogenic evolution, without any chance of further folding activity overlapping the foreland. In the case of the Alps, there is no reason why orogeny ought to continue northward or southward, since there is no sign of mass inequilibrium within the crust and mantle. Mountain chains like the Alps, that is, with negative gravity anomalies coinciding with the orographic contour lines, have reached their final geodynamic position (*Endlagengebirge* in the German terminology), whereas such chains as the Northern Apennines with shifted gravity anomalies cannot be in gravitational and geodynamic equilibrium. This is why in the latter case orogeny continues until gravity distribution is in full harmony with orography. Comparing the Alps and the Northern Apennines geologically, it is not difficult to recognize that the former are several million years older and therefore in an advanced geodynamic stage compared to the Apennines.

About 50 years ago, when gravity measurement started in the Swiss Alps, Heim (1921) thought the weight of the enormous mountain masses themselves to be the cause of the strong negative Bouguer anomaly found at that time in the southern part of Switzerland. Later, geophysicists as well as geologists looked for the so-called mountain roots (*Gebirgswurzeln* in the German terminology) as an interpretation of gravity deficiency of Alpine type. However, neither of these explanations appears adequate when the orographic boundaries do not coincide with those of the gravity anomaly. There are mountain chains without any coincidence with the gravity distribution. As pointed out by Vening-Meinesz (1934, 1948, 1950, 1964), Hess (1938), and others, it is especially in island arcs like Indonesia, the Aleutians, the Kuriles, the Kermadec-Tonga arc, and the northern and southern Antilles that gravity anomalies are commonly on the convex outer side of the arc, more or less far off the coastline of these mountainous islands.

In southern Europe and the Mediterranean, not only the Northern but also the Southern Apennines, the southern Carpathians, the Hellenides, and particularly the Peloponnese-Crete island arc (comprising also Kithera, Antikithera, Carpathos, and Rhodes) show more or less outward-shifted negative gravity anomalies. It is important to note that shifting of gravity anomalies does not depend on the manner in which arithmetic corrections are carried out in relation to the forms of relief or the height of the terrain above sea level. Thus, although Faye (free air) and Bouguer anomalies may indeed differ considerably from each other with respect to the *amounts* of the anomalies determined from corresponding points of observation, the *positions* of the anomalies in relation to the mountain chains (i.e., to the topographic pattern) are the same on either map.

Comparing these young Mediterranean orogens geologically, one may be led to the conclusion that shifting of negative gravity anomalies is most prominent where orogenic activity continues most strongly at the present time.

As an example of shifted gravity anomalies in the Mediterranean region, perhaps the best example in all of Europe, Fig. 1 shows the free air gravity distribution in the Aegean and Ionian seas and the island arc of Crete as measured by Fleischer (1964). The extensive negative anomaly south of the Peloponnese, Crete, and Rhodes is the most strongly marked in Europe, reaching -160 to -180 mgal. Adjacent to it there is a strong positive anomaly in the Aegean Sea north of Crete and centered around Santorin, one of the live volcanoes of the Mediterranean with activity as recent as 1956. The Peloponnese–Crete–Rhodes island arc follows closely the boundary between deficiency and excess of gravity, minimum and maximum being as far distant to the south as to the north. Here we have a typical Mediterranean orogen in the stage of an island arc, unique in Europe but comparable more or less to the well-known island arcs around the Pacific Ocean. Pointing to the fact that the Ionian gravity anomaly is the only one in the Mediterranean region situated almost completely within the foreland of an orogenic ridge, whereas the other shifted gravity

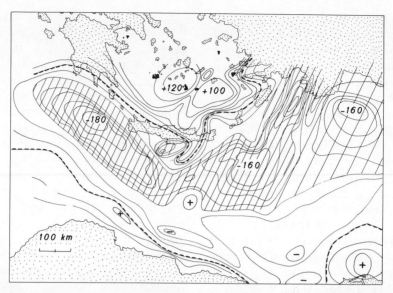

FIG. 1 Free-air gravity anomalies in the eastern Mediterranean measured by Fleischer (1964) with an Askania sea gravimeter (anomalies in milligals).

anomalies in the neighborhood take in more and more of the mountain chains themselves, the author stated in 1966 that the Peloponnese–Crete orogen must be considered as the one in the least-advanced stage of orogenic development in the entire Mediterranean and therefore may currently be in full orogenic activity.

Recently, the deep-sea drilling project of D/V Glomar Challenger carried out by the Scripps Institution of Oceanography (Ryan and Hsü, 1970) has brought strong new evidence that this assumption was correct: beneath the floor of the Hellenic trough, between the Peloponnese and Crete, Lower Cretaceous limestones some 120 million years old were discovered directly overlying soft oozes of Pliocene age. As far as known today, these are the youngest sediments within the entire Mediterranean fold belt to be involved in thrusting.

These newly acquired data all indicate for certain that recent crustal movements in the eastern Mediterranean, especially in front of the Peloponnese-Crete island arc, are of orogenic origin.

Shifted Gravity Anomalies and Shifting Orogenic Foredeeps

As pointed out, negative gravity anomalies of young fold belts with recent orogenic activity do not coincide with orographic crustal heights but occupy greater or lesser parts of the foreland, that is, the area of foredeeps in front of mountain ridges. If we keep in mind that there are "mountain roots" without any mountainous orography as well as negative gravity anomalies without isostatic uplift, some even showing in part antiisostatic subsidence, we can say that neither the hypothetical mountain roots (depressed into the upper mantle by the weight of the mountains) nor the principle of isostasy can explain this remarkable shifting of gravity anomalies. In this case, the causes and effects seem to be changed, for some, if not all, of these shifted anomalies originate primarily within the foreland far from the true mountain ridge. Consequently, we must assume that negative gravity anomalies may be caused by an anisostatic, that is, orogenic subsidence and the accumulation of diagenetically less condensed young sediments, much along the lines of the down-buckling hypothesis of Vening-Meinesz (1934). Later on, when the anisostatic forces of subsidence become weak, isostatic uplift begins to carry the day. As the orogenic front advances toward the foreland, the former zone of subsidence begins to rise little by little, so that zones of gravity deficiency and the zone of uplift fall increasingly in with one another, until they finally coincide, as in the case of the Alps.

Shifted gravity anomalies therefore require anisostatic (orogenic) forces disturbing the mass equilibrium within the crust and upper mantle. They also show the wandering of orogenic zones, that is, zones of subsidence, compression, uplift, and lateral distension. The lack of agreement between gravity distribution and orography can be explained only in terms of a shifting of orogenic activity, especially of orogenic foredeeps. One can put together a series of neighboring Mediterranean orogens with increasing conformity between gravity and orographic patterns. Such a series would start with the Peloponnese-Crete orogen, whose negative gravity anomaly lies totally within the foreland, continuing by way of the Hellenides, the southern Carpathians, and the Apennines, where the negative anomalies comprise increasingly the outer parts of the mountain ranges, and ending in the Alps with their high degree of orographic and gravitational conformity. This series may be considered as a continuous genetic sequence of successive evolutionary stages.

In other words, shifted gravity anomalies within the foreland of young mountain chains or island arcs show orogenic forces remaining active today and mark the direction and intensity of future mountain-building activity. Gravity research enables us to localize present-day orogenic activity as well as discover the trends and behavior of recent orogenic processes. From the outer to the inner parts of the active orogen, we can distinguish, side by side (1) an external zone of undisturbed but slowly sinking foreland, (2) a zone of strong orogenic subsidence, usually with little sedimentation and sometimes, especially in an earlier stage of development, accompanied by strong geosynclinal volcanism, (3) a zone immediately adjacent to the orogenic front where large amounts of flysch-type sediments are deposited, and almost simultaneously overthrust and folded, (4) a zone of orogenic uplift showing dilatation tendencies rather than the lateral compression that characterizes the flysch trough, and (5) another zone of subsidence producing a broad backdeep accompanied, at the outer margin, by remarkable extensional fracture and graben tectonics and strong volcanic activity, while the inner girdle of these active orogens is normally invaded by large granitic intrusions (provided that continental crust is available to supply sufficient granitic magma) and subjected to regional metamorphism owing to intensified heat flow from below.

During orogenic evolution in time and place, all these tectonic zones show a slow but continual change of position with regard to the epeirogenic foreland. However, for orogenic fronts to advance, it is necessary for foreland troughs to subside. Where such fronts abut directly against tables or shields of folded Precambrian continental crust they may be stopped more or less abruptly by epicontinental platforms that have been highly elevated by epeirogenic uplift instead of tectonically depressed. In times of geocracy as today, when the continents are extended as far as possible and the shelf margins show their smallest dimensions, active orogenic fronts occur only where they are not prevented from advancing by adjacent elevated continental blocks. Like waves on an ascending beach, orogenic fronts with continentward movement die out upon approaching an epicontinental shield; these orogenic fronts could be called *epeiropetal*, in contrast to *epeirofugal* fronts whose advancing direction points to the open sea.

Therefore recent active orogenic fronts lie mainly beneath sea level, limiting geologic field observation to a few small upheavals in the form of lonely islands or archipelagos. Thus a study of the actualistic setting of orogenic fronts, that is, in terms of actuotectonics, must be based primarily on oceanographic and geophysical research. As we explore the geology and structural patterns of the islands, we can see that some of these upheavals may be attributed to the zone of orogenic uplift immediately behind the true front, whereas others belong to the volcanic belt along the outer margin of the backdeep or to the plutonic center behind the volcanic island arc, composed of highly metamorphosed sediments intruded by mostly granitoid plutons. The wide spaces between these few upheavals can be studied only by using the methods of modern oceanography.

Since only epeirofugal orogenic fronts have a chance to advance and continue with active mountain building today, whereas epiropetal fronts were halted in the course of Late Tertiary or even Early Tertiary time when geocracy began to increase, the degree of spreading and the number of recently active orogenic fronts is very restricted. Moreover, some of the well-known circum-Pacific island arcs consist of only the volcanic belt (the zone of so-called *Rückseiten-Vulkanismus* of the German geologic terminology), and the other zones of the active orogen are inaccessible by reason of the deep ocean. As may be easily seen, oceanic orogenic fronts far off the continents will perish by underfeeding, for in order to develop they require a balanced proportion of detrital contributions supplied by terrigenic erosion. In the absence of sedimentary fill capable of folding, isostatic uplift cannot produce mountains. On the other hand, epeiropetal orogenic fronts may be paralyzed by overfeeding of their foredeeps, which inhibits the free advance of the front toward the shield. In tectonic as in human history, gluttony as well as hunger is harmful!

In the Mediterranean region there was a well-balanced relation between orogenic subsidence and sedimentary supply during long periods of geologic history, so that many well-developed mountain chains could come into being. There is only one fully active orogenic island arc left and available for geologic research into recent orogenic activity, the previously mentioned Peloponnese-Crete orogenic arc. Therefore we shall look at it with special interest.

The Peloponnese-Crete Island Arc as an Example of Recent Alpinotype Mountain Building

As noted, this part of Greece connecting Europe and Asia Minor may be considered, tectonically speaking, as the youngest still active part of the Mediterranean fold belt. If anywhere, this is where we shall see how orogeny takes place. Let us consider the largest island within this arc, the isle of Crete, known for its early Minoan culture (the first high culture in Europe after the Stone Age).

This great and mountainous island belongs to the orogenic zone of uplift, as shown by the numerous outcrops of highly elevated marine sediments of Late Tertiary age. The young uplift seems to be restricted more or less to the truly mountainous region, for several Minoan rock tombs cut near the coast in both the northern and southern slopes (for instance, near Megaron Nirou about 12 km east of Iraklion, and at Matalla Bay south of Phestos) have been submerged beneath the sea during the last 3000 or 4000 years. Possibly, this submergence is also due to eustatic rise of sea level caused by the regression of glaciers around the world.

Late Tertiary sediments uplifted tens to hundreds of meters above sea level are of molasse-type, composed of fine-grained sandy or marly sediments carrying in some places a rich marine fauna mainly of lamellibranchiata (pelecypods). Locally, late Miocene evaporites such as gypsum may be interbedded with nonfossiliferous, extremely dessiccated sediments.

The Hellenic trough immediately south of the Peloponnese-Crete island arc goes down to 5000 m below sea level and represents the orogenic foredeep, whereas the Aegean Sea north of Crete, locally almost 3000 m deep, may be considered as the backdeep trough. Starting with the Saronian Gulf at Aigina and Methana, the backland volcanic belt extends by way of Milos, Komolos, Antimilos, the well-known isle of Thira (Santorin, showing strong volcanic activity in historic time), Anaphi, Nisyros, and southwestern Kos, to Patmos, near the coast of Asia Minor. These islands form an arc slightly convex to the south and are composed of volcanic material of mainly Late Tertiary age. The inner zone of regional metamorphism and plutonic rocks is represented by the Cycladic archipelago, formerly known as a part of the so-called Attic-Cycladian crystalline massif. When geologic research started in Greece, geologists thought these Attic-Cycladian metamorphic and intrusive rocks to be of Prealpine, that is, Paleozoic or even pre-Paleozoic age. However, the granites of Tinos, Mikonos, Ikaria, Seriphos, and Naxos as well as the recrystallized metamorphic rocks, especially the marbles of Paros and some of the neighboring islands, show little or no Alpine fabric. Although Hercynian or Alpine fabrics in sediments could have been blotted out by late Alpine recrystallization, the lack of Alpine fabrics in plutons can only mean that intrusion occurred after the main Alpine orogeny. Intrusion as well as recrystallization therefore seem to be of Tertiary age. This zone of regional metamorphism and plutonic activity corresponds in an excellent way to the accompanying volcanic belt nearby. Thus we have a complete series of tectonic zones of an active orogenic front, consisting of foredeep trough, zone of uplift, backdeep trough, and a volcanic and plutonic belt (or zone of *Rückseitenvulkanismus* and *Rückseitenplutonismus*) with regional metamorphism.

On the geologic map of Greece (1:500,000) (Renz, Liatsikas, and Paraskevaidis, 1954) Crete is seen as part of a crystalline massif, named Peloponnese-Crete, in view of the occurrence of semicrystalline rocks such as sericite schists, phyllites, and crystalline limestones, both in the southern Peloponnese and in the westerly and easterly regions of Crete. On the other hand, Aubouin and others (1960–1963) consider this part of the Peloponnese (called Tripolitsa zone within the series of Hellenic tectonic zones) as a ridge zone which, together with the Gavrovo ridge, divides an outer miogeosynclinal "Ionian" trough from an inner eugeosynclinal "Pindos" trough.

For the most part, Fig. 2 is based upon these isopic zones outlined by Aubouin and his co-authors. The Vardar (Axios) zone, marked mainly by basic and ultrabasic igneous rocks, comes to an end in the northwestern part of the northern Cyclades. The Pelagonian and Attic-Cycladian crystalline massifs are divided from each other by a transverse zone of axial depression covered by Mesozoic and Tertiary sediments. The Subpelagonian ophiolite zone, with shales and chert in fine alternations (called "Schiefer-Hornstein" complex) and with abundant peridotite intrusives, reaches only as far as the eastern Peloponnese. In its place, the Peloponnese-Crete (Tripolitsa) zone with semimetamorphosed schists, phyllites, and crystalline limestones begins in the central Peloponnese and continues southeastward as far as the isle of Elassa, directly east of Crete.

The Parnas-Ghiona zone is exposed in the mountain ranges near Delphi and composed of Triassic dolomites, Megalodon limestone, mainly calcareous Jurassic and Cretaceous rocks (limestones with ammonites, algae, corals, snails, and rudists of the Upper Cretaceous), and Upper Cretaceous to Lower Tertiary shales and sandstones, followed by nummulite limestone and flyschlike sediments. It is restricted to continental Greece and the Peloponnese south of the Sperchios River.

As can be seen, there exist important differences between the Hellenides *sensu stricto*

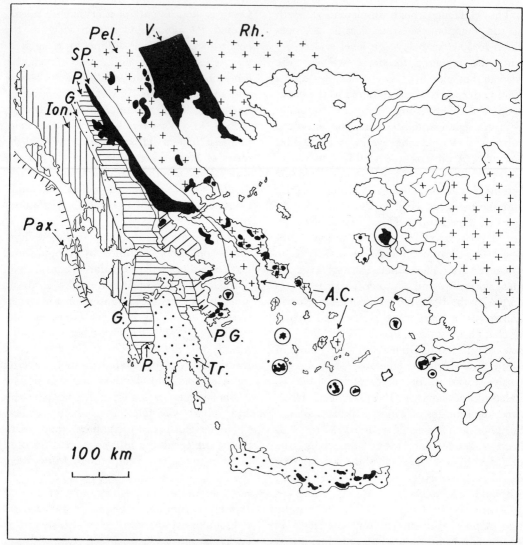

FIG. 2 Tectonic and facies zones of the Hellenides after Aubouin and others (1960-1963). (Pax.) Paxos foreland zone; (Ion.) Ionian outer miogeosynclinal zone; (G.) Gavrovo ridge; (P.) Pindos outer eugeosynclinal zone; (Tr.) Tripolitsa central miogeosynclinal zone (later partly overthrust by P.); (P.G.) Parnas-Ghiona inner miogeosynclinal zone; (SP.) Subpelagonian zone (later chert and ophiolite nappe); (Pel.) Pelagonian zone; (A.C.) Attic-Cycladian crystalline massif; (V.) Vardar innermost eugeosynclinal zone; (Rh.) Rhodope crystalline hinterland massif. Black indicates distribution of volcanic rocks. Encircled areas are Late Tertiary to Recent volcanoes.

and the young, active orogenic arc connecting the central and southern Peloponnese with Crete. In spite of a few similarities, decisive evidence tells its own tale: both the Santorin volcanic belt and the negative gravity anomaly are restricted to the Peloponnese-Crete island arc and do not continue farther to the northwest. As we shall see immediately below, not only the Hellenides, but the Peloponnese-Crete orogen as well possesses a eugeosynclinal trough of its own. Neighboring orogens with their own independent eugeosynclinal furrows should be carefully distinguished and called by their own names.

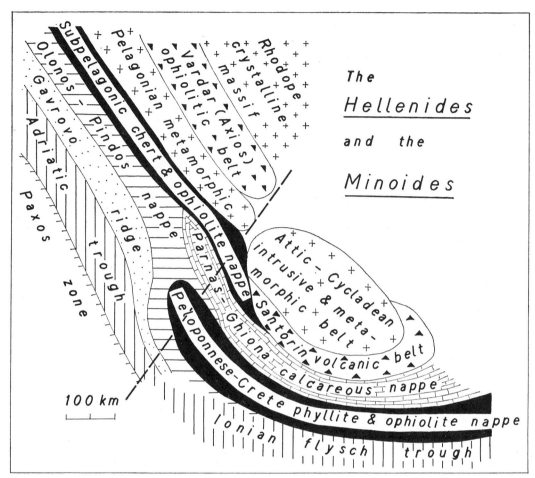

FIG. 3 Comparison of isopic sedimentary zones in the Hellenides (*sensu stricto*) and in the Minoides (the Peloponnese-Crete island arc).

For this reason the new term "Minoides" (instead of "Peloponnese-Crete orogen") is introduced in juxtaposition to the term Hellenides, which may be restricted as shown in Fig. 3.

The Hellenides and the Minoides both belong to the "Dinaridian" epeirofugal orogenic belt (relative to the eastern European platform that constitutes the real nucleus of the European continent), and are divided by an approximate line of separation reaching from the center of the island of Euboia by way of the Isthmus of Korinthos and the central Peloponnese to the western Ionian sea in front of Pylos (Fig. 3). The Minoides, named after the Minoan culture of Crete, extend across the same area as this culture, which was evidently brought to an abrupt half around the fifteenth century BC by the seismic and volcanic activity associated with the current Minoidian orogeny.

Central Crete: Geologic Framework, Overthrusting, and the Succession of Nappes

Let us consider the central part of Crete in the midst of the zone of orogenic uplift, as shown in Fig. 4. Mountains up to nearly 2500 m above sea level and made up of upper Paleozoic to Upper Cretaceous limestones

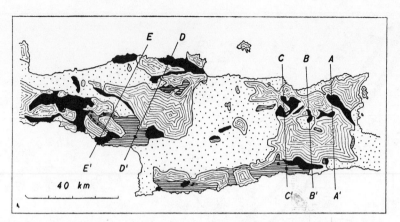

FIG. 4 Sketch map of central Crete with nappe units of Psiloritis (Ithi Ori, to the left, crossed by section lines E and D) and of Dikte (Lassithi Ori, to the right, crossed by section lines C, B, and A). Black: ophiolite-bearing phyllites at the base of the calcareous nappes. Horizontally hatched: flysch-like sediments of Early to Late Tertiary age. Dotted: neoautochthonous Late Tertiary to Quaternary sediments of molasse type.

rise above foothills mainly consisting of Tertiary flysch and molasse. Special attention should be paid to the phyllite outcrops of the so-called Peloponnese-Crete crystalline massif, shown by the black areas in Fig. 4. Deep road cuts show that these phyllites contain great amounts of ophiolitic material composed of basic and ultrabasic igneous rocks of eugeosynclinal origin. In many places this phyllite-ophiolite complex shows great similarity to the Schistes Lustrés series of the western Alps and other typically eugeosynclinal sediments and igneous rocks of the Alpine orogenic belt. As pointed out by Wurm (1950), black slates, quartzites, conglomerates, phyllitic limestones, and Triassic dolomites and gypsum are combined with this phyllite-ophiolite series in western Crete. In the Western Alps, sedimentation in the Penninic facies zone started with a carbonate-evaporite series of Triassic age, and overlying ophiolite-bearing calcareous phyllites announce a more or less abrupt change in facies conditions. Quite probably, the ophiolite-bearing phyllites of Crete also represent a Mesozoic eugeosynclinal sedimentary and volcanic series.

Wurm (1950) believes that the contact between the phyllites and the overlying limestone masses is of tectonic origin and in all likelihood caused by nappe tectonics. The overlying limestones belong to the so-called Tripolitsa zone of the central Peloponnese (Renz, 1950), reaching 500–1000 m in thickness and starting with Upper Triassic dolomites. Wurm (1950) was uncertain whether the phyllite-ophiolite series, supposedly of Early Triassic age, was the true stratigraphic base of the overlying limestones. Subsequently, Kuss (1963) found Permian fusulinids in Crete, proving that limestone sedimentation in the Tripolitsa zone of Crete started much earlier. Today we may assume that the phyllite-ophiolite series of central Crete has been overthrust by upper Paleozoic to Cretaceous dolomites and limestones as a result of far-reaching nappe tectonics. The cross sections of Fig. 5 pass through the Dikte Mountains (A, B, C) and Mount Psiloritis (D and E) (cf. Fig. 4). Invariably, a nappe unit consisting of limestones and dolomites is in tectonic contact with underlying dark, ophiolite-bearing phyllites (shown in black in Fig. 5).

There is, in addition, a thrust plane between the phyllite-ophiolite series and under-

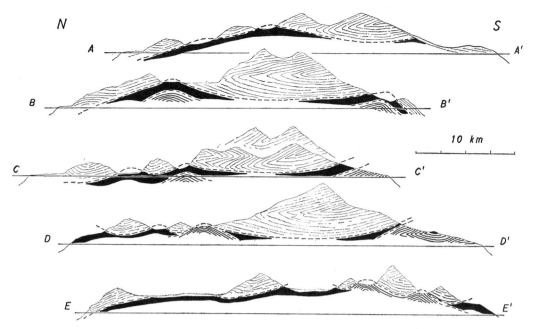

FIG. 5 Geologic cross sections through central Crete along the section lines of Fig. 5. From top to bottom: calcareous nappes of miogeosynclinal origin; mylonitic gliding horizon comprising ophiolite-bearing phyllites of eugeosynclinal origin (black); and overthrust Ionian flysch.

lying flyschlike sediments. Exposures along the southern slope of the Dikte Mountains show that the upper calcareous nappe with the phyllite-ophiolite series at the base (representing a mylonitic gliding horizon) slid into a Late Tertiary flysch-bearing foredeep (Fig. 5).

Most probably, nappe transport was of a gravitational nature, for blocks of every kind and size (limestones, ophiolites, and metamorphic rocks ranging from a few centimeters to hundreds of meters in diameter) can be found among the flysch sediments. Metamorphic rocks of greenschist and glaucophane schist facies alternate with unmetamorphosed flysch sediments (Wurm, 1950). Wedge-shaped particles enclosed near the thrust plane show that the nappe units came from the north, that is, from the area between Crete and the Attic-Cycladian crystalline massif. The tectonically superposed series in central Crete each had to be derived from its own primary sedimentary regions, located side by side with different facies characteristics. In the south, the Late Tertiary flysch trough (called Ionian flysch trough in Fig. 3) is represented today by the orogenic zone of uplift, that is, the Minoidian island arc. A eugeosynclinal, ophiolite-bearing furrow primarily of Mesozoic age was located near the northern edge of the autochthonous flysch zone. Still farther north lay the miogeosynclinal facies belt of compact upper Paleozoic to Cretaceous limestones and dolomites.

In comparing these parallel isopic sedimentary zones in the Minoides with those in continental Greece, especially with the Hellenides *sensu stricto*, we have to take into account that the Tripolitsa zone of the central Peloponnese has been overthrust by the fillings of the Olonos-Pindos furrow, which is as yet unknown in Crete. Apparently, the Olonos-Pindos trough comes to an end between the Peloponnese and Crete. Whereas the Tripolitsa and Parnas-Ghiona limestone sequences are separated from each other where an Olonos-Pindos furrow existed, they were quite possibly linked farther east, although they are difficult to tell apart because of their great similarities.

However, there are also some differences between these two facies zones. In the Tripolitsa zone of Crete no bauxites have been found such as exist along mid-Cretaceous emersion horizons in the Parnas-Ghiona zone of the Peleponnese, nor do we find the characteristic red ammonitic limestones known elsewhere by such names as Bulog, Asklepeion, Hallstatt, *ammonitico rosso*, or red Gault facies. Evidently, the bauxites are restricted to areas of local uplift during Early to mid-Cretaceous time and are totally absent in all neighboring regions that underwent epeirogenic subsidence. The red ammonitic limestones of the Parnas-Ghiona zone are, for the most part, neither of great stratigraphic nor of great lateral extent. Thus one may assume that the calcareous nappes of central Crete formed originally in a sedimentary zone which was the southeastward continuation of the Parnas-Ghiona zone and lay next to the Tripolitsa sedimentary zone, but which did not develop an inter-reef Hallstatt facies and was not subjected to Early to mid-Cretaceous uplift.

As far as the phyllite-ophiolite series of central Crete is concerned, one cannot speak of a crystalline massif in the true sense of the word. The newly acquired data, including those of D/V Glomar Challenger, all suggest young crustal movements of eugeosynclinal orogenic origin. Hence the conclusions of Wurm (1950) have been verified (translated as follows): "According to their composition and structure, the Cretan–Peloponnesian metamorphic rocks form part of an alpinotype geosyncline and stand in contrast to the other Aegean crystalline areas."

Thus the geodynamic comparison of the Minoides, Hellenides, Dinarides, Carpathians, Apennines, and Alps is well justified, as pointed out by the author in 1966 and carried out in detail by the research program "Geodynamics of the Mediterranean" of the Deutsche Forschungs-Gesellschaft.

The Gravity Field and the Mechanics of Nappe Transport

If there really is a recently active orogen in the eastern Mediterranean containing thrust sheets of late to post-Pliocene age, let us consider how orogeny and nappe transport may be supposed to function.

It seems evident that nappe transport could not be carried out in a single stage, but that two or more stages are involved. At first, the calcareous nappes, originally located around the southern border of the Attic-Cycladian massif, glided southward to the phyllite-ophiolite-bearing eugeosynclinal trough. One may assume that the miogeosynclinal reef platforms of Tripolitsa limestone initially rose above the deeply subsided eugeosynclinal furrow, so that a considerable slope was available for gravity gliding. Probably this happened during the Early Tertiary, accompanied by flysch sedimentation in the foredeep in front of the thrust sheets. Hence the subsiding troughs were filled not only by sedimentation but also by gliding tectonics. But gliding of nappes had to come to an end when the thrust masses reached the foreland border of the eugeosynclinal furrow and could not be renewed except by a shifting of foredeeps and zones of uplift. The eugeosynclinal trough, until then a zone of considerable subsidence, began to rise while the Ionian zone to the south sank. As a result, the sedimentary and tectonic fill of the eugeosynclinal trough became unstable again and began to glide down the slope between uplifted eugeosynclinal and subsiding Ionian flysch zone. As before, flyschlike sediments were laid down in the foredeep in front of the nappes, sometimes of wildflysch type, with blocks and sliding masses up to hundreds of meters in diameter that moved down the slope in front of the advancing nappes.

We see then, that the Minoidian island arc is composed of (1) an autochthonous flysch or wildflysch complex at the base, (2) overriding nappes consisting of at least two tectonically superposed sedimentary facies, and (3) normally deposited Upper Tertiary neoautochthonous molasselike sediments, covering both the autochthonous and the overriding rocks. Recently the Ionian flysch trough has been lifted up as well, and thrusting due to an orographic slope may occur in a restricted zone along the southern border of the island arc where the surface descends from 2500 m above to nearly 500 m beneath sea level. One cannot exclude the possibility that flyschlike sediments may be deposited today within this foredeep furrow south of the island arc, whereas sedimentation within the zone of orogenic uplift is of molasse type as it was in Late Tertiary and Quaternary time. Already during nappe transport, molasselike sediments seem to have been deposited on top of overthrust rock masses, as is known from the Eastern Alps ("Gosau conglomerate") and from the Northern Apennines ("Tongriano" of the internal Apennines). The Upper Tertiary molasselike sediments above the thrust masses of central Crete likewise belong to these Gosau-type deposits and therefore do not mark an early termination of orogenic activity. Flysch sedimentation differs from molasse sedimentation mainly with respect to depth relative to sea level; during orogeny, flysch and molasse may be deposited side by side, the one within deeply subsided troughs, the other upon ridges resulting from orogenic uplift. Certainly, the study of Recent sediments in the foredeep south of Crete is of special interest with regard to geology as well as oceanography.

As mentioned earlier, the zone of thrusting shifted outward as the foredeeps and zones of uplift were shifting to the south. This wandering of orogenic activity calls to mind the early work of Van Bemmelen done 40 years ago in the East Indies, where he recognized the same outward shifting of orogenic activity in space and time (1932; see also 1965). Therefore we may assume that advancing orogenic fronts play an essential role in geodynamics. Advancing of orogenic fronts also causes the movement of nappes, as pointed out previously. In 1965, the author compared such nappe transport by shifting foredeeps with surf-riding, where lateral transport is similarly due to wavelike subsidence and uplift. Minoidian nappe transport shows great similarity: there is at present no great "geotumor" behind the island arc that would permit gravity gliding from the border of the Attic-Cycladian massif all the way to southern Crete, and we may conclude from Neogene sedimentation around the Aegean that such an extensive hypothetical geotumor did not exist there during Tertiary time either. On the contrary, the Aegean Sea goes down to almost 3000 m beneath sea level and represents a typical backdeep trough, filled during the Late Tertiary partly by fresh water, partly by brackish, and partly by marine deposits. Like waves on the surface of the water, shifting zones of uplift and sinking continue to produce slopes between crustal heights and depths that make the driving of nappes possible. As may be verified in central Crete, the overthrust calcareous masses of the Dikte Mountains and Mount Psiloritis (Fig. 5) are not isolated portions of a once coherent nappe covering the entire island and the surrounding area and pushed forward by tectonic forces from a hypothetical root zone, but rather represent individual nappe units detached long ago and driven along independently by gravitational forces. Only gravity is capable of driving two such isolated thrust sheets and all the other smaller ones around them.

If one compares the direction of nappe transport with the gravity distribution map shown in Fig. 1 it may be observed that the nappes are moving, geophysically speaking, from the center of a positive gravity anomaly toward the outer zone of strong negative gravity anomalies, that is, from regions of gravitational heights to those of gravitational depths. All kinds of tectonic transport, such as folding and nappe movement, are capable of restoring the gravitational equilibrium in the upper mantle and lower crust which had been disturbed by orogenic activity. Tectonic movement therefore seems to continue as long as strong gravitational differences between neighboring orogenic zones exist. Today, the eastern Mediterranean, and especially the Minoidian island arc, must be considered as snapshots of a tectonic evolution that extended over a long period of geologic time and which will continue in the future until gravitational equilibrium is achieved. The gravitational inequilibrium of today is caused by the different structural positions of the mantle-to-crust discontinuity on the two sides of the orogenic front line, the discontinuity being high where gravity exceeds the normal, and low or absent where negative gravity anomalies exist. Instead of a far-reaching slope of an hypothetical "geotumor" on the surface of the Earth, there exists a characteristic slope of the Mohorovičić discontinuity, reaching from a depth of at least 10–15 km behind the orogenic front to perhaps 40–50 km beneath the foredeep trough. This important slope at the base of the crust, buried and counterbalanced in some measure by different crustal thicknesses, produces unstable layering conditions within the upper crust and therefore tectonic movement from the inner to the outer orogenic zones. As the zone of negative gravity anomalies shifts outward in space and time, the thrust sheets wander (in the case of epeiropetal fronts) similarly from the inner border of the miogeosyncline to the outer front of a stable epicontinental shield undergoing epeirogenic uplift and therefore able to halt the shifting of gravity anomalies, subsidence, and tectonic transport. In case of epeirofugal ones, shifting will continue, perhaps, on an oceanic scale.

Local differences in structural height of the transition zone between the mantle and the crust may be caused by the slow but continual change of geothermal and other conditions within the deep interior of the earth. Today, we are not able to decide whether the changing heat output of the earth's interior proved by the different heat flow values measured on a global scale is due to different amounts of radioactive materials in crustal rocks, to changes in heat radiation and conduction, or to gravitational compaction of the core and lower mantle. But, returning to the question raised at the outset, whether orogenic activity shows periodicity or continuity during geologic time, the newly acquired data all indicate, in spite of short-lived earthquake shocks and volcanic eruptions, a slow shifting of gravity anomalies, suggesting more or less continuous orogenic movements during nearly all of Tertiary and Quaternary time.

References

Aubouin, J., 1965, *Geosynclines*: Developments in Geotectonics, v. 1: Amsterdam–London–New York, Elsevier, 335 p.

———, Brunn, J. H., Celet, P., Dercourt, J., Godfriaux, I., and Mercier, J., 1960–1963, Esquisse de la géologie de la Grèce: *Livre à la Mémoire du Prof. P. Fallot*, Mém. Soc. Géol. France, v. 2, p. 582–610.

Fleischer, U., 1964, Schwerestörungen im östlichen Mittelmeer nach Messungen mit einem Askania-Seegravimeter: *Deutsch. Hydrograph. Zeitschr.*, v. 17, no. 4, p. 153–164.

Gilbert, G. K., 1890, Lake Bonneville: *U.S. Geol. Surv. Monogr.*, v. 1, 438 p.

Godfriaux, I., 1967, Panorama de la géologie de l'Olympe (Grèce): *Bull. Soc. Belge Géol. Paléontol. Hydrol.*, v. 76, no. 1–2, p. 114–124, geol. map 1:166,660 (1968).

Heim, A., 1921, *Geologie der Schweiz*, v. 2, *Die Schweizer Alpen*: Leipzig, C. H. Tauchnitz, 1018 p.

Hess, H. H., 1938, Gravity anomalies and island arc structure: *Proc. Am. Phil. Soc.*, v. 79, p. 71.

Kuss, S. E., 1963, Erster Nachweis von permischen Fusulinen auf der Insel Kreta: *Praktika Akad. Athen*, v. 38, p. 431–436.

———, 1965, Eine pleistozäne Säugetierfauna der Insel Kreta: *Ber. Naturf. Ges. Freiburg i. Br.*, v. 55, p. 271–348.

Renz, C., 1955, *Stratigraphie Griechenlands*: Athens, Institute Geol. and Subsurf. Research, 637 p.

———, 1957, *Ergänzungsheft und Tektonische Übersichtskarte*: Athens, Institute Geol. and Subsurf. Research, 55 p.

———, Liatsikas, N., and Paraskevaidis, I., 1954, Geologic map of Greece, 1:500,000: Institute Geol. and Subsurf. Research, Athens.

Ryan, W. B. F., and Hsü, K. J., 1970, *Glomar Challenger*

deep-sea drilling project: Univ. Calif., San Diego, Scripps Inst. Oceanography Release no. 155, p. 1–7.

Stensen, N., 1669, *De solido intra solidum naturaliter contento*: Florence, 131 p.

Stille, H., 1924, *Grundfragen der vergleichenden Tektonik*: Berlin, Borntraeger, 443 p.

Suess, E., 1885, *Das Antlitz der Erde*, v. 1: Prague, F. Tempsky; Leipzig, G. Freytag, 778 p.

Temple, P. G., 1968, Mechanics of large-scale gravity sliding in the Greek Peloponnesos: *Geol. Soc. Am. Bull.*, v. 79, p. 687–700.

Van Bemmelen, R. W., 1932, De undatie-theorie: *Natuurk. Tijdschr. Ned. Ind.*, v. 92, no. 1, p. 89–242.

——, 1965, Phéonomènes géodynamiques: *Mém. Soc. Belge. Géol. Pal. Hydr.*, no. 8, p. 1–127.

Vening-Meinesz, F. A., 1934, Gravity and the hypothesis of convection currents in the Earth: *Proc. Kon. Akad. Wetensch. Amsterdam*, v. 37, no. 2, p. 37.

——, 1948, Gravity expeditions at sea: *Neth. Geodetic Comm.*, v. 4, p. 1–233.

——, 1950, Earth's crust deformations in geosynclines: *Proc. Kon. Ned. Akad. Wetensch. Amsterdam*, ser. B., v. 53, p. 27–46.

—— 1964, *The Earth's crust and mantle*: Amsterdam, Elsevier, 124 p.

Wunderlich, H. G., 1965, Zyklischer Bewegungsablauf beim Vorrücken orogener Fronten und der Mechanismus des Deckschollentransports nach dem Surf Ridingprinzip: *Geol. Mijnb.*, v. 44, p. 440–457.

——, 1966, *Wesen und Ursachen der Gebirgsbildung*: Mannheim, Bibliogr. Inst., 367 p.

Wurm, A., 1950, Zur Kenntnis des Metamorphikums der Insel Kreta: *Neues Jahrb. Geol. Paläont. Mh.*, Stuttgart, p. 206–239.

——, 1955, Geologische Beobachtungen im Asterussia-Gebirge auf der Insel Kreta: *Elleniki Geol. Etiria*, v. 2, p. 80–87.

WILLIAM M. TURNER

Hydrotechnics
Albuquerque, New Mexico

The Cyprian Gravity Nappe and the Autochthonous Basement of Cyprus[1]

Detailed geological mapping of western Cyprus by the writer disclosed for the first time the allochthonous character of an extensively developed unit of highly deformed rock. The allochthonous rock unit was assigned to the Cyprian gravity nappe during the mapping program and is discussed in considerable detail by Turner (1971). The major concerns of this paper are the basement autochthon and the overlying allochthon, or Cyprian gravity nappe, and their composition, structural character, and relationships.

Inasmuch as the Cyprian gravity nappe is interpreted to have moved by gravity tectonics, a field of structural geology pioneered by R. W. Van Bemmelen, and because Van Bemmelen himself was interested in the petrogenesis of the Troodos Massif (Van Bemmelen, 1956), it seems appropriate that results of the writer's recent work in Cyprus be included in the present volume.

Regional Geologic Setting

Cyprus lies within the eastern Mediterranean and is part of the foreland border zone of the Taurus Mountain system of southern Turkey (Fig. 1). The Taurus Mountains are

---

[1] I am indebted to Mr. Y. Hji Stavrinou, Director of the Geological Survey Department of Cyprus, for providing field transportation and laboratory facilities during the geological mapping program. I am also grateful to the many scientists with whom I have discussed the problems of Mediterranean geology and in particular I would like to acknowledge Drs. T. Thayer, L. Frakes, L. Picard, and H. Klemme. Drs. A. Rosenzweig, H. Klemme, and D. Supkow have made many suggestions for the improvement of this paper. I accept responsibility for the interpretations presented.

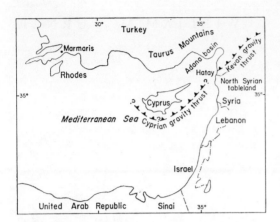

FIG. 1 Regional map of the eastern Mediterranean area.

the eastward continuation of the Dinaric Alps into Asia.

Southern Turkey was a broad continental shelf area during most of the Mesozoic. In the Late Cretaceous a sequence of shallow water sediments and basic and ultrabasic igneous rocks of Triassic and Jurassic age underwent regional uplift, structural deformation, and gravity sliding onto a low-lying border zone. In southern Turkey the gravity slide is represented by the Kevan gravity nappe, which moved south and overthrust the North Syrian tableland (Arabian foreland). This gravity nappe has great masses of ophiolite at the line of thrust (Rigo de Righi and Cortesini, 1964). A second phase of southward directed compression and movement is recorded in southeastern Turkey by the Elazig gravity nappe, which corresponds to most of the Tauric orogenic belt in southeast Turkey. The Elazig gravity nappe moved south in the late middle Miocene and overrides parts of the older Kevan gravity nappe.

Both of these tectonic units have homologs in Cyprus. The Kevan gravity nappe is homologous to the Cyprian gravity nappe. Strong southerly thrusting in the Kyrenia mountains and general southerly displacement of the entire Cyprus block in the late middle Miocene probably make the entire Cyprus block the structural homolog of the Elazig gravity nappe.

Cyprus is broadly divisible from north to south (Fig. 2) into the following zones, all of which contain some rock attributed to the Cyprian gravity nappe:

1. A northern zone of late Miocene to Holocene shelf sediments underlain by the same highly deformed rocks that make up the Kyrenia Mountains to the south.
2. The intensely folded, overturned, and imbricately thrust Kyrenia Mountains formed in late middle Miocene time of early Campanian to middle Miocene carbonate, flysch, and conglomerate with immense olistolites of Permian (?) marble and limestone and minor amounts of allochthonous pillow lava, serpentinite, syenite, and rhyolite, all remnants of the Cyprian gravity nappe or homotaxial olistolites.
3. The broad Mesaoria Plain with more than 9000 ft (2800 m) of strongly folded and faulted early Campanian to middle Miocene tuffaceous graywacke, chalk, and flysch, overlain by less deformed late Miocene to Holocene marl and calcarenite. The middle Miocene flysch indicates the area is an extension of the Adana basin of southeastern Turkey (Weiler, 1965).
4. The Troodos igneous massif, which is overlapped by early Campanian to Holocene sediment and overthrust by up to 3000 ft (900 m) of serpentinized ultrabasic rock.
5. The southern folded and faulted foothill belt of autochthonous early Campanian sediments. Neoautochthonous Maestrichtian to Holocene sediments rest above the Cyprian gravity nappe, which is most extensively preserved in this area.

Lithologic Units of the Autochthon

Most of the following discussion is based on a reorganization and reinterpretation by the present investigator of the autochthonous igneous basement rocks and associated

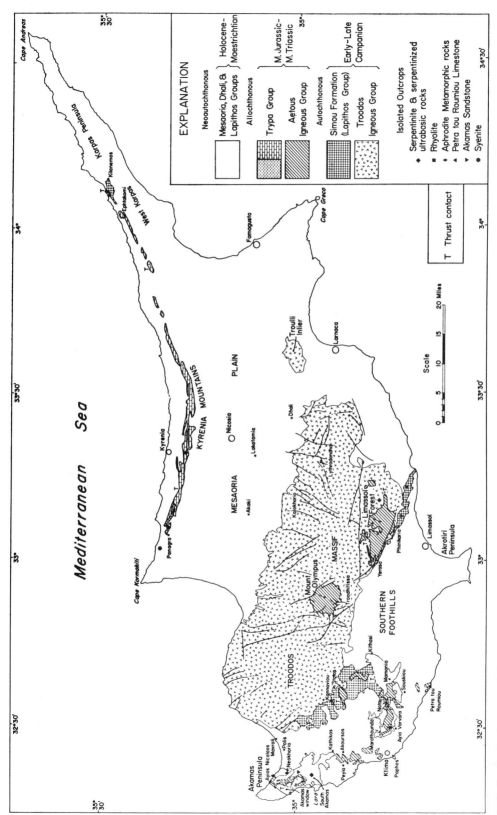

FIG. 2 Geologic map of Cyprus showing outcrop areas of autochthonous, allochthonous, and neoautochthonous rocks.

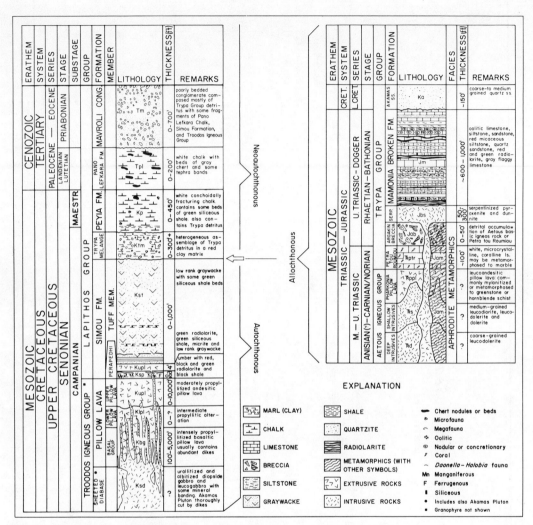

FIG. 3 Stratigraphy of the Mesozoic rocks of Cyprus.

sediments. The igneous rocks are placed in the Troodos Igneous group, whereas tuffaceous graywacke associated with autochthonous extrusive rocks is placed in the Simou formation of the Lapithos group. The rock relations of the autochthon are presented in Fig. 3 and the petrology of the component rock units is presented in Table 1.

Troodos Igneous Group

The Troodos Igneous group is composed of extrusive and intrusive rocks of Late Cretaceous age (Santonian to early Campanian). The extrusive rocks consist of a shallow water accumulation of basaltic and andesitic pillow lava more than 13,000 ft (4000 m) thick. The pillow lava was altered into three mappable alteration facies: the basal group, the Lower pillow lava, and the Upper pillow lava, following the emplacement of intrusive basic rocks into the pillow lava. These units were considered by all earlier workers, with the possible exception of Wilson (1959), as volcano-stratigraphic units.

The intrusive rock consists of a large

gabbroic pluton called the Akamas pluton, which was emplaced into the pillow lava sequence in the late Campanian. The Akamas pluton and the pillow lava were subsequently intruded by innumerable north–south trending dikes of intermediate composition, and ranging from 1 to 3 ft thick, such that the intrusive contact of the pluton with the pillow lava is commonly obscured. It is visible, however, at several places in the Akamas Peninsula (Fig. 4). The density of dikes is so great that in most areas the Akamas pluton is a minor component of the terrane. Where the terrane is composed of more than 85 percent dikes it is mapped as Sheeted diabase.

In the Troodos area the Akamas pluton is mapped as an insular body of gabbroic rock surrounded by topographically lower Sheeted diabase terrane (Fig. 5). Parts of the Akamas pluton are mapped as much smaller insular areas of gabbro in both Sheeted diabase and pillow lava terrane. Although the latter occurrences have been referred to as post-Upper pillow lava intrusive by Gass (1959), this is a misnomer.

TABLE 1 Petrology of the Troodos Igneous Group[a]

Unit	Rock Type	Composition of Original Plagioclase	Comments
Pillow lava	Basaltic and andesitic pillow lava usually highly altered; the highest degree of alteration, the Basal Group pillow lava, may contain epidosite	An_{30}–An_{70}	Highly vesicular with vesicles commonly more than 2 mm in diameter
Akamas pluton	Originally diopside gabbro and leucogabbro now altered to uralitic and albitic gabbro	An_{41}–An_{87}	Rocks of the Akamas pluton from both the Akamas Peninsula and the Troodos massif are similar compositionally and texturally, and exhibit the same kind of alteration; small amounts of modal quartz are not uncommon
Sheeted diabase	Akamas pluton host rock plus more than 85 volume percent dike rock		
Dikes	Dike rock examined is diorite, meladiorite, quartz-bearing meladiorite, leucogabbro, and andesine anorthosite	An_{32}–An_{72}	They contain more quartz and are less calcic than the gabbro of the Akamas pluton; there is no major compositional or tectural difference in the dike rocks whether they are from the Akamas pluton, the sheeted diabase, or the pillow lava; usually the dikes have withstood the effects of alteration
Granophyre	Quartz-rich augite leucoquartz gabbro, quartz-rich augite-quartz gabbro, barkevikite leucoquartz diorite, and quartz diorite have been observed	An_{28}–An_{50}	The total outcrop area of granophyre is small compared to the area occupied by other extrusive and intrusive rocks of the Troodos Igneous group

[a] Terminology from Streckeisen (1967).

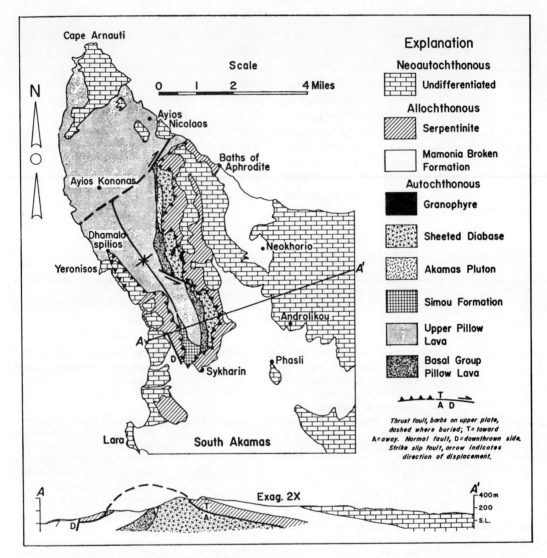

FIG. 4 Geologic map of the Akamas Peninsula of western Cyprus.

Geologic maps of Cyprus show north–south trending sinuous dike trends which pass from pillow lava into and across the Akamas pluton, indicating a post-Akamas pluton age for most of the dikes and suggesting a common mode of origin.

Late stage granophyre occurs as minor intrusives into all older intrusive and extrusive members of the Troodos Igneous group.

Lapithos Group

The only autochthonous member of the Lapithos group is the Simou formation, which is primarily a unit of tuffaceous graywacke, more than 1000 ft (300 m) thick and composed of presumably pyroclastic material ejected into a marine environment during the final phase of extrusive igneous activity. The base of the Simou formation is interbedded

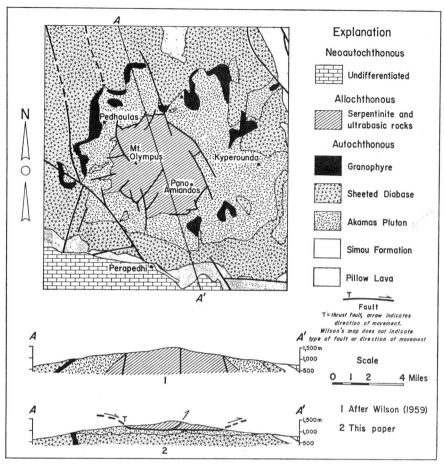

FIG. 5 Geologic map of the Mount Olympus area (adapted from Wilson, 1959).

with pillow lava of the Troodos Igneous group, indicating penecontemporaneity of the Simou with the youngest pillow lava of the Troodos Igneous group. Paleontologic evidence (Turner, 1971) indicates an early to late Campanian age for the Simou.

Lithologic Units of the Allochthon

Since the beginning of geologic investigation of Cyprus by Gaudry in 1854, the occurrence of thin-bedded radiolarite, brown calcareous siltstone, quartz sandstone, and other rock types including ultrabasic and basic igneous rocks has puzzled geologists. These rocks were allocated to the Trypanian series by Bellamy and Jukes-Browne (1905); this was changed to the Trypa group by Henson, Browne, and McGinty (1949). The Trypa group has been further divided by Turner (1971) into the Aetous Igneous group and the Trypa group, *sensu stricto*, which now includes only sedimentary and metamorphic rocks. Recent geologic mapping by the writer has shown that these rocks belong to an allochthonous unit, named the Cyprian gravity nappe, which was emplaced into the Cyprus area in the Late Cretaceous (late Campanian). The nappe was a more extensive structural unit in Cyprus than it is presently, much of it having been eroded. Erosional remnants of the nappe are composed of up to and exceeding 2000 ft (600 m) of intensely deformed sedimentary, igneous, and metamorphic rocks and are found from the Karpas Peninsula of eastern Cyprus to the Paphos district of western Cyprus (Fig. 2), where they occur most extensively.

Aetous Igneous Group

The Aetous Igneous group consists of the extrusive and intrusive basic and ultrabasic rocks of the Cyprian gravity nappe. The petrology of these rocks is summarized in Table 2. The extrusive rock unit is highly brecciated basaltic and/or andesitic Pharkonia pillow lava of unknown thickness. The Pharkonia is Triassic (Anisian to Carnian-Norian) in age based upon paleontologic evidence (Turner, 1971). The intrusive rocks are essentially (1) diorite, which is intrusive into the Pharkonia and which is divided into shallow and deep intrusives on the basis of mineralogy, texture, and structure and (2) serpentinite and ultrabasic rocks. The serpentinite and ultrabasic rocks occur at the base of the Cyprian gravity nappe and vary from about 300 ft (90 m) thick at localities in the Akamas Peninsula to possibly more than 3000 ft (900 m) thick at Mount Olympus. The serpentinite and ultrabasic rocks are also much younger than the other rocks of

TABLE 2 Petrology of the Aetous Igneous Group[a]

Unit	Rock Type	Composition of Plagioclase	Comments
Pharkonia pillow lava	Augite-olivine basalt to diopside leucoandesite	An_{18}–An_{53}	The Pharkonia pillow lava is similar to rock described by Bear (1959, 1961) from the Kyrenia Mountains; rocks are commonly very green due to the formation of chlorphaeite from glass; hematite may also be abundant; this unit has been metamorphosed into hornblende schist in places
Shallow intrusive	Medium to coarse-grained pyroxene bearing glass-rich leucodolerite, leucodiorite, and diorite	An_{30}–An_{40}	This unit differs from the Pharkonia in a coarser grain size, the paucity of hematite, the presence of some glass, and the absence of flow morphology
Deep intrusive	Coarse grained leucodiorite and diorite	An_{25}–An_{50}	The coarser grain size and total absence of glass distinguish this unit
Serpentinite and ultrabasics	Medium to coarse grained pyroxenite, pegmatitic pyroxenite, wehrlite, lherzolite, dunite, and serpentinite	An_{70+}	Serpentinite is composed of varying proportions of antigorite after enstatite and chrysotile after olivine with minor amounts of magnetite; the petrography of the serpentinites and ultrabasics whether from the Troodos massif, the Akamas Peninsula, or from isolated outcrops surrounded by Trypa group sediment, is similar

[a] Terminology from Streckeisen (1967).

the Aetous Igneous group which they intrude. In general, rocks of the Aetous Igneous group are visible only where the overlying Trypa group has been eroded away.

The serpentinite is considered Early Jurassic (Liassic) in age because it invaded Argakin breccia of the Trypa group, which is Late Triassic or Liassic in age, but apparently not the overlying Mamonia Broken formation which is Middle Jurassic (late Bathonian) in age.

Trypa Group

Sedimentary and metamorphic rocks of the Cyprian gravity nappe are placed in the Trypa group, which ranges up to about 1000 ft (305 m) in total thickness. The formations of the group from oldest to youngest are Petra tou Roumiou limestone, Argakin breccia, Aphrodite metamorphics, Mamonia Broken formation, and Akamas sandstone.

The Petra tou Roumiou limestone is a massive, pure white, porcellaneous, partly coralline limestone of Late Triassic (Carnian–Norian) age, as indicated by its fauna (Henson, Brown, and McGinty, 1949, p. 10). It ranges up to 50 ft thick and, where the contact is preserved, it conformably overlies the Pharkonia pillow lava.

Argakin breccia is an accumulation of angular detritus of greatly varying clast size and composition. The unit may vary from essentially igneous detritus to Petra tou Roumiou limestone detritus. It is interpreted as Liassic in age because it contains Late Triassic rocks and is overlain conformably by the Mamonia Broken formation of Middle Jurassic (late Bathonian) age.

The Aphrodite metamorphics include all metamorphic rocks in Cyprus. These include epidote-hornblende schist, talc-cummingtonite schist, quartz-mica schist, marble, and metaradiolarite which formed in Liassic time from metamorphism of the Aetous Igneous group, Petra tou Roumiou limestone, and Argakin breccia. Where serpentinite abuts metamorphic terrane the grade of metamorphism decreases away from the serpentinite, indicating that the metamorphic event was associated with the emplacement of serpentinite at temperatures ranging from 400 to 600° C and at pressures between 1 and 2 kbar.

The Mamonia Broken formation of Middle Jurassic (late Bathonian) age occurs primarily in the Paphos district of western Cyprus and comprises most of the sedimentary rock of the Cyprian gravity nappe. It is an intensely deformed sequence of radiolarite, calcareous sandstone, and *Daonella*- and *Halobia*-bearing flaggy, commonly recrystallized limestone. It contains no exotic blocks and functions as a rock-stratigraphic unit in the sense of Hsü (1968). The Mamonia Broken formation is 909 ft (277 m) thick in its thickest measured section and lies unconformably above all older allochthonous rocks and their metamorphic equivalents. The Mamonia was deposited in a very shallow sea and, at times, the depositional area was under subaerial conditions as indicated by reptile tracks and mud cracks in a slab of brown Mamonia siltstone[2] (D. Baird, personal communication, 1970).

The Akamas sandstone, of probable Early Cretaceous age, is the youngest and stratigraphically highest rock unit of the Cyprian gravity nappe. It occurs in the Akamas Peninsula and near the village of Parakklisha in the Limassol area. This medium- to coarse-grained quartzitic sandstone attains a maximum thickness of about 100 ft (30 m) in the Akamas Highland and rests with angular unconformity on the highly deformed Mamonia Broken formation and the serpentinite.

Lithologic Units of the Neoautochthon

Lapithos Group (Continued)

The Trypa melange is the first neoautochthonous sediment deposited above the nappe and is the only neoautochthonous unit

---

[2] Specimen 3208 in collection of the Hebrew University of Jerusalem.

described here. It is composed of red montmorillonitic clay, which encloses a chaotic assortment of exotic blocks derived from the Trypa and Aetous Igneous groups by gravity sliding and erosional mass wasting. No horizontal gradation has been observed between the melange and rock of the nappe. The Trypa melange occurs widely in western and southern Cyprus where it overlies unconformably the nappe or the Simou formation and ranges in thickness from wedge edges to more than 200 ft (60 m). In places it is unconformably overlain by the Peyia chalk of Maestrichtian age.

Structure

Tectonic Contact

Erosion through the Cyprian gravity nappe exposes the tectonic contact at the base of the nappe. This contact can be traced around many large erosional outliers of allochthonous Pharkonia pillow lava and serpentinite in deep, wide river valleys of western Cyprus. In the Akamas Peninsula the tectonic contact is continuously traceable for more than 13 miles (20 km) (Fig. 4).

The contact pattern of allochthonous serpentinite with the overriden autochthon (Fig. 4) is serrate as a consequence of the contact's dip and headward erosion of rivers. The dip of the tectonic contact is highly variable. On the eastern side of the Akamas Peninsula (Fig. 4) dips range from 16 to 90° to the east and average between 30 and 60°. At the southern terminus of the Akamas Peninsula, the tectonic contact surface dips very steeply to the south, and on the western side of the window dips range from 13 to 40° westerly. The dip of the tectonic contact generally diminishes away from the Akamas window (Fig. 4).

Quaternary marine planation has removed most of the nappe along the west coast of the Akamas Peninsula. Near Dhamalospilios only a narrow finger of serpentinite remains in a pre-Maestrichtian structural or erosional trough. On the eastern side of the serpentinite finger the tectonic contact dips westerly, whereas on its western side it dips easterly.

The tectonic contact between the serpentinite and overriden autochthonous rock is sharply defined, although a narrow zone of rock up to 5 ft (2 m) wide on either side of the contact may be intensely sheared and brecciated. No fragments of overridden rock were located within the serpentinite and no apophyses of serpentinite were found penetrating autochthonous rock.

Observations made in the Mount Olympus area indicate that the contact between the ultrabasic rocks and underlying altered gabbroic rock of the Akamas pluton dips toward the central part of Mount Olympus. The contact seems relatively well defined on the north, south, and east sides of the Mount Olympus ultrabasic area. On the western side, however, the relationships are not as clear and it appears as though the area has been considerably deformed.

Internal Structure of the Nappe

The Mamonia Broken formation, which comprises most of the sedimentary rock of the nappe, is intensely deformed into tight chevron folds, recumbent isoclinal folds, and crinkle folds. Contorted stratigraphic sections hundreds of feet thick may be overturned. The rocks of the Aetous Igneous group as well as the minor components of the nappe, except the Akamas sandstone, are commonly brecciated.

In the Polis-Kathikas area of western Cyprus (Fig. 2) nearly all measured fold axes in the Cyprian gravity nappe have a mean axial trend of north–south with deviations up to 20° east and west of this trend. The fold axes plunge either north or south at angles ranging from 21 to 38°. The structure of the Aphrodite metamorphics type section is an antiform with an axial direction conformable to the fold trends indicated.

Faults within the Cyprian gravity nappe are manifold but difficult to trace owing to erosion and intense deformation. Some faults bring Aetous Igneous group rocks into contact with Trypa group rocks.

Discussion

Mechanism of Emplacement

Earlier interpretations considered the ultrabasic rocks of the Akamas Peninsula, the Limassol Forest, and the Central Troodos as the ultrabasic zone of a differentiated pluton (Wilson, 1959) (Fig. 5, Section 1) or as a diapiric emplacement (Pantazis, 1967).

In the case of the Central Troodos it follows that either the antiformal warping of the area, required to reveal the ultrabasic zone of the Akamas pluton, or diapirism would have had to occur following the intrusion of the Akamas pluton into the pillow lava and following the injection of the numerous dikes. Either structural event would have produced profound structural deformation, which would be indicated by a distinct change in the attitude of the dikes. However, there is no major change in the attitude of the dikes as they trend across the Troodos massif.

In addition, a local Bouguer gravity anomaly centered over the Central Troodos indicates that a local mass deficiency occurs over the center of the serpentinite mass (Gass and Mason-Smith, 1963). Such a mass deficiency would hardly seem likely under Wilson's or Pantazis' hypothesis, because the serpentine mass in both cases would be expected to continue downward to even more basic rocks of still higher density, thereby providing a mass excess. Figure 5, Section 2, presents the interpretation of the structure of the Troodos Massif incorporating the relationships between units of the Troodos Igneous group as they are now understood, and suggests that the ultrabasic occurrence in the Central Troodos is a klippe resting above the upper zone of a relatively dike-free part of the gabbroic Akamas pluton.

Furthermore, the geologic conditions in the areas of major serpentinite occurrence are fundamentally different and several mechanisms of emplacement would have to be invoked. The earlier workers were not of the opinion that the different serpentinite occurrences and related ultrabasic rocks were in any way different petrographically or that their origin was due to anything but a single cause or episode. Many subsequent workers have viewed the serpentinite within Trypa terrane and serpentinite on Mount Olympus and in the Limassol Forest as essentially different in origin. Gass (1960, p. 27), however, writes:

> Abundant small irregular masses of bastite-serpentinite occur in association with the rocks of the Mamonia (Broken[3]) Formation. There seems to be no reason why these small bodies are not part of the main ultrabasic emplacement and no evidence has been forthcoming to suggest that there were two periods of ultrabasic intrusion.

The writer's work supports Gass's interpretation in that it appears that large and small masses of serpentinite not presently associated with the Trypa sediment belong to the Aetous Igneous group and represent erosional outliers. This is particularly evident in the case of the smaller serpentinite masses near Nata and Kannaviou, which rest above the Simou formation (Fig. 2).

With regard to the Limassol Forest serpentinite (Fig. 2), Morel (1962) suggested that the contact between the serpentinite and rock of the Troodos Igneous group is a thrust contact. Morel's observations in the Limassol Forest area suggest that, in places, the contact has been folded and that the folds have general east–west oriented axes. The infolding of the serpentinite may have given rise to the diapiric hypothesis.

At present no allochthonous sedimentary rock is associated with the serpentinite and

---
[3] Author's word.

ultrabasic rocks of the Central Troodos as it is with sepentinite and ultrabasic rocks in other parts of Cyprus. However, evidence for the presence of allochthonous sedimentary rock in the Central Troodos in former times is found in the neoautochthonous Mavroli conglomerate of late Eocene to Oligocene age. The Mavroli conglomerate has a circum-Troodos distribution and is composed of sedimentary and igneous detritus derived from the Cyprian gravity nappe and some clasts of early Eocene Pano Lefkara chalk and gabbro from the Akamas pluton. The conglomerate occurs at Mavroli and Statos in western Cyprus, and a provenance in the area of the present Troodos Massif is indicated by the fact that no other high ground exists which could have been a source for the gabbro clasts from the Akamas pluton. This evidence further supports the present writer's interpretation that the serpentinite and ultrabasic rocks in the Central Troodos are an erosional remnant of a once more extensive nappe cover, the upper sedimentary part of which has been removed by erosion and has accumulated as the Mavroli conglomerate.

The nappe is interpreted to have moved by gravitational sliding from its root zone because, as a structural unit, it is incompetent and could not have transmitted the stress required to move by mechanical thrusting. If we consider only that portion of the Cyprian gravity nappe whose erosional remnants we can see within the confines of Cyprus, we can imagine a plate about 70 km wide and 80 km long in a direction parallel to the movement, and with an original thickness ranging from about 400 to more than 1000 m. The consistent north–south orientation of fold axes within the nappe over large areas suggests that at least the observable material moved as a single plate from its root zone. If a motive force were applied continuously at the rearward edge of the nappe, the pressure would be more than 3.2 kbar, which exceeds the crushing strength of the rock.

There is no particular reason for believing the nappe moved in a submarine environment. Partial topographic inversion of the Akamas syncline and the formation of kaolin from the Simou formation indicates that the overriden surface had been subjected to weathering and subaerial erosion immediately prior to arrival of the nappe. Further, the Trypa melange is interpreted as the product of subaerial mass wasting processes, formed shortly after emplacement of the nappe. The proposed origin for the Trypa melange is supported by the formation of similar material at the present time by mass wasting of the allochthonous rock. Since subaerial conditions existed just before and immediately following nappe emplacement, it seems likely that similar conditions existed during nappe emplacement as well.

The regional gradient of the overridden area may have been very slight, but deformation of the area immediately prior to arrival of the nappe suggests that locally the area may have presented considerable relief. In the Akamas Peninsula the nappe encountered the Akamas syncline from which part of the sedimentary core had been removed by erosion. The presence of overridden kaolinitized Simou formation in the core of the Akamas syncline suggests that not much of the sedimentary material could have been removed by the emplacement of the nappe. The irregularity of the overridden surface is still visible as it affects the present distribution of the nappe and the dip of the tectonic contact.

The rock comprising the Cyprian gravity nappe appears to correspond to the Perdeso and Cermik units of the Kevan gravity nappe of southeastern Turkey and possibly included rock similar to the Hezan unit (Rigo de Righi and Cortesini, 1964). The radiolarite, pillow lava, and serpentinite of the Cyprian gravity nappe are also possibly correlative with the radiolarite-spilite complex and ultrabasic rock of the Marmariş area (Tatar, 1968) of southwestern Turkey and the Hatay area (Dubertret, 1934) northeast of Cyprus. Consequently, the source of the nappe is most likely north of Cyprus where the same suite of rocks occurs over large areas. The nappe is interpreted to have moved at the same time and with the same sense of motion as the Kevan gravity nappe. Furthermore, it is evident that the nappe moved at a rapid rate because only a small hiatus exists between cessation of Simou deposition in the late Campanian and commencement of overlying Peyia deposition in the Maestrichtian.

Contortion and intense folding of radiolarite particularly has caused speculation that deformation was due to preconsolidational slumping; however, competence and lack of attenuation of other interbedded sedimentary units suggest that deformation was postconsolidational. The Akamas sandstone is relatively undeformed and unconformably overlies the Mamonia Broken formation, indicating that most of the internal structure of the nappe formed before deposition of the Akamas sandstone, which is allochthonous and considered to be Early Cretaceous in age. The internal structure of the nappe, then, was formed during a Late Jurassic or Early Cretaceous east–west compressional episode.

The predominant north–south orientation of fold axes within the nappe appears to be at odds with the north to south direction of nappe movement. That is, a north to south movement might be expected to create east–west oriented folds in the nappe. This would occur if a significant compressional regime were responsible for an initial southward thrust of the nappe or if a compressional regime existed within the moving nappe. The observed situation therefore indicates that (1) a compressional stress did not initiate movement of the nappe and (2) a compressional regime did not exist within the nappe as it moved. The absence of east–west oriented compressional structures supports further the suggestion that the nappe was severed from its root zone by gravity and that it moved downslope under the influence of gravity. There is no reason why a freely sliding plate should have an internal compressional regime.

It is difficult to determine, however, whether the observable plate underwent any rotation during its transport, and the foregoing discussion has assumed that it has not. Vine and Moores (1969) have recently suggested a post-Cretaceous 90° anticlockwise rotation of the entire Cyprus block on the basis of paleomagnetic vector orientations. Structural and stratigraphic evidence from Cyprus would preclude this possibility. It can be demonstrated that the major north–northwest trending block faults of Cyprus developed late in the late Campanian as a result of relaxation of the preceding compressional regime. These same faults are oriented along the Erythrean trends of Shalem (1954) and were reactivated in the late middle Miocene. Furthermore, paleocurrent studies by Weiler (1965) of the middle Miocene Kythrea flysch from the Mesaoria Plain of Cyprus indicates that no rotation of the Cyprus block has occurred since that time.

Sequence of Events

Between the Middle Triassic and at least the Middle Jurassic (Bathonian) the area initially occupied by the Cyprian gravity nappe was at times under shallow water and at times exposed to subaerial conditions. The rocks of the area were subjected to at least three phases of major deformation. The first phase of deformation occurred in the Triassic (Rhaetian?) and affected the basic igneous basement and the overlying Petra tou Roumiou limestone and contributed to the formation of the Argakin breccia. The second phase was associated with the emplacement of serpentinite and ultrabasic rocks into the Triassic rocks and probably occurred in the Liassic. The third phase of deformation occurred in the Late Jurassic or Early Cretaceous and created the north–south oriented folds in the Mamonia Broken formation.

In the area in which the present Cyprus basement formed, east–west compression occurred in the early upper Campanian and created broad folds in the igneous basement and conjugate shear fractures which offset the fold axes. Subsequent relaxation of the compressional regime created major block faults and the Cyprus basement was formed into basins and ranges. Elevation of the basement exposed the area to subaerial erosion and pillow lava above the Akamas pluton was removed in both the Akamas Peninsula and the Central Troodos (Fig. 6, top section). This phase of deformation corresponds in time to the Late Cretaceous Alpine orogenic phase and uplift north of Cyprus.

In the Late Cretaceous the Cyprian gravity nappe became separated from its root zone (Fig. 6, top section) under the influence of gravity and, by the beginning of the Maestrichtian, it had moved to its present position above the Cyprus basement (Fig. 6, middle section).

At the close of the Cretaceous and possibly during the early Paleocene, following the emplacement of the nappe and the deposition of the Peyia chalk, southward-directed compression created east–west oriented folds in Cyprus. This period of deformation may have caused infolding of the nappe into the pillow lava basement near Limassol. It may also

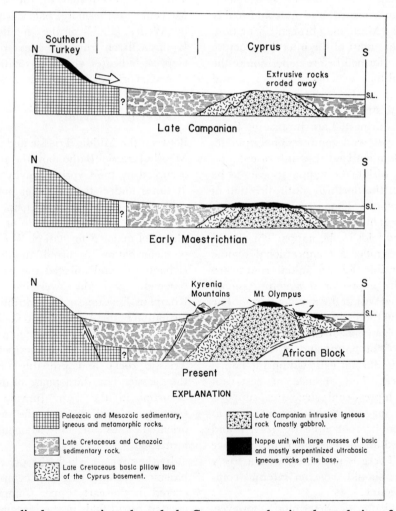

FIG. 6 Generalized cross sections through the Cyprus area showing the evolution of the Cyprian gravity nappe.

have caused deformation of the tectonic contact between the serpentinite and ultrabasic klippe in the Central Troodos area and the underlying basic rock of the Akamas pluton.

During the middle Eocene to early Miocene interval uplift of the Troodos area is recorded by the accumulation of the neoautochthonous Mavroli conglomerate. Regional uplift of the entire Cyprus block began in the late middle Miocene as the result of southward thrusting of the entire Cyprus block over a thick sequence of Paleozoic, Mesozoic, and Cenozoic sedimentary rocks belonging to the African block (Fig. 6, bottom section); uplift continued until the end of the early Pliocene (Tabianian). Relaxation of this compressional episode may have caused the Mediterranean crust between Cyprus and Turkey to have foundered. Erosion of Cyprus since the early Pliocene is largely responsible for the present morphology and rock outcrop patterns of Cyprus.

References

Bellamy, C. V., and Jukes-Browne, A. J., 1905, *The geology of Cyprus*: Plymouth, W. Brendon & Son, 72 p.

Dubertret, L., 1934, Sur la structure de la côte orientale de la Méditerranée: (France), *Office Nat. Combustibles Liquides, Ann.*, v. 9, no. 1, p. 150–152.

Gass, I. G., 1959, The geology of the Dhali area: *Cyprus, Geol. Surv. Dep. Mem. 4*, Nicosia, Geological Survey Dep., 116 p.

———, 1960, The geology of the Akamas Peninsula, in *Ann. Rept. Geol. Surv. Dep. for 1959*: Nicosia, Geological Survey Dep., p. 19–29.

———, and Masson-Smith, D., 1963, The geology and gravity anomalies of the Troodos Massif, Cyprus: *Phil. Trans. Roy. Soc. London*, Ser. A, v. 255, no. 1060, p. 417–467.

Gaudry, G., 1862, Géologie de l'île de Chypre: *Mém. Soc. Géol. Fr.*, 2nd ser., v. 7, p. 149–314.

Henson, F. R. S., Browne, R. V., and McGinty, J., 1949, A synopsis of the stratigraphy and geologic history of Cyprus (with discussion): *Quart. Jour. Geol. Soc. London*, v. 105, no. 417, p. 1–41.

Hsü, K. J., 1968, Principles of melanges and their bearing on the Franciscan–Knoxville paradox: *Bull. Geol. Soc. Am.*, v. 79, p. 1063–1074.

Morel, S. W., 1962, Preliminary account of the geology of the Parakklisha area, Limassol District, in *Ann. Rept. Geol. Surv. Dep. for 1961*: Nicosia, Geological Survey Dep., p. 24–27.

Pantazis, Th., 1967, The geology of the Pharmakas-Kalavasos area: *Cyprus Geol. Surv. Dep. Mem. 8*, Nicosia, Geological Survey Dep., 190 p.

Rigo de Righi, M., and Cortesini, A., 1964, Gravity tectonics in foothills structure belt of Southeast Turkey: *Am. Assoc. Petrol. Geol. Bull.*, v. 48, p. 1911–1937.

Shalem, N., 1954, Red Sea and Erythrean distrubances: *19th Geol. Cong.*, Algiers, 1952, Sect. 15, v. 17, p. 223–231.

Streckeisen, A. L., 1967, Classification and nomenclature of igneous rocks (A final report of an inquiry): *Neues Jahrb. Mineral., Abh.*, v. 107, p. 144–240.

Tatar, Y., 1968, *Geologie und Petrographie des Marmariş Gebietes* (S.W. Turkei): Ankara, M.T.A.

Turner, W. M., 1971, Geology of the Polis-Kathikas area, Cyprus: Doctoral Dissertation, Dept. Geol., Univ. New Mexico, Albuquerque, New Mexico, 430 p.

Van Bemmelen, R. W., 1956, The geochemical control of tectonic activity: *Geol. Mijnb.*, N.S., v. 18, no. 4, p. 131–144.

Vine, F. J., and Moores, E. M., 1969, Paleomagnetic results for the Troodos Igneous Massif, Cyprus (Abs. GP13): *Am. Geophys. Union Trans.*, v. 50, p. 131.

Weiler, Y., 1965, The folded Kythrea Flysch: Doctoral Dissertation, Hebrew University, Jerusalem, Israel, March 1965, 85 p. plus two appendices.

Wilson, R. A. M., 1959, The geology and mineral resources of the Xeros-Troodos area: *Cyprus Geol. Surv. Dep. Mem. 1*, Nicosia, Geological Survey Dep., 135 p.

PART 3

Papers on the Orogenic Systems in North America

A. J. EARDLEY[1]

Department of Geological and Geophysical Sciences
University of Utah
Salt Lake City, Utah

Introduction to Part 3: Tectonic Divisions of North America

Location of Areas Described in Relation to Tectonic Divisions

The new tectonic map of North America (King, 1969) is taken as a guide in constructing Fig. 1, which shows the principal divisions of the continent, each of which is briefly described in the figure caption. The divisions recognized are platforms and foldbelts. The articles in Part 3 are concerned only with the foldbelts of Phanerozoic time. The widest and longest foldbelt is the Cordilleran (O) along the western margin of the continent. Along the eastern margin is the Appalachian foldbelt (L), which probably continues around the southern margin (LO). It is overlapped from the Atlantic Ocean side by the Atlantic and Gulf Coastal Plain (C). Along the north perimeter is the Innuitian foldbelt (K). The broad stable interior is in two parts, the platform deposit on the Precambrian foldbelts (B) and the exposed Precambrian rocks, the Canadian Shield (PC). MR is a section of B called the Middle Rocky Mountains, which is noted for a group of ranges that are generally separated by intermountain basins. It includes the Colorado Plateau. All of the ranges are individual uplifts.

Three of the articles deal with areas in the Appalachian foldbelt, five deal with areas in the Cordilleran foldbelt, two with the Middle Rockies, and one with the Ouachita Mountains along the south of the continent in the Appalachian foldbelt. The general locations of these areas are shown in Fig. 1. An interesting comparison between the areas in terms of structural synthesis and tectonic

---

[1] Deceased November 7, 1972

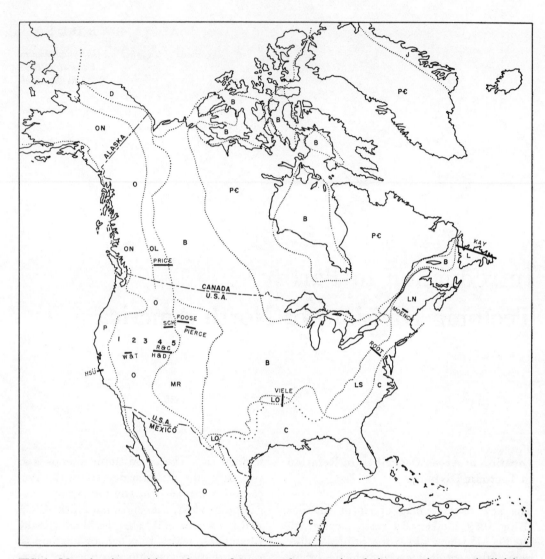

FIG. 1 Map showing positions of areas of monograph reports in relation to major tectonic divisions of North America. Taken from the new *Tectonic Map of North America* (King, 1969). Letters correspond to symbols on King's map, with the exception of MR, which stands for Middle Rocky Mountains. (O) Cordilleran foldbelt, mainly of Mesozoic age (numbers 1-5 relate to Fig. 2); (ON) Nevadan; (OL) Laramide; (P) Pacific foldbelt, late Mesozoic and Cenozoic age; (PV) principal basalt field of mid-Tertiary age; (B) platform deposits on Precambrian foldbelts; (PC) exposed Precambrian foldbelts constituting the Canadian shield; (L) Appalachian foldbelt including the Ouachita foldbelt (Paleozoic age); (LN) Northern Appalachians; (LS) Southern Appalachians; (LO) Ouachita foldbelt; (K) Innuitian foldbelt, including the Swerdrup Basin and Parry Islands and Cornwallis foldbelt (Paleozoic); (J) East Greenland foldbelt (Paleozoic); (D) Arctic Coastal Plain; (C) Atlantic and Gulf Coastal Plain; (MR) Middle Rockies uplifts and intermountain basins, including Colorado Plateau; (R & C) Roberts and Crittenden; (H & D) Hose and Daneš; (W & T) Wright and Troxel; (SCH) Scholten.

evolution as viewed by the various authors may be made by referring to the conceptual diagrams accompanying each article.

Cordilleran Foldbelt (O)

The Cordilleran foldbelt extends from Alaska to Central America, ranges from 650 to 1600 km wide, and is composed of several subdivisions that evolved from mid-Paleozoic to the Present (see Fig. 1). The widest extent of the Cordilleran division is in the western United States where it is composed of the following five orogenic complexes, from west to east: the Nevadan, the Sonoma, the Antler, the western division of the Sevier, and the eastern division of the Sevier (Fig. 2). The ages and main characteristics of each belt of orogeny are given in the caption of Fig. 2. Superimposed on the five divisions, from easternmost California to central Utah, is a

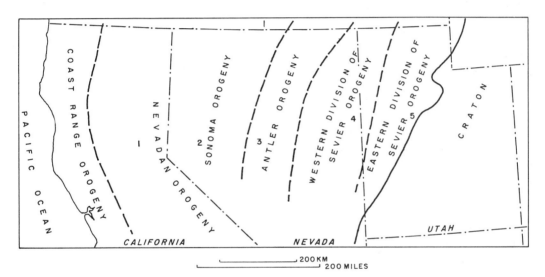

FIG. 2 Orogenic belts of the western Cordillera. *Coast Range orogenic belt:* Cenozoic, Cretaceous, Jurassic and possibly Triassic rocks and events related to the collision of Pacific and North American plates. *Nevadan orogenic belt:* Jurassic and Cretaceous rocks and events, noted for the great batholithic intrusions. Probably related to subduction of the Pacific plate under the North American plate. Boundary diffuse with the Sonoma orogenic belt. *Sonoma orogenic belt:* Lower and upper Paleozoic eugeosynclinal deposits deformed by the Sonoma orogeny during Permian time; overlapped by Triassic deposits from the west, and deformed again by middle Mesozoic orogenies. *Antler orogenic belt:* Lower eugeosynclinal deposits deformed in middle Paleozoic time and carried eastward in thrust sheets over miogeosynclinal deposits. The principal structure is the Roberts thrust, which is overlapped on the east by upper Paleozoic deposits, whose source was the orogenic belt. *Western division of Sevier orogenic belt:* Upper Precambrian and lower Paleozoic miogeosynclinal deposits, upper Paleozoic clastic wedges from the Antler orogenic belt and minor Mesozoic deposits, deformed mostly in Jurassic and Cretaceous time. This is the region of a great décollement, with younger strata thrust over older. The infrastructure rocks are metamorphosed, but the supracrustal rocks not. *Eastern division of the Sevier orogenic belt:* Miogeosynclinal deposits of late Precambrian and Paleozoic ages. Pennsylvanian and Permian deposits locally very thick. This is the "older on younger" thrust belt and regarded as the compressional segment complementary to the attenuated segment of the western division of the Sevier orogenic belt by Roberts and Crittenden and Hose and Daneš. Mesozoic and Early Tertiary sediments were deposited along the east flank and over the craton, with those of the Late Jurassic, Cretaceous, and Early Tertiary having accumulated to great thicknesses in a foredeep to the orogenic belt. These were also overrun in places by the advancing thrust sheets. Data taken from papers of this symposium and from King (1969).

system of extensional faults that resulted in a large region of internal drainage, known as the Basin and Range province. This faulting event, called the Basin and Range orogeny, has been evolving since mid-Tertiary time, accompanied by much acidic and later basaltic volcanism, from about 40 million years ago to the Present (McKee, 1971; Hedge and Noble, 1971).

The papers by Roberts and Crittenden and by Hose and Daneš (this volume[2]) deal with the Sevier belt, and the paper by Wright and Troxel deals with the Basin and Range orogeny.

The major Columbia and Snake River basalt fields lie north of the Basin and Range province and range in age from Miocene to Pleistocene. A belt of Cenozoic andesitic volcanoes lies near the Pacific coast west of the Columbia basalts.

The Nevadan orogeny with its great batholiths has been related to the collision of the Pacific and American plates, with subduction of the Pacific plate under the western edge of the American. The Basin and Range province has recently been analyzed to have been formed by a change from subduction and island arc generation to one of interarc spreading. It is proposed that this was accomplished by the rising of the subduction slab in the upper mantle to the lithosphere as a vast diapir, thereby creating lateral expansion and vertical rise of the lithosphere (Scholz and others, 1971). By a regional study of the chemical composition of the volcanic rocks of the province, Lipman and others (1971) postulate that there were two subduction slabs that developed in imbricate arrangement, each rising as diapirs to the lithosphere and creating the interarc expansion. The older belts of compressional orogeny, the Sonoma, the Antler, the western Sevier, and the eastern Sevier, have not yet been clearly related to the collision processes of the American and Pacific plates (Roberts and Crittenden).

The Cordilleran foldbelt in western Canada is composed of the Nevadan (ON) and the Laramide (OL) orogenic belts. The Nevadan may have elements of the Sonoma and Antler belts extending from the south, and the Laramide appears to be chiefly the equivalent of the Sevier. The tectonics of the Canadian Rockies is the subject of the fascinating paper by Price.

The segments connecting the Laramide belt of the Canadian Rockies with the Sevier belt is the site of the large Idaho batholith. The belt of deformation along its north and east side is the subject of the accompanying paper by Scholten.

The Cordilleran foldbelt in Alaska is bordered on the north by the Arctic Coastal Plain and on the south and west by the Pacific foldbelt. In Mexico on the south the Cordilleran foldbelt has a western division, the Nevadan, in the peninsula of Baja California, and an eastern division, the Laramide, which is not very well understood. The extension of the Basin and Range province into Mexico is also somewhat enigmatic.

Middle Rockies Uplifts and Basins (MR)

East of the deformed miogeosyncline in the western United States and occurring in the platform or shelf deposits of Paleozoic age is a group of sizable uplifts and basins, referred to as the Middle Rockies. The uplifts measure 100 to over 200 km in length and are about half as wide. They are generally oval shaped and are separated by wide basins, but in places they are crowded together with only limited depressions. These crowded separations are generally marked by folds and thrusts. The uplifts range in amplitude from

---

[2] In subsequent references to articles in *Gravity and Tectonics* the words "this volume" are omitted and the authors' names are not followed by the year of publication.

3000 to 10,000 m or more, and they came into existence in Late Cretaceous and Early Tertiary time. The largest uplifts have been eroded to their Precambrian cores and now are marked by high glaciated mountains. The intermountain basins are the sites of Early and mid-Tertiary sediments derived from the uplifts. Extensive Tertiary volcanic accumulations in scattered locations add mountainous masses to those of the uplifts.

Those uplifts that have risen more than 5000 m above the adjacent basins are not only asymmetrical but have developed upthrusts on the steep side. The uplifted rocks have tended to spread or slide toward the basins in various manifestations of gravity-caused movement. This is the subject of the accompanying well-documented papers by Foose and Pierce.

Pacific Foldbelt (P)

The Pacific foldbelt is an accretion to the western edge of the North American continent in Late Mesozoic and Cenozoic time and appears to have been formed on oceanic crust. It is characterized by Cenozoic orogenic activity, seismicity, and volcanism (King, 1969). It has a basement of eugeosynclinal rocks that is younger than that of the main Cordillera. These features are well described in the accompanying paper by Hsü. The Pacific belt is involved in a complex change from the compressional stresses of subduction to those of northwesterly horizontal translation through the evolution of transform faults.

Appalachian Foldbelt (L)

The Appalachian foldbelt extends along the east margin of the continent from Newfoundland on the northeast to Alabama on the southwest. It is partly covered by the Atlantic Coastal Plain sediments, which nearly bisect it in the New York region. Its widest exposure is 650 km, and it undoubtedly extends eastward to the continental margin under the Coastal Plain and shelf sediments. The foldbelt had mainly a Paleozoic evolution. It is fairly regular in plan and its longitudinal divisions maintain their identity for long distances. It is composed of a succession of salients and recesses, each salient being about 650 km long. The salients arc toward the continent. The Appalachian foldbelt has been divided transversely into a northern and a southern division (King, 1969), with the northern extending from New York City through New England, the Maritime provinces and Newfoundland, and consisting of the three northern salients. The southern division in the eastern United States consists of two salients.

The orogenies of the Appalachian foldbelt are as follows: the Grenville of late Precambrian age, the Taconian of late Ordovician age, the Acadian of Late Devonian age, and the Allegheny orogeny of Pennsylvanian age. Post-Acadian rifting extends into the Late Triassic. The Acadian orogeny was the climactic one of Paleozoic age, at least in the northern Appalachians. The Appalachian belts of orogeny are more difficult to show on maps than the Cordilleran foldbelts, because the Appalachian orogenies were variable in intensity and partly or largely overlapping in character.

Both miogeosynclinal and eugeosynclinal deposits are present. Those of Cambrian age are miogeosynclinal, were derived from the west and northwest (continental), and were deposited as sand-carbonate facies grading easterly into shale facies. This was a continental shelf-type deposit. Orogenies then brought about a striking change in the source of at least the miogeosynclinal deposits, and Ordovician, Silurian, and Devonian sediments were derived from the east to be spread in great clastic wedges over the previously deposited Cambrian sediments. Deposition in the correlative eugeosynclinal belt was accompanied or interrupted by the

Taconian and Acadian orogenies, and thus the eugeosynclinal deposits have proved very complex (Zen and others, 1968).

Northern Appalachians (LN)

The northern Appalachian division, in New England especially, has been well studied and is the classical locality for the Taconian and Acadian orogenies. The Taconian structures are allochthonous masses along the west side of the foldbelt, whereas the dominant Acadian structures are mantled gneiss domes and intrusive plutons. Two belts of mantled gneiss domes extend longitudinally in the eastern or eugeosynclinal part of the foldbelt.

The cores of the domes are mainly gneisses ranging from quartz diorite to granite and the mantling gneisses include amphibolites, mica schists, quartzites, and calc-silicate rocks of Paleozoic age. It has been concluded that the domes are a late tectonic feature, and that they have been preceded by the formation of at least three giant nappes. The style of the deformation accompanying much of the metamorphism in New England suggests plastic, almost fluidlike behavior, and it seems likely that lithostatic pressures equivalent to an overburden of 15 km were sufficient to have caused the metamorphism (Thompson and Norton, 1968). The plutonic rocks of the principal magmatic invasion (New Hampshire Plutonic Series) appear in most areas roughly contemporaneous with, or slightly younger than, the main regional metamorphism. The high-grade metamorphic zones show a striking correlation with these plutons.

Precambrian rocks emerge in the Northern Appalachian foldbelt in a chain of uplifts a short distance away from the western front of the belt. These are the Long Range of Newfoundland and the Green Mountains and Hudson River Highlands of New England. Another chain on the southeast of the foldbelt includes the Precambrian rocks of the Avalon Peninsula of Newfoundland, Cape Breton Island, southern New Brunswick, and some exposures in southeastern New England. The Precambrian rocks have been worked over to various degrees by the Paleozoic orogenies, including the Avalonian event with an age of 600–550 million years.

The Merrimack synclinorium discussed by Moench lies between the eastern zone of Precambrian uplifts and the belts of mantled gneiss domes in Maine and New Hampshire.

Newfoundland consisted of three divisions in Cambrian and Ordovician time, the Western Platform, the Central Volcanic or Mobile Belt, and the Avalon Platform. In both platform divisions the Cambrian and Ordovician strata were little deformed. The Central Volcanic or Mobile Belt has been interpreted as an oceanic trench above a subduction zone, adjacent to the border of an early Atlantic ("Protacadia") Ocean (see Kay). Island arc formation in the Central Mobile Belt was accompanied by uplift, and from the uplift a vast, gravity-impelled sheet slid northwesterly in Chazyan time over the Western Platform of Cambrian and Early Ordovician sediments. Other gravity slide masses can be inferred in the complex of the Central Mobile Belt. The Eastern or Avalon Platform has Cambrian and Ordovician fossils of European affinity.

In terms of plate tectonic theory Bird and Dewey (1970) postulate that in late Precambrian time a continuous North American/African continent began to separate with the development of a graben. Continued distension in Cambrian time led to an American accreting plate margin, and thus to shelf sedimentation. By early Ordovician time the Protacadia Atlantic had expanded to a maximum width. Then a zone of mid-ocean expansion came into existence, and subduction and plate loss on the edge of the American continent was initiated. By early to medial Ordovician time an island arc system had developed (the Taconian orogeny in eugeosynclinal sediments) and allochthonous thrusting followed incident to the consumption of the oceanic crust. As the continent of Africa approached North America the sediments of the oceanic crust were scraped off as it descended and finally the collision of the two continents resulted in the Acadian orogeny.

But the Atlantic must open up again, and the Triassic block faulting is believed by Bird and Dewey to be related to the initial stages of this second separation.

Southern Appalachian (LS)

The Southern Appalachians consist, longitudinally, of the Cretaceous to Recent Coastal Plain and shelf province, the crystalline Piedmont province, the Blue Ridge province, and the Valley and Ridge province. The crystalline Piedmont is essentially a continuation southwestward of the Acadian orogenic belt of New England, but it is less well exposed and known geologically than the metamorphic and igneous region of the Northern Appalachians. The Blue Ridge is a continuation of the western belt of Precambrian uplifts of western New England and is made up both of late Precambrian and Cambrian sedimentary rocks. The Valley and Ridge province is a fold and thrust system of the miogeosyncline. Its deformation is mostly late Pennsylvanian, but the orogeny had phases as early as Mississippian. A system of multiple imbricate faults in the south, with thrusting toward the continent, gives way at the surface to the classical folds of Pennsylvania farther north. The accompanying discerning paper by Root deals with the fold system in Pennsylvania.

Ouachita Foldbelt (LO)

The Ouachita foldbelt extends along the south margin of the continent in a series of salients and recesses, like those of the Appalachian foldbelt, but of greater amplitude. It is covered primarily by coastal plain deposits. The sedimentary rocks are of Paleozoic age, and especially the formations of Devonian and Pennsylvanian age are similar to those of the southernmost Appalachians. Folding and thrusting of late Pennsylvanian age are characteristic and generally directed toward the continent. A series of foredeep basins front the foldbelt. The accompanying paper by Viele deals with the tectonic evolution of the Ouachita Mountains section of the general foldbelt; here Viele postulates a nappelike core structure. Oil-well drilling has defined the position of the front (toward the continent) of much of the foldbelt under cover of coastal plain sediments, but only minor indications of the older crystalline belts of orogeny exhibited in the Appalachians have been found oceanward of the Ouachita Mountains.

East Greenland Foldbelt (J)

The East Greenland foldbelt is mainly an early Paleozoic feature and a result of the late Silurian Caledonian orogeny. It extends from Scoresby Sound at the seventieth parallel northwestward, partly under the ice to a possible junction with the Innuitian foldbelt. The bulk of the geosynclinal accumulation consists of upper Proterozoic sediments as much as 16,000 m thick. These are followed nearly conformably by rather thin Cambrian and Ordovician strata, and in the north by Silurian. All are miogeosynclinal in character. All strata were strongly deformed, and thrusting was toward the Greenland Shield and its platform cover (King, 1969).

Innuitian Foldbelt (K)

The Innuitian foldbelt extends across the northern part of the Arctic Islands of Canada, and into Peary Land of northern Greenland, where it probably meets the East Greenland foldbelt (Thorsteinsson and Tozer, 1961). Its eastern and western ends plunge beneath the Arctic Ocean and its western part is covered by Mesozoic and Tertiary strata.

Most of the sedimentary rocks are of miogeosynclinal character and of Ordovician and Upper Devonian age, about 6000 m thick. They are mostly carbonates with minor clastics and evaporites. In the Parry Islands they have been deformed in long symmetrical east–west folds. Here the folding is later than the Devonian and earlier than the unconformably overlying middle Pennsylvanian strata. On Cornwallis Island, however, the strata were folded along north–south axes between Silurian and Middle Devonian time.

A limited exposure of eugeosynclinal rocks extends across northern Ellesmere Island to the tip of Axel Heiberg Island. The strata are thicker than those of the miogeosyncline and have yielded a few Ordovician and Silurian fossils. Basal exposures of schists and gneisses may be Precambrian.

References

Bird, J. M., and Dewey, J. F., 1970, Lithosphere plate: Continental margin tectonics and the evolution of the Appalachian orogen: *Geol. Soc. Am. Bull.*, v. 81, p. 1031–1060.

Hedge, C. E., and Noble, D. C., 1971, Upper Cenozoic basalts with high Sr^{87}/Sr^{86} and Sr/Rb ratios, southern Great Basin, western United States: *Geol. Soc. Am. Bull.*, v. 82, p. 3503–3510.

King, B., 1969, The tectonics of North America, and Tectonic Map of North America, *U.S. Geol. Survey Prof. Paper 628*, 94 p.

Lipman, P. W., Prostka, H. J., and Christiansen, R. L., 1971, Evolving subduction zones in the western United States, as interpreted from igneous rocks: *Science*, v. 174, p. 821–825.

McKee, E. H., 1971, Tertiary igneous chronology of the Great Basin of western United States—Implications for tectonic models: *Geol. Soc. Am. Bull.*, v. 82, p. 3497–3502.

Scholz, C. H., Barazangi, M., and Sbar, M. L., 1971, Late Cenozoic evolution of the Great Basin, western United States, as an ensialic inter-arc basin: *Geol. Soc. Am. Bull.*, v. 82, p. 2979–2990.

Thompson, J. B., Jr., and Norton, S. A., 1968, Paleozoic regional metamorphism in New England and adjacent areas, *in* Zen, E-an, and others, eds., *Studies of Appalachian Geology: Northern and maritime*: New York, Interscience, p. 319–328.

Thompson, J. B., Jr., Robinson, P., Clifford, T. N., and Trask, N. J., Jr., 1968, Nappes and gneiss domes in west-central New England, *in* Zen, E-an, and others, eds., *Studies of Appalachian Geology: Northern and maritime*: New York, Interscience, p. 203–218.

Thorsteinsson, R., and Tozer, E. T., 1961, Structural history of the Canadian Arctic Archipelago since Precambrian time, *in* Raasch, G. O., ed., *Geology of the Arctic*, v. 1, Toronto, Ontario, Toronto Press, p. 339–360.

Zen, E-an, White, W. S., Hadley, J. B., and Thompson, J. B., Jr., 1968, *Studies of Appalachian geology: Northern and maritime*: New York, Interscience, 475 p.

MARSHALL KAY

Department of Geology
Columbia University
New York

Tectonic Evolution of Newfoundland[1]

The island of Newfoundland (Fig. 1) (Williams, 1967a), 130,000 km² in size, lies at the northeastern end of the Appalachian Mountain system. The traditional interpretation of the early Paleozoic geography envisioned a central "borderland," called New Brunswickia, with troughs to the east and west (Schuchert and Dunbar, 1934). This was superseded by the theory of geosynclines and island arcs (Kay, 1942, 1951), according to which Newfoundland was given a threefold division into the Western platform, the Central Volcanic or Mobile belt, and the Avalon platform (Williams, 1964; Kay, 1966, 1967).

The early Paleozoic structural history has been rather well determined in both the western and eastern structural belts, where fossiliferous Cambrian and Ordovician are relatively little deformed. The Central Mobile belt is separated into blocks by high angle faults, some of them transcurrent. The stratigraphic sequence has been ascertained in some of the blocks, but until the relative movements of contiguous blocks are determined, the reconstruction of structural history is tenuous.

Two particularly significant structural interpretations have been developed. Exotic rocks and thrust sheets (Johnson, 1941; Kay, 1945) on the west coast are thought to have been gravity slides (Rodgers and Neale, 1963; Kay, 1966) during the Ordovician Taconian orogeny. On the other hand, an area of bouldery mudstone in central Newfoundland, called the Dunnage melange by Kay and Eldredge (1968) and Kay (1970), is interpreted as representing an oceanic

---

[1] The field studies in central Newfoundland have been supported by grant GA-15350 of the National Science Foundation. The igneous rocks intruded into the Dunnage melange were studied by D. V. Manson (Manson and Kay, 1972) of the American Museum of Natural History, and the conodonts in carbonate rocks were identified by Stig Bergstrom of The Ohio State University.

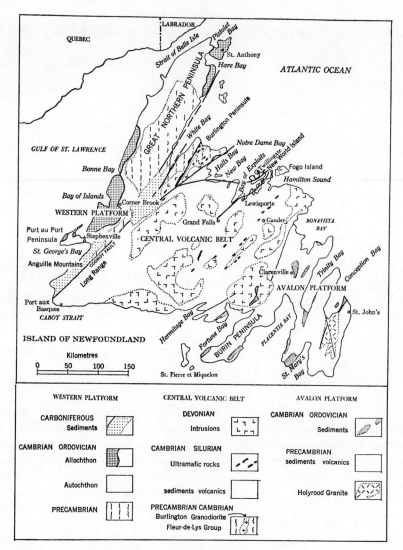

FIG. 1 Geologic map of the island of Newfoundland.

trench above a subduction zone on the border of the closing Protacadic or "Proto-Atlantic" Ocean (Dewey, 1969; Bird and Dewey, 1970; Horne, 1970).

The principal structures formed in the Ordovician Taconian deformation are widespread and involved several phases of disturbance, the effects continuing into the Silurian. The Devonian of central Newfoundland was affected by the Acadian Disturbance, with strong folding and batholithic intrusion. The record of the late Paleozoic Alleghanian Disturbance is largely in the west, where Carboniferous sedimentary rocks were folded (Belt, 1969). High-angle faulting in the late Paleozoic may have continued into later eras but is inadequately dated.

This paper describes the structure of several fault-bounded blocks and belts of Newfoundland and discusses interpretations that have been presented. The record has been summarized briefly by Poole (1967),

Williams (in Poole and others, 1970), and Rodgers (1970). More comprehensive summaries are in a memoir on North Atlantic geology (Kay, 1969a).

Western Belt

The Western belt has a basement of Precambrian metamorphic and igneous rocks exposed mainly in the core of the Great Northern Peninsula (Fig. 1), difficult of access and relatively little known (Clifford, 1969). They have been referred to the Grenvillian, on the basis of very few analyses, giving K/Ar dates of 900–960 million years (Wanless, 1970). The basement is succeeded unconformably by a succession of Lower Cambrian to middle Ordovician (Llanvirnian) sandstone, dolomite, limestone, and shale (Whittington and Kindle, 1969) (Figs. 2 and 3).

Structurally above the autochthon is the parautochthon, with contemporaneous Cow Head carbonate boulder conglomerates, with blocks that slid eastward from a carbonate bank into a muddy deeper basin floor (Kindle and Whittington, 1958; Kay, 1966; Rodgers, 1968), the whole subsequently gliding westward beneath allochthon. This overlying *Humber-Hare Bay allochthon* is composed of lower thrust sheets with Cambrian and Ordovician graywacke, argillite, and lava, and higher sheets of ophiolite-ultramafic rocks (Williams, 1971). On the west coast of Port au Port Peninsula the thrust sheets are succeeded unconformably by middle Ordovician Long Point limestone and shale (Kay, 1969c) and Silurian and Devonian shale and limestone.

The presence of thrust sheets in western Newfoundland was proposed in 1940 because carbonate rocks and graptolite-bearing shales of the same age are in adjoining sections and because of the analogy with similarly contrasting facies interpreted as allochthon and autochthon in western New England and eastern New York (Johnson, 1941; Kay, 1945). The hypothesis was substantiated by Rodgers and Neale (1963), who introduced the concept of gravity sliding and logically attributed the ultramafic rocks to a source above similar intrusives in the Burlington Peninsula, nearly 100 km to the east. On the other hand, Bird and Dewey (1970) present

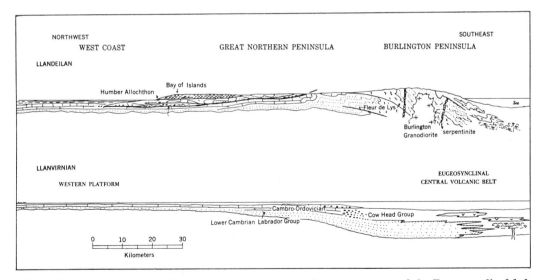

FIG. 2 Restored sections of the Western platform and the western part of the Eugeosynclinal belt prior to and following the Bonnian phase of the Taconian orogeny within the Ordovician Period.

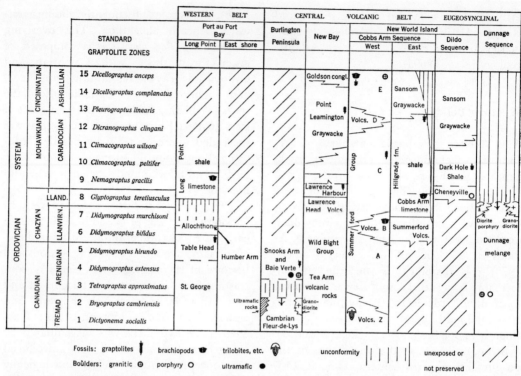

FIG. 3 Ordovician correlations from the Western platform through the Central Volcanic or Eugeosynclinal belt (recent corrections in dates of some intrusions are noted in the text).

a section with more conventional thrust faults rising from the flank of a tectonic welt.

Kay (1966, 1969c) and Stevens (1970) have elaborated the gravity slide interpretation. Such slides must be attributed to a rise of the sea floor of the youngest allochthonous sediments to such elevation as to permit their gliding nearly 100 km westward. The rocks had been deposited shortly before, so presumably they were only partially consolidated; such argillaceous sediments must have remained rather ductile. If the gravity slides moved westward from a welt raised in the midst of the depositional trough, perhaps there should have been similar slides that moved eastward into the central belt. This question will be considered subsequently.

The time of gliding has been defined as the Bonnian phase of the Taconian orogeny. The youngest rocks known in the autochthon (Fig. 3) and in the Humber Arm allochthon of western Newfoundland are graptolite-bearing shales of early Llanvirnian age (Morris and Kay, 1966) (Zone 6, Fig. 3). Thus the sediments were raised and the muds derived from the tectonic land were carried westward during Llanvirnian and Llandeilian or about Chazyan time. Any complementing structures on the east should be of that age span.

The Long Point group and succeeding sedimentary rocks on the west coast were affected by minor folding, low-angle faulting, and extensive late Paleozoic warping and high-angle faulting. Some faults are synchronous with Carboniferous deposition, whereas others are younger. A graben-rift system developed along the eastern border of the Western belt in the late Paleozoic comparable to the Midland Valley belt in Scotland and Ireland, which was nearby prior to continental drift.

In summary, metamorphic rocks of the Western belt were deformed several times during the Precambrian, perhaps in Grenvillian time. The Bonnian phase of the Taconian Disturbance produced gravity slides emplaced far to the west as allochthonous masses during the Chazyan Epoch. Subsequent faulting and folding accompanied and followed Carboniferous deposition.

Fault Blocks of the Central Belt

On the northeast coast, the Central Mobile belt is separable into several fault-bounded blocks, most having well-determined stratigraphic sequences and structures. The faults are thought to be transcurrent, notably the Lukes Arm–Lobster Cove fault and the Reach fault (Fig. 4). They will be described as the Fleur-de-Lys, Burlington, Halls Bay-Twillingate, New World Island, Dunnage, Loon Bay, Change Islands, and Gander Lake blocks.

The *Fleur-de-Lys block* (Fig. 2) contains the Fleur-de-Lys group (Fuller, 1941), a dominantly siliceous sedimentary rock probably belonging to the Precambrian and Lower Cambrian (the latter a greatly thickened offshore equivalent of that west of White Bay in the Great Northern Peninsula); it is intruded, or structurally overlain, by serpentinized ultramafic rocks (Kennedy and Phillips, 1971; Dewey and Bird, 1971). The block lies in an orthotectonic belt, one of severe deformation and metamorphism as compared to the rocks in the paratectonic belt to the east (Dewey and Kay, 1968). Several deformational phases are recorded. If the ultramafic rocks in the upper sheets of the west coast allochthon had roots in the Burlington Peninsula terrane, they must have been intruded by medial Ordovician time, though they could have been remobilized later (Neale, 1957). Southward, a large area of granite gives Devonian K/Ar ages of 358 to 384 million years (Wanless, 1970).

The Burlington block to the east has extensive exposure of Burlington granodiorite (Fig. 1). Interpretations of the sequence within the terrane have varied, but in general it consists of volcanic rocks, radiolarian chert,

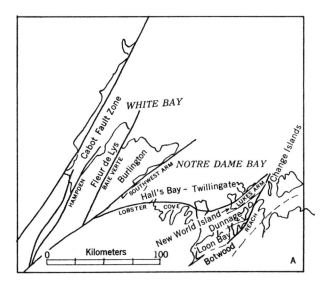 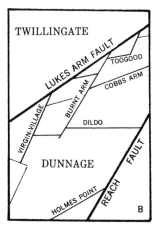

FIG. 4 (A) Outline map of northeastern Newfoundland showing structural blocks and main faults. (The Loon Bay intrusive has recently been dated as Devonian.) (B) Diagram of the faults between the Lukes Arm and Reach faults.

and argillite with lower Ordovician (Arenigian) graptolites; underlying ultramafic rocks are intrusive into or thrust upon Fleur-de-Lys metasediments (Neale and Nash, 1963; Phillips, Kennedy, and Dunlop, 1969; Church and Stevens, 1971; Dewey and Bird, 1971; Kennedy and Phillips, 1971). Conglomerate and sandstone overlie the Burlington granodiorite.

The *Halls Bay-Twillingate* or *Lush's Bight block* (Horne and Helwig, 1969; Dewey and Bird, 1971) has abundant lavas and fragmental volcanic rocks, some of which are of early Ordovician age (Dean, 1970), but others are probably much older. The Twillingate granodiorite that intrudes the terrane 75 km farther east has strong penetrative deformation and is of unknown age, perhaps Precambrian. The Lobster Cove and Lukes Arm fault along its southeastern border is transcurrent and may have very great displacement, possibly scores of kilometers.

The *New World Island blocks* (Fig. 4) are multiple, divided by high-angle faults separating, southeastward, the Toogood, Cobbs Arm, and Dildo sequences of Ordovician lavas, tuffs, limestone, and argillites. The more northwesterly sequences are unconformably beveled by early Silurian conglomerate and graywacke but the more southeasterly ones are conformably overlain by late Ordovician graywacke and Silurian graywacke and conglomerate (Kay, 1969, 1970). Silurian conglomerate has large boulders of volcanic rocks in the lower part and polymictic assemblages higher up, where plutonic boulders do not have penetrative deformation (Helwig and Sarpi, 1969). The high-angle faults are of several generations.

The *Dunnage block* contains the widespread Dunnage melange (Kay and Eldredge, 1968; Horne, 1969; Kay, 1970, 1972), lying between the Dildo fault on the north and the Loon Bay batholith and the Holmes Point fault on the south; eastward the Dunnage is cut off by the northeast-trending Reach fault with a displacement of scores of kilometers. The melange contains cleaved argillite, interbeds and slides of graywacke, pillow lava and tuff, and mudstones with boulders composed predominantly of graywacke in addition to lava, dolomite, and, more rarely, plutonic rocks. An interbed of limestone in volcanic rocks near the southern margin of outcrop contains Middle Cambrian trilobites and brachiopods (Kay and Eldredge, 1968). Scores of observations of the facing of lava pillows and graywacke cross-bedding are variable in azimuth. Some pillows at the north face northward, but the great majority of scattered observations have southward facing. A few lithic bands are continuous for hundreds of meters, and one of polymictic conglomerate continues for 1 km. Structural analysis shows two periods of deformation.

The Dunnage block is distinctive in the abundance and variety of intrusions, mostly of rock types known only in this belt. Stocks of Coaker diorite porphyry with many square kilometers exposure are widespread; the porphyry passes into mafic phases, some with abundant serpentinite and gabbro xenoliths.[2] The Dunnage has many intrusions, such as are unknown in the blocks to the north. Two small stocks, 1 km in diameter, are strongly zoned and surrounded by hornfels aureoles. Small diabase stocks and dikes are widespread; lamprophyre dikes are like those dated as mid-Mesozoic in the Twillingate block (Poole, 1967, p. 42).[3] The Loon Bay granodiorite batholith intrudes the margin of the Dunnage toward the southwest, but eastward, the Holmes Point fault bounds the Dunnage to its junction with the Reach fault.

The *Loon Bay block* has the Loon Bay batholith of granodiorite or tonalite in an

---

[2] The Coaker (Causeway) xenolith phase has yielded K/Ar dates of 428 ± 13 million years and 435 ± 13 million years, and Rb/Sr dates of 454 million years (using a decay constant of 1.47×10^{11} yr^{-1}) or 480 million years (using a decay constant of 1.38×10^{11} yr^{-1}), indicating an Ordovician age (Manson and Kay, 1972).

[3] Similar intrusions in the Loon Bay block have K/Ar ages of 165 ± 5 million years.

area of 60 km². On the south, northwest of the Reach fault, the batholith intrudes a north-facing sequence of a few kilometers of schistose volcanic rocks, including pillow lavas, succeeded by hundreds of meters of manganiferous chert, silty argillite, and graywacke flysch. A late Ordovician age (440 million years) has been attributed to it by Williams (1949b); subsequently made K/Ar determinations gave Devonian ages of 372 ± 10 million years and 365 ± 10 million years (Manson and Kay, 1972). Contrary to published sections, it is this sequence that adjoins the Dunnage melange on the southeast rather than the ultramafic rock-bearing Gander Lake sequence.

Southeastward of the great transcurrent Reach fault, relationships become decreasingly known in what is broadly the *Change Islands-Gander Bay block*. The fault is nearly parallel to the trend of the Silurian sedimentary and fragmental volcanic rocks of the Change Islands sequence (Eastler, 1969) and the southwesterly continuing Botwood sequence (Williams, 1967b) for more than 100 km. Along the fault are narrow horses or slivers of granite as much as 1 km long, and hornfels and trap, dissimilar to other rocks of the area. The fault truncates and terminates the successive blocks to the northwest until it meets the Lukes Arm fault beneath the open Atlantic. It is a very great transcurrent fault. The rocks of the Silurian Change Island sequence are gently folded and, though cleaved, are little metamorphosed. Across the Indian Island thrust fault on the southeast, rocks on the northern margin of the Gander Lake sequence are similar but show ductile deformation; local polymictic conglomerate contains slightly flattened granodiorite clasts and extremely deformed chert pebbles.

The *Gander Lake block* (Jenness, 1963; Anderson and Williams, 1970; Williams, 1963) extends southeastward to a broad belt of batholithic granite intrusions and late Precambrian sedimentary rocks that belong to the Avalon platform; the boundary probably is faulted (Fig. 1). The Gander Lake group is estimated to be 8 or 10 km thick, but neither the stratigraphy nor the structure is adequately known. The terrane is dominantly argillaceous, containing Ordovician graptolites and shelly faulules. Silurian occurs to the west and southwest. Local bands of conglomerate contain dioritic boulders. Volcanic fragmental rocks and flows are few and only tens of meters thick, some being associated with beds having Ordovician fossils. Plutonic rocks were exposed long before the intrusion of batholithic rocks of dated Devonian age (Wanless, 1970; Williams, 1969). Diorite stocks with penetrative deformation have yielded Rb/Sr dates of about 600 million years, indicating a late Precambrian or Early Cambrian age (Fairbairn and Berger, 1969). This age, if valid, suggests that there was an adequate source of plutonic pebbles in welts or tectonic lands within the Gander Lake belt. The Gander River ultramafic belt, 200 km long and up to 50 km wide, consists of lenses of serpentine, peridotite, and associated rocks (Fig. 1). The lenses lie along the structural trend, but they have no obvious stratigraphic or structural control. The intrusions have been considered relevant to theories on the origin of ancient ocean trenches (Hess, 1939, 1955). Though they have been related to the Dunnage melange (Dewey, 1969), they are on the opposite side of the great Reach fault.

Source of the Western Allochthon and Search for the Eastern Counterpart

The gravity slides of the west coast should have eastern counterparts if they slid from a welt of ductile geosynclinal rocks at the Burlington Peninsula (Kay, 1966). The gliding took place on the west coast after early Llanvirnian (Zone 6) and before late Chazyan (Llandeilan Zone 8) (Figs. 2 and 3). Structures were reported to be west-facing in the western and east-facing in the eastern part of the peninsula, so this seemed a reasonable place for a welt of ductile rocks.

No counterpart record of eastward sliding is reported on the east side of the peninsula itself. Much farther east, the Dunnage melange seemed a possible representative, assuming the Middle Cambrian fossils to be in a slide within the melange. The Fleur-de-Lys group, the lower Ordovician Snooks Arm group and similar Baie Verte group, and the underlying ultramafic rocks of the Burlington Peninsula are reasonably similar to those of the allochthon. If the melange were the counterpart, there should be a record in the structure and stratigraphy of each area between it and the Burlington Peninsula. The rocks northward from the melange to the Lukes Arm fault in the New World Island sequences are well dated by good faunal control that spans the stage of the westward sliding, yet they show no suitable discontinuities in the stratigraphy. There is no place for a mid-Ordovician slide to pass through the sequence and no suggestion that the Bonnian phase of the Taconian orogeny had an effect in introducing terrigenous sediments; in fact, in the Cobbs Arm sequence, Llandeilan and Caradocian sediments are limestone and argillite.

There remains the possibility that transcurrent faults such as the Lukes Arm fault have such great displacement that the sequences to the southeast have no significance in resolving the matter; perhaps they are like the transcurrent faults in California in bringing rocks that were proximal in the Mesozoic to far separated positions. Until the displacements are understood, there is no more probable validity to a Paleozoic restoration across Newfoundland on present geography than of one across California disregarding the displacements there! The alternate interpretation would be that the allochthonous masses moved exclusively westward down the slope of a west-tilted block whose east face produced conglomerates in adjacent rift blocks (Kay, 1966).

It is conceivable that the Burlington and Halls Bay–Twillingate blocks are part of an originally continuous belt, and one can speculate that the Dunnage block was farther along such a belt, with preservation of the subduction zone. Events in the Dunnage are of similar age to those in the Burlington Peninsula. These are problems that are under intensive study. Until they are solved, much of what is written about the interrelations of structural blocks in the Central Volcanic belt of the early Paleozoic is speculation. All of the foregoing tectonic blocks between the White Bay faults and the Gander Lake block on the northeast coast of Newfoundland are cut out along the south coast—the Gander Lake block meets the Cabot fault zone in southwestern Newfoundland (Williams and others, 1970).

Avalon Belt

Southeastern Newfoundland, like the northwestern part of the island, was a platform during the early Paleozoic in contrast to the great mobility of the Central Volcanic belt (Williams, 1964a; Kay, 1967). The platform, called the Avalon belt (Fig. 1), has three principal rock suites: a late Precambrian succession of several groups of sedimentary and volcanic rocks, a latest Precambrian batholith of Holyrood granite intruding the oldest group, and an unconformably overlying sequence of basal Cambrian sandstone and conglomerate, followed by Cambrian and lower Ordovician shaly rocks plus some impure limestone and a single lava flow (McCartney, 1967). Cambrian trilobites are more similar to those in northwestern Europe than to those in most of North America (Hutchinson, 1962).

The Precambrian basement is broken by faults, some of which may have defined graben and horsts that controlled Cambrian deposition and later deformation. The Cambrian and Ordovician in the eastern block of Conception Bay (Fig. 1) are gently dipping, but the Cambrian rocks in more westerly areas are folded into synclinal troughs and basins. In northwestern basins, axial planes dip westward on the west and eastward on the east (Jenness, 1963). Axial plane dips seem variable in the other troughs.

Meguma Belt

The continental shelf extends 500 km eastward from the Avalon Peninsula. The only observations have been on the Virgin Rocks from diving operations (Lilly, 1965). The rocks are metasediments like those of the Precambrian Avalon belt. Some 800 km to the southwest the continuation of the Avalon belt can be recognized in Nova Scotia, where several thousand meters of terrigenous sediments, principally graywacke, belong to the latest Precambrian and earliest Paleozoic Meguma group (Schenk, 1970). As in the case of the Virgin Rocks metasediments, the Meguma beds resemble some of the Precambrian rocks of the Avalon Peninsula. Unless there have been unrecognized, complex transcurrent movements of the Meguma block, there should be a continuation of these beds under the Grand Banks and eastward to the continental shelf off eastern Newfoundland. Abundant primary structures in the Meguma beds seem to require a source to the southeast, and in present interpretations of the Atlantic Ocean history, the source of the Meguma and its southeastward continuation lies in the continental shelf of the Iberian Peninsula or northwest Africa. The data for the reconstruction of early Paleozoic paleogeography are inadequate.

Subduction and the Protacadic [4] Ocean

The early Paleozoic record of central Newfoundland has been related to plate tectonics (Isacks, Oliver, and Sykes, 1968; LePichon, 1968; Morgan, 1968). The Gander Bay ultrabasic belt was considered by Hess (1940) to be a product of deformation of a tectogene beneath an oceanic trench. Gunn (1947; see also Kay, 1951, p. 73), anticipating modern views, attributed such associations to subduction. The Dunnage melange has been compared (Kay, 1970) to that of the Franciscan of California (Hsü, 1968, 1969, 1973). The Franciscan was originally associated with a deep downbuckle or "tectogene" (Ernst, 1965, p. 905) and subsequently to a subduction zone (Hamilton and Myers, 1966; Hamilton, 1969; Ernst, 1970; Coleman, 1971; Yeats, 1968). Dewey (1969) and Horne (1969, 1970) attributed the Dunnage melange to a northwest-descending subduction zone below the lower Ordovician Summerford volcanic rocks of the Dildo sequence that formed a volcanic arc to the northwest. They assumed the Dunnage and Summerford to be intertonguing and stratigraphically continuous, but the Dildo fault separates the Dunnage from the volcanics (Kay, 1970); transcurrent movement may not have been great. In contrast to Franciscan, the Dunnage has not yielded high-pressure rocks such as glaucophane schist. The internal structure and facing of slabs and clasts are better known; intermediate intrusions are voluminous, but ultramafic rocks are known only in xenoliths.

In his original paper, Dewey (1969, p. 128) had the Dunnage melange in a trench above a subduction zone descending northwestward beneath the rocks of the Burlington block, thought to lie on an original oceanic crust; he later interpreted that block to have become a tectonic island with a marginal ocean basin on the northwest (Dewey and Bird, 1971, p. 1302). The significance of the Gander River ultramafic belt was uncertain (Bird, Dewey, and Kidd, 1971, p. 30). Church and Stevens (1971, p. 1465) postulated "underthrusting of the western continental margin

---

[4] The term "Protacadic" has been given for the early Paleozoic ocean basin that was the site of *Ta*conian, *Ca*ledonian and *Acad*ian deformation (Kay, 1971). Proto-Atlantic (Wilson, 1966) suggests an ocean that was genetically antecedent to the Atlantic; it has been applied, logically, to the early Mesozoic progenitor of the South Atlantic.

and a segment of adjacent oceanic crust beneath oceanic lithosphere" of the Burlington block, along a southeasterly dipping subduction zone, and subduction of oceanic crust "to the southeast beneath the Gander Lake zone." These suggestions did not take into account the significance of the Reach fault (Kay, 1970), nor the fact that the Loon Bay block has volcanic rocks, manganiferous chert, and graywacke, conventionally considered to be oceanic. The writer will not speculate further with regard to the Burlington–Fleur-de-Lys relationships. The fact that the Gander River volcanic belt and associated rocks trend into the Cabot fault zone in southwestern Newfoundland (Williams and others, 1970) emphasizes the problems of accounting for the lack of continuity of the several blocks on the north coast. The very exposure of the Dunnage is fortuitous, for none is preserved against the open Atlantic because of displacement along faults.

The writer interprets the Dunnage as having formed in Cambrian and early Ordovician time above a subduction zone with oceanic floor of the Loon Bay block to the south; the crust descended beneath volcanic rocks in an island belt on the north (Fig. 5). Huge pillow lava blocks could have come from flows on the north or the south; the latter, with associated manganiferous chert, seems more oceanic. It is tempting to assume that granodiorite clasts in the melange came from Twillingate or Loon Bay, but the former is gneissic and displaced from afar, and the latter intrudes granodiorite-pebble beds in the Dunnage. Thus the production of the melange preceded intrusion of the Loon Bay granodiorite, dated as late Ordovician. It also preceded intrusion of the Coaker porphyry. The middle Ordovician north of the Dunnage has boulders of porphyry like that intruding the Dunnage, but similar boulders are present in the Dunnage. Probably, part of the Dunnage is older than in contiguous present exposures.

The Halls Bay–Twillingate block north of Lukes Arm fault has moved transcurrently from uncertain distance, and original relationships northwestward are not established. The original terrane south or southeast of the Loon Bay block is not known; there is a lack of evidence as to the nature of the southeastern limit of oceanic crust. Placement of the Gander Lake blocks is uncertain. Oceanic suites are not recognized unless represented, perhaps remobilized, in the Gander River ultramafics.

The Ordovician rocks in blocks to the north had been deformed and eroded in the more northerly New World Island sequences by early Silurian time. Silurian volcanic and polymictic conglomerates coarsen northward, but granodiorite clasts are not gneissic as in exposures in the Twillingate block to the north; in any case, the latter was far away prior to the transcurrent Lukes Arm faulting. The Twillingate block has mafic pillow lavas older than the intrusive gneiss, as though oceanic prior to its being intruded. It is possible that the block belongs in the same original tectonic belt as the Burlington block.

The first appreciable volcanism southeast of the Reach fault is in the shallow-laid Silurian agglomerates and tuffs of the Change Islands–Botwood belt. It would be axiomatic that they be on the margin of a southeast-dipping subduction zone of oceanic crust to the northwest. This is speculative, unsupported by direct evidence.

Summary

The oldest deformation in western Newfoundland produced structures of the Grenville orogeny about 1 billion years ago. The Central Mobile belt illustrates the great complexity of history of an orogenic belt—a belt containing ultramafic suites attributed to oceanic crust at times and places, a melange above an assumed subduction zone, ultramafic, mafic, and intermediate intrusions, volcanic rocks thought to represent island arcs, and associated sedimentation. In the Burlington Peninsula, west-derived terrig-

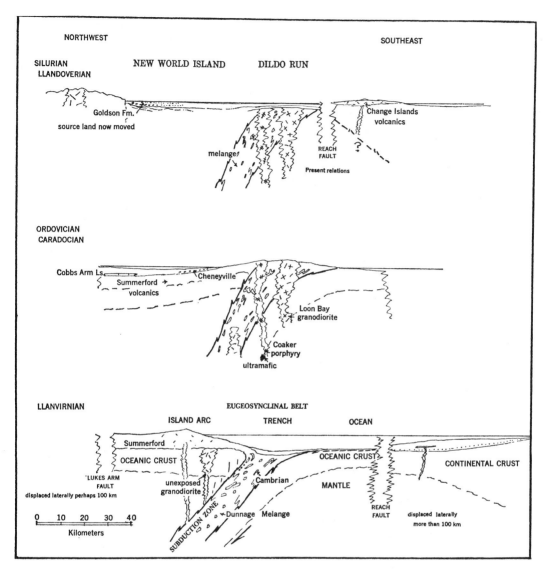

FIG. 5 Restored sections of the eastern part of the eugeosynclinal belt prior to and following Ordovician orogenies.

enous metasediments are overlain by lower Ordovician ultramafic and sedimentary rocks that may have been overthrust by obduction. Structures within the western part of the belt projected gravity slides onto the platform to the west during the middle Ordovician Bonnian phase of the Taconian orogeny.

Farther southeast, the Dunnage melange accumulated in a trench in Cambrian and earlier Ordovician time, with oceanic crust on the south. The melange soon was invaded by stocks and batholiths of intermediate abyssal and plutonic rocks. These were unroofed by late middle Ordovician. In latest Ordovician, sequences north of the subduction zone were folded and bevelled by earliest Silurian sediments, gaining coarse detritus from the erosion of rising highlands of volcanic and plutonic rocks to the north, their provenance now lost in the transport of the source rocks along the Lukes Arm transcurrent fault. On the southeast, the Reach fault lacks appreciable mafic flows such as might be expected within an oceanic crust but has scores of kilometers of strike-slip movement. The terrane beyond has the Gander River ultramafic lensing intrusions, which may have been remobilized. Subsequently the strong deformation in the Devonian Acadian Disturbance folded rocks throughout the region, and brought intrusions of granitic batholiths.

Carboniferous deformation produced rift structures, well shown in the west of the island; some folding of Carboniferous sediments followed. Later history involved the great displacement of crustal blocks along high-angle, commonly transcurrent faults.

The southeastern part of Newfoundland, the Avalon belt, was the site of deposition of several kilometers of Precambrian terrigenous mud volcanic sediments, and intrusion of late Precambrian granite. The succeeding Cambrian and lower Ordovician are only hundreds of meters thick, laid on a platform having some shallow fault-bordered troughs and horsts. The rocks were folded in the western part, probably also in the Devonian Acadian orogeny. The interpretation of the integration of the structural history of the structural blocks is still speculative and is under continuing investigation, including mapping and research on the petrology of igneous and sedimentary rocks, as well as their geochemistry and radiometry, to clarify the relation between plate tectonics, uplift, and gravity sliding.

References

Anderson, F. D., and Williams, H., 1970, Gander Lake (West Half): *Geol. Survey Can. Map 1195A.*

Belt, E. S., 1969, Newfoundland Carboniferous stratigraphy and its relation to the Maritimes and Ireland: *Am. Assoc. Petrol. Geol. Mem. 12*, p. 734–753.

Bird, J. M., and Dewey, J. F., 1970, Lithosphere plate continental margin tectonics and the evolution of the Appalachian orogen: *Geol. Soc. Am. Bull.*, v. 81, p. 1031–1060.

Bird, J. M., Dewey, J. F., and Kidd, W. S. F., 1971, Proto-Atlantic oceanic crust and mantle: Appalachian/Caledonian ophiolites: *Nature*, v. 231, p. 28–31.

Church, W. R., and Stevens, R. K., 1971, Early Paleozoic ophiolite complexes of the Newfoundland Appalachians as mantle-oceanic crust sequences: *Jour. Geophys. Res.*, v. 76, p. 1460–1466.

Clifford, P. M., 1969, Evolution of Precambrian massif of western Newfoundland: *Am. Assoc. Petrol. Geol. Mem. 12*, p. 647–654.

Coleman, R. G., 1971, Plate tectonic emplacement of upper mantle peridotites along continental edges: *Jour. Geophys. Res.*, v. 76, p. 1212–1222.

Dean, W. T., 1970, Lower Ordovician trilobites from the vicinity of South Catcher Pond, Northeastern Newfoundland: *Geol. Surv. Can. Paper 70.44*, 15 p.

Dewey, J. F., 1969, Evolution of the Appalachian/Caledonian orogen: *Nature*, v. 222, p. 124–129.

———, and Bird, J. M., 1971, Origins and emplacement of the ophiolite suite: Appalachian ophiolites in Newfoundland: *Jour. Geophys. Res.*, v. 76, p. 3179–3206.

———, and Kay, M., 1968, Appalachian and Caledonian evidence for drift in the North Atlantic, in Phinney, R. A., ed., *The history of the Earth's crust*: Princeton University Press, p. 161–168.

Eastler, T. E., 1969, Geology of Change Islands and eastern Notre Dame Bay, Newfoundland: *Am. Assoc. Petrol. Geol. Mem. 12*, p. 425–432.

Ernst, W. G., 1965, Mineral paragenesis in Franciscan metamorphic rocks, Panoche Pass, California: *Geol. Soc. Am. Bull.*, v. 76, p. 879–913.

Fairbairn, H. W., and Berger, A. R., 1969, Preliminary geochronological studies in northeastern Newfoundland: *Mass. Inst. Tech., 17th Rept. on Age Determinations*, p. 19–20.

Fuller, J. O., 1941, Geology and mineral resources of the Fleur-de-Lys area: *Newfoundland Geol. Survey Bull. 15*, 41 p.

Gunn, R., 1947, Quantitative aspects of juxtaposed ocean deeps, mountain chains and volcanic ranges: *Geophysics*, v. 12, p. 238–255.

Hamilton, W., 1969, Mesozoic California and the underflow of the Pacific mantle: *Geol. Soc. Am. Bull.*, v. 80, p. 2409–2430.

———, and Myers, W. B., 1966, Cenozoic tectonics of the western United States: *Rev. Geophys.*, v. 4, p. 509–549.

Helwig, J. A., and Sarpi, E., 1969, Plutonic-pebble conglomerates, New World Island, Newfoundland,

and history of eugeosynclines: *Am. Assoc. Petrol. Geol. Mem. 12*, p. 443–466.

Hess, H. H., 1939, Island arcs, gravity anomalies and serpentine intrusions—A contribution to the ophiolite problem: *17th Int. Geol. Congr., Rept.*, v. 2, p. 263–282.

———, 1940, Appalachian peridotite belt; its significance in sequence of events in mountain building (abst.): *Geol. Soc. Am. Bull.*, v. 51, p. 1996.

———, 1955, Serpentines, orogeny and epeirogeny: *Geol. Soc. Am. Spec. Paper 62*, p. 391–408.

Horne, G. S., 1969, Early Ordovician chaotic deposits in the central volcanic belt of northeastern Newfoundland: *Geol. Soc. Am. Bull.*, v. 80, p. 2451–2646.

———, 1970, Complex volcanic-sedimentary patterns in the Magog Belt of northeastern Newfoundland: *Geol. Soc. Am. Bull.*, v. 81, p. 1767–1788.

———, and Helwig, J. A., 1969, Ordovician stratigraphy of Notre Dame Bay, Newfoundland: *Am. Assoc. Petrol. Geol. Mem. 12*, p. 388–407.

Hsü, K. J., 1968, Principles of melanges and their bearing on the Franciscan–Knoxville problem: *Geol. Soc. Am. Bull.*, v. 80, p. 1063–1074.

———, 1969, Melanges of San Franciscan Peninsula—Geologic reinterpretation of type Franciscan: *Am. Assoc. Petrol. Geol. Bull.*, v. 51, p. 579–600.

———, 1973, Mesozoic evolution of the California coast ranges: A second look (this volume).

Hutchinson, R. D., 1962, Cambrian stratigraphy and trilobite faunas of southeastern Newfoundland: *Geol. Survey Can. Bull.*, v. 88, 156 p.

Isacks, B., Oliver, J. E., and Sykes, L. R., 1968, Seismology and the new global tectonics: *Jour. Geophys. Res.*, v. 73, p. 5855–5899.

Jenness, S. E., 1963, Terra Nova and Bonavista map areas, Newfoundland: *Geol. Survey Can. Mem. 327*, 184 p.

Johnson, H., 1941, Paleozoic lowlands of western Newfoundland: *New York Acad. Sci. Trans.*, ser. 2, v. 3, p. 141–145.

Kay, M., 1942, Development of the northern Alleghany synclinorium and adjoining regions: *Geol. Soc. Am. Bull.*, v. 53, p. 1601–1658.

———, 1945, Paleogeographic and palinspastic maps: *Am. Assoc. Petrol. Geol. Bull.*, v. 29, p. 426–450.

———, 1951, North American geosynclines: *Geol. Soc. Am. Mem. 48*, 143 p.

———, 1966, Comparison of the lower Paleozoic volcanic and non-volcanic geosynclinal belts in Nevada and Newfoundland: *Bull. Can. Petrol. Geol.*, v. 25, p. 579–599.

———, 1967, Stratigraphy and structure of northeast Newfoundland bearing on drift in North Atlantic: *Am. Assoc. Petrol. Geol. Bull.*, v. 51, p. 579–600.

———, ed., 1969a, North Atlantic geology and continental drift—a symposium: *Am. Assoc. Petrol. Geol. Mem. 12*, 1082 p.

———, 1969b, Silurian of northeast Newfoundland, in Kay, M., ed., *North Atlantic geology and continental drift—a symposium*: *Am. Assoc. Petrol. Geol. Mem. 12*, p. 414–424.

———, 1969c, Thrust sheets and gravity slides of western Newfoundland, *Am. Assoc. Petrol. Geol. Mem. 12*, p. 665–669.

———, 1970, Flysch and bouldery mudstone in northeast Newfoundland: *Geol. Assoc. Can. Spec. Paper 7*, p. 155–164.

———, 1972, Dunnage Melange and lower Paleozoic deformation in northeast Newfoundland: *24th Int. Geol. Congr., Rept.*, v. 3, p. 122–133.

———, and Eldredge, N., 1968, Cambrian trilobites in central Newfoundland volcanic belt: *Geol. Mag.*, v. 105, p. 372–377.

Kennedy, M. J., and Phillips, W. E. A., 1971, Ultramafic rocks of the Burlington Peninsula, Newfoundland: *Geol. Assoc. Can. Proc.*, v. 22, p. 35–46.

Kindle, C. H., and Whittington, H. B., 1958, Stratigraphy of the Cow Head region, western Newfoundland: *Geol. Soc. Am. Bull.*, v. 69, p. 315–342.

LePichon, X., 1968, Sea-floor spreading and continental drift: *Jour. Geophys. Res.*, v. 73, p. 3661–3705.

Lilly, H. D., 1965, Submarine examination of the Virgin Rocks area, Grand Banks, Newfoundland: Preliminary notes: *Geol. Soc. Am. Bull.*, v. 76, p. 131–132.

Manson, D. V., and Kay, M., 1972, Intrusions in the Dunnage melange subduction zone, northeast Newfoundland: *Am. Museum Nat. Hist. Novitates* (in preparation).

McCartney, W. D., 1967, Whitbourne map-area, Newfoundland: *Geol. Surv. Can. Mem. 341*, 135 p.

Morgan, W. J., 1968, Rises, trenches, great faults and crustal blocks: *Jour. Geophys. Res.*, v. 73, p. 1959–1982.

Morris, R. W., and Kay, M., 1966, Ordovician graptolites from the middle Table Head Formation at Black Cove, Port au Port, Newfoundland: *Jour. Paleont.*, v. 40, p. 1223–1229.

Neale, E. R. W., 1957, Ambiguous intrusive relationships of the Betts Cove-Tilt Cove serpentinite belt, Newfoundland: *Geol. Assoc. Can. Proc.*, v. 9, p. 95–102.

———, and Nash, W. A., 1963, Sandy Lake (east half), Newfoundland: *Geol. Survey Can. Paper 62-28*, 40 p.

Phillips, W. E. A., Kennedy, M. J., and Dunlop, G. M., 1969, Geologic comparison of western Ireland and northeastern Newfoundland: *Am. Assoc. Petrol. Geol. Mem. 12*, p. 194–211.

Poole, W. H., 1967, Tectonic Evolution of Appalachian region of Canada: *Geol. Assoc. Can. Spec. Publ. 4*, p. 9–51.

———, Stanford, B. V., Williams, H., and Kelley, D. G., *in* Douglas, R. J. W., ed., *Geology and mineral resources of Canada*: Geol. Survey Canada, Econ. Geol. Rept. 1, p. 228–364.

Rodgers, J., 1968, The eastern edge of North America during the Cambrian and lower Ordovician, *in* Zen, E-an, and others, eds., *Studies of Appalachian geology: Northern and maritime*: New York, Interscience, p. 141–149.

———, 1970, *The tectonics of the Appalachians*: New York, Interscience, 270 p.

———, and Neale, E. R. W., 1963, Possible "Taconic" klippen in western Newfoundland: *Am. Jour. Sci.*, v. 261, p. 213–230.

Schenk, P. E., 1970, Regional variation of the flysch-like Meguma Group (lower Paleozoic) of Nova Scotia compared to recent sedimentation off the Scotia shelf: *Geol. Assoc. Can. Spec. Paper 7*, p. 127–154.

Schuchert, C., and Dunbar, C. O., 1934, Stratigraphy of western Newfoundland: *Geol. Soc. Am. Mem. 1*, 123 p.

Stevens, R. K., 1970, Cambro-Ordovician flysch sedimentation and tectonics in west Newfoundland and their possible bearing on a Proto-Atlantic Ocean: *Geol. Assoc. Can. Spec. Publ. 7*, p. 165–178.

Wanless, R. K., 1970, Isotope age map of Canada: *Geol. Survey Can., Map 1256A*.

Whittington, H. B., and Kindle, C. H., 1969, Cambrian and Ordovician stratigraphy of western Newfoundland: *Am. Assoc. Petrol. Geol. Mem. 12*, p. 655–664.

Williams, H., 1963, Twillingate map area, Newfoundland: *Geol. Survey Can. Paper 63–36*, 30 p.

———, 1964a, The Appalachians in northeastern Newfoundland—A two sided symmetrical system: *Am. Jour. Sci.*, v. 262, p. 1137–1158.

———, 1964b, Notes on the orogenic history and isotope ages in Botwood map area, northeastern Newfoundland: *Geol. Survey Can. Paper 64–17*, pt. 2, p. 22–25.

———, 1967a, Island of Newfoundland: *Geol. Survey Can. Map 1231A*.

———. 1967b, Silurian rocks of Newfoundland: *Geol. Assoc. Can. Spec. Publ. 4*, p. 93–137.

———, 1969, Wesleyville, Newfoundland: *Geol. Survey Can., Map 1227a*.

———, 1971, Mafic-ultramafic complexes in western Newfoundland Appalachians and the evidence for their transportation: A review and interim report: *Geol. Assoc. Can. Proc.*, v. 24, p. 9–25.

Wilson, J. T., 1966, Did the Atlantic close and then re-open?: *Nature*, v. 211, p. 676–681.

Yeats, R. S., 1968, Southern California structure, sea-floor spreading, and history of the Pacific Basin: *Geol. Soc. Am. Bull.*, v. 79, p. 1693–1702.

ROBERT H. MOENCH

U.S. Geological Survey
Denver, Colorado[1]

Down-Basin Fault-Fold Tectonics in Western Maine, with Comparisons to the Taconic Klippe[2]

The Rangeley area is on the northwest limb of the Merrimack synclinorium (Fig. 1), which spans much of zone *B* of the Appalachian-Caledonian orogen, according to Dewey (1969a) and Bird and Dewey (1970, Fig. 1). According to the Bird–Dewey lithosphere plate model, all the lower and middle Paleozoic oceanic and trench sediments of zone *B* were deformed by compression when the proto-Atlantic Ocean closed during plate consumption and opposing continents collided, resulting in the Devonian Acadian orogeny. No one is likely to argue for a noncompressional origin of the tight folds with subvertical axial surfaces and penetrative slaty cleavage that characterize the structure of much of zone *B*. Compression, however, may be either a boundary force phenomenon controlled externally by crustal shortening or a body force phenomenon controlled principally by the influence of gravity on the mass being deformed.

Dewey (1969b, p. 331–333) recognized this problem in zone *B* of the British Caledonides, and he argued for boundary force crustal shortening. He pointed out that the "vast décollement" required by a body force model probably does not exist. Such a décollement might, however, be extremely difficult to recognize. In contrast, I have suggested that the fundamental early structure of the Merrimack synclinorium in western Maine was produced by gradual large-scale down-basin slump faulting, folding, and tectonic dewatering

---

[1] Publication authorized by the Director, U.S. Geological Survey.
[2] I am grateful to Wallace M. Cady and W. Bradley Meyers for their criticisms of an early version of this paper.

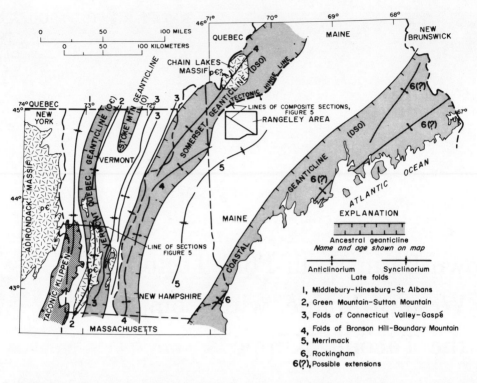

Fig. 1 Index map showing the tectonic and paleotectonic setting of the Rangeley area. (p C) Precambrian crystalline massif (queried where age conjectural); ages of geanticlines: (O C) Ordovician and Cambrian; (O) Ordovician; (DSO) Devonian, Silurian, and Ordovician

on the active slopes of a subsiding geosynclinal trough (Moench, 1970). Whether the bordering anticlinoria (Fig. 1) actually moved toward or away from one another is moot, but the mapped geometric relations between major folds and apparently normal premetamorphic faults indicate extension within the sedimentary mass when deformation began, followed by compression when the mass slumped against a barrier of more sediment in the trough.

Recognition of the principal premetamorphic faults is crucial to this interpretation. Although several formations are truncated along both sides of individual faults, in outcrop the faults are very inconspicuous (see Table 1; Moench 1970, Figs. 7, 8, 9, 12, 13).

Also fundamental to this interpretation is the controversial hypothesis of slaty cleavage formation by tectonic dewatering (Maxwell, 1962). My interpretation, if applied generally to the early structure of zone B, does not necessarily conflict with some modification of the plate model. It implies, however, that tight folds with penetrative slaty cleavage—commonly cited as evidence for orogenic compression—may originate as well in a deep trough undergoing subsidence or extension.

The intent of this paper is to illustrate how the Rangeley fault-fold structure and the Taconic klippen[3] of western New England (Fig. 1) may be different manifestations of a similar body force mass movement phenomenon, controlled by major paleotectonic

---

[3] Henceforth the plural term "Taconic klippen" refers to all of the more or less contiguous klippen in western New England (Fig. 1) and the faults or slides beneath them.

features. With the notable exception of Sales (1971), most current authors interpret the Taconic klippen as a succession of detachment gravity slides (see Cady, 1968; Zen, 1967). This interpretation is tentatively accepted here. The obvious differences between the Rangeley and Taconic structures are explained by reference to the different sedimentary histories and paleotectonic settings of both areas.

This paper is based on mapping by the writer in the Phillips and Rangeley quadrangles during 1961–1964 (Moench, 1971) and in the Rumford quadrangle in 1965 and 1970 by the writer and Carol T. Hildreth (unpub. mapping). Only the northernmost part of the Rumford quadrangle, mapped in 1965, together with the Phillips and Rangeley quadrangles are shown in Fig. 1 and 2. This area provides the simplest illustration of the pertinent stratigraphy and structural geometry. Structural complexity increases greatly toward the southwest in the Rumford quadrangle, because of crowding between major plutons.

Tectonic Setting and Structure of the Rangeley and Taconic Areas

The present expression of the major anticlinoria and synclinoria shown in Fig. 1 was attained during the Devonian Acadian orogeny. All apparently descended, however, from ancestral geanticlines and geosynclinal furrows that controlled both sedimentation and the origin of the earliest patterns of deformation. The Green Mountain–Sutton Mountain anticlinorium, for example, is considered a Devonian feature of tectonic overprinting that is obliquely superposed on the ancestral Vermont–Quebec geanticline, and on a complex pattern of early folds, slides, and foliation whose origin culminated in middle Ordovician time (Cady, 1968; 1969).[4]

The geanticline, according to Cady (1969, p. 19–21) was a positive tract of slowed deposition and stratigraphic convergence that persisted through early Paleozoic time. Prior to culmination of early deformation in middle Ordovician time, a long period of folding, sliding, and slate-grade metamorphism more or less accompanied sedimentation (Cady, 1968; 1969, p. 37–46, 53–55, 161–162). In western New England, according to Cady, early deformation was restricted to the sedimentary cover. It was a westward mass movement pattern, controlled largely by a persistent geanticline in the east, that produced dominantly recumbent folds overturned to the west, the Taconic klippen, and the early regional foliation. Late deformation and metamorphism in Devonian time involved the Precambrian basement. It produced flexural folds, slip cleavage, thrusts, and domes, which define the Green Mountain–Sutton Mountain anticlinorium and adjacent synclinoria (Cady, 1969, p. 59–72; Osberg, 1969).

Salient characteristics of the Taconic klippen (Fig. 1) are summarized in Table 1. Interpretation of the klippen as thin, flat-lying detachment thrust sheets has been amply discussed by Cady (1968, 1969), Zen (1961, 1967, 1969), and others, and needs little further discussion here. The most pertinent difference of opinion concerns the temporal and spatial relations of sliding to early recumbent folding, the formation of slaty cleavage, and metamorphism. According to Zen (1969), rocks of the Taconic area were probably recumbently folded, slaty-cleaved, and metamorphosed in late Ordovician to early Silurian (?) time after emplacement of the earliest klippen, and they were deformed and metamorphosed again during the Devonian Acadian orogeny.

---

[4] Zen (1967, p. 46–67) and others have suggested that the Vermont-Quebec geanticline did not exist but was instead a belt of starved deposition in a deep ocean basin east of a carbonate bank. Sliding that produced the Taconic klippen was caused, according to Zen (1967, Fig. 4), by abrupt uplift and bathymetric reversal along the site of deposition of the thin starved sequence in Middle Ordovician time. Stratigraphic evidence for the geanticline summarized by Cady (1968, p. 570; 1969, p. 19–21) is tentatively accepted here.

TABLE 1 Comparative Characteristics and Interpretations of the Taconic Klippen and the Early Fault-Fold Structure of the Rangeley Area

Characteristic or Interpretation	Taconic Klippen (from Cady, 1968, 1969; Zen, 1961, 1967, 1969)	Rangeley Fault-Fold Structure (from Moench, 1970)
Shape of transported bodies	Thin sheets above flat-lying detachment faults; 1400 m of strata preserved in klippen, somewhat less in each klippe	Thick wedges above curved normal faults that flatten in depth; 4500+ m of strata preserved in fault-fold cell above Hill 2808 fault
Internal and external structures	Recumbent isoclinal passive flow folds within and below klippen; large amplitude/wavelength ratio	Tight upright passive flow syncline-anticline pairs above Hill 2808 and Barnjum faults; fold amplitudes and plunges correlate with displacement along underlying fault, which truncates next underlying folds
	Folds overturned to west toward transport direction	Folds upright and overturned to northwest away from transport direction
	Slaty cleavage penetrates faults and is parallel to axial planes of early folds	Slaty cleavage penetrates faults and is subparallel to axial planes of early folds
Stratigraphic relations across faults	Older over younger most common; older may face downward	Younger over and southeast of older; younger may face northwest toward older
Fault surfaces or zones	Variably indistinct, gradational, sharp; soft sediment deformation; precleavage; local wildflysch-type conglomerate near base of lowest, oldest klippe	Sharp, firm walls; bedding truncated at low angles, otherwise not disrupted; fault zones ½ in. to a few feet wide, local precleavage disaggregation, and soft-sediment mixing; cataclasis in soda-granite on Barnjum fault
Relative movements in folds	Westward, overthrusting	Eastward underthrusting, or orthogonal
Leading edges	Toes exposed during sliding	No toes; faults disappear along bedding in depth and along strike
Trailing edges	Characteristics not known; possibly fragmented during sliding	Normal growth faults; dip 50–70° SE
Distances transported	Long; 40± km	Short; 3–7± km maximum each fault-fold cell
Rates and times of transport	Rapid; beds with "zone 12" graptolites transported and emplaced during interval of lower subzone of graptolite "zone 13" (Berry, 1970)	Very slow; basinward creep and faulting initiated during Silurian deposition; faulting, folding, tectonic dewatering culminated in Early Devonian time
Tectonic settings	Tectonic slope controlled tectonic transport and some sediment provenance	Tectonic slope controlled tectonic transport and some sediment provenance
	Thin strata of transported sheets slowly deposited over Vermont–Quebec geanticline (Cady, 1968)	Thick strata of slumped wedges rapidly deposited on trough side of tectonic hinge line
Tectonic stages	Westward transport initiated by maximum uplift in east	Southeastward creep initiated by critical sedimentary loading during relative tectonic quiet; culminated during maximum uplift in northwest
Sources of excess fluid pressure	External; possibly fluids expelled from compacting older sediments (Cady, 1968)	Internal; rapid sedimentary loading plus low hydraulic conductivity
Role of excess fluid pressure	Primarily weakening, permitting downslope movement, leading to rupture; some flotation effect during sliding on wet pelitic sediment	Primarily weakening, permitting downslope movement, leading to rupture; also permitting compactional formation of slaty cleavage

Cady, however, cited evidence that some folding, metamorphism, and slaty cleavage formation predated as well as postdated emplacement of the early klippen (Cady, 1968; 1969, p. 37–46, 53–55, 161–162).

In contrast with the Green Mountain–Sutton Mountain anticlinorium, the Merrimack synclinorium is not primarily a feature of tectonic overprinting. Although superposed deformation is intense in the synclinorium and in the adjacent Bronson Hill–Boundary Mountain anticlinorium in New Hampshire (Cady, 1969, p. 73, 74; Thompson and others, 1968), in Maine the synclinorium is dominated by early longitudinal folds, faults, and steeply dipping axial surface cleavage (Osberg and others, 1968; Hussey, 1968). Early deformation involved Ordovician, Silurian, and Devonian strata. It culminated in Early Devonian time and was closely followed by emplacement of the Devonian plutons. Late deformation and metamorphism were partly related to the plutons and were entirely unrelated to the gross trends of the synclinorium and adjacent anticlinoria (Moench, 1970, p. 1470–1475). Evidence that the Merrimack synclinorium coincides with an ancestral trough is provided, for example, by Boucot's (1968; 1969) paleogeographic maps of the Silurian and Devonian systems and by Pavlides and others' (1968) review of stratigraphic evidence for the Taconic orogeny.

Briefly, the synclinorium coincides with "the locus of the thickest sequence of Silurian sediment in the northern Appalachian region" (Boucot, 1968, p. 87), and with a narrower belt of "Taconic conformity," in which Ordovician strata pass gradationally into the thick Silurian clastic deposits (Pavlides and others, 1968). The tectonic hinge line shown in Fig. 1 is the approximate southeast limit of the Taconic unconformity in the Rangeley area (Boone and others, 1970). The unconformity is again recognized far to the southeast on the coastal geanticline (Pavlides and others, 1968, Fig. 5-1).

The structural framework of the Rangeley area (Figs. 2, 3; Table 1) is defined by the mapped distribution of major stratigraphic units (see Moench, 1971). The stratigraphy of the area (see Fig. 2, Explanation) is thus fundamental to the interpretation. Despite greenschist to amphibolite facies metamorphism, sedimentary features are well preserved, and the stratigraphic order was established by abundant use of evidence of sedimentary tops, especially across exposed contacts. Although identifiable fossils have not been found in the mapped area, they have been found in some equivalent formations a few miles to the north and northwest (Harwood and Berry, 1967; Moench and Boudette, 1970) and in probable equivalents of the Silurian (?) formations 30–60 miles to the east and southeast (Osberg and others, 1968; Ludman, 1971). For practical purposes, the major formations are thus biostratigraphic units.

An exception is the southern sequence (Fig. 2), which is correlated in part with the Seboomook (?) formation north of the Blueberry Mountain fault on the basis of lithologic similarity and dissimilarity with rocks below the Seboomook (?). The shapes and variable plunges of the major folds (Figs. 2, 3) accord with abundant evidence of sedimentary tops and with the orientations of small folds and early cleavage-bedding intersections. The premetamorphic faults shown in Fig. 2 were delineated by truncations of map units, locally by evidence that sedimentary tops face toward older rocks across a dislocation, and by observed characteristics of exposed faults (Table 1; see also Moench, 1970, p. 1475–1487). Some bedding-plane faults within formations may have escaped recognition. The internal subdivision of formations is too detailed and consistent, however, to permit the existence of major folded thrusts within the mapped area that might correlate with the mapped premetamorphic faults. Moreover, independent evidence for folded thrusts is lacking—for example, downward-facing folds or map units or consistent zones of disrupted bedding.

Two principal deformations are recognized in the Rangeley area. The early deformation

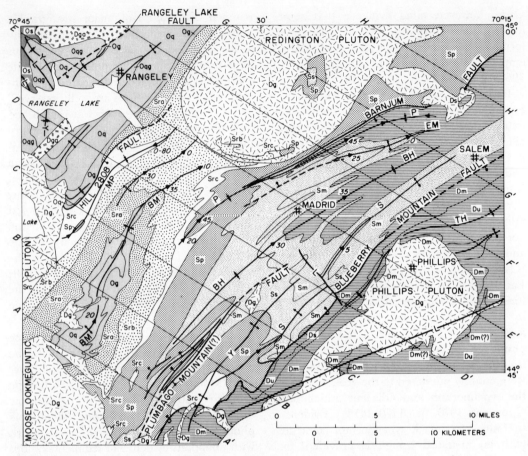

FIG. 2 Geologic map of the Rangeley area, and lines of sections of Figs. 3 and 4.

is expressed by several mapped and inferred longitudinal premetamorphic faults and by large, tight passive flow folds (Donath and Parker, 1964). The early cleavage is defined largely by pervasive parallel orientation of mica. It crosses the premetamorphic faults. Although the cleavage is subparallel to the axial surfaces of the early folds, the present expression of the cleavage appears to be slightly younger than the folds (Moench, 1970, p. 1487). The late deformation is expressed by locally major flexural folds that are obliquely superposed on the older structures and by slip cleavage and slip cleavage-derived schistosity. These features are most conspicuous near the plutons or near the postmetamorphic faults.

Two metamorphic events above slate grade are recognized as well. Although isograds of only the younger event conform to known plutons, porphyroblasts of both events are overprinted on the early slaty cleavage or schistosity (Moench, 1970, p. 1473–1475, Figs. 4, 8, 9; Guidotti, 1970). These relationships permit restoration of the structural framework to its approximate form prior to emplacement of the plutons (Figs. 3, 4).

The major prepluton structure of the Rangeley area is separable into large fault-fold units, or cells (Fig. 4). Where unaffected by younger deformation the major faults dip 50–70° SE. As we shall see shortly, the faults are inferred to flatten in depth. Since younger rocks are consistently downthrown on the

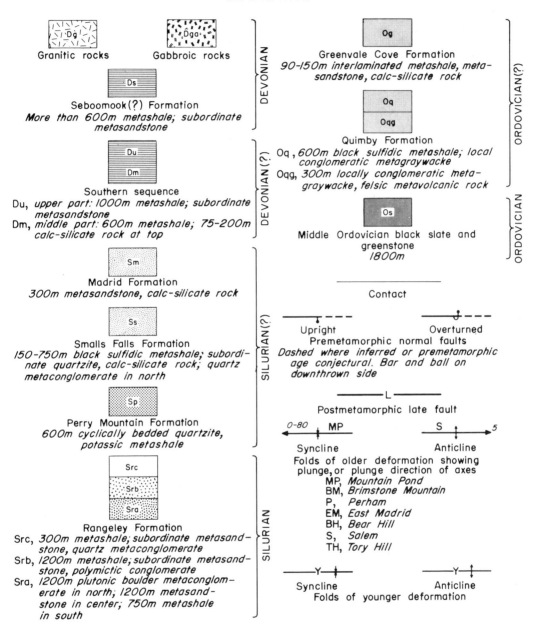

EXPLANATION

southeast against older rocks on the northwest, the faults are considered normal faults. Geometric relations between faults and folds are best illustrated by the Hill 2808 fault, Mountain Pond syncline, and Brimstone Mountain anticline (Figs. 2–4). Note that the syncline plunges southwest in the direction of increasing apparent displacement along the fault, and opposite to the northeast plunge of the anticline. Structural relief in the syncline-anticline pair thus increases in the direction of increasing apparent normal displacement

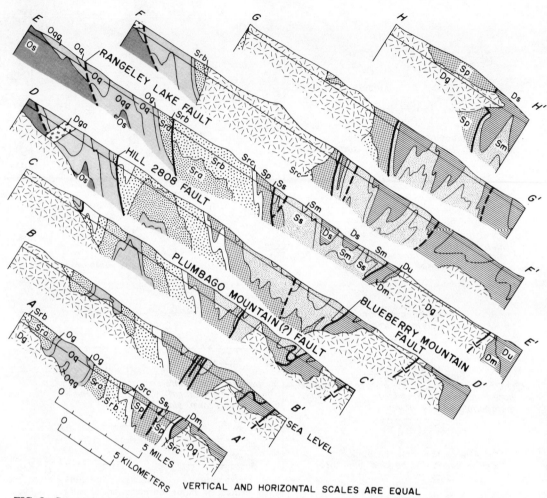

FIG. 3 Structure sections across the Rangeley area. For explanation see Fig. 2.

along the fault. Moreover, this plunge contrast is greatest near the area where displacement along the fault increases most abruptly. Conversely, the folds apparently disappear in the opposite direction where the fault apparently ends. Owing to the Redington pluton, evidence for these disappearances is lacking. Structural relief is smallest, however, along section E–E' (Figs. 3 and 4), near the inferred or approximate end of the fault.

The Barnjum fault, Perham syncline, and East Madrid anticline illustrate similar relations, but opposite in direction. Both folds plunge northeast, but the syncline plunges more steeply than the anticline. The anticline enlarges and structural relief in the syncline–anticline pair increases northeastward, in the direction of increasing apparent normal displacement along the Barnjum fault. The folds are thus geometrically tied to apparent normal movements along the faults.

All the major faults are shown in Fig. 4 to flatten in depth toward the southeast. This inferred flattening is based on relationships between "reverse drag" and curvature in depth of normal faults (Hamblin, 1965; Cloos, 1968). Owing to this curvature, strata of the hanging-wall mass bend downward to fill open space that would appear during movement if the mass were rigid. Depending

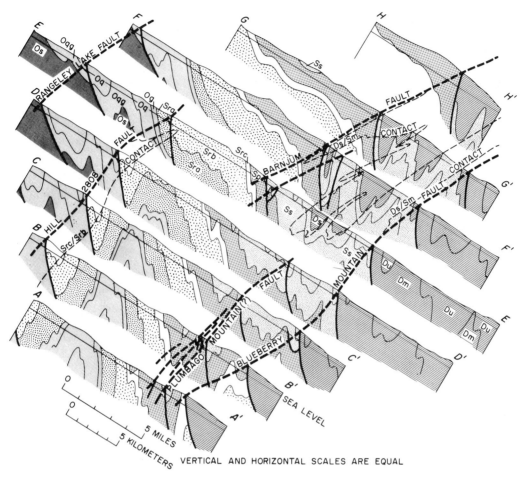

FIG. 4 Approximate prepluton restored structure sections across Rangeley area. For explanation see Fig. 2.

on the properties of materials involved, the same process might produce a graben, or a broad syncline, whose northwest limb above a southeast-dipping fault would actually be normal drag. Direct evidence for flattening was found during recent mapping (Moench and C. T. Hildreth, unpub. data) in the southwest part of the Rumford quadrangle and adjacent areas south of the area of Fig. 2. There, another major premetamorphic fault has been delineated and tentatively named the Plumbago Mountain fault. The probable extension of this fault is named in Figs. 2 and 4. Where the Plumbago Mountain fault is delineated to the south, it has been deformed by major late folds. The geometry of the late folds indicates that the early fault originally dipped southeast and extended subhorizontally under a large area to the southeast of the original fault line.

Comparative Structural Evolution of the Rangeley Area and Western New England

The Taconic klippen and associated early folds of western New England and the fault-fold structures of the Rangeley area are here considered different manifestations of gravitational mass movement (Cady, 1968; Moench, 1970). The klippen were thin, detached sheets that slid rapidly westward over long distances during rather short periods of time, whereas the Rangeley fault-fold cells were thick wedges, attached in depth and along strike, that crept slowly southeastward over short distances during long periods of time (Table 1).

In addition, folds associated with the klippen are commonly recumbent and overturned toward the west in the direction of mass movement, whereas folds related to the Rangeley cells are upright or overturned toward the northwest, opposite the inferred direction of mass movement. Finally, the controlling paleotectonic slopes dipped west in western New England, but southeast in the Rangeley area. It should be remembered, however, that early deformation ended in western New England long before it is known to have begun in the Rangeley area, where a generally younger sequence is deformed. Uplift that apparently shed the klippen to the west in middle Ordovician time apparently also shed clastics to the east and southeast, which were then deformed contemporaneously with continued geanticlinal uplift, trough subsidence, and sedimentation. The principal stages of the evolution of the Rangeley area and western New England are summarized in Fig. 5.

Rangeley Area

Section I (Fig. 5) sets the stage for enactment of the earliest phase of the structural evolution of the Rangeley area. This section represents the time of deposition of about the middle of the Perry Mountain formation. This time was chosen because the earliest hint of deformation that can be related to known structural features is recorded by abrupt thickness and facies changes in the next younger Smalls Falls formation (Moench, 1960, p. 1469, 1470, 1490). The Smalls Falls is interpreted to have accumulated in a closed structural trough whose northwest margin was partly defined by the earliest expressions of the Barnjum fault and Brimstone Mountain anticline. The Perry Mountain, in contrast, is part of an extensive blanket of cyclically bedded metashale and quartzite that exhibits only gradual changes over long distances (Osberg and others, 1968). It was deposited probably rather slowly on a featureless, gently sloping ocean floor during relative tectonic quiet.

Deposition of the Perry Mountain was preceded, however, by rapid deposition of a sedimentary wedge as much as 17,000 ft (5.2 km)[5] thick, composed of the upper Ordovician (?) Quimby and Greenvale Cove formations and the lower Silurian (?) Rangeley formation. The great thickness of this wedge, together with its sedimentary features, indicates both very rapid sedimentation and low hydraulic conductivity of the mass. Excess fluid pressure may thus have been established by sedimentary loading (Moench, 1970, p. 1468–1469; Bredehoeft and Hanshaw, 1968). Owing to the weakening effect of excess fluid pressure, the wedge was unstable.

As shown in section I (Fig. 5) the Quimby-Rangeley wedge conformably overlies middle Ordovician and older strata. The wedge thins abruptly toward the Somerset geanticline in the northwest, where the Silurian beds generally represent a more landward facies relative to equivalent rocks of the synclinorium (Moench and Boudette, 1970). The upper part of the Rangeley formation and the Perry Mountain formation on the geanticline apparently unconformably overlap previously deformed older rocks (Moench and Boudette, 1970, Fig. 2; Harwood and Berry, 1967). The thinned sequence and the previously deformed older rocks of the geanticline were thus relatively stable.

---

[5] Measured thicknesses recomputed to assumed water content of 40 percent.

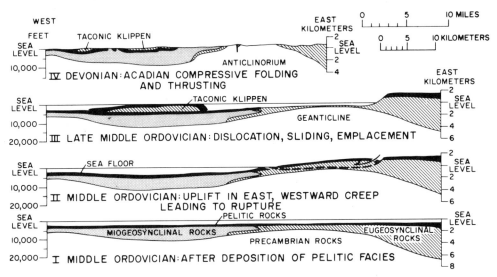
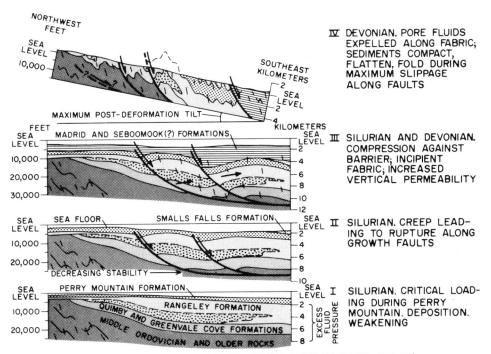

FIG. 5 Restored sections across the Rangeley area and western New England showing evolution of early structural features. Lines of sections are shown in Fig. 1.

When the weakened, unstable Quimby-Rangeley wedge was further loaded by the Perry Mountain blanket, and was tilted slightly, it began to creep glacierlike downslope (Fig. 5, section II). As the wedge pulled away from more stable rocks upslope, it ruptured by faulting. With continuing creep the wedge separated into probably several cells, each bounded on the northwest and below by longitudinal normal faults that flattened in depth. Owing to this curvature a shallow syncline formed on the downthrown side of each fault, in a manner similar to the formation of "reverse drag" (Hamblin, 1965; Cloos, 1968). A gentle anticline grew farther down-basin. Initially, the anticline was no more than the break in slope between the syncline and the basinward slope of the ocean floor.

The initial sedimentary wedge did not thin on a continental slope, but rather filled the trough between the Somerset and coastal geanticline (Fig. 1). The center of the trough thus contained a barrier of sediment fill. As the mass above each fault compressed under body forces against the barrier, it folded slightly; clay particles tended to reorient normal to compression and across bedding (Fig. 5, section III). Vertical permeability was improved slightly by the new fabric, and pore fluids began to migrate vertically.

As uplift in the northwest and subsidence in the trough continued, basinward slippage along established faults accelerated, body force compression against the barrier increased, and fabric improved. Pore fluids thus escaped vertically and were pressed out more rapidly as the degree of preferred orientation improved. Horizontal shortening required to produce the observed folds must have been accompanied by some vertical thickening. A large part of the shortening, however, would be permitted by a directional volume loss of at least 25 percent owing to vertical expulsion of water. If important differences in compactability and perhaps competence existed in the mass, this horizontal compaction could have produced folds. Such differences can be inferred from the stratigraphic column.

Shaly formations with rather high water contents, for example, would compact more and would transmit stress less effectively than sandy or conglomeratic units. Accordingly, gentle folds established by the early extensional mechanisms of stage II (Fig. 5) would have been tightened and amplified by the body force compressional pile-up during stages III and IV. Geometrically similar folds would have resulted where compactability and competence contrasts were low, whereas disharmonic folds would have resulted where these contrasts were relatively greater. Perhaps large-scale cleavage fanning across major folds as well as cleavage refraction from one bed to another (Moench, 1966, p. 1458) are expressions of differential compactability.

In addition, the fact that major folds tend to be overturned away from the inferred direction of movement (Figs. 4 and 5) might reflect greater horizontal compaction and shortening at depth than near the surface. In the Rangeley area, expulsion of pore fluid was probably assisted by thermal processes of diagenesis and low-grade metamorphism in the deeper levels of the system (Moench, 1970, p. 1488, 1493). The geothermal gradient may have been steep, because stage IV (Fig. 5) was closely followed by emplacement of the Devonian plutons and by amphibolite facies metamorphism.

Final stages of the process may thus be visualized partly as the directional collapse of a system of excess fluid pressure. During collapse and flattening, earlier structures were amplified, and pervasive early slaty cleavage was impressed on the mass. The driving force was the head within each cell caused by regional basinward tilting. Since this head must have been limited by the upward growth of anticlines in each cell, section IV is shown tilted about 12° SE after deformation. This tilt is excessive, but the amount of required tilt would have been lessened by contemporaneous erosion of the anticlines and deposition in the synclines.

Four major premetamorphic down-basin faults have been mapped in the Rangeley area. It is assumed that many more exist to the northwest and southeast, and that groups of premetamorphic faults extend the full length of the Merrimack synclinorium. In addition, approximate mirror images of the premetamorphic faults of the Rangeley area may exist on the southeast limb of the Merrimack synclinorium, although such faults have not yet been proved (see Moench, 1970, p. 1492). The total early structural pattern thus might reflect the final gravitational body force pile-up of many fault-fold cells and the interactions between them on both limbs of the synclinorium. The process would require a hindmost-bottommost décollement, shown tentatively in section IV (Fig. 5). Alternatively, regional boundary force compression might account for some tightening and cleavage formation, but not for the major fault-fold pattern.

Taconic Area

As shown in section I (Fig. 5), a thin sequence of Cambrian and Ordovician sediments—probably about 4500 ft (1.4 km) thick—was slowly deposited over the crest of the Vermont–Quebec geanticline (Cady, 1965). More than 20,000 ft (7 km) of clastic sediments and some volcanic rocks of the same age accumulated in the eugeosyncline to the east, and about 7000 ft (2 km) of carbonate-rich strata was deposited in the miogeosyncline to the west. In middle Ordovician time a thin blanket of rather impermeable pelitic sediments was laid down across the entire region.

Because sedimentation was slow over the geanticline, excess fluid pressure probably did not accumulate by sedimentary loading. The thin sequence probably compacted at a normal rate, and it probably was reasonably stable before the culmination of uplift. Meanwhile, large volumes of pore fluids were gradually expelled by compaction and by lowgrade metamorphism from the thick eugeosynclinal sequence in the east. These fluids moved westward parallel to bedding (and perhaps parallel to incipient fabric) toward the geanticline and were trapped beneath the impermeable pelitic blanket. With uplift of the geanticline and the eugeosyncline in Middle Ordovician time, an extensive west-dipping slope was established (Fig. 5, section II). Under the head established by this slope, excess fluid pressure accumulated in or near the base of the pelitic sequence on the geanticline.

The initial role of excess fluid pressure may have been to weaken rather than float the sedimentary blanket, much as postulated in the Rangeley area (Moench, 1970, p. 1490, 1493). If so, the following continuous sequence of events leading to folding and detachment sliding may be visualized:

1. Excess fluid pressure accumulated unevenly in the thin sedimentary cover on the geanticline. Fluid pressure adequate to float the blanket is not required.
2. Uneven weakening permitted uneven downslope creep. The predominant westward movement pattern produced asymmetrical longitudinal folds overturned to the west, but the uneven movement produced cross-folds as well (Cady, 1969, p. 42–46).
3. As the moving blanket pulled away from more stable material upslope, it ruptured (Fig. 5, section II). The rupture, a west-dipping normal fault, flattened abruptly and propagated westward at shallow depth along the zone of maximum shear strain between the moving blanket and the stable underlying rocks.
4. The fracture propagated farther west, where it was deflected to higher stratigraphic levels by the competent carbonate facies, and then it intersected the ocean floor.

When detachment was completed, the blanket slid independently westward on a cushion of wet pelitic sediments (Zen, 1967, p. 69) and came to rest in the trough (Fig. 5, section III). Material that cascaded from the leading, bulldozing edge of the blanket produced wildflysch-type conglomerate (Zen, 1967, p. 35–40). The trailing edge may have been fragmented during sliding.

In the absence of a barrier of sediment fill in the trough, folding was not a necessary result of sliding. Processes of mass movement leading to folding and sliding, however, were repeated at least twice (see Cady, 1968, p. 571; Zen, 1967, p. 16, 25) and were closely followed by additional westward mass movement, which produced the folds and slaty cleavage that are overprinted on the Taconic klippen and surrounding strata. Folding and sliding were initiated and largely completed during late middle Ordovician time (Zen, 1967, p. 40, 68–69; Berry, 1970). The earliest klippen, at least, were wet and plastic when they arrived, and were thus susceptible to slightly later formation of slaty cleavage by tectonic compaction.

Summary

Principal differences in style of the fault-fold structure of the Rangeley area and the older detachment klippen of western New England are attributed to the presence or absence of a sedimentary barrier in the receiving trough and to the thickness-width ratio of the sedimentary mass involved. These factors in turn were controlled by the geosynclinal setting. In the Rangeley area, a thick sedimentary wedge was deposited on the southeast or down-basin side of the tectonic hinge line (Fig. 1). The wedge crept basinward and separated into perhaps several fault-fold slump units or cells, each of which folded and compacted under its own head against the barrier of sediment in the trough. Perhaps owing to greater tectonic compaction and shortening at depth than near the surface, combined with mass movement by slippage on an underlying fault, the resulting folds within each cell tend to be overturned away from the inferred direction of movement.

In western New England, on the other hand, a thin sedimentary blanket was deposited on the crest of an ancestral geanticline. The blanket was tilted by regional uplift in the east and it moved downslope to the west. Westward movement, largely restricted to the extensive but thin blanket, did not involve slippage on a décollement surface. The movement produced longitudinal recumbent folds overturned to the west toward the direction of movement. Movement led, however, to rupture and detachment along flat-lying shallow depth faults that intersected the ocean floor along the leading edge of the moving blanket. In absence of a barrier, there was no internal compression due to pile-up during sliding. Since the earliest klippen, at least, were wet and plastic when emplaced, they were subject to slightly later compactional origin of slaty cleavage, possibly produced by continued sedimentary loading and westward mass movement. Proposed sources of excess fluid pressure were different in the two regions, but the initial role of fluid pressure in both regions was more to weaken than to float the sedimentary mass, and thus permit slow downslope creep.

The role of boundary force in Acadian compression is moot in the Rangeley area. The compression might have produced some tightening of the early fold pattern, but it cannot account for the major features of the early fault-fold pattern. It left no systematic pattern of tectonic overprinting. If Acadian compression did occur, it seems likely that it was largely transmitted through the bulk of the Merrimack synclinorium and that it was focused in western New England. There it produced the late superposed Green Mountain–Sutton Mountain anticlinorium and adjacent synclinoria (Cady, 1969; Osberg, 1969). Acadian compression might also have been focused in the weaker, hotter, higher rank metamorphic terrane of New Hampshire and southern New England, where tectonic overprinting is more conspicuous regionally than in western Maine (Cady, 1969, p. 73, 74; Thompson and others, 1968). Moreover, the general southwestward convergence of major tectonic and paleotectonic features (Fig. 1) might also be explained by the greater intensity of superposed compressive deformation in that direction.

References

Berry, W. B. N., 1970, Review of late Middle Ordovician graptolites in eastern New York and Pennsylvania: *Am. Jour. Sci.*, v. 269, p. 304–313.

Bird, J. M., and Dewey, J. F., 1970, Lithosphere plate-continental margin tectonics and the evolution of the Appalachian orogen: *Geol. Soc. Am. Bull.*, v. 81, p. 1031–1060.

Boone, G. M., ed., 1970, The Rangeley Lakes-Dead River basin region, western Maine: *New England Intercoll. Geol. Conf. Guidebook*, 246 p.

——, Boudette, E. L., and Moench, R. H., 1970, Bedrock geology of the Rangeley Lakes-Dead River basin region, western Maine, *in* Boone, G. M., ed., The Rangeley Lakes-Dead River basin region, western Maine: *New England Intercoll. Geol. Conf. Guidebook*, p. 1–24.

Boucot, A. J., 1968, Silurian and Devonian of the Northern Appalachians, *in* Zen, E-an, and others, eds., *Studies of Appalachian geology—Northern and Maritime*: New York, Interscience, p. 83–94.

——, 1969, Silurian-Devonian of Northern Appalachians-Newfoundland, p. 477–483 *in* Kay, M., ed., *North Atlantic geology and continental drift—A symposium: Am. Assoc. Petrol. Geol. Mem. 12*, p. 477–483.

Bredehoeft, J. D., and Hanshaw, B. B., 1968, On the maintenance of anomalous fluid pressures, I. Thick sedimentary sequences: *Geol. Soc. Am. Bull.*, v. 79, p. 1097–1106.

Cady, W. M., 1968, Tectonic setting and mechanism of the Taconic slide: *Am. Jour. Sci.*, v. 266, p. 563–578.

——, 1969, Regional tectonic synthesis of northwestern New England and adjacent Quebec: *Geol. Soc. Am. Mem. 120*, 181 p.

Cloos, E., 1968, Experimental analysis of Gulf Coast fracture patterns: *Am. Assoc. Petrol. Geol. Bull.*, v. 52, p. 420–444.

Dewey, J. F., 1969a, Evolution of the Appalachian-Caledonian orogen: *Nature*, v. 222, p. 124–129.

——, 1969b, Structure and sequence in paratectonic British Caledonides, *in* Kay, M., ed., *North Atlantic geology and continental drift—A symposium: Am. Assoc. Petrol. Geol. Mem. 12*, p. 309–335.

Donath, F. A., and Parker, R. B., 1964, Folds and folding: *Geol. Soc. Am. Bull.*, v. 75, p. 45–62.

Guidotti, C. V., 1970, Metamorphic petrology, mineralogy and polymetamorphism in a portion of N.W. Maine, *in* Boone, G. M., ed., The Rangeley Lakes-Dead River basin area, western Maine: *New England Intercoll. Geol. Conf. Guidebook*, p. B-2, 1–29.

Hamblin, W. K., 1965, Origin of "reverse drag" on the downthrown side of normal faults: *Geol. Soc. Am. Bull.*, v. 76, p. 1145–1164.

Harwood, D. S., and Berry, W. B. N., 1967, Fossiliferous lower Paleozoic rocks in the Cupsuptic quadrangle, west-central Maine, *in Geological Survey research 1967*: U.S. Geol. Survey Prof. Paper 575-D, p. D16–D23.

Hussey, A. M., II, 1968, Stratigraphy and structure of southwestern Maine, *in* Zen, E-an, and others, *Studies of Appalachian geology—Northern and Maritime*: New York, Interscience, p. 291–301.

Kay, M., ed., 1969, *North Atlantic geology and continental drift—A symposium: Am. Assoc. Petrol. Geol. Mem. 12*, 1082 p.

Ludman, Allan, 1971, A fossil-based stratigraphy in the Merrimack synclinorium, central Maine: *Geol. Soc. Am. Abst. with Programs*, v. 3, no. 1, p. 43.

Maxwell, J. C., 1962, Origin of slaty and fracture cleavage in the Delaware Water Gap area, New Jersey and Pennsylvania, *in* Engel, A. E. J., James, H. L., and Leonard, B. F., eds., *Petrologic studies—A volume in honor of A. F. Buddington*: Geol. Soc. America, p. 281–311.

Moench, R. H., 1966, Relation of S_2 schistosity to metamorphosed clastic dikes, Rangeley-Phillips area, Maine: *Geol. Soc. Am. Bull.*, v. 77, p. 1449–1462.

——, 1970, Premetamorphic down-to-basin faulting, folding, and tectonic dewatering, Rangeley area, western Maine: *Geol. Soc. Am. Bull.*, v. 81, p. 1463–1496.

——, 1971, Geologic map of the Rangeley and Phillips quadrangles, Franklin and Oxford Counties, Maine: *U.S. Geol. Survey Misc. Geol. Inv. Map I-605*.

——, and Boudette, E. L., 1970, Stratigraphy of the northwest limb of the Merrimack synclinorium in the Kennebago Lake, Rangeley, and Phillips quadrangles, western Maine, *in* Boone, G. M., ed., The Rangeley Lakes-Dead River basin region; western Maine: *New England Intercoll. Geol. Conf. Guidebook*, p. A-1, 1–25.

Osberg, P. H., 1969, Lower Paleozoic stratigraphy and structural geology, Green Mountain–Sutton Mountain anticlinorium, Vermont and southern Quebec, *in* Kay, M., ed., *North Atlantic geology and continental drift*: Am. Assoc. Petrol. Geol. Mem. *12*, p. 687–700.

Osberg, P. H., Moench, R. H., and Warner, J., 1968, Stratigraphy of the Merrimack synclinorium in west-central Maine, *in* Zen, E-an, and others, *Studies of Appalachian geology—Northern and Maritime*: New York, Interscience, p. 241–253.

Pavlides, L., Boucot, A. J., and Skidmore, W. B., 1968, Stratigraphic evidence for the Taconic orogeny in the northern Appalachians, *in* Zen, E-an, and others, *Studies of Appalachian geology—Northern and Maritime*: New York, Interscience, p. 61–82.

Sales, J. K., 1971, The Taconic allochthon—Not a detachment gravity slide: *Geol. Soc. Am. Abst. with Programs*, v. 3, no. 1, p. 53.

Thompson, J. B., Jr., Robinson, P., Clifford, T. N., and Trask, N. J., Jr., 1968, Nappes and gneiss domes in west-central New England, *in* Zen, E-an, and others, *Studies of Appalachian geology—Northern and Maritime*: New York, Interscience, p. 203–218.

Zen, E-an, 1961, Stratigraphy and structure at the north end of the Taconic Range in west-central Vermont: *Geol. Soc. Am. Bull.*, v. 72, p. 292–338.

———, 1967, Time and space relationships of the Taconic allochthon and autochthon: *Geol. Soc. Am. Spec. Paper 97*, 107 p.

———, 1969, Petrographic evidence for polymetamorphism in the western part of the northern Appalachians and a possible regional chronology: *Geol. Soc. Am. Abs. with Programs for 1969*, pt. 7, p. 297–299.

Zen, E-an, White, W. S., Hadley, J. B., and Thompson, J. B., Jr., eds., 1968, *Studies of Appalachian geology—Northern and Maritime*: New York, Interscience, 475 p.

SAMUEL I. ROOT

Pennsylvania Geological Survey
Harrisburg, Pennsylvania[1]

Structure, Basin Development, and Tectogenesis in the Pennsylvania Portion of the Folded Appalachians[2]

Recent extensive geologic mapping by the Pennsylvania Geological Survey and U.S. Geological Survey as well as geophysical studies and deep petroleum drilling in the Valley and Ridge and Appalachian Plateaus Province adduce on tectogenesis and basin development in the central and western part of Pennsylvania. This western part of the Appalachian geosyncline is regarded as a continuously subsiding basin throughout the Paleozoic, and deformed, principally by folding above décollement surfaces, during a single terminal Paleozoic orogeny. The significant conclusion of this paper is that during deformation, movement of the detached cover was up the basement slope—similar to décollement movements in analogous terrains such as the Jura Mountains and Southern Canadian Rockies.

Pennsylvania lies wholly within the Central Appalachian System (King, 1951). Within the system several physiographic provinces are recognized which correspond in considerable measure to distinct structural provinces. From southeast to northwest (Fig. 1) they are: (1) the Piedmont Province, which is characterized by Precambrian to Ordovician sedimentary, metamorphic, and plutonic

---

[1] Published by permission of the Director, Pennsylvania Geological Survey.
[2] I am indebted to R. A. Price and R. Scholten for many of the ideas, which were developed in the field and during discussions. Colleagues on the Pennsylvania Geological Survey—D. M. Hoskins, D. M. Lapham, D. B. MacLachlan, and A. A. Socolow—offered encouragement and many helpful ideas to improve this manuscript. W. R. Wagner kindly furnished data on thickness of the various rock units. A particular debt to D. B. MacLachlan is acknowledged for the many years he has counseled me in problems of Pennsylvania geology.

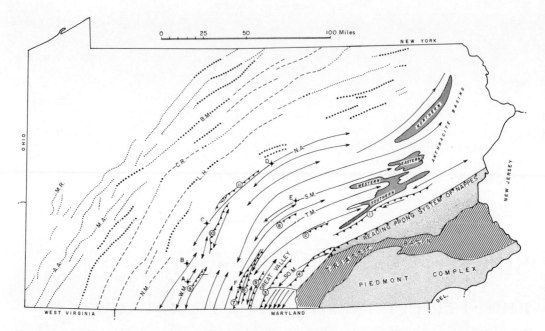

FIG. 1 Generalized structure map of Pennsylvania showing major anticlines, faults, and deep well locations in the Valley and Ridge-Appalachian Plateaus regions. Abbreviation for anticlines: (SOM) South Mountain; (TM) Tuscarora Mountain; (SM) Shade Mountain; (WM) Wills Mountain; (NA) Nittany anticlinorium; (NM) Negro Mountain; (LH) Laurel Hill; (CR) Chestnut Ridge; (BM) Boone Mountain; (MA) Murraysville anticline; (AA) Amity anticline; (MR) Mercer region. Anticlines shown by continuous line are in the Valley and Ridge. On the Appalachian Plateaus (after Gwinn, 1964) dashed lines indicate anticlines of greater than 800 ft structural relief, large dots indicate anticlines of 800-300 ft structural relief, small dots indicate anticlines of less than 300 ft relief. Major faults are numbered as follows: (1) Sweet Arrow; (2) Perry County; (3) Carbaugh-Marsh Creek; (4) Readings Bank-Piney Mountain; (5) North Mountain; (6) Path Valley; (7) Little Scrub Ridge; (8) Limestone Ridge; (9) Friends Cover; (10) Henrietta-Yellow Springs-Grazier Mill; (11) Birmingham. Deep wells located by letters: (A) Mary Martin No. 1; (B) S. Miller No. 1; (C) Rankey No. 1; (D) Mobil No. 1 Long; (E) Shell No. 1 Lost Creek; (F) No. 1 T. E. Nesbitt.

rocks complexly folded and faulted; (2) the Triassic Lowlands, composed of homoclinal northwest-dipping red beds with diabase dikes and sills; (3) the Reading Prong, which is structurally related to the Piedmont; (4) the Valley and Ridge Province, a folded and faulted terrain of Paleozoic sedimentary rocks and including, for purposes of this paper, the structurally related Precambrian crystalline-cored Blue Ridge Province (South Mountain anticlinorium), which forms a narrow belt adjacent to the western margin of the Triassic section; and (5) the Appalachian Plateaus, which is a terrain of low-amplitude folds that diminish to the west. The Valley and Ridge–Appalachian Plateaus constitute the folded Appalachians of this paper.

The Reading Prong and Piedmont terrains, which represent the eastern portion of the geosyncline, are considerably complicated by Alpine-type structure, polyphase deformation, metamorphism, and plutonism, so that the development of this portion of the Paleozoic Appalachian geosyncline is still somewhat enigmatic.

Structural Geology

The Valley and Ridge–Appalachian Plateaus terrain is some 125–150 miles wide across tectonic strike. This dimension approximates the width of the southern Canadian Rocky Mountains (Price and Mountjoy, 1970, p. 11), an area with many elements of structural similarity to this region. The principal difference between the Central Appalachians in Pennsylvania and the southern Canadian Rockies or, indeed, the Appalachians to the north and south is the paucity of faulting at the surface and the greater amount of folding in the Central Appalachians (Fig. 2). The sparsity of both steep and low-angle thrusts in the Central Appalachians is considered to imply less cover shortening than in the Southern Appalachians. This is based on the assumption that the Southern Appalachians (Fig. 2) are nearly like the Southern Canadian Rockies. The Rockies have been shortened 125 miles (200 km) (Price and Mountjoy, 1970), whereas in this paper only 50 miles (80 km) of shortening is proposed for the folded Appalachians.

It has been shown by Gwinn (1964, 1970) that the folds involve allochthonous Paleozoic rocks that have been transported northwestward along southeastward-inclined, nonoutcropping low-angle detachment (décollement) faults. According to Gwinn (1964), folds have formed in passive response to steplike upward shearing of detachment thrusts from one of the several glide zones present in the Paleozoic strata. Evidence for extending these detachment faults southeastward, even under the Precambrian rocks of the Blue Ridge (Fig. 3), has been presented by Root (1970).

Folds

Examination of the structure map (Fig. 1) indicates a fairly uniform wavelength of folds across this terrain. Nickelsen (1963, p. 16) recognized various orders of folds in the western part of the Valley and Ridge, but only the first-order folds merit consideration here. These are the largest folds and have a wave length of 7–11 miles (11–18 km). This spacing is consistent across the entire folded Appalachians, since Nickelsen observed that folds in the Appalachian Plateaus also occur in this category. The spacing is probably a reflection of the thickness of the dominant member of these folds (Currie and others, 1962, p. 669). In the eastern part of the area, on the South Mountain anticlinorium (Fig. 3), wavelengths of folds appear to be less because of digitation and faulting on the first-order fold as a consequence of greater shortening.

The lengths of the individual anticlinal folds vary considerably. Some of the folds are only about 35 miles (55 km) long, but most are considerably longer. The Wills Mountain anticline (Fig. 1) is a single structure extending from Pennsylvania southwestward well into Virginia, a distance of nearly 200 miles (320 km). Other anticlines have almost comparable lengths when viewed as *en echelon* compound structures.

The fold geometry of the Valley and Ridge does not conform to the classic concept of a concentric fold. As pointed out by Nickelsen (1963, p. 19), the folds have nearly planar limbs and curved hinges. It was also observed that at certain stratigraphic levels thickened hinges occur, resulting in a similar fold form. This was elaborated upon by Faill (1969), who noted that many folds change radically in profile along trend. Because of the fold geometry and flexural slip mechanism on bedding surfaces Faill interprets these structures as large kink bands. This observation is significant since kink band folds may form without the response to upward shearing that Gwinn (1964) postulates. Thus not all of the folds must have faults in their core (Fig. 4); however, as discussed subsequently, a considerable number of folds do contain faults in their core.

In the eastern part of the Valley and Ridge (South Mountain anticlinorium, Fig. 1), it is difficult to interpret fold geometry and mechanics. In this terrain of more intense deformation Cloos (1947) demonstrated a pervasive cleavage in which there was extension of oolites in carbonate rocks on the limb and extension of chlorite blebs and amygdules in the metavolcanics in the core. He interpreted the South Mountain anticlinorium as a large

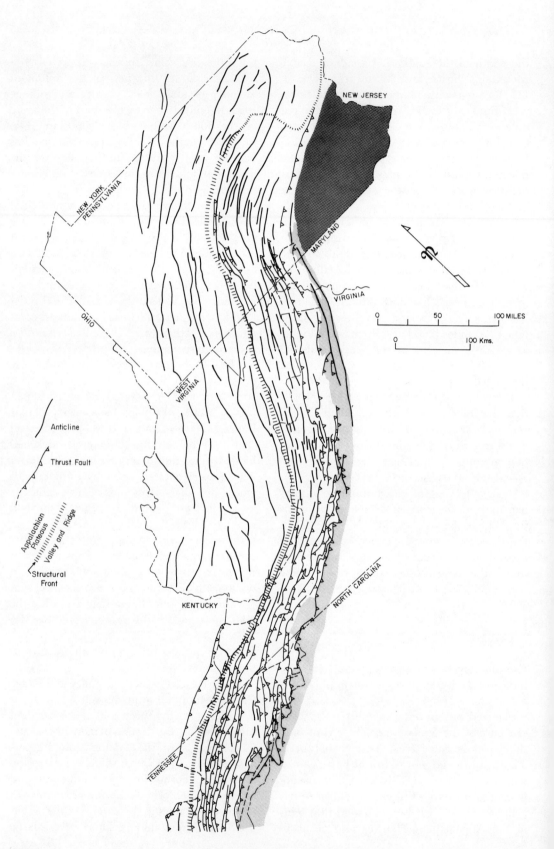

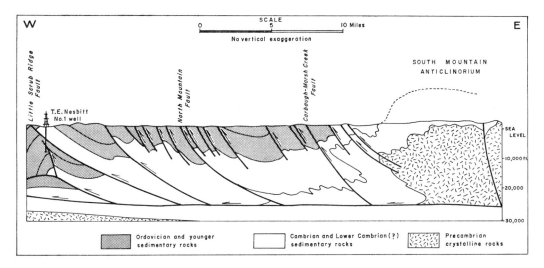

FIG. 3 Generalized cross section across the Great Valley. Approximate location extends east from the Little Scrub Ridge fault to the South Mountain anticlinorium (see Fig. 1).

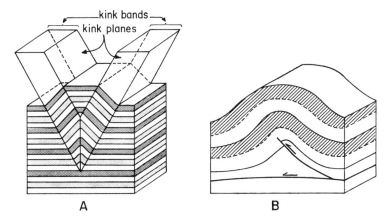

FIG. 4 Difference between kink band folds and folds formed above a décollement. (A) A fold in horizontal beds resulting from the juncture of two oppositely inclined kink planes; the kink bands are of equal width, and inclination but of opposite rotational sense (after Faill and Wells, in press). (B) A concentric fold developed above an upward shear from a décollement surface. Both types of folds developed in the folded Appalachians.

FIG. 2 Tectonic map of Central and Southern Appalachians. Note difference in tectonic style between the dominantly folded terrain of the Central Appalachians (Pennsylvania, West Virginia, and northern Virginia) and the dominantly thrust-faulted terrain of the Southern Appalachians. The structural front in northeastern Pennsylvania is shown by smaller hachures to denote a less distinct development than to the south. The light stippled pattern indicates the crystalline-cored Blue Ridge and the darker pattern indicates the Piedmont terrain and Triassic basins. (Modified from the 1962 U.S. Geol. Survey, Tectonic Map of the United States.)

"shear" fold wherein deformation is due to laminar flow on subparallel planes. Root (1968, 1970) concluded that the folds in much of the carbonate terrane are cylindrical flexural-slip folds and only near the core of this anticlinorium are flexure-flow and passive-slip folds present.

In the Plateaus region flexures are so broad and gentle that they are best analyzed on structural contour maps. These maps clearly indicate a geometry that is part of the Valley and Ridge fold system. In the eastern part of the area structure contours define some folds with broad hinges and planar limbs. These folds then may also approximate a kink band geometry. On the Negro Mountain anticline (Fig. 1) dips are about 3° on the crest and about 5–8° on the limbs. Farther west, dips on the limbs of the Laurel Hill anticline are up to 8°. On the Chestnut Ridge anticline maximum dips are 4° on the west limb and as much as 8° on the east limb. Where faulting is present, as on the Boone Mountain anticline, dips of 10–20° are locally attained. Still farther west dips decrease progressively so that, for example, limbs on the Murraysville anticline dip at 100–200 ft/mile (20–40 m/km) and on the Amity anticline they dip at 50 ft/mile (10 m/km). In the Mercer region, the most westerly area, structures are vague and limbs dip at only 20–40 ft/mile (4–8 m/km).

An idea of the progressive decrease of shortening of the cover from east to west is indicated by the progressive diminution of fold appression as shown by interlimb angles (Fig. 5). On the east side of the Great Valley (South Mountain anticlinorium) the interlimb angle is 50–70° with folds commonly overturned. On the west side of the Great Valley the interlimb angle is about 80°. In the central Valley and Ridge (Shade Mountain anticline) the interlimb angle is about 100° (Faill and Wells, in press); folds are not overturned but are commonly asymmetric. In the Appalachian Plateaus the interlimb angle is nearly at maximum (=180°) and folds, especially the smaller folds, commonly are symmetrical.

The progressive decrease in cover shortening is also reflected in structural relief of the folds, which diminish from possibly 35,000 ft (10,500 m) on the South Mountain anticlinorium in the southeast (Gwinn, 1964) to

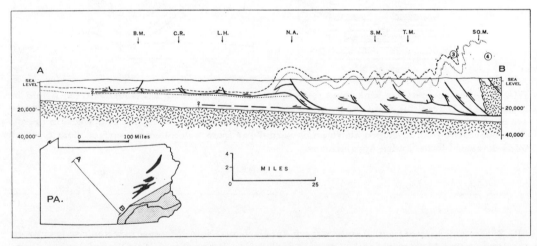

FIG. 5 Highly generalized cross section across the folded Appalachians. Dashed pattern indicates Precambrian rocks, dotted line indicates base of the Silurian, dashed line indicates base of the Devonian. Heavy solid line indicates system of faults. Line of section shown on inset map. Abbreviations referred to in Fig. 2.

7000–9000 ft (2100–2700 m) on folds in the central part of the Valley and Ridge to about 4500 ft (1400 m) on a typical fold on the Nittany Mountain anticlinorium in the western part of the Valley and Ridge (Fig. 5). Structural relief on the west side of the Nittany anticlinorium is considerably greater— 17,000–20,000 ft (5100–6000 m). This large figure is based on relief on Ordovician carbonate rocks which are repeated three times by faulting in the core of the fold (Moebs and Hoy, 1959; Gwinn, 1970, Fig. 4). Many of the major folds have thrust faults in the core so that these figures are truly comparable.

There is an abrupt decrease in structural relief to the west of the Nittany anticlinorium in the Appalachian Plateaus. This abrupt transition in structural magnitude has been termed the Appalachian structural front (Fig. 2) by Price (1931), who attributed it to an inherited line of weakness related to a Paleozoic depositonal trough. Gwinn (1964) postulated that the structural front marks the zone along which sole thrusts shear upward from a Cambrian to an Ordovician glide zone. These ideas have yet to be substantiated. For purposes of this paper the significant feature is that folds are continuous across the structural front; they share a common fold limb which is substantially unfaulted.

In the Plateaus area some of the anticlines close to the structural front, such as Chestnut Ridge and Laurel Hill, have as much as 2500 ft (750 m) of structural relief and Negro Mountain has as much as 3000 ft (900 m) of relief. Relief on anticlines diminishes progressively westward (Fig. 3), and the most westerly folds have relief of less than 300 ft (100 m).

Faults

Scrutiny of the structure map (Fig. 1) demonstrates that at present levels of exposure, this region is dominantly a fold terrane and that faulting is subordinate. This is in marked contrast to the Southern Appalachian region (Fig. 2), which is dominantly a fault terrane (e.g., Keith, 1923; Price, 1931; King, 1951; Gwinn, 1964). Keith (1923) ascribed the lack of faulting and dominance of folding in the Central Appalachians to the fact that quartzites, dolomites, and limestones comprise the lower part of the Paleozoic column and are overlain by a great thickness of shale and sandstone. These general proportions of the stratigraphic column are reversed in the Southern Appalachians, where there is an enormous development of thrust faulting and comparatively little folding. Implicit in this reasoning is a competency variation in the rocks. Gwinn (1964), on the other hand, concluded that the tectonic style of the Central Appalachians is entirely analogous to that of the Southern Appalachians but that erosion has progressed far enough to expose the forward toes of the thrust faults of the Southern Appalachians, whereas the toes of the Central Appalachian low-angle thrust sheets are still covered by an unfaulted blanket of younger sediments draped across the toes. Considering that many low-angle faults occur in lower Paleozoic rocks close to the Blue Ridge in the Southern Appalachians, whereas only a few low-angle faults occur in similar lower Paleozoic rocks close to the Blue Ridge in the Central Appalachians, Gwinn's (1964) assessment may be too simplistic.

All major faults exposed in the Valley and Ridge and Appalachian Plateaus of Pennsylvania, except those in the Anthracite region, are shown on Fig. 1. There are about a dozen such faults and they are typically moderate to steep thrusts with dips ranging from 40 to 70° southeast. They range from a length of 7 miles (11 km) and a stratigraphic displacement of only a few hundred feet for the Limestone Ridge fault (Conlin and Hoskins, 1962) to a length of 200 miles (320 km) for the North Mountain fault and an actual displacement possibly as great as 5 miles (8 km) for the Little Scrub Ridge fault. The faults are intimately related to the development of anticlines with fault and fold surface traces parallel. It has been proposed

(Gwinn, 1964, 1970; Moebs and Hoy, 1959; Root, 1970) that the steep thrusts are splays off major subsurface décollements. Folding and faulting appear to proceed concomitantly as part of the general shortening of this terrain.

The North Mountain fault is singularly significant. It furnishes direct evidence that some faults of the Southern Appalachians extend into the Central Appalachians and that structural mechanics are similar. This fault extends from near Natural Bridge, Virginia, to northeast of Chambersburg, Pennsylvania, where its definition is lost in an Ordovician shale terrane (Clark, 1970). The Carbaugh–Marsh Creek fault (Root, 1970, 1971) is of interest in that it swings normal to the regional trend and becomes a strike-slip fault at its northern end where displacement amounts to $2\frac{1}{2}$ miles (4 km).

The Birmingham fault (Moebs and Hoy, 1959; Gwinn, 1964, 1970) is of extreme importance because it is the only fault in Pennsylvania associated with a major, well-exposed décollement. It has a length of about 33 miles (53 km) and has been shown by drilling to be an east-dipping, steep thrust. This fault, which is developed in the core of the Nittany anticlinorium, is a minor fault splay off a major décollement termed the Sinking Valley fault. Erosion has developed a *fenster* known as the Birmingham window through the décollement. Moebs and Hoy (1959) calculated displacement along this décollement surface to be about 2 miles and displacement on the secondary Birmingham fault is considerably less.

A few minor faults have been mapped in the eastern part of the Plateaus area, and recent mapping by Edmunds (1968) and Glass (in press) reveal a considerable number of wrench faults, normal to regional structure, developed at the structural front. Considering the Plateaus area in a regional sense the amount of faulting and displacement by faulting is deemed to be insignificant (Fig. 5).

Numerous thrusts, reverse faults, low-angle folded thrusts, and tear faults have been mapped in the Anthracite basins in recent years (Wood and others, 1969; Wood and Bergin, 1970). The precise nature and magnitude of some of these low-angle folded thrust faults is still to be resolved; Hoskins (personal communication, 1971) has been unable to map their extension in the adjoining quadrangles on the west.

Subsurface studies in the Valley and Ridge reveal southeast-dipping steep thrusts related to faults mapped at the surface. Nearly all wells drilled to considerable depths pass through one or more steep thrusts into younger beds. On the west side of the Valley and Ridge the No. 1 Mary Martin, No. 1 S. Miller, and No. 1 Rankey wells (A, B, and C in Fig. 1) admirably illustrate this relation. The No. 1 Ray Sponaugle well drilled on the continuation of the same geologic feature (Wills Mountain anticline) as the No. 1 Mary Martin well, but about 110 miles (180 km) to the south, penetrated a thrust that dips 25° southeast and has a displacement of $1\frac{1}{2}$ to $2\frac{1}{4}$ miles ($2\frac{1}{2}$ to $3\frac{1}{2}$ km) (Perry, 1963).

Other wells that encountered thrusts in the cores of anticlines are in the central part of the Valley and Ridge. The Mobil No. 1 Long and the Shell No. 1 Lost Creek wells (D and E in Fig. 1) were drilled to depths of 15,662 ft and 10,036 ft, respectively; both passed through thrusts that repeated significant amounts of section (Gwinn, 1970). The No. 1 T. E. Nesbitt well (F in Fig. 1), in the eastern part of the Valley and Ridge commenced in Cambrian carbonates, but at 5500 ft it passed through a thrust into middle Ordovician shales. This thrust is the Little Scrub Ridge fault (Figs. 1 and 3), and this is the only case in which the fault drilled at depth extends to the surface. A dip of 40° southeast can be computed (Root, 1970) and a displacement of about 5 miles (8 km) is postulated.

In the Plateaus area, Gwinn (1964) has demonstrated that in the subsurface many of the anticlines are considerably faulted. Faulting was accomplished along imbricate thrusts that splayed from a décollement. In some instances, faults are symmetrically disposed along the flanks of the folds as a consequence of the space problem in the core of a concentric fold. The faults in the Plateaus area generally have a displacement of less than 1000 ft (350 m) and usually only of several hundred feet (Gwinn, 1964). Displacement on the faults diminishes to the west in harmony with decreasing structural relief. On a regional basis the total amount of movement on these faults is negligible.

Age of Deformation

Consideration of all recent data including isotopic analyses has led Rodgers (1967) and Root (1970) to conclude that major deformation in the folded Appalachians occurred as a terminal Paleozoic (Alleghanian) event. The immense Upper Devonian clastic wedge and succeeding coarser-grained Mississippian and Pennsylvanian clastic wedges reflect varying intensities of earlier orogenic activity in the Piedmont area. If the 340 million year date on the mylonite from a possible décollement in Ordovician shales (Pierce and Armstrong, 1966) is valid and significant, then some local tectonic activity may have occurred during latest Devonian. A major terminal Paleozoic orogeny is concluded to be the only obvious structural event because (1) beds from Precambrian to Pennsylvanian, and locally even Permian, are folded together and (2) regionally, the sedimentary rocks are essentially conformable from Precambrian to the Pennsylvanian/Permian except for a feather edge of the Ordovician (Taconic) unconformity.

It is not to be conceived that the orogeny was an almost instantaneous climactic event. For example, several deformation phases are recognized (Root, 1970) in the development of the South Mountain anticlinorium. This fold is then truncated by later thrust sheets, containing Piedmont rocks, which were emplaced subsequent to the folding and probably even subsequent to substantial erosion of the fold (MacLachlan, 1967). It may also be that the major folds and thrusts of the Valley and Ridge-Plateaus region did not develop simultaneously but rather developed sequentially due to progressive displacement of the type described by Price (1969) in the Canadian Rockies. The synorogenic upper Paleozoic sedimentary rocks reflect older orogenic events in the Piedmont, and what is observed in the Valley and Ridge-Plateaus area may be the youngest phases of a progressively westward-migrating orogeny—a concept mentioned early by Barrell (1914, p. 253).

Shortening of the Cover

Consideration of structural geology in the preceding sections indicates that the carpet of Paleozoic sediments has maintained structural integrity during deformation. There are no major steep thrusts that displace cover segments many miles; none of the steep thrusts has a displacement of more than several miles. Although folds and steep thrusts develop together, the folds appear to accommodate a somewhat greater part of the cover shortening. This system of coherent folds and faults is derived from major subsurface décollements along which an aggregate of 50 miles of displacement is postulated. These décollements do not shear across the sedimentary blanket but rather they gradually die out in bedding (Fig. 5) as displacement progressively decreases. Structural relief on the folds and displacement on faults progressively decrease northwestward as the décollements die out without emerging.

It is important to emphasize the regional structural integrity of the sedimentary carpet with its coherent fold and fault system. Calculations of cover shortening are greatly simplified in this type of terrane as contrasted with the Southern Appalachians where large thrust sheets must first be restored many miles to their original position prior to correction for folding and steep thrusting.

Many estimates of cover shortening have been made for the Valley and Ridge during the years. Chamberlin (1910) reviewed the estimates of shortening up to 1910 and calculated that the amount of shortening on a line across the middle of the Valley and Ridge amounted to 19 percent. Cloos (1940) calculated shortening in the Anthracite basins and demonstrated that the amount of shortening increased from an average of 3–5 percent in the Northern field to an average of 20 percent in the Eastern Middle field, 25 percent in the Western Middle field, and 35 percent in the Southern field. These calculations are derived from sinuous bed computations.

In recent years the equal area method has been used in two studies to estimate the amount of cover shortening in this region. By this method Dennison and Woodward (1963) calculated 30 miles (48 km) or 22 percent of shortening across the Valley and Ridge along a line from southern Pennsylvania to northeastern West Virginia. Gwinn (1970), on the basis of most recent data, calculated cover shortening across the southern part of the Valley and Ridge and concluded that shortening amounted to 47 percent or 50 miles (80 km). A more conservative calculation by Gwinn, based on the assumption that the rock strata are somewhat thicker than the values used in his initial computation, obtains a shortening of 29 percent, or 26 miles (41 km).

In the Plateaus area there is an abrupt decrease in the amount of cover shortening, so that it becomes insignificant. Calculations made on some of the larger folds by the sinuous bed method indicate a shortening of about 1 percent. Since dips of less than 8° correspond to shortening of less than 1 percent, it is obvious that most of the Plateaus area has been shortened less than 1 percent. This calculation, however, does not consider the many small faults in this region. Allowing for these faults, a reasonable estimate of cover shortening here is 2 percent or about $1\frac{1}{2}$ miles (2.5 km).

Assuming that Gwinn's (1970) estimate of 47 percent shortening in the Valley and Ridge and the estimate of 2 percent shortening in the Plateaus are valid, then the sedimentary rocks on the western flank of the South Mountain anticlinorium have been translated about 50 miles (80 km) northwestward from an original depositional site near present-day Baltimore. The amount of northwestward translation from the original site of sedimentary deposition decreases progressively, but not uniformly, away from South Mountain until in the Plateaus area the strata are at most only about 1 mile from their original site of deposition. The total shortening is not uniform along tectonic strike of this terrane, since translation is less across the Anthracite Region because of considerably diminished width of the Valley and Ridge belt (Fig. 1). Translation is maximal on the south and in the central parts, decreasing to the northeast along tectonic strike.

Basin Analysis

Studies at the systemic level of the amount of sedimentary infilling of the western part of the Appalachian geosyncline reveal certain basinal conditions of marked uniformity and persistence throughout the Paleozoic.

Sedimentary Thickness

The amount of sediment accumulated during each of the Paleozoic systems is shown in Fig. 6. Despite the complexity of sedimentary infill and change of major source areas there is a remarkable uniformity to the thickness patterns. At the systemic level all strata thicken from west to east. The individual thickness gradients across the area are relatively uniform in all systems except the Silurian. Maximum observed thickness for all systems is at the eastern limit of the study area; however, it is not known if this represents actual maximum thickness or if some units continue to thicken into the Piedmont region where the stratigraphy is not yet decipherable. The position of maximum accumulation at times was a site for open-marine

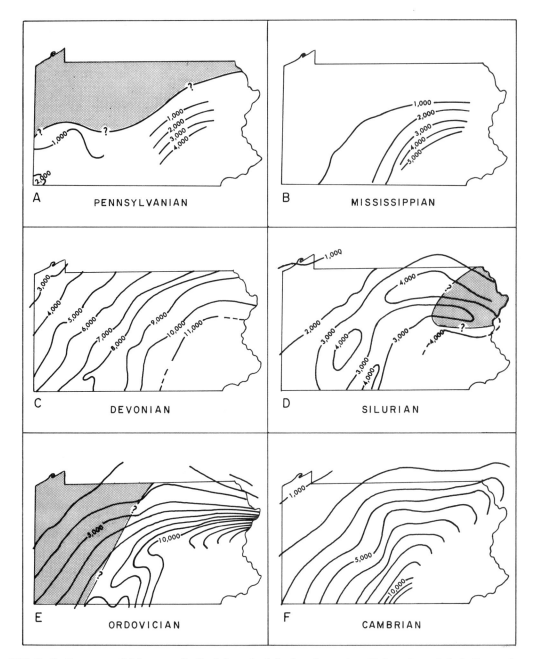

FIG. 6 Sedimentary thicknesses (in feet) deposited during the various Paleozoic periods in Pennsylvania. Pennsylvanian and Mississippian data modified from Colton (1970), Devonian data modified from Oliver and others (1967). Silurian, Ordovician, and Cambrian thicknesses compiled by W. R. Wagner and the author. Light stippled pattern shows approximate extent of recognized unconformities at base middle Ordovician (E), at base Silurian (D), and at base Pennsylvanian (A).

deposition and at other times a site of marginal-marine or fluvial deposition. Independent of the nature of the sediment accumulating or depth of water, maximum deposition occurred continuously in this area, indicating that this part of the geosyncline was subsiding during the entire Paleozoic as a consequence of localized primary crustal downwarping or exogeosynclinal development (Kay, 1951, p. 19). Eloquent testimony to this is the accumulation of nearly 48,000 ft (14,500 m) of virtually uninterrupted miogeosynclinal-shelf deposits (Figs. 6 and 7).

It should be noted that Figs. 6 and 7 are not corrected for postdepositional cover shortening. Corrections for shortening result in less closely spaced contours in the eastern part of the diagrams and the regional gradient is consequently more uniform. In Fig. 7, for example, where cumulative thickness attains a maximum value of 48,000 ft at the site of the present South Mountain anticlinorium, the true depositional thickness at this site, prior to shortening, was probably 32,000–36,000 ft (9,500–11,000 m). The 48,000 ft of strata were originally deposited in what is now the Baltimore area and have been shifted to their present position during Alleghanian shortening. For the general concepts developed in this paper shortening corrections need not be applied.

Tectonic Synthesis

Throughout the Paleozoic an essentially conformable sedimentary sequence accumulated; this thickens regularly southeastward across the folded belt, to a known maximum thickness of nearly 48,000 ft (14,500 m) at the limit of the region. The sedimentary cover was shortened above a relatively rigid basement during a single terminal Paleozoic (Alleghanian) orogeny. Shortening was accomplished by folding and faulting above décollements. Structural integrity and coherence were maintained and maximum shortening was about 50 miles (80 km).

Basement Slope

It is difficult to map the basement in this terrane. Especially in the east, where structures are most complex and there has been little deep drilling, the depth to the basement is inferred from some gravity and magnetic studies. For purposes of this paper basement is considered as relatively rigid Precambrian rocks below the Paleozoic cover. Recognizing that on a geophysical basis it is difficult to separate basement from some of the overlying dense sedimentary rocks, Kelley and others (1970) compiled a map (Fig. 8) on the present basement surface.

Since the basement now dips to the southeast (Figs. 3 and 8) and appears to have dipped southeast during Paleozoic deposition by about the same amount (Fig. 7), the central problem is whether the basement and the décollement faults that parallel it maintained a southeastward slope during Alleghanian deformation. It has been suggested that overthrust terranes may develop as a consequence of gravitational tectogenesis such as epidermal gliding (Van Bemmelen, 1954) or by pushing of wide thrust plates down a gentle slope in which high fluid pressures are involved (Rubey and Hubbert, 1959). However, as Price (1969) observes, the critical data of regional geology indicate that most low-angle thrust faults have always sloped in the wrong direction for gravitational gliding. Thus the question arises what the slope of the Appalachian basement was during Alleghanian deformation.

Basement Slope during Alleghanian Deformation

Among the factors to consider in determining basement slope during deformation is the maximum amount of uplift at the eastern edge of the detached cover and possible areas of tectonic denudation. If a subhorizontal component of resolved gravitational stress, induced solely by the mass of cover moving downslope, is concluded to be the driving

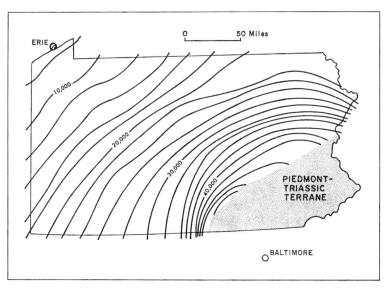

FIG. 7 Total accumulation of Paleozoic sediments on western flank of the Appalachian geosyncline. Thickness expressed in feet and not corrected for cover shortening.

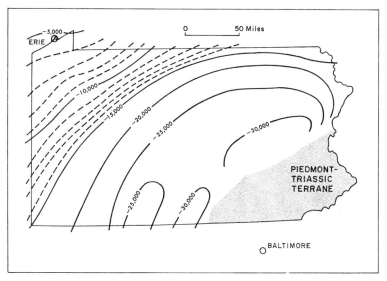

FIG. 8 Contours on surface of Precambrian igneous and metamorphic complex (from Kelley and others, 1970). Depth figures expressed in feet below sea level.

mechanism of Appalachian shortening (Gwinn, 1964, p. 897), the data of Rubey and Hubbert (1959) indicate that a slope of about 2° is required across the area. Assuming the effective width of the folded belt as about 60 miles (95 km), this implies that, at the time of sliding, basement was about 10,000 ft higher in the present area of the South Mountain anticlinorium than at the Appalachian structural front. The Paleozoic depositional record indicates that basement was initially 20,000 ft lower in the South Mountain area.

A minimum basement uplift of 30,000 ft must be generated at this locus to initiate gravity sliding. However, this is only part of the problem as the rocks in the South Mountain anticlinorium are deduced to have been translated about 50 miles (80 km) northwestward and there is no basis for presuming that they lay close to the detachment heel; the simple slide hypothesis requires that the southeastward basement rise continued for at least another 50 miles, and even then the crest of the uplifted system would not have been reached. Total basement uplift here is then several tens of thousands of feet higher than the 30,000 ft of uplift postulated for the area of the South Mountain anticlinorium. Uplift of this magnitude seems unreasonable in the nonmagmatic, essentially nonmetamorphic Valley and Ridge terrane.

It is true that the Piedmont, to the east, is in effect an area of considerable uplift. The area about Baltimore, which in the simple slide hypothesis is deduced to be the depositional site of the sedimentary rocks in South Mountain, is composed of intensively metamorphosed rocks with mantled gneiss domes and serpentinites, indicating that this is an uplifted terrain of orogenic basement rocks. However, many of the rocks are of the same age as the folded rocks of the Valley and Ridge and therefore this area cannot be a tectonically denuded uplift from which the Valley and Ridge rocks could have been derived by simple downslope gravity sliding. In fact, there is no known area in the Piedmont that can be regarded as a tectonically denuded uplift linked to the Paleozoic cover of the Valley and Ridge.

There is in reality little to indicate a sudden, temporary, large uplift in the southeast except the conceptual necessity to generate a slope for gravity sliding. However, it should not be construed that basement has been static; it has certainly moved up and down several thousands of feet during and after deposition of the Triassic (Faill, 1969) and in post-Triassic time (Owens, 1970).

I propose that basement in the folded Appalachians did not slope to the northwest during deformation but continued to slope to the southeast as it did prior and subsequent to deformation. It then remains to consider the possible processes involved in shortening of the folded Appalachians in light of these constraints. For this it is useful to consider analogous situations.

Geological and Experimental Analogues

Recent studies, based on considerable evidence from the southern Canadian Rockies, indicate that structures in these terranes were not developed as a consequence of strata gliding down an inclined fault surface (Price, 1969; Price and Mountjoy, 1970). Indeed in this area, which may be considered as an analog of the Central Appalachians, the sedimentary cover is believed to have moved up the basement slope and associated décollement fault slopes during deformation.

Price and Mountjoy describe a belt of detachment of supracrustal rocks adjacent to profound-orogenic metamorphic belts involving a mobilized basement. They conclude that lateral spreading, resulting from gravitational collapse of uplift in the mobilized basement zone, is the essential driving mechanism for displacement of the marginal cover rocks. The marginal cover is driven up the slope of the basement surface as the entire mass spreads.

Some experimental studies tend to confirm these concepts. Bucher (1956) simulated the role of gravity in laboratory experiments with stitching wax and showed that it flows and flattens under its own weight when raised to relatively great heights, or when slowly deformed with one part kept slightly warmer. In a series of centrifuged models of silicone and putty, Ramberg (1967, p. 108, Fig. 57) generated detachment folds that moved upslope from a laterally spreading central dome. On consideration of both the experimental and the rheological behavior of rocks, Carey (1962, p. 140) concluded that Appalachian décollements are simple upward and outward flow lines from a rising axial zone.

Evolution of the Folded Appalachians

The energy necessary to displace the sedimentary cover up the basement slope cannot have been contained within the sedimentary cover itself. It must be derived from the adjacent Piedmont terrane to the east—an area of repeated orogenic basement deformation and uplift. However, the juncture of the folded Appalachians and Piedmont terranes is largely obscured by younger Triassic sediments, so that interpretations concerning their relation are tenuous. There is clear evidence of multiple deformation in the Piedmont reflected in the development of at least three cleavages or S-surfaces (Freedman and others, 1964) and at least two periods of early and middle Paleozoic radiometrically dated thermal and/or tectonic events (Lapham and Root, 1971), whereas there only is a single later deformation apparent in the folded Appalachians. The possible relations between tectonic events occurring both with and between the two terrains are shown in Fig. 9, based in part on the work of Lapham (unpublished data, Pa. Geol. Survey), and are as follows:

Stage A. From Cambrian to middle Ordovician miogeosynclinal sedimenta-

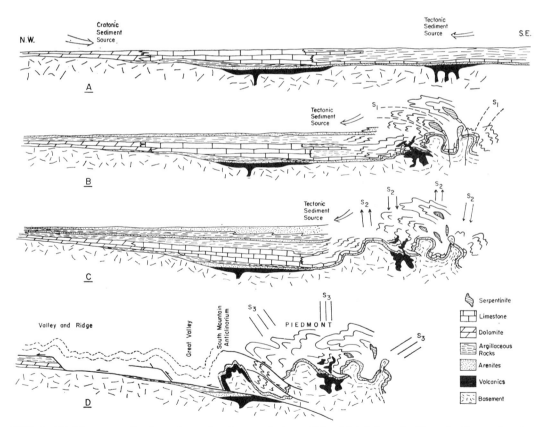

FIG. 9 Highly generalized sections showing evolution of the folded Appalachians in Pennsylvania. Sections have extreme vertical exaggeration and represent an initial width of a couple of hundred miles. (A) Development during middle Ordovician time. (B) Development at end of the Ordovician or "Taconic" time. (C) Development near end of the Devonian or "Acadian" time. (D) Development at end of Paleozoic or Alleghanian time.

tion occurs in the west part of the Appalachian geosyncline (folded Appalachians) and mio-eugeosynclinal sedimentation occurs to the east (Piedmont).

Stage B. Piedmont "proto-Taconic" and "Taconic" radiometric dates, terminating about 425 million years ago, reflect minor igneous intrusion with regional metamorphism, generation of nappes, large-scale thrusting away from the Piedmont core, development and/or mobilization of gneiss domes; development of S_1 cleavage, probably at this time, represents subhorizontal northwest transport. In what are now the folded Appalachians there is no evidence of related tectonism other than extrusive metabentonites and some local unconformities (Fig. 6), but the major sediment source lay to the east after early middle Ordovician time.

Stage C. Piedmont "Acadian" radiometric dates of 320–350 million years probably reflect regional uplift and erosion with only minor mineral recrystallization; the subvertical S_2 cleavage possibly developed at this time. Clastic wedges, in what are now the folded Appalachians, are the reflection of Piedmont tectonism.

Stage D. Sparse Alleghanian radiometric dates of 240–260 million years outside the Piedmont probably record cover shortening in the folded Appalachians and local development of cleavage in the South Mountain anticlinorium. There are no radiometric dates in the Pennsylvania Piedmont reflecting this tectonism, but a weak S_3 cleavage may indicate uplift at this time. Local thrusting of Cambro–Ordovician Piedmont rocks over the South Mountain anticlinorium (MacLachlan, 1967) is direct evidence of Piedmont participation in Alleghanian tectonism.

The absence of late Paleozoic radiometric dates and development of only weak S_3 cleavage indicates that during Alleghanian deformation the Piedmont was a relatively rigid terrane. Since there is no gap between the two terranes and locally the Piedmont is thrust over the folded Appalachians, it follows that despite its apparently rigid character the Piedmont also participated in northwestward-directed shortening of the folded Appalachians. This implies a model in which both uplift and northwestward movement of the relatively rigid Piedmont, during the Alleghanian deformation, caused the local sedimentary cover to rotate northwestward. By a combination of directed stress from the moving Piedmont block and lateral spreading within the sedimentary cover due to an unstable gravity distribution in the uplifted area the Paleozoic cover moved upslope on detachment surfaces, forming the folded Appalachians. The detachment surface probably begins to descend into the root zone of the Piedmont below the crystalline-cored South Mountain anticlinorium. This anticlinorium may be considered transitional in structural style between the Piedmont and folded Appalachians; it is characterized by a pervasive flow cleavage and greenschist facies metamorphism similar to parts of the Piedmont, but it bears the fold geometry and contemporaneity of the folded Appalachian.

Speculations beyond these general ideas await additional evidence on the relationships between the two terranes and, in turn, on plate tectonics.

References

Barrell, J., 1914, The Upper Devonian delta of the Appalachian geosyncline: *Am. Jour. Sci.*, v. 37, p. 225–253.

Bucher, W. H., 1956, Role of gravity in orogenesis: *Geol. Soc. Am. Bull.*, v. 67, p. 1295–1318.

Carey, S. W., 1962, Folding: *Alberta Soc. Petrol. Geol.*, v. 10, p. 95–144.

Chamberlin, R. T., 1910, The Appalachian folds of central Pennsylvania: *Jour. Geol.*, v. 18, p. 228–251.

Clark, J. H., 1970, Geology of the carbonate rocks in

western Franklin County, Pennsylvania: *Pennsylvania Geol. Survey, 4th ser., Prog. Report 180* (map with text).

Cloos, E., 1940, Crustal shortening and axial divergence in the Appalachians of southeastern Pennsylvania: *Geol. Soc. Am. Bull.*, v. 51, p. 845–872.

———, 1947, Oolite deformation in the South Mountain fold, Maryland: *Geol. Soc. Am. Bull.*, v. 58, p. 843–918.

Conlin, R. R., and Hoskins, D. M., 1962, Geology and mineral resources of the Mifflintown Quadrangle: *Pennsylvania Geol. Survey, 4th ser., Atlas 126*, 46 p.

Colton, G. W., 1970, The Appalachian Basin—Its depositional sequences and their geologic relationships, *in* Fisher, G. W., and others, eds. *Studies of Appalachian geology, Central and Southern*: New York, Interscience, p. 5–47.

Currie, J. B., Patnode, H. W., and Trump, R. P., 1962, Development of folds in sedimentary strata: *Geol. Soc. Am. Bull.*, v. 73, p. 665–674.

Dennison, J. M., and Woodward, H. P., 1963, Palinspastic maps of Central Appalachians: *Am. Assoc. Petrol. Geol. Bull.*, v. 47, p. 666–680.

Edmunds, W. E., 1968, Geology and mineral resources of the northern half of the Houtzdale 15′ Quadrangle, Pennsylvania: *Pennsylvania Geol. Survey, 4th ser., Atlas 85ab*, 150 p.

Faill, R. T., 1969, Tectonic development of the Triassic basin in Pennsylvania: *Geol. Soc. Am., Abst. with programs for 1969*, pt. 7, p. 273–275.

———, 1969, Kink band structures in the Valley and Ridge Province, central Pennsylvania: *Geol. Soc. Am. Bull.*, v. 80, p. 2539–2550.

———, and Wells, R. B., in press, Geology and mineral resources of the Millerstown 15′ Quadrangle, Pennsylvania: *Pennsylvania Geol. Survey, 4th ser., Atlas 136*.

Freedman, J., Wise, D. U., and Bentley, R. D., 1964, Pattern of folded folds in the Appalachian Piedmont along Susquehanna River: *Geol. Soc. Am. Bull.*, v. 75, p. 621–638.

Glass, G. B., in press, Geology and mineral resources of the Phillipsburg $7\frac{1}{2}$′ Quadrangle, Pennsylvania: *Pennsylvania Geol. Survey, 4th ser., Atlas 95a*.

Gwinn, V. E., 1964, Thin-skinned tectonics in the Plateau and northwestern Valley and Ridge Provinces of the Central Appalachians: *Geol. Soc. Am. Bull.*, v. 75, p. 863–900.

———, 1970, Kinematic patterns and estimates of lateral shortening, Valley and Ridge and Great Valley Provinces, Central Appalachians, South Central Pennsylvania, *in* Fisher, G. W., and others, eds., *Studies of Appalachian geology, Central and Southern*: New York, Interscience, p. 127–146.

Kay, M., 1951, North American geosynclines: *Geol. Soc. Am. Mem. 48*, 143 p.

Keith, A., 1923, Outlines of Appalachian structure: *Geol. Soc. Am. Bull.*, v. 34, p. 309–380.

Kelley, R. D., Lytle, W. S., Wagner, W. R., and Heyman, L., 1970, The petroleum industry and future petroleum provinces in Pennsylvania, 1970: *Pennsylvania Geol. Survey, 4th ser., Mineral Res. Report 65*, 39 p.

King, P. B., 1951, *Tectonics of Middle North America*: Princeton, N.J., Princeton University Press, 203 p.

Lapham, D. M., and Root, S. I., 1971, Summary of isotopic age determinations in Pennsylvania: *Pennsylvania Geol. Survey, 4th ser., Info. Circular 70*, 29 p.

MacLachlan, D. B., 1967, Structure and stratigraphy of the limestones and dolomites of Dauphin County, Pennsylvania: *Pennsylvania Geol. Survey, 4th ser., General Geol. Report 44*, 169 p.

Moebs, N. N., and Hoy, R. B., 1959, Thrust faulting in Sinking Valley, Blair and Huntingdon Counties, Pennsylvania: *Geol. Soc. Am. Bull.*, v. 70, p. 1079–1088.

Nickelsen, R. P., 1963, Fold patterns and continuous deformation mechanisms of the Central Pennsylvania folded Appalachians, *in Tectonics and Cambrian-Ordovician stratigraphy in the Central Appalachians of Pennsylvania*: Pittsburgh Geol. Soc., Appalachian Geol. Soc., Guidebook, p. 13–29.

Oliver, W. A., Jr., DeWitt, J., Jr., Dennison, J. M., Hoskins, D. M., and Huddle, J. W., 1967, Devonian of the Appalachian Basin, United States, *in* Oswald, D. H., ed., *International symposium on the Devonian System*, Alberta Society Petroleum Geologists, p. 1001–1040.

Owens, J. P., 1970, Post-Triassic movements in the Central and Southern Appalachians as recorded by sediments of the Atlantic Coastal Plain, *in* Fisher, G. W., and others, eds., *Studies of Appalachian geology, Central and Southern*: New York, Interscience, p. 417–428.

Perry, N. J., Jr., 1964, Geology of the Ray Sponaugle well, Pendleton County, West Virginia: *Am. Assoc. Petrol. Geol. Bull.*, v. 48, p. 659–669.

Pierce, K. L., and Armstrong, R. L., 1966, Tuscarora fault, an Acadian (?) bedding plane fault in Central Appalachian Valley and Ridge Province: *Am. Assoc. Petrol. Geol. Bull.*, v. 50, p. 385–389.

Price, P. H., 1931, The Appalachian structural front: *Jour. Geol.*, v. 39, p. 24–44.

Price, R. A., 1969, The Southern Canadian Rockies and the role of gravity in low-angle thrusting, foreland folding, and the evolution of migrating foredeeps: *Geol. Soc. Am. Abst. with program for 1969*, pt. 7, p. 284–286.

———, and Mountjoy, E. W., 1970, Geologic structure of the Canadian Rocky Mountains between Bow and Athabaska Rivers—A progress report: *Geol. Assoc. Can. Spec. Paper 6*, p. 7–25.

Ramberg, H., 1967, *Gravity, deformation, and the earth's crust as studied by centrifuged models*: New York, Academic Press, 214 p.

Rodgers, J., 1967, Chronology of tectonic movements in the Appalachian region of eastern North America: *Am. Jour. Sci.*, v. 265, p. 408–427.

Root, S. I., 1968, Geology and mineral resources of southeastern Franklin County, Pennsylvania: *Pennsylvania Geol. Survey, 4th ser., Atlas 119cd*, 118 p.

———, 1970, Structure of the northern terminus of the Blue Ridge in Pennsylvania: *Geol. Soc. Am. Bull.*, v. 81, p. 815–830.

———, 1971, Geology and mineral resources of northeastern Franklin County, Pennsylvania: *Pennsylvania Geol. Survey, 4th ser., Atlas 119ab*, 104 p.

Rubey, W. W., and Hubbert, M. K., 1969, Roles of fluid pressure in mechanics of overthrust faulting; II. Overthrust belt in geosynclinal area of western Wyoming in light of fluid-pressure hypothesis: *Geol. Soc. Am. Bull.*, v. 70, p. 167–206.

Van Bemmelen, R. W., 1954, *Mountain building*: The Hague, Martinus Nijhoff, 177 p.

Wood, G. H., Jr., Trexler, J. P., and Kehn, T. M., 1969, Geology of the west-central part of the southern Anthracite field and adjoining areas, Pennsylvania: *U.S. Geol. Survey Prof. Paper 602*, 150 p.

———, and Bergin, M. H., 1970, Structural controls of the Anthracite region, Pennsylvania, *in* Fisher, G. W., and others, eds., *Studies of Appalachian geology, Central and Southern*: New York, Interscience, p. 147–160.

G. W. VIELE

*Department of Geology
University of Missouri
Columbia, Missouri*

Structure and Tectonic History of the Ouachita Mountains, Arkansas[1]

Large recumbent folds of lower Paleozoic strata characterize the eastern part of the Ouachita folded belt of Arkansas and Oklahoma. A belt of gravity structures and imbricate thrust faults in Carboniferous rocks lies north of the folds. This paper describes these structures and others and attempts to relate them to the history of the folded belt as a whole.

Most of the new data come from several quadrangles immediately west and southwest of Little Rock, Arkansas (Fig. 1), areas that are typical of the Ouachitas in that they are easily accessible, but in which outcrops are scattered, generally being found in road and stream cuts. Were it not for this, the Ouachita Mountains of Arkansas might well constitute a classical locality for the study of Alpine-type tectonics.

The plan of this paper is first to describe the lithologic units and the individual tectonic provinces of the Ouachitas, then to describe the structure of the eastern Ouachitas, and finally to present a hypothesis for the tectonic history of the eastern Ouachita folded belt in Arkansas.

Regional Setting

The Ouachita Mountains constitute a belt of folded Paleozoic strata extending westward

---

[1] I have greatly benefited from unpublished information relayed by B. R. Haley of the U.S. Geological Survey and C. G. Stone of the Arkansas Geological Commission, as well as from their critique of the manuscript. Neither should, however, be held responsible for the conclusions presented in this paper, as they have made clear with vigor and good humor.

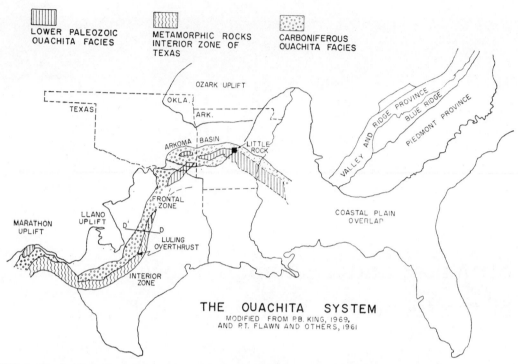

FIG. 1 Regional setting of Ouachita folded belt. Cross-section D'–D is Fig. 5.

across Arkansas from beneath the Mesozoic and Tertiary strata of the Gulf Coastal Plain (Fig. 1). In Oklahoma the belt turns southwestward and disappears once again beneath the Mesozoic-Tertiary cover. The Ouachitas form a continuous subsurface belt across Texas (Flawn and others, 1961), and rocks of similar lithology and age of deformation reappear in the Marathon uplift of southwestern Texas.

Well records indicate that, in the subsurface, the Ouachitas extend southeastward from central Arkansas toward the Appalachians, but although the folded belts closely approach one another our knowledge of their junction remains speculative. Seemingly the Ouachitas are the extension of the Paleozoic folded belts flanking eastern North America; thus they form the southern portion of a series of arcs that are convex toward the continent, maintaining the pattern of the Appalachian orogen relative to the craton.

Lithologic Units

Sedimentary Rocks

Figure 2 shows the major lithologic and structural units that compose the Ouachita folded belt of Arkansas and Oklahoma. Viewed in general, the stratigraphic column includes two lithofacies, each relatively uniform in overall aspect but each quite distinct from the other. The lowermost, exposed in a core known as the Benton-Broken Bow uplift, ranges in age from Ordovician to Mississippian and has an aggregate thickness of about 1000 m of strata, mostly Ordovician. From bottom to top the lower Paleozoic rocks comprise (1) dark oolitic limestones and clean quartzitic sandstone overlain by limestones and dark laminated shales, (2) a sandstone formation locally carrying cobbles and pebbles of granite, (3) a widespread sequence of black graptolitic shales interstratified with

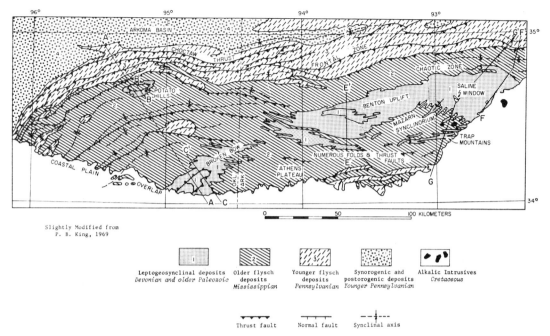

FIG. 2 Tectonic map of Ouachita folded belt in Arkansas and Oklahoma. Cross section lines A–C and E–G are in order of appearance in text (see Figs. 3, 4, 6, and 8).

thin beds of dark, silty, microcrystalline limestone, and (4) uppermost a sequence of siliceous shales, thin-bedded cherts, and thick beds of dense, light-colored radiolarian chert, known locally as novaculite. The depositional environment of these strata remains unresolved. Many recent workers (Park and Croneis, 1969; McBride, 1969; McBride and Thomson, 1970) suggest the strata represent a leptogeosynclinal environment (Trümpy, 1960), but others (Miser and Purdue, 1929; Goldstein and Hendricks, 1953; Berry, 1958; Folk, 1970) have interpreted the strata of the Ouachitas and equivalent strata in the Marathon region of Texas to be of shallow water or even subaerial origin.

The evidence for a starved, leptogeosynclinal basin may need reevaluation, for the measurements of original stratigraphic thickness may be substantially wrong. The rocks of the Benton-Broken Bow uplift are tightly folded and strongly cleaved. Ratios of major to minor axes of stretched pebbles on fold limbs are locally as high as 7:1; thus stratigraphic thicknesses on the strained limbs of folds are certain to be much less than the original thicknesses. If strain has reduced the thickness of the limbs by only one-half, the original thickness of the section would have been 2000 m.

Following deposition of the lower and middle Paleozoic sediments, rates of deposition rapidly increased and an aggregate thickness of more than 11,000 m of Carboniferous flysch accumulated, perhaps the thickest deposit of flysch in North America. The Mississippian portion of the flysch, as much as 5200 m thick, crops out throughout the Ouachitas south of the frontal thrust faults (Figs. 2 and 3). It wedges out northward (Cline, 1960), for geologists have not recognized it in the subsurface of the Arkoma Basin. As a whole shale is dominant, though graded beds of sandstone are common and predominate in the lower and middle portions of the stratigraphic section. Metamorphic rock fragments are common in the lower stratigraphic units but become less abundant upward. Tuffs and siliceous shales form important marker beds but represent an insignificant portion of the stratigraphic column.

Subaqueous slumps and slides are common in the upper units of the Mississippian flysch. Morris (1971), citing a general decrease in thickness of zones of disturbed bedding and an increasing number of distal turbidites toward the south, suggests most of the sliding was down the north slope of the basin. Walthall (1967), however, describing the Athens Plateau (Fig. 2), notes exotic blocks of graywacke and conglomerates containing metamorphic fragments, and from them he infers a southern source.

At the base of the Pennsylvanian is a predominantly argillaceous unit, the Johns Valley formation, containing abundant boulders of foreland carbonate strata. Dispersal of the boulders at least as far south as the area of the Benton-Broken Bow uplift was from fault-bounded ridges along the northern margin of the geosyncline (Miser, 1934; Shideler, 1970) (Fig. 3). Shideler (1970, p. 805) believes that transport was generally over distances of about 17 km southward away from the craton. This direction of transport for a wildflysch facies is opposite to that postulated for most geosynclines, but it has been documented for many years in the Ouachitas (Ulrich, 1927; Miser, 1934). Walthall (1967, p. 516) describes an opposite direction of transport for the Johns Valley exotics in the area of the Athens Plateau (Fig. 2). Here the exotic blocks are subgraywackes and graywackes, probably derived from a tectonically active welt on the south.

After deposition of the Mississippian and the basal Pennsylvanian strata, the northern limits of the area of geosynclinal deposition moved northward and as much as 6000 m of Pennsylvanian flysch accumulated in the area of the frontal zone and the Arkoma Basin. In

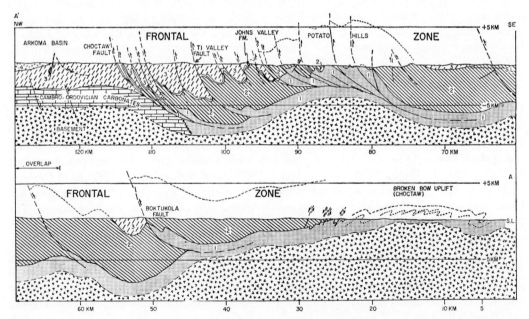

FIG. 3 Cross section of Oklahoma Ouachitas from Arkoma Basin to Broken Bow area. See Fig. 2 for location of A'–A and explanation of patterns and numbers. Figures 4A and 4B give preferred interpretation of Potato Hills and Broken Bow area. (Modified from Hopkins, 1968.)

the northern part of the Arkoma Basin, the Pennsylvanian flysch wedges out into shallow-water sandstones and is overlain by coal-bearing strata, but in the area of the present frontal belt it was a deep-water deposit (Stone, 1968; Cline, 1970; Chamberlain, 1971). The thickness south of the Benton-Broken Bow uplift is not known because the upper contact is not exposed, yet one can surmise thinning because of the probable rise of the uplift and, according to Ham and Wilson (1967, p. 384) and Chamberlain (1971), a northward shift of the depocenter. Walthall (1967, p. 518) describes associations of fossils from the Pennsylvanian of the Athens Plateau (Fig. 2) that were initially deposited in shallow water and brought to the final site of deposition by northward sliding down the slope of the geosyncline. Hendricks (1959, p. 54) describes conglomerates in the westermost outcrops (Fig. 2) of the Pennsylvanian flysch and postulates an uplift toward the south and east.

Cline (1960, 1970) and his students (Picha and Niem, 1970; Morris, 1971) describe the close affinities of the Carboniferous flysch of the Ouachitas with the flysch of the Alps and the Carpathians. They believe turbidity currents entering from both the north and south, some spreading over submarine fans, dispersed graded sands into a deep trough. Most of the paleocurrent trends are westward, suggesting the flows turned and continued their flow along a west-sloping axis toward even deeper water.

The ultimate source of the flysch is still a matter of debate. Some recent papers (Walthall, 1967; Briggs and Cline, 1967) suggest a southern source for the Mississippian flysch and the Johns Valley formation, but others favor more distant sources. For example, Morris (1971), describing the provenance of the middle and upper Mississippian flysch, suggests it came from the Illinois Basin or from the Southern Appalachians. Discounting the importance of a southern highland, he suggests that only volcanic islands and small tectonic lands lay to the south, and that they bounded and contained the sedimentary fill but did not contribute to it (Morris, 1971, p. 396–399). It may not be necessary to postulate any emergent areas on the south, for the submarine topography alone may have sufficed to limit the distribution of the flysch.

Igneous Rocks

Igneous rocks crop out in several areas of the eastern Ouachita Mountains. Dikes and sills, mostly of lamphrophyric and phonolitic composition, occur throughout the lower Paleozoic strata, and several alkalic stocks of Mesozoic age (Erickson and Blade, 1963, p. 5) intrude strata as young as Mississippian along a line roughly parallel to the edge of the Tertiary overlap. Sterling and Stone (1961, p. 103) contrast the Mesozoic age of the alkalic intrusives to the probable early Paleozoic age of several small, nickeliferous serpentinite masses embedded in the cherts and black slates of the leptogeosynclinal facies. Drill holes have passed completely through these masses and have re-entered black slates. The serpentinite masses appear to be caught in the noses of nearly recumbent folds, and they lie on, or close to, the trace of a large thrust fault (Figs. 6 and 7); originally they may have been gravity slide masses in the shales and cherts.

Tectonic Provinces

Arkoma Basin

The Arkoma Basin (Figs. 2 and 3) flanks the folded belt on the north; depths to basement may reach 5000–6000 m (Hopkins, 1968). Paleozoic strata, ranging from Cambro-Ordovician carbonates through the Pennsylvanian Morrow series are about 1500 m thick (Cline, 1968) and grade southward into the leptogeosynclinal (?) facies and the Mississippian flysch (Harlton, 1953; Ham, 1959; Decker, 1959). In the middle and late Pennsylvanian the basin dropped rapidly along south-dipping normal faults. Using well records, Buchanan and Johnson (1968) postulate as much as 1000 m of vertical offset of the basement across individual faults and a maximum subsidence of the basin of over 4000 m. Thick flysch sedimentation and faulting were simultaneous, as indicated by abrupt increases in thickness of individual units across faults (Haley and Hendricks, 1968; Merewether and Haley, 1968). The normal faults do not offset the uppermost Pennsylvanian beds.

Numerous broad, somewhat box-shaped synclines separated by relatively narrow anticlines constitute the dominant surface structures of the Arkoma Basin. Thrust faults surface in the crests of many anticlines and are known to underlie the folds (Reinemund and Danilchik, 1957; Merewether and Haley, 1961). The thrust faults offset the youngest Pennsylvanian deposits and are believed to offset or follow some of the earlier growth faults (Buchanan and Johnson, 1968).

Frontal Zone

The Choctaw thrust (Figs. 2 and 3) is generally taken to be the boundary between the Arkoma Basin and the frontal zone of the Ouachitas. In the frontal zone the basement slopes steeply southward, reaching depths as great as 10,000 m (Flawn, 1967). Cline (1960) believes the basement slope acted as a ramp for the thrust faults, causing them to turn upward toward the surface, but as is generally true in foreland belts, the basement is not involved in the thrusting. In the Oklahoma portion of the frontal zone, the imbricate thrust faults are associated with numerous northward-overturned folds, but at the surface the thrusts have steep to vertical dips. The thrusts effect a total cumulative northward movement of over 80 km (Hendricks, 1959, p. 50).

Perhaps the most controversial feature of the frontal zone has been the Potato Hills, an area where leptogeosynclinal (?) rocks immediately underlie the surface. Initially, Miser (1929) mapped the area as a window bounded by thrusts, whereas others have interpreted the structure to be an anticlinorium. Hopkins (1968) infers the presence of a broad basement uplift (Fig. 3) and depicts the thrusts on its north slope as having a gravity drive. A different interpretation, confirming Miser's view of the Potato Hills as a tectonic window (Fig. 4A) is given by Arbenz (1968), who cites more recent mapping and an analysis of fossil pollen from a well drilled by the Sinclair Oil Company in 1959. Arbenz suggests that initial, flat-lying thrusts were subsequently folded by thrusting at a lower level that emplaced recumbent folds of leptogeosynclinal (?) strata over the Mississippian flysch and are therefore the major faults of the region. It is not possible to say if the recumbent folds are anticlinal or synclinal; the cross section (Fig. 4A) arbitrarily shows them as anticlinal. Since Mississippian flysch underlies the major thrusts of the Potato Hills, there does not appear to be room for a major basement uplift. Also in evidence against an uplift is a 100 mgal negative gravity anomaly present in this area of the Ouachita frontal zone (Lyons, 1950; Howell and Lyons, 1959).

The frontal zone extends eastward into Arkansas, where folds and thrust faults form the most prominent surface structures. The main thrusts extend as far as the Tertiary overlap and probably underlie most of the folds in the Carboniferous flysch. Lying between the thrusts and folds and the Benton-Broken Bow uplift in the easternmost Ouachitas is a broad zone of upper Mississippian flysch, the Maumelle chaotic zone, in which the structures are completely disoriented. Reconnaissance work shows the chaotic zone is probably a mappable unit in upper Mississippian flysch that extends along the entire northern front of the Ouachitas in Arkansas (Haley and Stone, 1971, personal communication). Apparently it is absent in Oklahoma, for it is not mentioned in reports describing the area between the Potato Hills and the Broken Bow uplift (Shelburne, 1960; Cline, 1960).

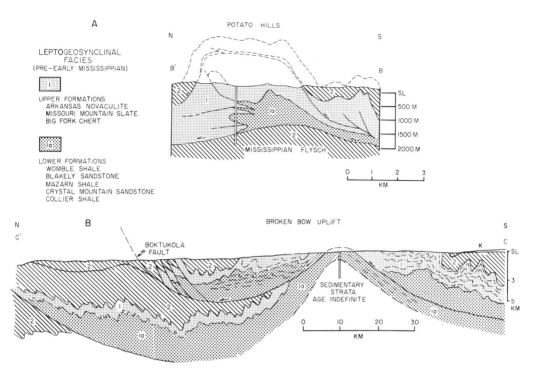

FIG. 4 (A) Cross section B'–B of Potato Hills (modified from Arbenz, 1968). (B) Cross section C'–C of Broken Bow area (modified from Miser, 1929). See Fig. 2 for locations.

Benton-Broken Bow Uplift and Interior Zone of Texas

The Benton-Broken Bow uplift, composed of rocks ranging from probable Late Cambrian to early Mississippian age, forms the core of the Ouachita Mountains. Its style of deformation is quite different from that of the imbricate thrust belt on the north, for in almost every outcrop the strata exhibit tight folds and pervasive cleavage. Toward the west these older Paleozoic strata are last seen at the surface in the Broken Bow area of Oklahoma, but numerous wells have demonstrated the presence of a low-grade metamorphic zone in the subsurface of Texas that loops around the Llano uplift and trends toward the Marathon uplift (Figs. 1 and 5). In general, the metamorphic grade is slightly higher than that of the Benton-Broken Bow uplift. Flawn (1961) describes the subsurface metamorphic rocks as belonging to an interior zone of the Ouachitas and contrasts them to the Carboniferous rocks of the frontal zone. Noting their lithologic similarity and structural position, he tentatively correlates the metamorphic rocks of the interior zone with the lower Paleozoic rocks of the Benton-Broken Bow uplift, as did earlier workers (Goldstein and Reno, 1952). Since the radiometric dates reflect only the late Paleozoic recrystallization, this assignment cannot be absolutely certain.

Flawn (1959, p. 22) has referred to the subsurface contact of the interior zone with the frontal zone in Texas as the "Luling overthrust front" and has with good insight suggested the possibility of large, overturned folds and possibly even nappes. The Luling overthrust front has not been recognized as such in the Ouachitas of Arkansas and Oklahoma, although evidence is accumulating

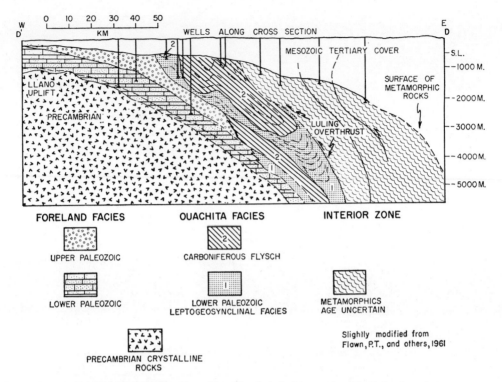

FIG. 5 Cross section D'–D of subsurface Ouachita belt in Texas. See Fig. 1 for location.

that the Benton-Broken Bow uplift is allochthonous.

The history of thought concerning the Broken Bow area (Fig. 2), which is also known as the Choctaw anticlinorium, is similar to that of the Potato Hills. Honess (1923) first described the southward-overturned folds of the pre-Carboniferous strata in the Broken Bow area but also noted that the oldest strata were only broadly folded. Citing the map of Honess, Miser (1929) suggested the structural discordance could be explained if the rocks in the core formed a window eroded in the upwarped Boktukola thrust fault (Fig. 4B). In later years other workers (Pitt, 1955; Misch and Oles, 1957; Ham, 1959; Hopkins, 1968) took the view that the Broken Bow uplift is an autochthonous anticlinorium developed over a basement high (Fig. 3). A different view emerges from a brief study by Bacon (1967), who reports that cleavage and axial surfaces in the Broken Bow are subhorizontal and indicative of recumbent folding, thus strengthening the hypothesis by Viele and Stone (1964) that major recumbent folds are present in the Ouachitas. Short reconnaissance trips to the Broken Bow area have confirmed these observations. As in the Potato Hills, drilling partially resolved the discussion, for in 1970 a well in the southwestern corner of the postulated Broken Bow window terminated in black slates of indeterminate age at a depth of about 3000 m. If future studies of the well samples reveal a post-Cambrian age, the evidence will be strong that the entire Broken Bow area is allochthonous.

The Athens Plateau Zone

For the purposes of this paper, all Carboniferous strata south of the Benton-Broken Bow uplift are described under this heading. Thus the zone includes the Mazarn synclinorium (Fig. 2), which contains more than 3500 m

(Lee, 1965, p. 41) of folded and faulted Mississippian flysch and is flanked north and south by tightly folded lower Paleozoic rocks. The flysch exhibits incipient metamorphism along the northern and eastern flanks of the basin, but the metamorphic grade decreases from north to south. Folds on the north side of the synclinorium are overturned southeastward; those on the south side, in the Trap Mountains (Fig. 2), are overturned northward or exhibit vertical axial surfaces, indicating the gross form of the synclinorium is that of an inverted fan fold (Purdue and Miser, 1923). Many folds and thrusts (Haley and Stone, 1971, personal communication) occur south of the Trap Mountains, but the tightness of folding decreases, and where last exposed unmetamorphosed Carboniferous strata dip homoclinally toward the south beneath the overlying sediments of the Gulf Coastal Plain.

The Athens Plateau proper lies southwest of the Mazarn synclinorium and is largely underlain by unmetamorphosed Carboniferous flysch, which here is as much as 7900 m thick (Ham and Wilson, 1967, p. 384). Walthall's (1967) maps and cross sections of the Athens Plateau show numerous folds and thrusts, which decrease in number toward the south. In the northern part of the plateau, near the Benton-Broken Bow uplift, northward-overturned anticlines broken by steep, south-dipping reverse faults constitute the dominant structures. Similar faults occur in the southern part, but the strata dip dominantly southward.

The Eastern Ouachita Mountains

Structure-Leptogeosynclinal Facies

Although Miser (1929) recognized thrust faults in the lower Paleozoic rocks in Oklahoma, his maps and cross sections of Arkansas show a broad anticlinorial structure composed of folds overturned toward both the north and the south (Fig. 6A). Miser did not

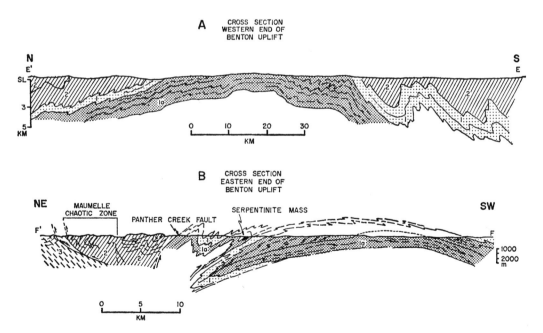

FIG. 6 (A) Cross section E'–E of western Arkansas Ouachitas (modified from Miser, 1929). (B) Cross section F'–F of eastern Ouachitas. See Figs. 2 and 4 for locations and legend.

publish maps or cross sections of the easternmost Ouachitas, but by extension the concept of an anticlinorium became applied to the entire range (Eardley, 1962, p. 227).

In an upright anticlinorial fold, axial surfaces and cleavage planes should be nearly vertical or fan about the vertical. Therefore, to test the hypothesis, attitudes of bedding and cleavage and of the axial surfaces of numerous small folds were collected throughout the eastern Ouachitas. They turned out to be all either flat-lying or only gently arched. Consequently, it was suggested (Viele and Stone, 1964; Viele, 1966) that the eastern end of the Benton-Broken Bow uplift comprised one or more large, broadly arched recumbent folds or, as Americans often use the term in a restricted sense, nappes, which traveled from south to north, diving toward the north. My original cross sections, although broadly correct, were premature, and further work has shown the structure is much more complex than this.

Superimposed on the recumbent folds is a second fold system of northeast–southwest trend that is essentially parallel to the southeastward-overturned folds on the north side of the Mazarn synclinorium. Locally along this north side, a second-generation crenulation cleavage has formed. Also superimposed on all earlier structures is a broad arching along the general trend of the Benton-Broken Bow uplift. The subsequent fold episodes resulted in reclined folds (Turner and Weiss, 1963, p. 118–120); only rarely does one find a hinge line parallel or even somewhat parallel to the strike of the axial surface. In Fig. 7 representative attitudes of axial surfaces and the rake of the hinge lines on the axial surfaces illustrate the variation in attitude of the folds.

Projected into a vertical cross section (Fig. 6B), the folds occur in the following general sequence: (1) on the extreme right of the section a packet of up-facing, nearly-recumbent folds, overturned toward the northeast, is indicated by the stratigraphic section and the direction of rotation of the mesoscopic folds; (2) a zone underlain by the overturned limb of a recumbent fold, the axial surface lying above the surface of the ground; (3) in the right center of the section a broad zone on the upper limb of a packet of down-facing folds, the axial surface having re-entered the ground; and (4) in the left center of the section, a packet of tightly digitated folds that are up-facing and overturned toward the southwest.

Confirmation of my earlier idea that the lower Paleozoic of the easternmost Ouachitas represents the core of a nappe (Viele, 1966, p. 264) came with the discovery by Boyd Haley of the U.S. Geological Survey and Charles Stone of the Arkansas Geological Commission of a window along the Saline River (Fig. 2) west of the map area of Fig. 7. Mississippian Stanley shale crops out structurally beneath the lower Paleozoic rocks. In the Stanley the direction of rotation of recumbent folds and the refraction of cleavage planes suggests an upper limb of a second, lower-lying recumbent fold, having its anticlinal hinge toward the north.

Structure—Carboniferous Strata

Although its stratigraphic position varies along trace, the Panther Creek thrust (Figs. 6B and 7) generally separates lower Paleozoic from upper Paleozoic strata. Northeast and north of the fault, in a belt 4–8 km wide, the flysch of the Mississippian Jackfork formation exhibits the same fold geometry and cleavage as the lower Paleozoic digitate folds on the south. North of this folded and cleaved zone, however, the structure changes completely, for here we are in the Maumelle chaotic zone mentioned previously, and numerous examples of deformation by gravity sliding may be seen in the Jackfork. Along the south side of Lake Maumelle (Fig. 7) contorted and convoluted shales, in which structures are completely disoriented, carry numerous isolated sandstone bodies of varying size. Locally the shale is diapiric into the sandstone masses, which may be rounded or blocky or intricately folded.

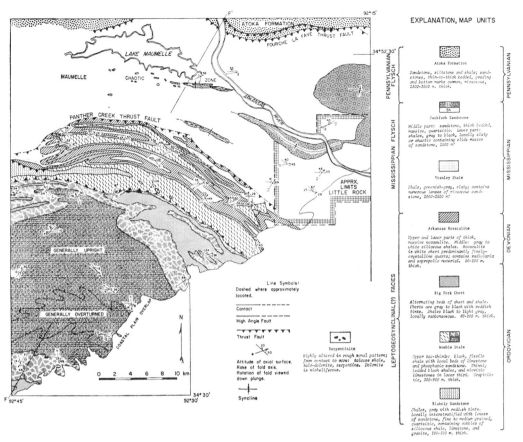

FIG. 7 Geological map of eastern Ouachita Mountains. Compiled by Viele from unpublished maps by Sterling, Stone, and Holbrook, Arkansas Geological Commission. See Fig. 6B for cross section F–F'.

In contrast to these bodies are thick-bedded, massive, regularly folded beds of sandstone that overlie chaotic shales west and north of Little Rock. The folds are mostly tight, infolded synclines slightly overturned southward that open and become less inclined farther north. Where the contact between the massive sandstones and the shales is exposed, it is demonstrably faulted, and locally in the underlying shales tension fractures, filled with syntectonic quartz veins (Miser, 1943; Engel, 1952), indicate relative northward movement of the sandstones. Down-facing cascade folds are exposed just south of the Fourche La Fave thrust (Fig. 7). Excellent sole markings and graded beds allow no misinterpretation of the direction of facing, and the folds demonstrate northward overturning. Thrusts that moved northward mark the contact between the Mississippian and Pennsylvanian flysch.

Tectonic History and Role of Gravity

Figure 8, a series of schematic cross sections, illustrates a sequence of events that could account for the structure of the Ouachita Mountains as it is known at present. Data from several areas are projected into the sections so that they represent a north-south composite section.

Section A shows the region after deposition of the Mississippian and early Pennsylvanian

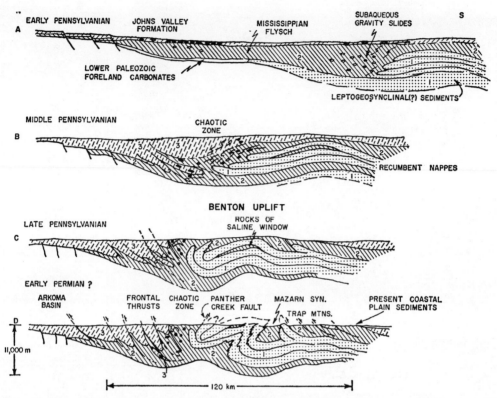

FIG. 8 Sketches of cross sections along line G'–G illustrating a possible scenario for formation of eastern Ouachitas. Sketches are diagrammatic. See Fig. 2 for location of G'–G and explanation of patterns and numbers. Further explanation in text.

flysch. In the north these strata thinned abruptly at the site of the present Choctaw thrust but had their full thickness of over 5000 m across the area of the present frontal zone and Benton-Broken Bow uplift. Both the north and south slopes of the basin probably had contributed subaqueous slumps and slides of Mississippian flysch, whereas the early Pennsylvanian wildflysch, containing exotics of foreland strata, had spread from fault blocks along the northern margin of the basin.

Earlier discussions (Walthall, 1967; Morris, 1971) have not attempted to link these slides of Carboniferous strata to the emplacement of the nappes composed of lower Paleozoic leptogeosynclinal (?) strata. Yet if one accepts the concept that the Benton-Broken Bow uplift is allochthonous, then one must consider the palinspastic position of the nappes in Early Carboniferous time. The deep subsidence of the basin that received the Mississippian flysch, the presence in the flysch of metamorphic pebbles, and the gravity slides themselves all bespeak tectonic activity. Probably, the nappes of lower Paleozoic strata were moving in the subsurface from south to north, but they had not yet reached the location of the present Benton-Broken Bow uplift. If the nappes were already at the site of this uplift prior to Mississippian deposition, neither the gravity slides in the lower flysch units nor the late Pennsylvanian thrusts of the frontal belt (Hendricks, 1959) could be related to nappe transport. Moreover, if movement of the nappes was restricted to the late Pennsylvanian, concurrent with the thrusting, then too the gravity slides in the lower flysch would necessarily be independent of nappe transport. It seems more probable that tectonism and sedimentation were synchronous; the question is in what way?

One possibility is that as the nappes traveled northward in the subsurface they arched, thereby lifting the covering strata into a submarine high. Northward sliding of masses of Mississippian flysch into the basin was simultaneous with sedimentation and tectonic transport. Subsidence of the area of the present frontal belt and later of the Arkoma Basin was rapid and probably associated with the northward shifting of a foredeep. The basement sank to depths of 10,000 m and the nappes followed this structural slope downward and northward. In the eastern Ouachitas, Mississippian strata slid off the nappes into the foredeep to form the Maumelle chaotic zone. Shales, already containing gravity-slide masses, were re-elevated as the foredeep shifted north and flowed again downslope, carrying beds of sandstone with them. Owing to their thickness and relative lack of ductility, the sandstones did not disaggregate. Some were infolded into the underlying shales and others cascaded down the slope to form down-facing folds. A free upper surface is inferred.

Section B (Fig. 8) shows the Ouachitas after deposition of the Pennsylvanian flysch composing the Atoka formation, and it also illustrates the difficulty of exactly dating the sliding. The Atoka of the frontal belt does not contain large slumped masses and Atokan rocks have not been recognized in the Maumelle chaotic zone. If the sliding predated Atoka deposition, the Atoka should rest unconformably on older rocks, and, indeed, Hendricks (1959, p. 54) describes an overlap of Atoka on lower Pennsylvanian and Mississippian strata in the far western Ouachitas. Evidence for an emergent area, however, is not definitive, and some workers strongly doubt its existence (Haley and Stone, 1971, personal communication). There are other possibilities. At the time of Atoka deposition the present frontal belt may have lain north of the deep axis of the trough, and therefore this area did not receive slides from the south. It is also possible that gravity-slide deposits in the Atoka lie buried in the Maumelle chaotic zone or that they were eroded from the area of the present Benton-Broken Bow uplift. To maintain the concept of simultaneous tectonism and sedimentation, the cross section arbitrarily shows gravity slides in Atokan rocks and buried Atokan rocks in the Maumelle chaotic zone.

Throughout the eastern Ouachitas thrust faults form the contact between the Mississippian and Pennslyvanian flysch. Since horizontal compressive stress cannot be transmitted for any great distance in a thick pile of water-saturated shale, a spreading mechanism associated with the diving nappes and gravity slides is indicated (Fig. 9). The mechanism is

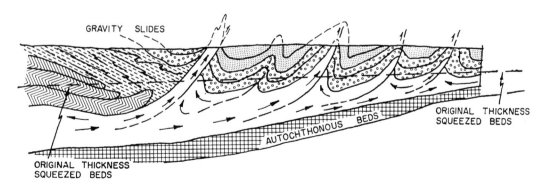

FIG. 9 Imbricate thrusting by loading and spreading of a décollement zone. Should the stratigraphic displacements of individual faults be summed to derive total shortening?

similar to that proposed by Price and Mountjoy (1970) for the Canadian Rocky Mountains. Cross sections B and C (Fig. 8) show the nappes diving into the foredeep and overriding the Carboniferous flysch. Sliding of similar Carboniferous strata from the backs of the nappes and concurrent sedimentation added to the load, the combined weight of all acting as a plunger that squeezed the basin fill northward and upslope. The slip surfaces in this squeezed mass are the thrusts. Growth of the thrusts began with the initial loading of the basin fill and continued throughout Carboniferous time as movement of the nappes, gravity slides, and sedimentation added increments to the load. The fact that the thrusts cut the youngest Carboniferous strata means only that thrusting continued into late Carboniferous time, not that this was the sole time of faulting.

If gravity loading drove the thrusts of the Ouachita frontal belt, the thrusts must have formed under the present Maumelle chaotic zone and north edge of the Benton-Broken Bow uplift. How then do we reconcile the estimated shortening in the frontal belt of over 80 km (Hendricks, 1959, p. 50)? This estimate assumes the stratigraphic displacement on each fault is cumulative. But if the individual faults form in pinched and spreading bodies of shale, the sum of their displacements need not be the measure of total shortening. Translation of the thrust sheets is not by simple overriding of individual rock sheets moving along vanishingly thin surfaces of shear but instead by internal strain within a thick décollement zone. The thrusts not only follow a décollement zone; they exist because of its pinching and flow (Viele, 1966, 1970). The mechanism allows the total horizontal translation at the rear of an imbricate thrust belt to approach minimal values.

We turn now to the nappes of lower Paleozoic leptogeosynclinal (?) strata. Cross section C (Fig. 8) shows growth of involutions along the northeastern margin of the Benton-Broken Bow uplift. This illustrates one of the most puzzling aspects of the structure of the Ouachitas. The regional arcuate pattern (Fig. 1), seismic profiles, and recent drilling in the interior zone of Texas (Rozendal and Erskine, 1971) as well as in the Potato Hills and Broken Bow areas, the thrusting of the frontal zone, and fold rotations throughout broad belts of lower Paleozoic strata all indicate northward or westward movement of allochthonous sheets toward the craton. Nevertheless, along the northeastern margin of the Benton-Broken Bow uplift and throughout the western Ouachitas, the folds are overturned toward the south and southeast, indicating an opposite sense of movement. A late stage of backfolding and faulting of stratigraphically upright nappes does not suffice as an explanation for this paradox, because in the eastern Ouachitas it does not explain the northeast-southwest sequence of up-facing, down-facing, and up-facing folds (Figs. 6B and 7).

In addition to this sequence other lines of evidence lead to the hypothesis of involution in the northeastern Ouachitas. The rocks of this area exhibit the strongest cleavage, the tightest folding, and the greatest recrystallization, an anomalous pattern for the continental side of a folded belt. The tightly folded and cleaved Mississippian flysch in the Saline window suggests the possibility of lower-lying allochthonous sheets. West of Little Rock the Panther Creek fault (Fig. 7) represents a major structural break, for it generally separates the lower Paleozoic strata from the upper Mississippian flysch. It is not exposed, but all other planar structural elements, including numerous smaller faults dip northeast in this area. Therefore a generally north or northeast dip is inferred for the fault. But this implies either a normal fault down to the north, which would be unique in this area, or a major thrust to the south, which would be opposite to the general sense of tectonic transport. Only if one infers a southward rotation does the fault make sense.

The involutions are consistent with the

hypothesis of gravity loading by nappes diving toward the north. The lower nappes are flattened and rotated by the weight of the higher ones carrying their load of Carboniferous flysch. The concept of involution of the frontal nappes finds its origin in Argand's (1916) classical cross sections of the Pennine nappes.

The last cross section (section D, Fig. 8) shows how all the tectonic features were accentuated by the end of the Pennsylvanian and perhaps during the early Permian (Hendricks, 1959). Continued northward movement and subsequent uplift had arched the Benton-Broken Bow uplift; folds along the northern margin of the Mazarn synclinorium were overturned southward; the frontal nappes and perhaps the chaotic zone were forced farther upward and rotated toward the south; and the continued spreading of the thrusts had now offset the youngest Pennsylvanian strata in the Arkoma Basin. These later deformations also accentuated the reclined attitudes of the folds throughout the eastern Ouachitas. On the south side of the Mazarn synclinorium the Trap Mountains had been thrust into position and still farther south other thrusts had offset the Carboniferous strata. It is tempting to speculate on the existence of yet another system of lower Paleozoic nappes in the subsurface of southern Arkansas.

Summary

The Ouachita orogeny included several styles of deformation. Initially, under a thick sedimentary cover, the northward gliding of the nappes and their internal strain increased the length of the original bedding perhaps several times. Diving nappes and chaotic gravity sliding of sediment from their backs loaded a basin-fill of soft shale, causing thrusts in the shale to spread up the north slope of the basin. Higher nappes plunging on the backs of lower ones formed southward-rotated involutions. Only the latest stages of the orogeny were primarily compressional, causing shortening and arching of the entire folded belt.

Notwithstanding the multiple deformations, the structures of the lower Paleozoic strata are ordered. In any given area they exhibit penetrative patterns probably as a result of deformation under load. They are confined deformations. Structures directly attributable to superficial gravity sliding are limited to the Maumelle chaotic zone along the northern margin of the folded belt. These structures are disordered, exhibiting only a general sense of northward transport. Notwithstanding their extent and excellent development, they represent a second-order feature of the orogeny, as do the thrusts of the frontal belt. The first-order structures were the subsurface nappes bending and moving down the structural slope of a geosynclinal foredeep.

References

Arbenz, J. K., 1968, Structural geology of the Potato Hills, Ouachita Mountains, Oklahoma, *in* Cline, L. M., ed., *Geology of the western Arkoma Basin and Ouachita Mountains, Oklahoma*: Okla. City Geol. Soc. Guidebook, p. 109–121.

Argand, E., 1916, Sur l'arc des Alpes Occidentales: *Eclog. Geol. Helv.*, v. 14, p. 145–191.

Bacon, J., 1967, A fabric study of the core of the Oklahoma Ouachita Mountains: Tulsa Univ., unpublished manuscript, 10 p.

Berry, William, B. N., 1958, Depositional environment of the Ordovician succession in the Marathon region, Texas: *Jour. Sed. Petrol.*, v. 28, p. 389–405.

Briggs, G., and Cline, C. M., 1967, Paleocurrents and source areas of late Paleozoic sediments of the Ouachita Mountains, southeastern Oklahoma: *Jour. Sed. Petrology*, v. 37, p. 985–1000.

Buchanan, R. S., and Johnson, F. K., 1968, Bonanza gas field—A model for Arkoma Basin growth faulting, *in* Cline, L. M., ed., *Geology of the western Arkoma Basin and Ouachita Mountains, Oklahoma*: Okla. City Geol. Soc. Guidebook, p. 75–85.

Chamberlain, C. K., 1971, Bathymetry and paleocology of Ouachita geosyncline of southeastern Oklahoma as determined from trace fossils: *Am. Assoc. Petrol. Geol. Bull.*, v. 55, p. 34–50.

Cline, L. M., 1960, Stratigraphy of the late Paleozoic

rocks of the Ouachita Mountains, Oklahoma: *Okla. Geol. Survey Bull. 85*, 113 p.

———, 1966, Late Paleozoic rocks of Ouachita Mountains, a flysch facies, *in* Field conference on flysch facies and structure of the Ouachita Mountains: Kansas Geol. Soc. Guidebook, 29th, p. 91–111.

———, 1968, Comparison of main geologic features of Arkoma Basin and Ouachita Mountains, Southeastern Oklahoma, *in* Cline, L. M., ed., *Geology of the western Arkoma Basin and Ouachita Mountains, Oklahoma*: Okla. Geol. Soc. Guidebook, p. 63–74.

———, 1970, Sedimentary features of Late Paleozoic flysch, Ouachita Mountains, Oklahoma: *Geol. Assoc. Can. Special Paper 7*, p. 85–101.

Decker, C. E., 1959, Correlation of lower Paleozoic formations of the Arbuckle and Ouachita areas as indicated by graptolite zones, *in* Cline, L. M., and others, eds., *The geology of the Ouachita Mountains—A symposium*: Dallas and Ardmore Geol. Soc. Guidebook, p. 92–96, 1959.

Eardley, A. J., 1962, *Structural geology of North America*: New York, Harper and Row, 743 p.

Engel, A. E. J., 1952, Quartz crystal deposits of western Arkansas: *U.S. Geol. Survey Bull. 973*, p. 173–260.

Erickson, R. L., and Blade, L. V., 1963, Geochemistry and petrology of the alkalic igneous complex at Magnet Cove, Arkansas: *U.S. Geol. Survey Prof. Paper 425*, 95 p.

Flawn, P. T., 1959, The Ouachita structural belt, *in* Cline, L. M., and others, eds., *The geology of the Ouachita Mountains—A symposium*: Dallas and Ardmore Geol. Soc. Guidebook, p. 20–29.

———, chm., 1967, Basement map of North America, between latitudes 24° and 60° N.: *Am. Assoc. Petrol. Geol.; U.S. Geol. Surv.*, scale 1:5,000,000.

———, Goldstein, A., Jr., King, P. B., and Weaver, C. E., 1961, The Ouachita system: *Texas Univ. Pub. 6120*, Austin, 410 p.

Folk, R. L., Evidence of peritidal origin for part of the Caballos Novaculite (Devonian radiolarian chert), Marathon Basin, Texas (abst.): *Geol. Soc. Am., Progr. Ann. Meetings*, v. 2, p. 552–553, 1970.

Goldstein, A., Jr., and Hendricks, T. A., 1953, Siliceous sediments of Ouachita facies in Oklahoma: *Geol. Soc. Am. Bull.*, v. 64, p. 421–442.

———, 1962, Late Mississippian and Pennsylvanian sediments of Ouachita facies, Oklahoma, Texas, and Arkansas, *in* Branson, C. G., ed., *Pennsylvanian system in the United States*: Am. Assoc. Petrol. Geol., p. 385–430.

Goldstein, A., Jr., and Reno, D. H., 1952, Petrography and metamorphism of sediments of Ouachita facies: *Am. Assoc. Petrol. Geol. Bull.*, v. 36, p. 2275–2290.

Haley, B. R., and Hendricks, T. A., 1968, Geology of the Greenwood Quadrangle, Arkansas–Oklahoma: *U.S. Geol. Survey Prof. Paper 536-A*, 15 p.

Ham, W. E., 1959, Correlation of pre-Stanley strata in the Arbuckle–Ouachita Mountain Region, *in* Cline, L. M., and others, eds., *The Geology of the Ouachita Mountains—A Symposium*: Dallas and Ardmore Geol. Soc. Guidebook, p. 71–86.

———, and Wilson, J. L., 1967, Paleozoic epeirogeny and orogeny in the Central United States: *Am. Jour. Sci.*, v. 265, p. 332–408.

Harlton, B. H., 1953, Ouachita chert facies, southeastern Oklahoma: *Am. Assoc. Petrol. Geol. Bull.*, v. 37, p. 778–796.

Hendricks, T. A., 1958, Interpretation of Ouachita Mountains of Oklahoma as autochthonous folded belt: Preliminary report: *Am. Assoc. Petrol. Geol. Bull.*, v. 42, p. 2757–2786.

———, 1959, Structure of the frontal belt of the Ouachita Mountains *in* Cline, L. M., and others, eds., *The geology of the Ouachita Mountains—A symposium*: Dallas and Ardmore Geol. Soc. Guidebook, p. 44–56.

Honess, C. W., 1923, Geology of the southern Ouachita Mountains of Oklahoma: *Okla. Geol. Survey Bull. 32*.

Hopkins, H. R., 1968, Structural interpretation of the Ouachita Mountains, *in* Cline, L. M., ed., *Geology of the western Arkoma Basin and the Ouachita Mountains, Oklahoma*: Okla. City Geol. Soc. Guidebook, p. 104–108.

Howell, J. V., and Lyons, P. L., 1959, Oil and gas possibilities of the Ouachita Mountains *in* Cline, L. M., and others, eds., *The geology of the Ouachita Mountains—A symposium*: Dallas and Ardmore Geol. Soc. Guidebook, p. 57–61.

King, P. B., 1964, Further thoughts on tectonic framework of the southeastern United States, *in* Lowry, W. D., ed., *Tectonics of the Southern Appalachians*: Blacksburg, Virginia Polytechnic Inst., Dept. of Geol., Memoir 1, p. 5–31.

———, 1969, compiler, Tectonic map of North America: *U.S. Geol. Survey Map*.

———, 1969, The tectonics of North America—A discussion to accompany the tectonic map of North America, scale 1:5,000,000: *U.S. Geol. Survey Prof. Paper 628*, 94 p.

Lee, M. A., 1965, Structural analysis of the Mazarn synclinorium: Univ. of Missouri, unpubl. M.A. dissertation, 56 p.

Lyons, P. L., 1951, A gravity map of the United States: *Tulsa Geol. Soc. Digest*, v. 18, p. 33–43.

McBride, E. F., 1964, Stratigraphy and sedimentology of the Woods Hollow Formation (Middle Ordovician), Trans-Pecos, Texas: *Geol. Soc. Am. Bull.*, v. 80, p. 2287–2302.

———, 1970, Stratigraphy and origin of Maravillas formation (Upper Ordovician), West Texas: Am. Assoc. Petrol. Geol. Bull., v. 54, p. 1719-1745.

———, and Thomson, A., 1970, The Caballos Novaculite, Marathon Region, Texas: Geol. Soc. Am. Spec. Paper 122, 129 p.

Merewether, E. A., and Haley, B. R., 1969, Geology of the Coal Hill, Hartman, and Clarksville Quadrangles, Johnson County and vicinity, Arkansas: U.S. Geol. Survey Prof. Paper 536-C, 27 p.

Miser, H. D., 1924, Structure of the Ouachita Mountains of Oklahoma and Arkansas: Okla. Geol. Survey Bull., v. 50, 30 p.

———, 1934, Carboniferous rocks of Ouachita Mountains: Am. Assoc. Petrol. Geol. Bull., v. 18, p. 971-1009.

———, 1943, Quartz veins in the Ouachita Mountains of Arkansas and Oklahoma, their relations to structure, metamorphism, and metalliferous deposits: Econ. Geol., v. 38, p. 91-118.

———, and Purdue, A. H., 1929, Geology of the DeQueen and Caddo Gap Quadrangles, Arkansas: U.S. Geol. Survey Bull. 808, 195 p.

Misch, P., and Oles, K. F., 1957, Interpretation of Ouachita Mountains of Oklahoma as autochthonous folded belt: Preliminary report: Am. Assoc. Petrol. Geol. Bull., v. 41, p. 1899-1905.

Morris, R. C., 1971, Stratigraphy and sedimentology of Jackfork Group, Arkansas: Am. Assoc. Petrol. Geol. Bull., v. 55, p. 387-402.

———, 1971, Classification and interpretation of disturbed bedding types in Jackfork flysch rocks (upper Mississippian), Ouachita Mountains, Arkansas: Jour. Sed. Petrol., v. 41, p. 410-424.

Park, D. E., and Croneis, C., 1969, Origin of Caballos and Arkansas Novaculite Formations: Am. Assoc. Petrol. Geol. Bull., v. 53, p. 94-111.

Picha, F., and Niem, A. R., 1970, Distribution and extension of some flysch deposits, Ouachita Mountains, Arkansas and Oklahoma (abs.): Geol. Soc. Am. Progr. Ann. Meetings, v. 2, p. 653.

Pitt, W. D., 1955, Geology of the core of the Ouachita Mountains in Oklahoma: Okla. Geol. Survey Circular 34, 34 p.

Price, R. A., and Mountjoy, E. W., 1970, Geologic structure of the Canadian Rocky Mountains between Bow and Athabasca River—A progress report: Geol. Assoc. Can. Spec. Paper 6, p. 7-25.

Purdue, A. H., and Miser, H. D., 1923, Description of the Hot Springs district: U.S. Geol. Survey Geol. Atlas, Folio 215.

Rozendal, R. A., and Erskine, W. S., 1971, Deep test in Ouachita structural belt of central Texas: Am. Assoc. Petrol. Geol. Bull., v. 55, p. 2008-2017.

Shelburne, O. B., 1960, Geology of the Boktukola syncline, southeastern Oklahoma: Okla. Geol. Survey Bull., v. 88, 84 p.

Shideler, G. L., 1970, Provenance of Johns Valley boulders in Late Paleozoic Ouachita facies, Southeastern Oklahoma and Southwestern Arkansas: Am. Assoc. Petrol. Geol. Bull., v. 54, p. 789-805.

Sterling, P. J., and Stone, C. G., 1961, Nickel occurrences in soapstone deposits, Saline County, Arkansas: Econ. Geol., v. 56, p. 100-110.

Stone, C. G., 1968, The Atoka formation in the southeastern Arkansas Valley, Arkansas (abs.): Geol. Soc. Am. Spec. Paper 121, p. 413.

Trümpy, R., 1960, Paleotectonic evolution of the central and western Alps: Geol. Soc. Am. Bull., v. 71, p. 843-908.

Turner, F. J., and Weiss, L. E., 1963, Structural analysis of metamorphic tectonites: New York, McGraw-Hill, 545 p.

Van Bemmelen, R. W., 1960, New views on East-Alpine orogenesis: Rept. 21st Sess. Int. Geol. Cong., pt. 18, p. 99-116.

Viele, G. W., 1966, The regional structure of the Ouachita Mountains of Arkansas, a hypothesis, in Field conference on flysch facies and structure of the Ouachita Mountains: Kansas Geol. Soc. Guidebook, 29th, p. 245-278.

———, 1970, Displacement by flow in imbricate thrust fault belts (abs.): Geol. Soc. Am. Progr. Ann. Meetings, v. 2, p. 302.

———, and Stone, C. G., 1964, Structural geology of the Ouachita Mountains, Benton–Little Rock area, Arkansas (abs.): Geol. Soc. Am. Spec. Paper 82, p. 212.

Walthall, B. H., 1967, Stratigraphy and structure, part of Athens Plateau, southern Ouachitas, Arkansas: Am. Assoc. Petrol. Geol. Bull., v. 51, p. 504-528.

Weaver, C. E., 1960, Possible uses of clay minerals in search of oil: Am. Assoc. Petrol. Geol. Bull., v. 44, p. 1505-1518.

K. JINGWHA HSÜ

Swiss Federal Institute of Technology
Zürich, Switzerland

Mesozoic Evolution of the California Coast Ranges: A Second Look

The innovations of the sea-floor spreading and plate-tectonics theories led us to return to the golden age of Alpine tectonics. When rereading the giants of Alpine geology (e.g., Argand, 1922; Staub, 1928), one will find expressions such as "rifting of plates" or "collision of continents" not uncommon in these classic writings. The conservatism during the middle decades of this century reflected the influences of those who had difficulty in accepting the evidence of crustal shortening (e.g., Jeffreys, 1929; Bucher, 1933). In the search for an alternative many of us found refuge in vertical tectonics and gravitational sliding. This school of thought was propounded by Ampferer (1906) and Haarmann (1930), nursed to maturity by Van Bemmelen (1932, 1954, 1960) and Beloussov (1962). Some theoretical basis for the gravity-sliding hypothesis was provided by Hubbert and Rubey (1959). The onrushing of the "new global tectonics" bandwagon, however, seems to crush all opposition, so that as convenor of a Penrose Conference on Tectonics, Dickinson (1970) was moved to remark: "It was notable that (during the conference) no discussion of thrust relations . . . emphasized gravitational gliding."

Van Bemmelen, to whom this volume is dedicated, himself has been quick to accept new evidence of continental drift and sea-floor spreading. He has attempted a synthesis to harmonize the "fixistic" and "mobilistic" schools of thoughts in his writings of the last decade (Van Bemmelen, 1965, 1967, 1969). He recognized that the plate-tectonics hypotheses are not incompatible with ideas of vertical tectonics, and that they do not necessarily refute the significance of gravitative forces. His flexibility and his restraint are admirable. Deep-sea drilling revealed, for example, that the spreading of the

Atlantic floor has been accompanied by vertical movements amounting to thousands of meters (Hsü and Andrews, 1970). And a newest hypothesis on plate-tectonics postulates gravitative forces as the driving mechanism for plate movements (Jacoby, 1970, 1973), somewhat in the fashion propounded by Van Bemmelen in 1965.

The evolution of my own thinking has wittingly or unwittingly followed the steps of Van Bemmelen. I met him at Neuchâtel in 1966, where I gave a paper on a working

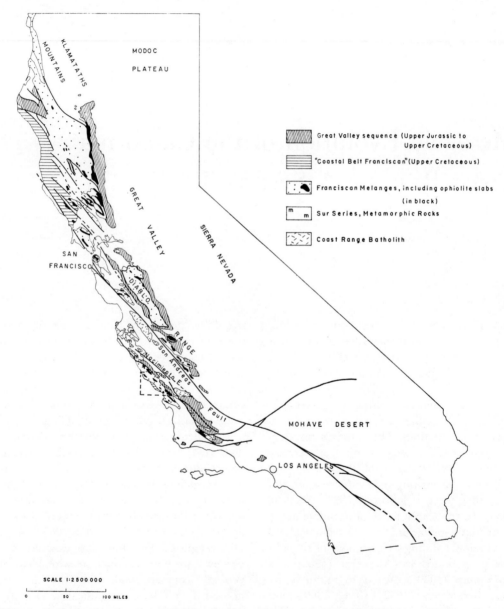

FIG. 1 Geologic sketch map of California Coast Ranges, showing distribution of major Mesozoic units. Location of the area covered by Figs. 2 and 5 is shown by stippled lines.

hypothesis on the Mesozoic history of California Coast Ranges in the Wegmann Symposium (Hsü, 1967). I was then much attracted to the idea of invoking gravitational sliding to explain the origin of the Franciscan mélange. Having been schooled in "fixism," I greatly admired the Haarmann–Van Bemmelen concept of primary and secondary tectogenesis. Yet Van Bemmelen attempted to persuade me on the merit of the mobilistic evidence of continental drift. I was skeptical then and was not fully convinced of the validity of sea-floor spreading until I served on the shipboard staff of the Deep-Sea Drilling Project's Leg III, Dakar–Rio de Janeiro. The acquisition of this new insight eventually gave me fresh ideas on the origin of the Franciscan mélange and its puzzling relation to the Great Valley sequence; I, too, began to search for a synthesis.

The purpose of this paper is to present my current interpretation of the Mesozoic history of the California Coast Ranges. This second effort represents a considerable departure from the original version I presented six years ago at Neuchâtel. The emphasis is now shifted to invoke actualistic models of tectonics and sedimentation. Furthermore, I have managed since to learn more about the mechanics of overthrust faulting, and this new knowledge permits me to attempt a clarification of the role of gravity in the genesis of mélanges.

Franciscan Stratigraphy

Franciscan rocks form the backbone of the California Coast Ranges (Fig. 1). These Mesozoic rocks are characterized by a general destruction of original junctions, whether igneous structures or sedimentary bedding, and by the shearing down of the more ductile material until it functions as a matrix in which fragments of the more brittle rocks float as isolated lenticles or boudins. This style of deformation is developed on a great scale. In a country of such tectonic fragmentation, the larger masses, which range up to 20 miles long, constitute mappable units. On the Geologic Map of California, scale 1:250,000, the mappable tectonic fragments appear as if embedded in a homogeneous and structureless matrix, the "undifferentiated Franciscan" (Fig. 2). However, this country rock itself is built up of interdigitating lenticular masses of all sizes, composed of graywackes, chert, limestones, pillow lava, diabase, gabbro, serpentinite, and their metamorphosed

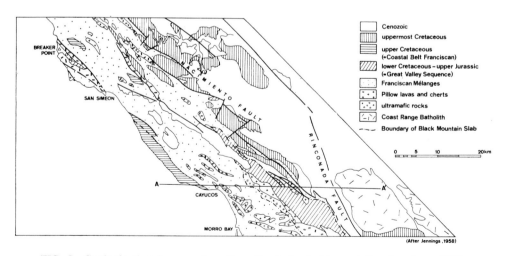

FIG. 2 Geologic sketch map of southern Santa Lucia Range (after Jennings, 1959).

equivalents (glaucophane and lawsonite schists).

Blocks of sedimentary rocks within the Franciscan mélanges have been correlated faunally and lithologically to the Late Jurassic (Tithonian) and Cretaceous sequence on the west side of the Great Valley of California; this is the so-called Great Valley sequence (Bailey and others, 1964). However, the mélanges also include unfossiliferous rocks which are older than Tithonian. I have presented evidence to reconstruct the Franciscan stratigraphy in a series of papers (Hsü, 1967, 1968, 1969; Hsü and Ohrbom, 1969). As exemplified by Table 1, which shows the Mesozoic stratigraphy of the southern Santa Lucia Range, this interpretation differs from the orthodox approach by emphasizing that the Franciscan ophiolites are in part older than the Great Valley sequence.

The Franciscan ophiolite includes basalt tuff, pillow breccia, pillow basalt, diabase, very fine-grained gabbro, and ultramafic rocks. The basalts are locally spilitic. However, the average Franciscan basaltic magma is a normal tholeiitic magma (Bailey and others, 1964, p. 51). The ophiolitic extrusives with a petrology similar to that of the tholeiite found in the oceanic ridge regions today (Moores, 1969) could represent the product of sea-floor spreading (Thayer, 1969; Dickinson, 1970). Such a model of genesis implies that the Franciscan ophiolites represent a homotaxial unit, including the products of sea-floor spreading west of the present Pacific margin, and that these rocks may range in age from early Mesozoic to Late Cretaceous. Fragmentary evidence does support such an interpretation, for metamorphosed pillow basalts yielding metamorphic ages of 130–150 million years (Lee and others, 1964; Suppe, 1969) must have been emplaced long before Late Jurassic, yet basalts containing Cenomanian limestone xenoliths are apparently Late Cretaceous in age (Hsü and Ohrbom, 1969).

The Franciscan sedimentary rocks include radiolarian cherts, pelagic limestones, and several types of graywackes. The radiolarian cherts are invariably associated with greenstones or altered basalt tuff and represent the topmost member of the "Steinmann trinity" (Hsü and Ohrbom, 1969). The genesis of the cherts was related to blooms of radiolaria during Late Jurassic, which resulted in the

TABLE 1 Mesozoic Stratigraphy of Santa Lucia Range

	Franciscan Mélange	Great Valley Sequence
Uppermost Cretaceous (Campanian and younger)		Asuncion formation
San Mateo Unit		
Upper Cretaceous (Campanian and older)	Type III graywacke	
Lower Cretaceous and Uppermost Jurassic (Tithonian)	Type II graywacke with local volcanic rocks	Marmolejo formation Knoxville formation
San Francisco Unit		
Pre-Tithonian	Type I graywacke Radiolarian chert Ophiolite suite, incl. basalt tuff, pillow basalt, gabbro-diabase, and ultramafic rocks (metamorphosed in part)	

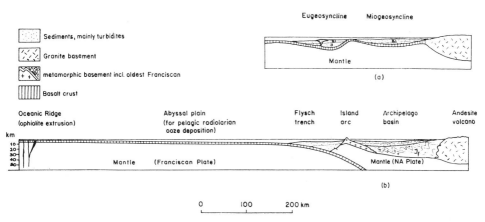

FIG. 3 Reconstruction of the Franciscan eugeosyncline (a) as the outer member of an eu-miogeosynclinal couple, traditional model (after Bailey and Blake, 1969) and (b) as a segment of the Pacific Ocean, plate-tectonics model.

widespread occurrence of radiolarites in the Tethyan and Circumpacific regions (Grunau, 1965). The absence of calcareous plankton may indicate deposition below the calcite-compensation depth, probably in an abyssal hill province between an oceanic ridge and a deep trench on the west margin of Mesozoic North America (Fig. 3b).

Three types of graywackes have been recognized in the mélanges (Table 1). Type I is characterized by an absence of potash-feldspars as detrital grains. This type of graywacke is widespread in the Coast Ranges. It is largely unmetamorphosed in the Santa Lucia Range (Hsü, 1969), but it has been metamorphosed to include such high-pressure minerals as jadeite and aragonite in the Diablo Range (McKee, 1962; Ernst, 1965). Sedimentary structures, where preserved, include graded bedding. These flyschlike sediments were obviously deposited by turbidity currents in deep-sea basins. Two other types of graywacke were found in Franciscan mélanges, either as formations in allochthonous slabs or thrust slices or as tectonic fragments too small to be mapped. Both of these types are characterized by the presence of "Franciscan detritus" as clasts. Also present is potash-feldspar ranging from a trace to 10 percent in type II graywacke to 10–20 percent in type III graywacke. The presence of potash-feldspar detritus is a characteristic of the Cretaceous sandstones of California, deposited when the granitic batholiths were being unroofed. Aside from such lithological evidence, the few fossils that have been found confirm that types II and III graywackes are mainly Cretaceous (Hsü, 1969; Hsü and Ohrbom, 1969; Page, 1970). These sediments constitute the part of the Franciscan that is correlative to the Great Valley sequence.

In reconstructing the Franciscan stratigraphy, I have refrained from the traditional view that the Franciscan rocks were emplaced in an eugeosyncline (Fig. 3a). The term commonly connotes a narrow trough of sedimentation with concurrent submarine volcanism. However, if we adopt an actualistic model, the Franciscan "eugeosyncline" would consist of broad reaches of an oceanic realm, hundreds or even thousands of kilometers wide, stretching from a mid-ocean ridge across an abyssal hill province to a marginal trench (Fig. 3b).

Great Valley Sequence

In addition to the Franciscan, another late Mesozoic sequence is present in the Coast

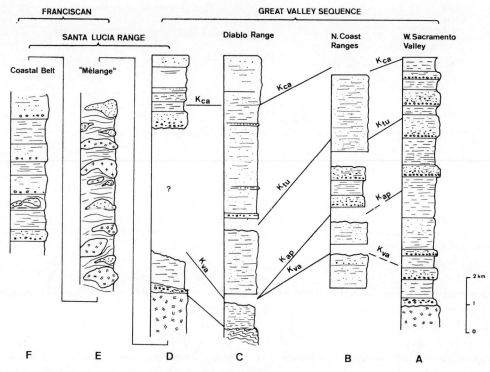

FIG. 4 Schematic stratigraphic columns: K_{va} = Valanginian, K_{ap} = Aptian, K_{tu} = Turonian, K_{ca} = Campanian. Sections D, E, F are tectonically superposed in the Santa Lucia Range.

Ranges. This sequence is best developed on the west side of the Sacramento or Great Valley (Fig. 1) and was thus referred to as the Great Valley sequence (Bailey and others, 1964). Here it includes an almost unbroken succession of latest Jurassic (Tithonian) to Late Cretaceous (Maestrichtian) strata (Fig. 4A). These mainly Cretaceous turbidite sandstones and shales are typical sediments of the flysch facies. They are largely coeval to the types II and III graywackes of the Franciscan, and they were deposited in similar sedimentary environments. In the Coast Ranges the Great Valley succession is less complete (Fig. 4B, C, D). In the Santa Lucia Range, for example, the Great Valley-type rocks fall into two separate age categories, separated by a large time gap and represented by the latest Jurassic to earliest Cretaceous (Tithonian–Valanginian) Knoxville–Marmolejo formations and the latest Cretaceous (Campanian–Maestrichtian) Asuncion formation (Hsü, 1969; Page, 1970).

Since the Cretaceous Franciscan and the Great Valley sequence are both flysch turbidites, a distinction of the two is based upon two sets of criteria: (1) the Great Valley rocks do not constitute broken formations or mélanges, and (2) they do not include radiolarian cherts or ophiolites.

Relation between Franciscan and Great Valley Sequence

Summary of Observations

The relation between the Franciscan and the Great Valley sequence is the problem of the Coast Range tectonics. The following salient facts are now recognized:

1. A fault commonly intervenes where the Great Valley sequence and the Franciscan

rocks are in contact (Page, 1966).
2. The Franciscan rocks are commonly severely sheared, whereas the Great Valley rocks show little disruption of stratal continuity (Hsü, 1968).
3. The base of the Great Valley sequence—the Knoxville formation—on the west side of Sacramento Valley (Fig. 4A) and in the Santa Lucia Range (Fig. 4D), is superposed upon an ophiolitic basement (Bailey and Blake, 1969; Hsü, 1969). In contrast, the Great Valley rocks overlie Franciscan jadeitic graywackes in the Pacheco Pass area of the Diablo Range (Fig. 4C).
4. The Franciscan rocks are largely unfossiliferous. The paleontologically dated Franciscan rocks are mainly type II and III graywackes and pelagic limestones. The fossiliferous rocks are on the whole coeval to the Great Valley sequence (Irwin, 1957; Hsü and Ohrbom, 1969); the unfossiliferous ophiolites and their metamorphosed equivalents are at least in part older than Tithonian (Hsü, 1968).
5. Radiometric dates indicate two or more episodes of Franciscan metamorphism (Coleman, 1967; Suppe, 1969; Ernst, 1970): a Jurassic episode, between 130 and 150 million years, for the unfossiliferous glaucophane-schists in the Diablo Range and its northern extension; and a Cretaceous episode, between 110 and 130 million years, for the sparsely fossiliferous lawsonite-schists in the Northern Coast Ranges. A younger (75–80 million year) event involved only the albitization of older glaucophane-schists (Keith and Coleman, 1968).
6. Experimental evidence indicates that the jadeitic and aragonitic Franciscan rocks must have been metamorphosed under relatively low-temperature, high-pressure conditions (Ernst, 1965; Brace and others, 1970).

This set of facts has led to two somewhat different schools of thought on the tectonic evolution of California Coast Ranges.

The Hypothesis of a Coeval Eugeosynclinal and Miogeosynclinal Couple

The fact that all fossils that have been found so far in the Franciscan range from latest Jurassic (Tithonian) to Late Cretaceous in age seems a strong argument in favor of the hypothesis advanced by Irwin in 1957, that the Franciscan and the Great Valley sequence were coeval and do not include rocks older than Tithonian. However, the assignment of a time range of deposition to all the rocks in a mélange on the basis of the oldest and youngest fossils found in such a mélange can be wrong (Hsü, 1968, p. 1067), because such a tectonic mixture often includes exotic blocks of igneous and metamorphic rocks which constituted the original basement of the fossiliferous sediments.

The Franciscan mélange includes metamorphic rocks which yielded a Late Jurassic metamorphic age. Those rocks must have been emplaced before the deposition of the earliest Great Valley sediments in Tithonian. Yet this fact was ignored by Bailey, Irwin, and Jones (1964), who further developed the hypothesis of a coexisting eugeosynclinal and miogeosynclinal pair (Fig. 3a). Their paleogeographic reconstruction has formed the basis of tectonic interpretation by many current workers. Others, including myself, have taken exception to this oversimplified interpretation because a synthesis of Coast Range geology must account for its pre-Tithonian history.

Hypothesis of Franciscan Rocks as a Polykinematic Mélange

The relations between the Franciscan and Great Valley rocks may be interpreted as follows:

1. The fault between the Great Valley and the Franciscan rocks is not a large peel-thrust involving 50–100-km displacement between the strata immediately above and below the fault. This fault represents the

contact between an internally undisturbed Great Valley sequence and the pervasively sheared Franciscan mélange.
2. The major crustal displacement did not take place along this, or any other, *single* surface of discontinuity, but along innumerable shear surfaces within the Franciscan mélange.
3. The lowest unit of the Great Valley sequence has been deposited unconformably upon different Franciscan basement types—upon a Franciscan ophiolitic complex on the west side of the Sacramento Valley, and in the Santa Lucia Range, but upon a pre-Tithonian metamorphic Franciscan basement in parts of the Diablo Range (Fig. 4).
4. Fossiliferous Franciscan rocks are indeed coeval to the Great Valley sequence, but at least some of the unfossiliferous Franciscan rocks were emplaced and deposited prior to Tithonian. The Franciscan has a pre-Tithonian history.
5. There have been two generations of high-pressure metamorphism: a pre-Tithonian glaucophane-schist metamorphism and a Cretaceous lawsonite-schist metamorphism.
6. The glaucophane-schists were formed under a high overburden pressure but were squeezed upward during a pre-Tithonian deformation.

The detailed arguments leading to my conclusions have been given in a series of papers (Hsü, 1967, 1968, 1969a, 1970a; Hsü and Ohrbom, 1969). Individually, these interpretations are rather straightforward and noncontroversial. However, the sum of these ideas indicates a Late Jurassic deformation of an older, unfossiliferous Franciscan, at about the same time as the destruction of the Sierra Nevada geosyncline by the Nevadan orogeny. This last conclusion has been disputed by others (e.g., Bailey and Blake, 1969), even though the evidence of pre-Tithonian deformation is clear. Subsequent to this Jurassic orogeny, deformation and metamorphism took place during the Cretaceous. In fact, much of the mélange structures we now see in the Coast Ranges originated during this later orogeny.

Franciscan Structures

Franciscan includes both large slabs or thrust slices and undifferentiated mélanges. Mélange structures are well displayed along the seashore or at new highway cuts. Where the Franciscan rocks include only interbedded graywackes and shales, the graywacke layers commonly have pinch-and-swell appearance, resulting from the stretching of brittle graywacke beds. Angular boudins, commonly lenticular and bounded in part by a complementary set of shear fractures, were produced when graywacke beds were disrupted by extension. Such shear fractures, indicating extension parallel to bedding, are rare or absent in the shaly matrix surrounding the boudins. The shales were deformed ductily by flowage, which induced further stretching and bodily rotation of the boudins. On a somewhat larger scale, small thrust planes subparallel to bedding are commonly observed. The strata between form slabs ranging from a few decimeters to many meters thick. Some such slabs have been stretched and sheared on a minute scale as described previously. Others show practically no internal strain. Minor recumbent folding of the slabs is present, commonly in the form of drag-folding accompanying small thrust faults.

The nature of the mélange terrane renders mapping difficult. A regional framework can be deciphered if large allochthonous slabs are associated with a mélange. This relation is exemplified by the geology of the southern Santa Lucia Range.

Figure 2 shows that the part of the Santa Lucia Range west of the Nacimento-Rinconada fault zone is underlain mainly by Franciscan and Great Valley rocks. Those rocks constitute three tectonic units (Fig. 5):

1. An allochthonous slab of Great Valley

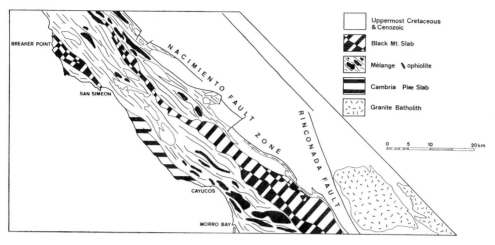

FIG. 5 Tectonic map of southern Santa Lucia Range.

rocks and their ophiolite basement—the Black Mountain slab.
2. An undifferentiated Franciscan mélange, including many ophiolite-chert and graywacke blocks.
3. An allochthonous slab of Coastal Belt Franciscan—the Cambria Pine slab.

The Black Mountain slab (Hsü, 1969) is 35 km long and has an outcrop width of about 10 km. It consists of Knoxville and Marmolejo strata and the underlying Franciscan ophiolite complex. The slab must have been nearly horizontal during the Miocene. Now the slab and the Miocene shale have been folded into a sharp syncline with considerable fracturing and shearing on the eastern limb of this syncline (upper diagram of Fig. 6). When the syncline is unfolded and the sheared-off limb restored, the Black Mountain slab is 20 or 30 km wide (Fig. 6G). The Jurassic ophiolite of the Black Mountain slab has been superposed tectonically on the Franciscan mélange, which includes mainly Cretaceous type II graywacke. The steeply dipping chaotic terrane has an outcrop width of about 10 km (Fig. 5), but its original extent must have been much wider. West of the mélange zone is another large slab, the so-called Cambria Pine slab; this allochthonous slice includes mainly an Upper Cretaceous formation (type III graywacke). Locally, the graywacke beds have been broken and mixed with ophiolites to form mélanges (Hsü, 1969).

Such a threefold tectonic division could also be applied to unravel the complex structure of the Northern Coast Ranges. Figure 1 shows that the highest tectonic unit there is present on the edge of the northern Great Valley, also known as the Sacramento Valley, where the Great Valley sequence overlies depositionally an ophiolite slab. On the Pacific Coast is the Coastal Belt Franciscan, another large allochthonous plate. Sandwiched between is the main body of the Franciscan mélange of the northern Coast Ranges, which encloses several smaller allochthonous slabs of the Cretaceous sedimentary rocks.

Geologic Evolution of Western North America

The geologic evolution of the Franciscan mélanges has been interpreted in terms of plate tectonics (e.g., Crowell, 1968; Hamilton, 1969; Swe and Dickinson, 1969; Bailey and Blake, 1969; Ernst, 1970). Not taking

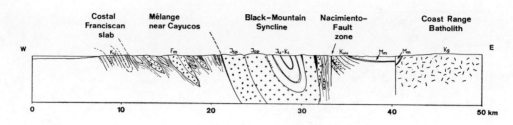
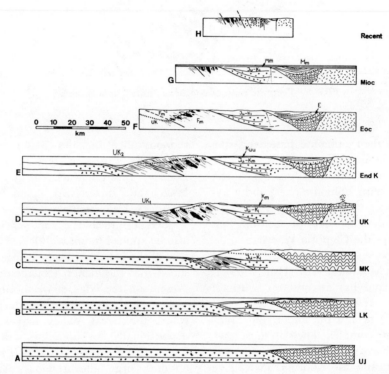

FIG. 6 Schematic diagrams showing the structural evolution of southern Santa Lucia Range and the genesis of a Franciscan mélange. (See text for detailed explanation.) Top diagram is a cross section across the Santa Lucia Range, identical to diagram H but on a larger scale.

into account the pre-Tithonian history, the model adopted by current workers assumes deformation along an ancient Benioff zone west of the Mesozoic North American continent. This interpretation left several puzzling facts unresolved, such as the presence of high-pressure metamorphic rocks in Franciscan mélanges and the existence of a large "pressure-discontinuity" at the contact between the Franciscan and Great Valley rocks of the Diablo Range. For these and other reasons it is necessary to postulate a more complicated history (Hsü, 1971), involving the following four stages as illustrated by Fig. 7:

1. *Early Mesozoic Rifting.* During this stage, a part of the continental crust—Salinia—was split away from the North American continent. Ophiolites were emplaced along an

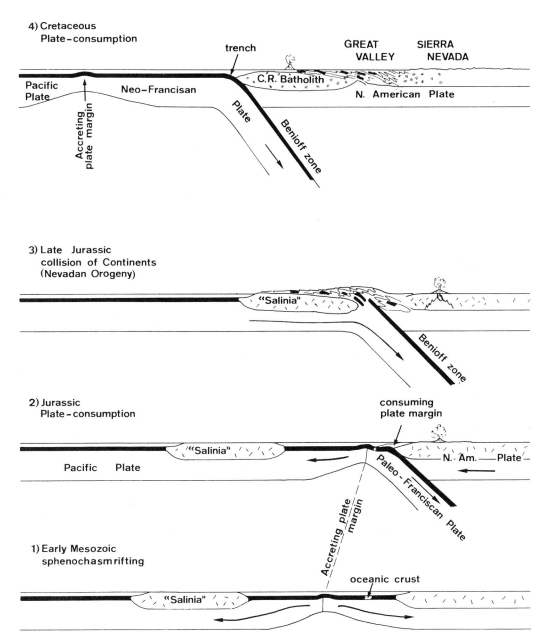

FIG. 7 Mesozoic evolution of western North America, according to the plate-tectonics model. (See text for detailed explanation.)

accretional ridge in this rift and were later buried under oceanic sediments. The ridge represented a plate margin, separating new crust welded to the Pacific plate from that welded to the North American plate.

2. *Jurassic Plate Consumption.* As the North American continent drifted westward during the Jurassic, the continental crust overrode oceanic crust and a trench was to mark the position of a new plate margin. The oceanic crust west of the trench belonged to the Paleo-Franciscan plate, which was plunged under the trench along a Jurassic Benioff zone. Type I Franciscan graywackes were deposited in this trench. Partial melting of the descending plate resulted in Late Jurassic volcanism, which represented the surface expression of the Sierra Nevada batholith.

3. *Late Jurassic Collision of Continents.* During the Late Jurassic, the Paleo-Franciscan plate and that part of the Pacific plate east of Salinia were completely consumed, resulting in the collision of the North America and the microcontinent Salinia. The Nevadan orogeny is an expression of this collision, when deeply buried metamorphic and mantle rocks were squeezed out and partly thrust over Salinia.

4. *Cretaceous Plate Consumption.* Beginning in latest Jurassic (Tithonian) and continuing till Late Cretaceous or Early Tertiary, a trench existed along the west side of the newly welded North American plate. A Neo-Franciscan plate was present between a spreading ridge and the North American plate. Ophiolites were extruded at the ridge, radiolarites accumulated in an abyssal hills province, and flysch-graywackes (types II and III) were poured into the trench, spilling over onto abyssal plains. The Neo-Franciscan rocks are coeval to the Great Valley sequence, which was deposited in deep basins behind or east of the island arc. Mixing of those and older rocks along a newly located Benioff zone under the trench produced the mélanges. Meanwhile, Cretaceous andesitic volcanoes were present in Salinia, as this former microcontinent was converted by magmatic activities into the Coast Range batholith.

Mechanism of Mélange Formation

The origin of a mélange belt tens or hundreds of kilometers wide and several kilometers thick is a great puzzle. Such a large body could not have been deposited as a sedimentary body during a single catastrophic event. The Franciscan mélange is obviously a tectonic mixture. Current theories relate the pervasive shearing of the mélange to tectonic movements along an ancient Benioff zone. But how is this achieved? Ernst (1970, Fig. 3) postulates underthrusting along a Benioff zone which dipped more than 30° under the continent and portrays this zone as the tectonic contact between the Great Valley and the Franciscan. This tectonic contact is, however, more or less parallel to the overlying Cenozoic sediments in many places, as my mapping in the Santa Lucia Range showed (Hsü, 1969a). Taking this fact into account, Ernst's model implies a 30° rotation of the Benioff zone after the underthrusting. If this has been the case, the eastern edge of the mélange zone—the lowest Great Valley rocks west of Sacramento Valley—must once have been buried 75 km below the surface and 30 km below the continental Moho. These implications are contrary to observations; the unmetamorphosed Great Valley rocks could never have been buried so deeply.

An alternative hypothesis is to assume that the mélange was produced at a migrating juncture under an ocean trench, an idea casually referred to by Hamilton (1969). The dip of this juncture—the so-called Benioff zone—near the surface must have been less than 30°, perhaps only 10° or less, as is the case now under the Aleutian Trench (Pflaker, 1969). The westward migration of such gently dipping Benioff zones could best account for a mélange zone over a hundred kilometers wide.

Figure 6 illustrates the genesis of the mainly

Cretaceous Franciscan mélange of the Santa Lucia Range:

1. During latest Jurassic (Tithonian) time, a marginal trench was created and a new Benioff zone was formed west of the Salinia.
2. Latest Jurassic and Early Cretaceous sediments were deposited on an ophiolitic oceanic crust in this trench.
3. The plate boundary was then moved west during Early Cretaceous. The body of rocks between the latest Jurassic and the Early Cretaceous Benioff zones constitute the Black Mountain slab. This slab was now welded to the Salinia block and formed a coastal range. Type II graywacke and sedimentary breccia were poured into a westward-migrating trench during the middle Cretaceous. Ophiolites were detached from an oceanic crust, Franciscan rocks slumped down from the adjacent coastal range, and newly deposited Cretaceous turbidites and olistostromes were sheared, broken, and mixed to form the main mélange body.
4. Deposition of type III graywacke and generation of mélange continued during the Late Cretaceous. Partial melting of the descending plate resulted in andesitic volcanism, which represented the surface expression of a deep-seated plutonic process that changed Salinia to the Coast Range batholith.
5. The consuming plate margin moved west of the Coastal Belt Franciscan. Andesitic debris removed from the top of the Coast Range batholith formed the main constitution of the uppermost Cretaceous conglomerates. These and finer clastics of the Asuncion formation constitute the basal "postorogenic" sediments of the Coast Ranges.
6. The western part of the Santa Lucia Range was uplifted during Early Tertiary. The Franciscan debris removed from that area was carried westward to form offshore deposits.
7. The rocks of the Santa Lucia Range were folded during Miocene. Middle and late Miocene sediments were then deposited upon bevelled Franciscan and granitic basements.
8. Deformation during Plio-Pleistocene caused the synclinal structure of the Black Mountain slab and the numerous faults of the region.

Role of Gravity in Tectonics

Role of Gravity in Mélange Genesis

I once was inclined to regard the Franciscan mélange as the product of gravity sliding; this interpretation was based mainly on comparison of its tectonic style with the *argille scagliose* of the Apennines (e.g., Hsü, 1965). I found the argument that tectonic mélanges like "the *argille scagliose* are too weak to have moved as a sheet by a push on one side" (Page, 1963, p. 669) very convincing. Furthermore, gravity sliding seemed to me then the only adequate mechanism to account for the very strange fact that the Great Valley rocks are allochthonous in the Coast Ranges but are autochthonously attached to their various basements in the Great Valley itself (Hsü, 1968). Nevertheless, I disagreed with Maxwell (1959, p. 2711) and have found no evidence that Franciscan mélange, like *argille scagliose*, could represent a gigantic submarine slide of cohesionless sediments. I emphasized the distinction between a tectonic *mélange* and an *olistostrome*. The former results from *subterranean* deformation of consolidated rocks under an overburden, whereas the latter defines a sedimentary deposit originated from a *submarine* slide. Thus I have compared the shearing of a mélange under a large allochthonous slab to the deformation of a groundmoraine under a glacier (Hsü, 1968). I was always somewhat uncomfortable with my own interpretations: the mélange deformation of the Franciscan indicates considerable shearing strain, yet the allochthonous slabs of the Great Valley rocks did not seem to have gone very far from their original site of deposition to induce such magnitudes of shearing.

During the last few years, a revolutionary change occurred in our thinking on tectonics. The plate-tectonics hypotheses provided a mechanism for Franciscan underthrusting which neatly explains its puzzling relation with the Great Valley sequence. Meanwhile, I have also learned more about the mechanism of overthrust tectonics and the processes that may lead to the formation of mélanges.

Considering the inclined junction between the relatively overthrusting North American plate and the underthrusting Neo-Franciscan plate, the wedge that rides over the underthrust mass is the toe, and the resistance to overthrusting produced by this toe is called *toe effect* (Raleigh and Griggs, 1963).

If the main thrust movement is horizontal and the toe has a triangular cross section (Fig. 8), the force F_2 required to push the triangular toe up a slope θ is expressed by (Hsü, 1969b):

$$F_2 \cos \theta = \tau_0 A_i + (1 - \lambda) \tan \phi \\ \times (\rho \cdot g \cdot V \cdot \cos \theta + F_2 \sin \theta) \\ + \rho \cdot g \cdot V \cdot \sin \theta \quad (1)$$

where τ_0 = cohesive strength, or initial shear strength
A_i = area of inclined thrust plane above the toe = $x \cdot y / \cos \theta = 2V/z \cos \theta$
V = volume of the toe = $\frac{1}{2} x \cdot y \cdot z$
x = length of toe
y = width of toe
z = thickness of toe
ρ = bulk density (average) of the thrust block
g = gravitational acceleration
ϕ = angle of internal friction
λ = pore-pressure to overburden-pressure ratio along the thrust

After thrust movement along a plane has taken place, the cohesive strength along this plane may be weakened, especially if the thrust plane is a brittle fracture (Hsü, 1969b).

On the other hand, an increasingly larger volume of the thrust mass will have to be pushed up the toe; consequently, the toe effect will be increasingly larger so that the thrust movement either is completely stopped by this resistance or must seek a new break behind the old toe (Raleigh and Griggs, 1963).

The minimum force F_1 required to push up an enlarged old toe is expressed by:

$$F_1 \cos \theta = \tau_0' A_i + (1 - \lambda) \tan \phi \\ \times (\rho \cdot g \cdot V \cdot \cos \theta + F_1 \sin \theta) \\ + \rho \cdot g \cdot V \cdot \sin \theta \\ + \tau_0' A_h \\ + (1 - \lambda) \tan \phi \cdot \rho \cdot g \cdot \cos \theta \cdot \Delta V \quad (2)$$

where τ_0' = initial shearing resistance to gliding along pre-existing fracture plane, or residual cohesive strength
A_h = area of the horizontal thrust plane between the old and new toes = $\Delta x \cdot y = \Delta V/z$
ΔV = volume difference between the old and new toes = $(\Delta x) \cdot y \cdot z$
Δx = horizontal separation between old and new toes

The force F_2 required to move a new triangular toe up along a new break is the same as that given by Eq. 1.

The movement will take place along a new toe when

$$F_2 = F_1 \quad (3)$$

Substituting Eqs. 1 and 2 into Eq. 3, we have:

$$\tau_0 \cdot \frac{2V}{z \cdot \cos \theta} = \tau_0' \cdot \frac{2V}{z \cdot \cos \theta} \\ + \left[\frac{\tau_0'}{z} + (1 - \lambda) \tan \phi \cdot \rho \cdot g \cdot \cos \theta \right] \Delta V \quad (4)$$

Solving for the volume ratio, we have:

$$\frac{\Delta V}{V} = \frac{2(\tau_0 - \tau_0')}{[\tau_0' + (1 - \lambda) \tan \phi \cdot \rho g z \cdot \cos \theta] \cos \theta} \quad (5)$$

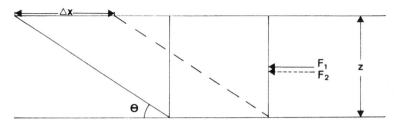

FIG. 8 The toe effect. The inclined portion of a thrust plane makes an angle θ with the horizontal. ΔX is the spacing of imbrication, and z the thickness of the thrust plate. F_1 is the force required to push a toe up an existing thrust plane. F_2 is the force required to push a new toe up along a new thrust plane (shown by the dashed line).

The spacing of the imbrication is expressed by $\Delta x/x$, which is the distance between two successive inclined thrust planes, divided by the length of the triangular toe. Substituting the linear ratio for the volume ratio in Eq. 5, we have:

$$\frac{\Delta x}{x} = \frac{(\tau_0 - \tau_0')}{[\tau_0' + (1 - \lambda) \tan \phi \cdot \rho g z \cdot \cos \theta] \cos \theta} \quad (6)$$

The two limiting cases given by Eq. 6 are as follows:

1. When $\tau_0' \to \tau_0$, $\Delta x/x$ approaches zero, or $x/\Delta x$ approaches infinity. In other words, thrusting without a reduction of cohesive strength would lead to a very densely spaced zone of imbrication, forming a tectonic mélange.
2. When $\tau_0' \to 0$ and $\lambda \to 1$, $\Delta x/x$ approaches infinity. That is, if an overthrust is displaced along a brittle fracture, with a total loss of cohesion and adhesion, and if the frictional resistance along the thrust plane has been rendered null by an extremely high pore pressure, there should be no imbrication.

In addition to the effects of relative cohesive strength and of pore pressure is the role of the overburden pressure, $\rho \cdot g \cdot z$, which is directly related to the thickness of the thrust plate, z. Equation 6 shows that a large overburden pressure because of a thicker thrust plate tends to increase the frictional resistance along the old thrust plane and thus reduce the ratio $\Delta x/x$. Furthermore, a larger overburden pressure favors deformation by flowage, which tends to reduce the difference between τ_0 and τ_0', and thus indirectly also leads to a denser spacing of imbrication. Factors ρ and θ in Eq. 6 vary within a small range of values, and their roles are secondary.

The detailed numerical results of my analysis of the toe effect have been presented elsewhere (Hsü, 1970) and will not be repeated here. Nevertheless, the foregoing discussion clearly indicates that the style of deformation characteristic of the tectonic mélanges is not necessarily typical of gravity-sliding tectonics. *The toe effect of thrusting under compression can also produce a mélange with densely spaced thrust planes, if the thrust is a zone of flowage.* The tectonic style of the Franciscan leaves little doubt that its predominantly pelitic matrix has been sheared ductily. In contrast, the Great Valley sequence has been deformed by thrust surfaces which represent deformation along brittle fractures. Consequently, the toe effect has been to produce densely spaced shear fractures in the Franciscan, leading to the formation of a mélange, and to produce widely spaced shear fractures in the Great Valley sequence, separating the various allochthonous slabs.

To summarize, I do not believe that gravity sliding is the only explanation for the tectonic style of mélange deformation. The features we observe in the Franciscan are consistent with the hypothesis that the mélange resulted from the disintegration of an oceanic plate, as it was dragged down along an ancient Benioff zone. Nevertheless, a secondary role must be assigned to gravity-induced downslope movements. The mélange includes many sedimentary breccia beds. The possibility must be considered that Franciscan mélange represents largely sheared olistostromes. In any case, sedimentary processes such as submarine sliding may have been largely responsible for the introduction of exotic elements. For example, the Jurassic glaucophane-schist blocks in the Santa Lucia Range were most likely sedimented in a Cretaceous trench as a component of olistostromes prior to their subsequent tectonic mixing with the sheared and broken graywacke beds.

Role of Gravity in Plate Tectonics

In this paper I have invoked the plate-tectonics hypothesis to interpret the tectonic evolution of the California Coast Ranges. Yet unresolved is the nature of the driving force for large horizontal displacements of lithospheric plates. Mantlewide convection cells have been proposed as a mechanism, but this assumption leads to serious difficulties (Knopoff, 1969; Orowan, 1969). Therefore, Elsasser (1967), Isacks and Molnar (1969), and Jacoby (1970) proposed that plate movements result from gravitational instability of the earth's upper mantle.

In their study of mantle earthquake mechanisms, Isacks and Molnar (1969) found "clear evidence for down-dip extensional stress within (lithospheric) slabs" at 300–500-km depths in several of the Circumpacific regions. The stress system could be accounted for by a model that postulates sinking of a lithospheric plate under its own weight. The predominance of down-dip orientation of principal stresses further encouraged the idea that "gravitational body forces on the down-going slabs are important forces in determining the stress within the lithosphere and may be important forces in driving the global system of plate movements" (Isacks and Molnar, 1969, p. 1121). Later work by Katsumata and Sykes (1969, p. 5946) afforded further evidence that earthquake mechanisms "are related to the state of stress within a single down-going plate rather than to the relative motion of two large plates." Furthermore, one of their profiles shows a nearly vertical distribution of the earthquake hypocenters for depths between 200 and 700 km in the Mariana region. This peculiar orientation can be explained if the plate is a gravitationally sinking slab.

The hypothesis of sinking slabs cannot explain the movements of overriding plates such as the westward drift of the Americas. However, this major shortcoming has been remedied by Jacoby (1970), who postulated that those plates were being pushed by gravitationally induced diapiric movement under a mid-ocean ridge. The basic idea of this new treatise was contained in the 1965 paper by Van Bemmelen on mega-undations as a cause of continental drift.

I am not sufficiently informed to judge the merits of these geophysical theories. However, I believe we should be reminded that gravity is always with us, and the role of gravitative forces cannot be ignored even if we accept the postulate of plate tectonics.

Looking Back after Six Years

When I first prepared my talk for the Wegmann Symposium, the California Coast Ranges represented just one of many poorly investigated mountain ranges; the puzzling Franciscan was compared to the blind men's elephant in an Indian saga. During the last few years we have witnessed an outburst of interest in the Franciscan and a frenzy of activity in the California Coast Ranges. New interpretations tend to unite and harmonize old contradictory or even hostile views. There is much similarity between the interpretations I have presented in this paper and the conclusions reached by others working, or not working, in the Coast Ranges (e.g., Crowell, 1968; Hamilton, 1969; Bailey and Blake, 1969; Ernst, 1970; Dewey and Bird, 1970). Perhaps we could indeed hold up the Franciscan geology as a model of eugeosynclinal sedimentation and underthrusting tectonics (Hsü, 1971).

My working hypothesis, when it was first presented, was centered on the theme that the Franciscan is largely a tectonic mélange. The recognition of its mélange structure released us from the suffocating confines of the eugeosyncline dogma. No longer do we need to view the Franciscan as a conformable sequence deposited in a long, narrow trough that was also the site of concurrent volcanism. Instead, the mélange could well have been a small remnant of a segment of the Mesozoic Pacific Ocean, which has been largely plunged into the mantle under the North American continent. Needless to say, as we are all children of our time, this second attempt to clarify a long-standing paradox is much tainted with an emphasis on plate tectonics and horizontal movements. It is interesting to note, however, that the swing of the pendulum has begun to move in an opposite direction. Theoreticians speculating on the motor of plate motion are leading us back to gravity tectonics.

References

Ampferer, O., 1906, Über das Bewegungsbild von Faltengebirgen: *k.k. Geol. Reichsanstalt, Jahrb.*, v. 56, p. 539–622.

Argand, E., 1922, La Tectonique de l'Asie: *13ème Congr. Géol. Int.*, v. 1, p. 171–372.

Bailey, E. H., and Blake, M. C., 1969, Late Mesozoic sedimentation and deformation in western California: *Geotektonika*, pt. 3, p. 17–30; pt. 4, p. 24–34.

Bailey, E. H., Irwin, W. P., and Jones, D. L., 1964, Franciscan and related rocks, and their significance in the geology of western California: *Calif. Div. Mines Bull. 183*, 177 p.

Beloussov, V. V., 1962, *Basic problems in geotectonics*: New York, McGraw-Hill, 816 p.

Brace, W. F., Ernst, W. G., and Kallberg, R. W., 1970, An experimental study of tectonic overpressure in Franciscan rocks: *Geol. Soc. Am. Bull.*, v. 81, p. 1325–1338.

Bucher, W. H., 1933, The deformation of the Earth's crust: Princeton, N.J., Princeton University Press, 518 p.

Coleman, R. G., 1969, Glaucophane schists from California and New Caledonia: *Tectonophysics*, v. 4, p. 479–498.

Crowell, J. C., 1968, Movement histories of faults in the Transverse Ranges and speculations on the tectonic history of California, *in* Dickinson, W. R., and Grantz, A., eds., *Proc. Conference on Geologic Problems of San Andreas Fault System*: Stanford California, Stanford Univ. Publ., Geological Sciences, 374 p.

Dewey, J. F., and Bird, J. M., 1970, Mountain belts and the new global tectonics: *J. Geophys. Res.*, v. 225, p. 521–525.

Dickinson, W. R., 1970, The new global tectonics: *Geotimes*, v. 15, no. 4, p. 18–22.

Elsasser, W. M., 1967, Convection and stress propagation in the upper mantle: *Princeton Univ. Tech. Rep. 5*, p. 223–246.

Ernst, W. G., 1965, Mineral parageneses in Franciscan metamorphic rocks, Panoche Pass, California: *Geol. Soc. Am. Bull.*, v. 76, p. 879–914.

———, 1970, Tectonic contact between the Franciscan mélange and the Great Valley sequence—Crustal expression of a late Mesozoic Benioff zone: *J. Geoph. Res.*, v. 75, p. 886–901.

Grunau, H. R., 1965, Radiolarian cherts and associated rocks in space and time: *Eclog. Geol. Helv.*, v. 58, p. 157–208.

Hamilton, W., 1969, Mesozoic California and the underflow of Pacific mantle: *Geol. Soc. Am. Bull.*, v. 80, p. 2409–2430.

Haarman, E., *Die Oszillationstheorie*: Stuttgart, Ferdinand Enke, 260 p.

Hsü, K. J., 1965, Franciscan rocks of Santa Lucia Range, California, and the argille scagliose of the Apennines, Italy: A comparison in style of deformation: *Geol. Soc. Am. Spec. Paper 87*, p. 210–211.

———, 1967, Mesozoic geology of the California Coast Ranges—A new working hypothesis, *in* Schaer, J., ed., *Etages tectoniques*: Neuchâtel, à la Baconnière, p. 279–296.

———, 1968, Principle of mélanges and their bearing on the Franciscan-Knoxville paradox: *Geol. Soc. Am. Bull.*, v. 79, p. 1063–1074.

———, 1969a, Preliminary report and geologic guide to Franciscan mélanges of the Morro Bay—San Simeon area, California: *Calif. Div. Mines Geol. Spec. Publ. 35*, 46 p.

Hsü, K. J., 1969b, Role of cohesive strength in the mechanics of overthrust faulting and of landsliding: Geol. Soc. Am. Bull., v. 80, p. 927–952.

———, 1970, Cohesive strength, toe effect, and the mechanics of imbricated overthrusts: Proc. 2nd Intern. Rock Mech. Conf., Paper 3–36, 4 p.

———, 1971, Franciscan mélanges as a model for geosynclinal sedimentation and underthrusting tectonics: J. Geoph. Res., v. 75, p. 1162–1170.

———, and Andrews, J. E., 1970, History of South Atlantic basin, in Maxwell, A. E., and others, Initial Reports of the Deep Sea Drilling Project, v. 3, p. 464–467.

Hsü, K. J., and Ohrbom, R., 1969, Mélanges of San Francisco Peninsula—Geologic reinterpretation of type Franciscan: Am. Assoc. Petrol. Geol. Bull., v. 53, p. 1348–1367.

Hubbert, M. K., and Rubey, W. W., 1959, Role of fluid pressure in mechanics of overthrust faulting: Geol. Soc. Am. Bull., v. 70, p. 115–166.

Isacks, B., and Molnar, P., 1969, Mantle earthquake mechanisms and the sinking of the lithosphere: Nature, v. 223, p. 1121–1124.

Jacoby, W. R., 1970, Instability in the upper mantle and global plate movements: J. Geoph. Res., v. 75, p. 5671–5680.

———, 1973, Gravitational instability and plate tectonics (this volume).

Jeffreys, H., 1929, The Earth: Cambridge, England, Cambridge University Press, 278 p.

Jennings, C. W., 1959, Geologic map of California: California Div. Mines 1/250,000.

Katsumata, M., and Sykes, L. R., 1969, Seismicity and tectonics of the western Pacific: Izu–Mariana–Caroline and Ryukyu–Taiwan regions: J. Geoph. Res., v. 74, p. 5923–5948.

Keith, T. C., and Coleman, R. G., 1968, Albite-pyroxene-glaucophane schist from Valley Ford, California: U.S. Geol. Survey Prof. Paper 600-C, p. 13–17.

Knopoff, L., 1969, Continental drift and convection, in The Earth's crust and upper mantle: Geophys. Monog. 13, Washington, D. C., Am. Geoph. Union, p. 683–693.

Lee, D. E., Thomas, H. H., Marvin, R. F., and Coleman, R. G., 1964, Isotopic ages of glaucophane schists from the area of Cazadero, California: U.S. Geol. Survey Prof. Paper 473-D, p. 105–107.

Maxwell, J. C., 1959, Turbidite, tectonic and gravity transport, northern Apennine Mountains, Italy: Am. Assoc. Petrol. Geol. Bull., v. 43, p. 2701–2719.

McKee, B., 1962, Widespread occurrence of jadeite, lawsonite, and glaucophane in central California: Am. Jour. Sci., v. 260, p. 596–610.

Moores, E. M., 1969, Petrology and structure of the Vourinos ophiolite complex of northern Greece: Geol. Soc. Am. Spec. Paper 118, 74 p.

Orowan, E., 1969, The origin of the oceanic ridges: Sci. Am., v. 214, p. 103–119.

Page, B. M., 1963, Gravity tectonics near Passo della Cisa, northern Apennines, Italy: Geol. Soc. Am. Bull., v. 74, 655–672.

———, 1966, Geology of the Coast Ranges of California: Calif. Div. Mines Geol. Bull. 190, p. 255–276.

———, 1970, Sur-Nacimiento fault zone of California: Continental margin tectonics: Geol. Soc. Am. Bull., v. 81, p. 667–690.

Pflaker, G., 1969, Tectonics of the March 27, 1964 Alaska Earthquake: U.S. Geol. Survey Prof. Paper 543-I, 74 p.

Raleigh, C. B., and Griggs, D. T., 1963, Effect of the toe in the mechanics of overthrust faulting: Geol. Soc. Am. Bull., v. 70, p. 115–166.

Staub, R., 1928, Der Bewegungsmechanismus der Erde: Berlin, Borntraeger, 270 p.

Suppe, J., 1969, Times of metamorphism in the Franciscan terrain of the northern Coast Ranges, California: Geol. Soc. Am. Bull., v. 80, p. 135–142.

Swe W., and Dickinson, W. R., 1970, Sedimentation and thrusting of late Mesozoic rocks in the Coast Ranges near Clear Lake, California: Geol. Soc. Am. Bull., v. 81, p. 165–188.

Thayer, T. P., Peridotite-gabbro complexes as keys to petrology of mid-oceanic ridges: Geol. Soc. Am. Bull., v. 80, p. 1515–1522.

Van Bemmelen, R. W., 1932, De Undatie-theorie: Natuurk. Tijdschr. Ned. Indië, v. 92, p. 85–242-373–402.

———, 1954, Mountain building: The Hague, Martinus Nijhoff, 177 p.

———, 1960, Zur Mechanik der ostalpinen Deckenbildung: Geol. Rund., v. 50, p. 474–499.

———, 1963a, Mega-undations as cause of continental drift: Geol. Mijnb., v. 9, p. 320–333.

———, 1967, The importance of the geonomic dimensions for geodynamic concepts: Earth Sci. Reviews, v. 3, p. 79–110.

———, 1969, The Alpine loop of the Tethys zone: Tectonophysics, v. 8, p. 102–113.

L. A. WRIGHT

Department of Geosciences
The Pennsylvania State University
University Park, Pennsylvania

B. W. TROXEL

California Division of Mines and Geology
Sacramento, California

Present address: Geological Society of America
Boulder, Colorado

Shallow-Fault Interpretation of Basin and Range Structure, Southwestern Great Basin

Geologists have long recognized that many of the ranges of the Basin and Range province of the North American Cordillera consist of tilted fault blocks. They also have accumulated abundant evidence that the tilting began in Cenozoic time and is attributable to rotation along normal faults that bound the blocks. The nature of the master fault surfaces upon which the blocks have rotated, however, has remained conjectural, because (1) their traces ordinarily lie hidden beneath the alluvium that flanks the ranges, and (2) adequate control in plotting their downdip configurations has been difficult to establish. Indeed, from region to region within the Basin and Range province, one observes such a wide range in the lengths and widths of individual blocks and in the angles of tilt that it is difficult or impossible to conclude whether a given block should be considered typical or atypical of the province.

Evidence of the overall form of a hidden bounding fault exists in the angle of Cenozoic tilting detectable in the overlying block, the attitudes and traces of normal faults of smaller displacement in the exposed part of the block, the shape of the range margin if it can be inferred to parallel the bounding fault, and the subsurface configuration of the bordering basins as determined by geophysical techniques.

That many of the master normal faults are curved in plan view is strongly suggested by the arcuate traces of a large proportion of those that are exposed and by the arcuate margins of most of the ranges that consist of tilted blocks (Moore, 1960). In vertical dimension the master faults are more difficult

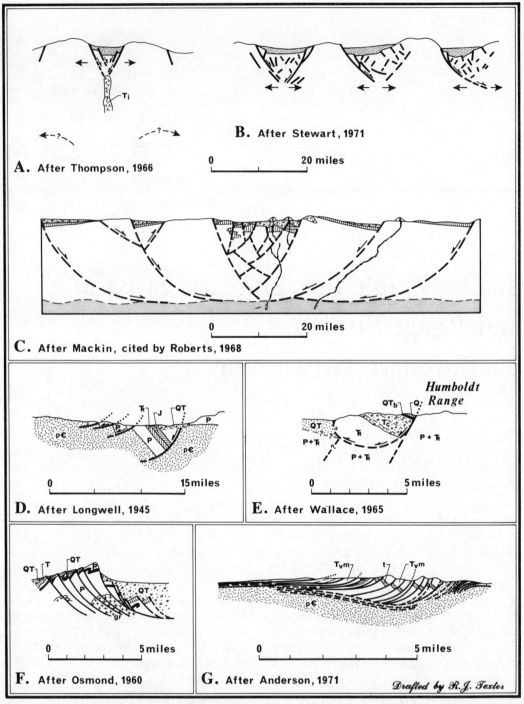

FIG. 1 Cross sections through actual, idealized, or hypothetical areas of the Great Basin, showing various interpretations of the downdip configurations of normal faults; modified from the indicated sources. (A, B) Idealized cross sections, after Thompson (1966) and Stewart (1971), illustrating deep-seated horst and graben structure. Major faults penetrate crust to depths of 10 miles or more. Blocks are gently tilted if at all. Faults maintain steep to moderate dips. Extension of deeper parts of crust is featured by intrusion of bodies of igneous rock (pattern of crosses) and/or by solid flow. (G) Hypo-

to envision, their lower parts being rarely shown in published cross sections. Virtually all cross sections, however, represent them as curved, becoming less steep with depth, and as dipping at nearly right angles to the angle of tilt of the overlying rotated blocks. The faults thus have been commonly viewed as spoon-shaped, concave to the blocks above them, and analogous to landslides of the rotated slump type. This interpretation has been based in part upon the fact that flattening can be observed on fault planes that can be traced through topography of considerable relief, as in the vicinity of the Grand Wash and Hurricane Cliffs near the boundary between the Great Basin and the Colorado Plateau about 150 miles east of the area shown in Fig. 2 (Longwell, 1945; Hamblin, 1965). It also is supported by observations that the tilt, as in rotated slumps, is ordinarily toward the convex sides of the blocks (Moore, 1960). Moreover, where the angle of rotation and the attitude of the underlying surface can be determined, the angle between them, indeed, commonly approximates 90°.

Cross sections that represent the lower as well as the upper parts of actual or hypothetical Basin and Range master normal faults show widely differing interpretations of the degree of flattening and the depths to which the fault planes extend (Fig. 1). In the central, northern, and eastern parts of the Great Basin the Cenozoic tilting of major blocks rarely exceeds 20° and the bounding faults, where exposed at the surface, ordinarily dip at nearly vertical angles (Gilluly, 1963, p. 158; Stewart, 1971, p. 1027). These faults generally have been viewed as maintaining steep dips to depths of 5 to 15 miles or more while forming deeply seated horsts and graben (Figs. 1A, 1B). This "deep-fault" interpretation is supported by geophysical data accumulated by various workers and recently summarized by Stewart (1971); it is geometrically compatible with relatively wide and gently tilted fault blocks. A variation of the "deep-fault" interpretation was proposed by Mackin (1960; cited by Roberts, 1968, p. 103), who envisioned Basin and Range normal faults as related to centers of volcanism and becoming essentially horizontal at depths of about 20 miles (Fig. 1C).

Normal faults have been shown on still other Basin and Range cross sections as flattening at much shallower depths (Figs. 1D, 1F, and 1G). Some of the steeply tilted blocks have been interpreted as cradled in unrooted, trough-shaped fault planes in the manner of Fig. 1E (Wallace, 1965). In this concept they assume the full form of a rotated slump. To date this interpretation appears to have been applied only to relatively small blocks no more than a few miles long, which can be viewed as second- or

---

thetical cross section (after Mackin, cited by Roberts, 1968), illustrating possible relationship between volcanism and normal faulting. Master faults bound gently tilted blocks and flatten to horizontal at depths of 15 to 20 miles. Vertical lines = bodies of extrusive and intrusive igneous rock; dotted pattern = Tertiary and Quaternary sedimentary rocks. (D) Cross section, after Longwell (1945), from Virgin Valley, Nevada, to Colorado Plateau, Arizona, showing blocks steeply tilted along shallow normal faults that flatten at different depths: p C = Pre-cambrian crystalline complex; P,   , J, and QT = Paleozoic, Triassic, Jurassic, and Tertiary-Quaternary sedimentary rocks, respectively. (E) Cross section, after Wallace (1965), across part of Humboldt Range, Nevada, showing interpretation of fault block as unrooted slump: P,   , Tc, QT, and Q = Paleozoic, Triassic, Tertiary, Tertiary-Quaternary, and Quaternary sedimentary rocks respectively; QT_b = Tertiary-Quaternary basalt. (F) Hypothetical cross section, after Osmond (1960), of a typical Basin Range, showing normal faults that tend to flatten at about the same, relatively shallow depth: P = Paleozoic sedimentary rocks; gr = Mesozoic or Tertiary granite rock; T = Tertiary sedimentary and volcanic rocks; QT = Tertiary and Quaternary sedimentary rocks. (G) Cross section (after Anderson, 1971), Eldorado Mountains, Nevada, showing extreme distension of layered volcanic rocks by normal faults arranged in shingle fashion and merging at very shallow depth: p C = Precambrian crystalline complex; $T_v m$ = mixed Tertiary volcanic rocks containing tuff layer (t) and subordinate sedimentary rocks in lower part.

third-order features formed by slumping off the uptilted sides of larger blocks. A similar interpretation may prove applicable, however, to some of the more steeply tilted blocks as large as an entire mountain range. The master normal faults of the Basin and Range province therefore may well range widely between those that break most of the thickness of the crust and shallow, canoe-shaped surfaces in which rootless fault blocks are cradled. The shallow type may include slumps in the true sense of the word as well as features related to deeper crustal extension.

Shallow-Fault Model Evidenced in Southwestern Great Basin

A "shallow-fault" interpretation seems particularly applicable to many of the ranges of the southwestern part of the Great Basin where fault blocks commonly have tilted through angles of 45–80°. This region has been referred to as structurally unlike the rest of the Great Basin, primarily because it is also featured by elongate northwest-trending valleys that mark the positions of zones of high-angle faulting against which the normal faults terminate (Figs. 2 and 3). Like the remainder of the Great Basin, however, the southwestern part has been extended along normal faults that generally trend northward to northeastward. The pervasiveness of such faulting constitutes evidence that the deformation of the area of the Great Basin, beginning in mid-Tertiary time, has been accomplished in a uniformly oriented stress field. The southwestern part appears simply to have been extended to a greater degree than the rest. The northwest-trending zones of high-angle faulting seem best interpreted as localized along zones of weakness established in Precambrian time (Wright and Troxel, 1966; Wright and Williams, 1970).

Features compatible with a shallow-fault, large-extension model are well displayed in the group of ranges that lies southwest of the Northern Death Valley–Furnace Creek fault zone, the more southerly of this group being shown in Figs. 2 and 3. Each of the ranges is fault bounded and is composed of a basement of earlier Precambrian crystalline rocks and a cover of well-layered rocks composed variously of later Precambrian, Paleozoic, Mesozoic, and Cenozoic units. Dips in the layered cover are generally eastward and mostly moderate to steep. That the dips are attributable largely to Cenozoic tilting is evidenced in the observation that the Cenozoic units, both sedimentary and volcanic, are widespread and generally dip in the same direction and to about the same degree as the older, layered units upon which they rest. In blocks that contain pre-Tertiary, Tertiary, and Pliocene–Pleistocene units, one can observe progressively gentler dips up-section. In several areas, notably in the Black Mountains and the Panamint Range, the crystalline basement has been invaded by Mesozoic and Tertiary plutons and by swarms of Tertiary dikes.

All but a few of the north-northwest to northeast trending faults within the area of Figs. 2 and 3 are normal in sense of movement, and all but a small proportion of the normal faults can be observed or inferred to dip westward. Where measurable in surface exposures, dips on the latter generally are less than 60° and commonly less than 45°.

Exposures of the contact between the layered cover and the underlying complex frequently show evidence of severe movement. At several localities exposures of the contact appear to mark master normal faults that dip gently westward beneath east-tilted blocks. Cross sections through two of these localities, one on the east side of the Alexander Hills at the southern end of the Nopah Range, and the other low along the east flank of the Panamint Range, are shown in Figs. 4A and 4B.

The fault surface at each locality separates an underlying complex, composed mostly of Precambrian crystalline rocks, from a block composed mostly of later Precambrian and Cambrian sedimentary units. Each of the two surfaces has been interpreted as part of a regional thrust fault, the Amargosa thrust (Noble, 1941; Hunt and Mabey, 1966, p. A129), but various lines of evidence suggest that they are separate, low-angle normal faults (Wright and Troxel, 1969).

Regardless of the relationship these fault surfaces bear to one another, they clearly constitute the planes on which the overlying fault blocks have moved by rotation. Since the blocks contain numerous normal faults of small displacement, to the virtual exclusion of other Cenozoic deformational features, their rotation apparently was accompanied by internal extension.

The east-tilted block that forms the Alexander Hills rests on a fault surface that can be traced as an undulating but essentially horizontal feature for as much as 2 miles east of the main mass of the hills. On the east side, among exposures of Tertiary intrusive and extrusive volcanic rocks, the surface dips gently westward beneath the block where it is presumed to penetrate the complex at a low angle (Fig. 4A). This locality thus affords an excellent three-dimensional view of a surface of movement that underlies a moderately to steeply tilted Basin Range. The Alexander Hills form the most southerly of a group of linked fault blocks that constitute the Nopah Range which, in turn, closely resembles the nearby Resting Spring Range. These similarities in structural setting suggest that (1) many of the normal faults, and especially those of relatively small displacement in the internal parts of the two ranges, terminate at the basement-cover contact, and (2) those that penetrate the basement are moderately to gently dipping.

The fault surface along the east face of the Panamint Range (Fig. 4B) occurs in a terrane that, in many respects, is a large-scale counterpart of the terrane of the Alexander Hills. This west-dipping fault also separates an underlying complex of Precambrian crystalline rocks from east-dipping sedimentary units of late Precambrian and Cambrian age. There, too, bodies of intrusive and extrusive Tertiary igneous rock are exposed, but much more abundantly than in the Alexander Hills.

Hunt and Mabey (1966) propose that the low-angle fault of the Panamint Range is part of a regional thrust, formed in Cretaceous or Jurassic time and later fragmented by Basin and Range normal faulting. They observe, however, that thrusting may be related to the emplacement of the bodies of Tertiary igneous rock in that some of the low-angle faults have offset Tertiary volcanic rocks. They also note that Tertiary extrusive rocks, exposed above the low-angle fault, dip 15–20° eastward and about 20° less steeply than the older sedimentary rocks upon which they rest. They therefore conclude that the dip of the volcanic rocks was produced by a tilting of the entire range, including the fault surface and following the thrusting.

We favor an alternate interpretation, holding that the low-angle fault is simply the gently dipping lower part of the master normal fault along which the units above the fault have rotated eastward in Cenozoic time. Thus the tilting of the overlying layered rock need not have appreciably affected the dip of the fault. This interpretation permits the elimination of hypothetical thrusting along the fault surface before the inception of normal faulting, and it is compatible with the evidence for a degree of contemporaneity between the faulting and the Tertiary volcanism. It also explains the difference in dip between the Tertiary rocks and the older sedimentary rocks as an effect of continued rotation of individual blocks and slices.

The bodies of Tertiary intrusive rock exposed on the lower east face of the Panamint Range consist mostly of dike swarms and small, irregular plutons and are mostly acidic in composition (Hunt and Mabey, 1966). They are much more abundant in the crystalline rocks beneath the fault than in the overlying block. The dike swarms in the former strike northward, are in sharp contact with the intruded metamorphic rocks, and are estimated by Hunt and Mabey to occupy

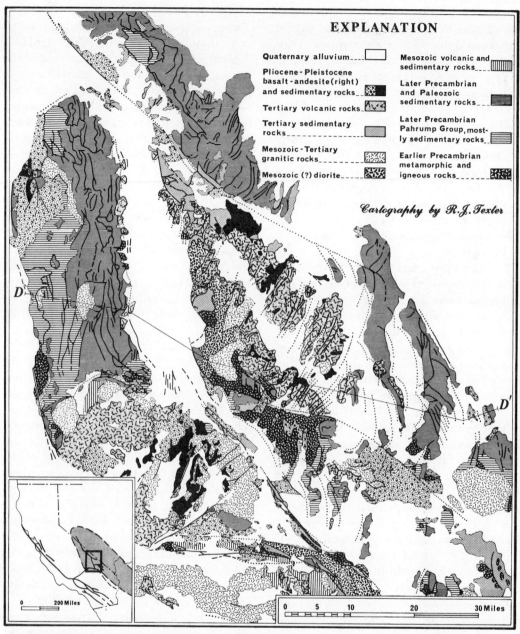

FIG. 2 Geologic map of the region of central and southern Death Valley. Shaded pattern on inset index map indicates area in which Cenozoic fault blocks generally tilt at steeper angles than they do in rest of Great Basin. Data from various sources acknowledged on Death Valley Sheet, Geologic Map of California (1958; and revision in preparation, 1972) and Trona Sheet, Geologic Map of California (1963).

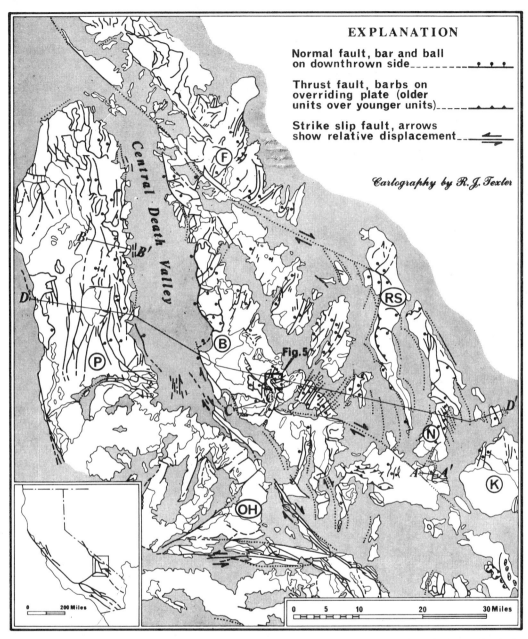

FIG. 3 Map of region of southern and central Death Valley showing principal faults and relative displacements: shaded pattern = Quaternary alluvium, locally deformed; blank areas = pre-Quaternary rocks. All displacement along normal faults probably has occurred within Cenozoic time. Thrust faulting in Resting Spring and Nopah Ranges probably pre-dates normal faulting. Lines of cross sections A–A', B–B', C–C', D–D', refer to Figs. 4 and 6. (P) Panamint Range; (B) Black Mountains; (F) Funeral Mountains; (RS) Resting Spring Range; (N) Nopah Range; (K) Kingston Range; (OH) Owlshead Mountains.

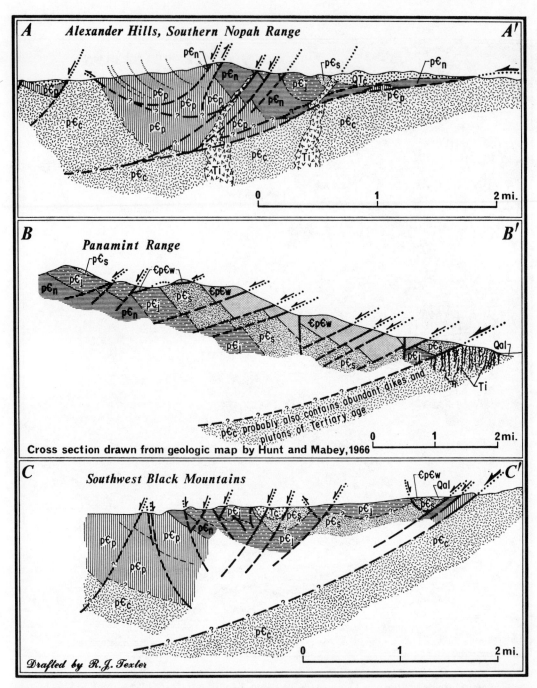

FIG. 4 Cross sections through localities, shown in Fig. 3, where gently west-dipping normal faults of large displacement are exposed. At each locality the fault surface separates a Precambrian crystalline complex from an overlying block composed mostly of later Precambrian and Cambrian sedimentary units which, in general, dip moderately to steeply eastward. Normal faulting dominates internal structure of each block. (pCc) Precambrian crystalline complex; (pCp) later Precambrian Pharump Group, undivided; (pCn, pCj, pCs) later Precambrian Noonday Dolomite, Johnnie Formation, and Stirling Quartzite, respectively; (Cp Cw) later Precambrian-Cambrian Wood Canyon formation; (Ti) Tertiary intrusive igneous rocks; (T$_c$) Tertiary sedimentary rocks; (QT$_c$) Tertiary and Quaternary sedimentary rocks; (Qal) Quaternary alluvium.

20 percent or more of the volume of the complex beneath the fault. They thus record a strong east–west dilation of the crystalline rocks that appears to have occurred contemporaneously with some of the movement on the low-angle fault and on the normal faults of smaller displacement within the overlying block.

The east-tilted fault block of Fig. 4C forms part of the west flank of the southern Black Mountains and is contained in the area in which Noble (1941) first noted the structurally disordered bodies of rock that he named the "Amargosa chaos." Although the bodies of chaos were originally interpreted as segments of a regional thrust plate, we have interpreted them as having formed, in part, on the undersides of tilted fault blocks (Wright and Troxel, 1969). Evidence for this interpretation can be seen in the cross section. The deformed rocks immediately above the principal fault qualify as chaos by Noble's usage, but the rocks higher in the block are less deformed than the term implies. The crudely synclinal structural feature into which the east-dipping units of the tilted block can be traced seems best viewed as an effect of drag along the fault. Consequently, at this

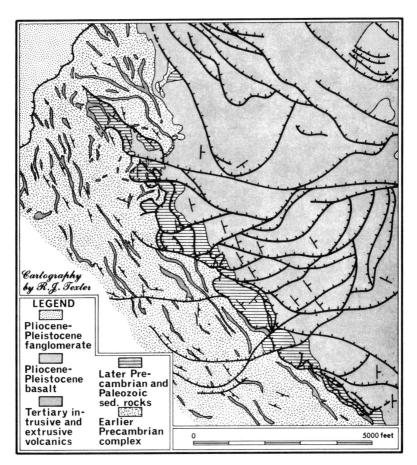

FIG. 5 Geologic map of a part of the southern Black Mountains, showing tendency of normal faults (hachured heavy lines) to terminate at or near the contact between the crystalline complex and a cover composed of Precambrian and Paleozoic sedimentary rocks and Cenozoic volcanic and sedimentary rocks. Location of area is shown in Fig. 3.

locality, too, one observes evidence that a low-angle fault has served as the surface of rotation for an east-tilted block. Although there the fault and tilted block may be second-order features formed as a result of sliding off the face of a larger, less steeply tilted block in the manner of Fig. 1*E*, their geologic setting so nearly resembles that of the two blocks in sections 4*A* and 4*B*, that we suggest an analogous relationship, all three resulting from large-scale slumping. At other places, the contact between the complex and the layered cover displays a different kind of deformational pattern, this one featured by numerous normal faults of small displacement that cut the cover, each tending to flatten and to join tangentially, forming a zone of movement at or near the contact. These features are especially well exposed at the locality of Fig. 5, high in the Black Mountains. Here, too, the contact has been described as a segment of the Amargosa thrust (Noble, 1941), but we ascribe the chaotic appearance of the cover largely to the normal faulting. We also observe, as did Noble, that few of the faults that do cut the complex penetrate it more than a few hundred feet, and that the complex is intimately intruded by a swarm of Tertiary dikes.

Regional Cross Section Through Central Death Valley

Most of the features that we have described as characterizing the Cenozoic deformation of the central Death Valley region are shown in the cross section of Fig. 6. We feel that the upper part of the cross section, based on surface observations, accurately shows that the master normal faults penetrate the crystalline complex. The normal faults of smaller

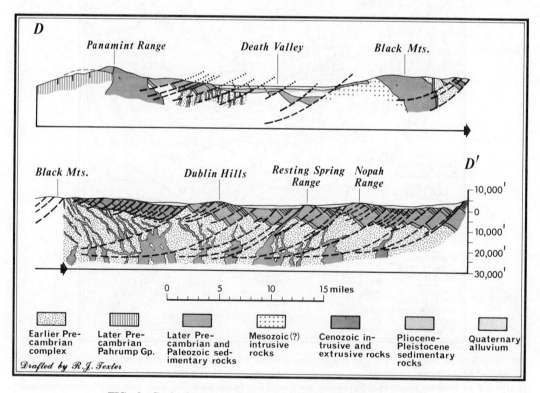

FIG. 6 Geologic cross section along line *D–D'* of Figs. 2 and 3.

displacement, on the other hand, tend to terminate near the contact between the complex and the layered cover. The lower part of the cross section is, of course, hypothetical, and it involves the assumption that the master faults, like the shallower ones, flatten with depth. If so, the evidence that the master faults dip moderately to gently in their upper parts strongly suggests that they flatten to the horizontal at depths of 5000–30,000 ft in the western half of the section. The various observable features in the eastern half suggest that the master faults there become horizontal at a depth range of 15,000–30,000 ft. We thus interpret all of the normal faults as being relatively shallow features, caused by severe crustal extension.

If the layered rocks of the eastern half of the cross section are restored to the horizontal, an extension of 30–50 percent of the original width is indicated. We suggest that the faults give way with depth to dike swarms and plutons as the principal phenomena related to the extension. A complete correlation between extension and igneous activity, however, can be questioned because (1) the extension seems as pronounced in the eastern part of the cross section as it is in the Black Mountains where bodies of Tertiary igneous rocks are more abundantly exposed and (2) the tilting of the blocks has continued following the last major volcanic episode. We therefore suggest that in the part of the crust that lies beneath the normal faults the extension is featured by both the emplacement of igneous bodies and solid flow.

References

Anderson, R. E., 1971, Thin skin distension in Tertiary rocks of southeastern Nevada: *Geol. Soc. Am. Bull.*, v. 82, p. 43–58.

Gilluly, J., 1963, The tectonic evolution of the western United States: *Geol. Soc. London Quart. Jour., Spec. Paper 80*, 69 p.

Hamblin, W. K., 1965, Origin of "reverse drag" on the downthrown side of normal faults: *Geol. Soc. Am. Bull.*, v. 76, p. 1145–1164.

Hunt, C. B., and Mabey, D. R., 1966, Stratigraphy and structure, Death Valley, California: *U.S. Geol. Survey Prof. Paper 494-A*, 162 p.

Longwell, C. R., 1945, Low-angle normal faults in the Basin-and-Range province: *Am. Geophys. Union Trans.*, v. 26, p. 107–118.

Mackin, J. H., 1960, Structural significance of Tertiary volcanic rocks in southwestern Utah: *Am. Jour. Sci.*, v. 258, p. 81–131.

Moore, J. G., 1960, Curvature of normal faults in the Basin and Range province of the western United States, *in* Geological Survey research, 1960, *U.S. Geol. Survey Prof. Paper 400-B*, p. B409–B411.

Noble, L. F., 1941, Structural features of the Virgin Spring area, Death Valley, California: *Geol. Soc. Am. Bull.*, v. 52, p. 941–1000.

Osmond, J. C., 1960, Tectonic history of the Basin and Range province in Utah and Nevada: *Mining Eng.*, v. 12, p. 251–265.

Roberts, R. J., 1968, Tectonic framework of the Great Basin: *UMR Jour.*, no. 1, Univ. Missouri Rolla, p. 101–119.

Stewart, J. H., 1971, Basin and Range structure: A system of horsts and grabens produced by deep-seated extension: *Geol. Soc. Am. Bull.*, v. 82, p. 1019–1044.

Thompson, G. A., 1967, The rift system of the western United States, *in* The world rift system-International Upper Mantle Commission Symposium: Ottawa, Can. Geol. Surv. Paper 66–14, p. 280–290.

Wallace, R. E., 1965, Possible origin of homoclinal structure in West Humboldt Range by landslide-like mechanism, an illustration, *in First Basin and Range Geol. Field Conf. Guidebook*, Mackay School of Mines.

Wright, L. A., and Troxel, B. W., 1967, Limitations on strike-slip displacement along the Death Valley and Furnace Creek fault zones, California: *Geol. Soc. Am. Bull.*, v. 78, p. 933–950.

———, 1969, Chaos structure and Basin and Range normal faults: evidence for a genetic relationship (Abs.): *Geol. Soc. Am. Spec. Paper 121*, p. 580–581.

Wright, L. A., and Williams, E. G., 1970, Precambrian reef complex, Death Valley; Evidence for tectonic control of facies patterns and for antiquity of a major fault: *Geol. Soc. Am. abs. with programs*, v. 2, p. 727.

RALPH J. ROBERTS
M. D. CRITTENDEN, JR.
*U.S. Geological Survey
Menlo Park, California*[1]

Orogenic Mechanisms, Sevier Orogenic Belt, Nevada and Utah[2]

The nature of orogenic mechanisms has been a major geologic problem for more than 100 years. After the demise of the contraction theory, orogeny was discussed primarily in terms of the geotectonic (geosynclinal) cycle. Recently the validity of this concept has been challenged, and many geologists have turned to ocean-floor spreading as an exclusive driving mechanism for orogeny in the western Cordillera (Wilson, 1968; Hamilton, 1969, 1970; Dewey and Bird, 1970; Coney, 1970, 1971; Dickinson, 1969, 1970b; Roeder and Nelson, 1970; Burchfiel and Davis, 1971). We agree that ocean-floor spreading is a powerful mechanism but feel that an assist from the mechanisms of the geotectonic cycle is needed to control and localize the uplift, metamorphism, and plutonism observed in the Sevier belt.

In this paper we propose to examine some of the late Paleozoic and early Mesozoic events in the western Cordillera that give insight into mechanisms of uplift and deformation in the Sevier orogenic belt; in addition, we will consider structural features within the belt related to gravitative transport that followed uplift.

Hose and Daneš (1973) emphasize later aspects of the orogeny, especially the development and movement of tectonic plates impelled by gravity from the Sevier belt into western and central Utah as earlier outlined

---

[1] Publication authorized by the Director, U.S. Geological Survey.
[2] Critical reviews of the manuscript by D. H. Whitebread, R. Scholten, and K. A. De Jong are gratefully acknowledged. We wish to thank R. L. Armstrong, B. C. Burchfiel, G. A. Davis, L. T. Grose, W. Hamilton, R. K. Hose, R. H. Moench, M. Mudge, W. G. Pierce, R. A. Price, J. Riva, and R. G. Yates for helpful discussions.

by Roberts and others (1965), Hose and Daneš (1967), and Hose and Blake (1969).

Summary of Phanerozoic Geologic History

The tectonic evolution of western North America during Phanerozoic time proceeded through three principal phases: (I) an orthogeosynclinal phase that lasted from late Precambrian through middle Late Devonian; (II) an orogenic-late geosynclinal phase from Late Devonian through Jurassic; and (III) an orogenic postgeosynclinal phase during Late Cretaceous and Tertiary.

The Sevier orogeny and the related Nevadan orogeny were the culminating events of phase II of the evolution of the western continental margin (Table 1) (Roberts and others, 1958, 1965; Roberts, 1969, 1972).

The following discussion will center on the zone along the fortieth parallel because the geosyncline was broader here, and the depositional environments and structural development of the western Cordillera are more clearly shown than to the north and south (Fig. 1).

Initial deposition in the orthogeosyncline (phase I) took place in three principal facies belts (Fig. 2A): an eastern miogeosynclinal belt of shallow-water carbonate rocks deposited on sialic crust; a western eugeosynclinal belt of deep-water clastic and volcanic rocks deposited on simatic crust; and a transitional belt at the early Paleozoic continental margin that separates these two major facies and includes representatives of both. Sedimentation in the orthogeosyncline lasted from late Precambrian through middle Late Devonian, amounting to 15,000–25,000 ft (4600–7700 m) in the miogeosynclinal zone, and to 40,000 ft (12,800 m) or more in

Table 1 Major events in the Development of the Cordilleran Fold Belt

	Phase	Depositional events	Orogenic events	Time span	Marginal tectonics
III Post-geosynclinal	III	Basaltic volcanism / Andesite-rhyolite volcanism / Local basins	Basin-Range	Late Miocene to present (16-0 m.y.) / Oligocene to Miocene (40-15 m.y.) / Eocene to Oligocene (60-30 m.y.)	Plate tectonics dominant along western continental margin (Dewey and Bird, 1970)
		Rocky Mountain Basins	Laramide	Latest Cretaceous and early Tertiary (75-40 m.y.)	
			Coast Range	Late Cretaceous and early Tertiary (110-50 m.y.)	
		Pacific trough		Cretaceous (130-60 m.y.)	
			Sevier (late stage)	Late Cretaceous and early Tertiary (110-50 m.y.)	
		Rocky Mountain-Sonoran troughs		Cretaceous (130-60 m.y.)	
II Orogenic - late geosynclinal	IIc	Local basins and troughs in eastern California - western Nevada and in western Utah - eastern Nevada	Sevier (early stage) / Nevadan	Early Jurassic to Late Cretaceous (170-110 m.y.) / Middle and Late Jurassic (167-132 m.y.)	Cordilleran-type margin / Plate tectonics minor
				Early Triassic to Late Jurassic (220-140 m.y.)	
	IIb		Sonoma	Late Permian and Early Triassic (240-220 m.y.)	
		Havallah sequence in western trough; Carlin sequence and other Mississippian to Permian Formations in eastern trough		Late Pennsylvanian to Late Permian (300-225 m.y.)	
	IIa		Antler	Late Devonian through Middle Pennsylvanian (340-300 m.y.)	
I Orthogeosynclinal	I	Cordilleran orthogeosyncline		Late Precambrian into Late Devonian (550-360 m.y.)	Orthogeosynclinal-type margin

FIG. 1 Index map of the western United States, showing position of the 40th parallel and section A–A', Fig. 8.

the eugeosynclinal zone. The site of maximum sedimentation lay astride the boundary between the eugeosynclinal and transitional facies near the base of the continental slope.

In latest Devonian time phase II was initiated by uplift and deformation along this boundary, heralding the Antler orogeny. This culminated in the development of the Roberts Mountains thrust, which carried deep-water offshore sediments and volcanics of the eugeosynclinal and transitional facies eastward as much as 100 miles (160 km) across shallow-water sediments of the miogeosyncline (Roberts, 1968, p. 106) (phase IIa, Fig. 2B). Clastics from the orogenic belt were shed into flanking troughs formed in an oceanic environment to the west and in an epicontinental environment to the east. The Sonoma orogeny in late Permian and Early Triassic time succeeded middle Pennsylvanian to early Permian sedimentation and volcanism in the western trough (phase IIb,

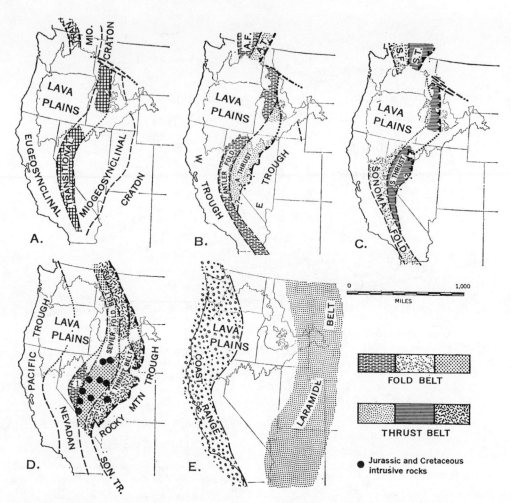

FIG. 2 Major phases in Cordilleran geotectonic cycle. (A) Phase I, Facies in Cordilleran orthogeosyncline, late Precambrian to Late Devonian time; (B) Phase IIa, Antler orogeny, Late Devonian through middle Pennsylvanian; (C) Phase IIb, Sonoma orogeny, late Permian and Early Triassic; (D) Phase IIc, Nevadan-Sevier orogeny, Jurassic, and Cretaceous [(1) designates zone of minor deformation and intrusive activity]; (E) Phase III, postgeosynclinal, Cretaceous, and Tertiary.

Fig. 2C), whereas only local uplift occurred in the eastern trough. The Nevadan orogeny followed continued volcanism and sedimentation in the western trough and was characterized by large-scale intrusive activity from Late Triassic into Cretaceous time (210–80 million years) (Bateman and Eaton, 1967; Evernden and Kistler, 1970; Kistler and others, 1971; Shaw and others, 1971), though its principal development is considered to be Middle and Late Jurassic (167–132 million years) by Lanphere and others (1968) (Fig. 2D). The partly correlative Sevier orogeny in the eastern trough was the site of moderate igneous activity during Jurassic and Early Cretaceous time (Roberts and others, 1965; Armstrong, 1968a, 1968c; Riva, 1970).

The Nevadan and Sevier belts are separate and distinct in Nevada but merge to the north and south. During middle Mesozoic orogeny the intervening Antler belt, designated (1) in Fig. 2D, was a zone of only minor deformation and intrusive activity. Clastic rocks from flanking highlands accumulated in local basins, and the region was probably an intermontane highland.

With the ending of the orogenic–late geosynclinal phase, a new regime involving thrusting of the Pacific plate under the western continental margin dominated subsequent developments in Late Jurassic and Cretaceous time (Bailey and others, 1964; Irwin, 1964; Bailey and Blake, 1969; Hsü, 1968; Dickinson, 1969, 1970a, 1970b). These events ushered in the orogenic–postgeosynclinal phase III. Thereafter, the Coast Range orogenic belt developed in the Pacific trough, the Sevier orogeny passed through its waning stages, dominated by gravity tectonics, and the block uplifts of the Laramide belt developed in the Rocky Mountain–Sonoran trough (Fig. 2E). The Basin and Range province formed still later in the zone between the Coast Range and Laramide belts during late Cenozoic time.

This brief summary indicates that in the course of Phanerozoic time the western Cordillera developed symmetrically outward from the Antler belt through successive Paleozoic, Mesozoic, and Cenozoic depositional-orogenic phases. During this time span the Cordillera grew westward onto oceanic crust while the continental crust on the east was thickened and stabilized.

These three major phases define a non-repeating scheme of continental development that lasted for more than 550 million years. Each major phase records a separate and distinct chapter in the evolution of the Cordilleran fold belt involving very different sedimentary environments, structural development, and marginal tectonics.

Sedimentation Immediately Preceding Sevier Orogeny

During the Antler orogeny in early Mississippian time, deposition at first took place mainly in troughs adjacent to the Antler belt (Fig. 3A). During Pennsylvanian time the locus of most rapid subsidence shifted eastward into central and northwestern Utah (Fig. 3B).

Sedimentation during Permian time in general followed the Pennsylvanian patterns (Fig. 3C). Locally in northwestern Utah more than 20,000 ft (6200 m) of clastic deposits accumulated; in central-western Utah another deep within the basin received more than 11,000 ft (2750 m) of sediments. These deeps were on either side of the Cortez–Uinta axis, a positive feature at times during the late Paleozoic and Mesozoic (Fig. 4).

The record of early Mesozoic sedimentation in the eastern Great Basin is fragmentary, but locally several hundreds to a few thousand feet of Triassic and Jurassic clastic deposits and limestone accumulated (Clark, 1964; Silberling and Roberts, 1962; Stanley and others, 1971).

A composite isopach map (Fig. 3D) shows that 7000–25,000 ft (2200–7700 m) of late Paleozoic and early Mesozoic sediments were deposited in the eastern Great Basin. These sediments accumulated mostly in shallow-marine environments, indicating that sedimentation was largely controlled by the rate of subsidence. In early Mesozoic time subsidence slowed and the region remained nearly at sea level until Early Jurassic time, when uplift locally began in the west. In Fig. 4 a palinspastic reconstruction of the basins has been made, based on the assumption that the glide plates in the frontal zone have moved as much as 70 miles (112 km) eastward. Crosby (1969) postulated movements of comparable magnitude in the overthrust belt in southern Idaho. Tooker (1970) has suggested that glide plates in Utah may have moved even greater distances.

Sevier Orogeny

Terminology, Age, and Extent

The name Sevier, which was first proposed by Harris (1959), was later restricted by

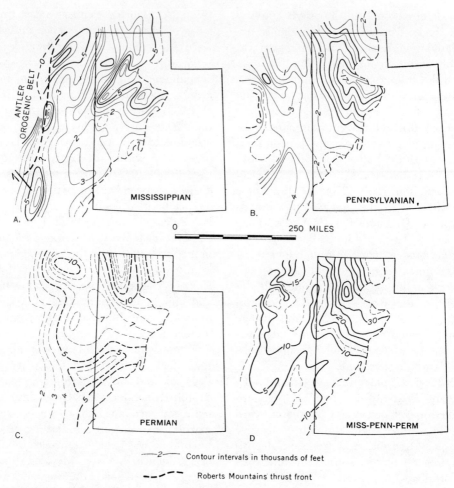

FIG. 3 Late Paleozoic basins in Nevada and Utah. (A) Mississippian; (B) Pennsylvanian; (C) Permian; (D) Combined isopachs, Mississippian-Permian time.

Armstrong (1963, 1968a) to the Utah-Nevada portion of the zone of thrust faults and folds on the east flank of the Cordilleran fold belt of King (1969b). However, the hinterland (to be defined shortly) from which the thrust or glide plates moved was also an integral part of the Sevier orogenic belt, and the term is therefore redefined by us to include this zone.

Two or more phases of deformation have been recognized in different parts of the Sevier belt by Misch and Hazzard (1962), Hose and Blake (1969), Thorman (1970), and Compton (1971). In contrast, Bally and others (1966), Armstrong (1968a), and Price and Mountjoy (1970) regard orogeny essentially as a single-stage, long-continuing process.

The onset of the orogeny was dated long ago by Spieker (1946), who showed on the basis of first appearance of clastics in central Utah that orogenic movements in the west began in Late Jurassic time and continued through Late Cretaceous. Armstrong (1968a, 1971) has summarized available geochronologic information on the age of the Sevier orogeny (Fig. 5). In addition, Adams and others (1966), Burchfiel and Davis (1971),

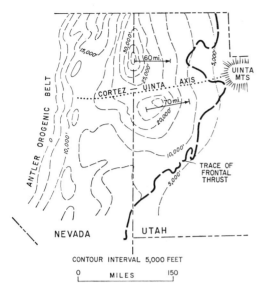

FIG. 4 Palinspastic map, showing reconstruction of late Paleozoic basins, Nevada and Utah.

and Stevens (1969) have shown that thrust faults in southern Nevada and southeastern California are cut by intrusive rocks 165 and 200 million years old or older, implying deformation of pre-Middle Jurassic age. In central Nevada, Ferguson and Muller (1949) recognized folding and thrusting of Middle Jurassic age east of the Nevadan orogenic belt. In northeastern Nevada Coats and others (1965) and Riva (1970) have described intrusive rocks about 150 million years old that cut folds in Paleozoic and Mesozoic rocks. In the Ruby–East Humboldt, Toiyabe, Cortez, and East Ranges, Nevada, intrusive rocks have been dated 143–168 million years (Kistler and Willden, 1969; Willden and Kistler, 1969; Silberman and McKee *in* Roberts and others, 1971).

These orogenic and intrusive epochs are part of a continent-long belt of disturbance that extended from southeastern California into Canada and Alaska. Like the Antler and Sonoma belts, the Sevier belt comprises two zones—a western fold belt (the hinterland of this report) and an eastern thrust belt (Fig. 2D). It is evident that orogeny began in Late Jurassic in the southern and western part of the Sevier belt and continued through at least Late Cretaceous (Crittenden,

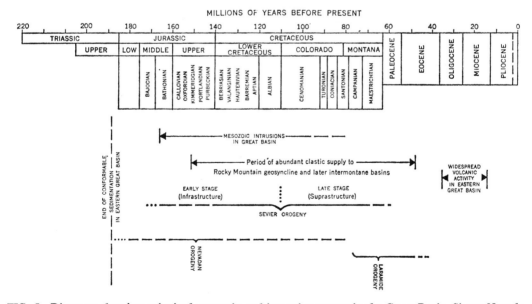

FIG. 5 Diagram showing principal orogenic and intrusive events in the Great Basin, Sierra Nevada, and Rocky Mountains (after Armstrong, 1968a).

1969). In the extension of the Sevier belt in southern Wyoming, however, thrusts continued to develop at the eastern margin into early Eocene (Oriel and Armstrong, 1966). Bally and others (1966) and Price and Mountjoy (1970) have postulated a similar chronology in the southern Canadian Rockies.

Basic Features

Whatever models are invoked to explain the development of the Sevier orogenic belt, the following basic physical features must be accounted for: (1) an eastern foreland belt of imbricate "older-on-younger" thrusts carrying the rocks of the geosyncline over the craton; (2) an allochthonous medial zone of tabular to gently folded rocks, occupying much of western Utah; (3) a hinterland comprising a little-deformed suprastructure, the western extension of zone 2, separated by a widespread décollement from an infrastructure characterized by high-grade metamorphism, penetrative deformation, and plastic flow (Armstrong and Hansen, 1966). In general, the suprastructure is characterized by low-angle "younger-on-older" thrusts and high- to moderate-angle normal faults that end at the décollement surface, all implying vertical attenuation and horizontal extension. In contrast, the infrastructure is characterized by complex deformation, locally at least, nappelike folds, and extreme thickening and thinning of individual units. Wherever the temporal relations of these two structural levels can be determined, the structures of the suprastructure appear to postdate (because they truncate) those of the infrastructure. This relation is borne out also by isotopic dating of synorogenic or postorogenic intrusives within the hinterland. For this reason, the following discussion separates the events into two phases, an early phase concerned largely with the infrastructure and a later phase concerned largely with the suprastructure.

Early-Stage Deformation—Infrastructure. The early-stage deformation of the Sevier belt was characterized by polyphase penetrative deformation, low-angle faulting, plutonism, and metamorphism. In the few places where the infrastructure has been brought to the surface, the patterns of deformation are extremely complex.

Imbricate thrusts, recumbent folds, and penetrative deformation characteristic of the infrastructure have been described by Compton (1971) in the Raft River Range, Utah (Fig. 6A). Deformation in three substages is recognized: an early set of folds overturned to the west and northwest is followed by another set overturned toward the north and northeast. Both sets of folds and related thrusts are truncated by the décollement which separates them from little-deformed rocks of the Oquirrh formation (Pennsylvanian), which form the suprastructure.

Similar complexity has been described by Armstrong (1968c, 1970) in the Albion Range, Idaho. Though still not clearly understood, the structures are an extension of those in the Raft River Range and show comparable degrees of folding, thrusting, and remobilization of basement rocks within the infrastructure.

A second large area of infrastructure is exposed in the Ruby–East Humboldt Range and the adjoining Wood Hills in northeastern Nevada. Howard (1971) has shown recumbent nappes and extreme tectonic attenuation of sedimentary units in an area of sillimanite-grade metamorphism and migmatization. Synorogenic granite just south of this area has been dated as Jurassic (Kistler and Willden, 1969).

In the nearby Wood Hills, Nevada (Fig. 6B), Thorman (1970, p. 2430–2443) has shown that northwestward overturning dominates the infrastructure, both during and after metamorphism. Later structures show that the higher structural plates of the suprastructure were emplaced by southeastward transport.

Dover (1969) has described a domal area near Ketchum, Idaho (Fig. 6C), which exhibits many features of Great Basin infrastructure. Lower plate metamorphic rocks

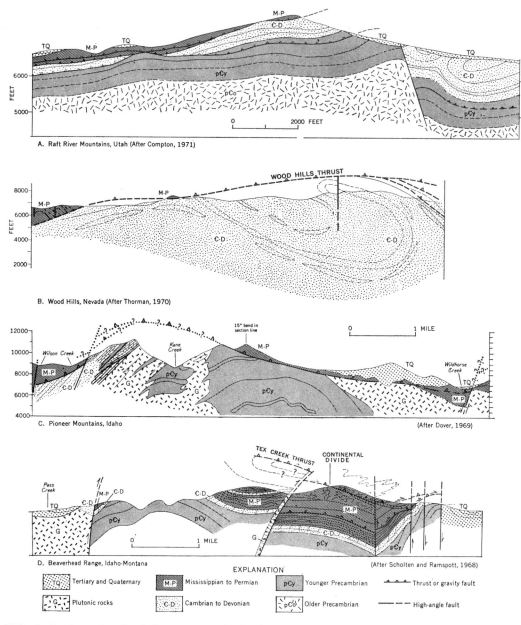

FIG. 6 Sections showing infrastructure, Sevier belt, Utah, Nevada, Idaho, and Montana.

are strongly folded, cut by imbricate faults, and invaded by gneissic rocks of the Pioneer Mountains pluton. An unmetamorphosed upper plate is inferred to have been transported northeastward after emplacement of the pluton.

Early-stage deformation and metamorphism in eastern Nevada and western Utah are considered by Misch (1960) and Misch and Hazzard (1962) to have taken place at considerable depth. They postulate compressive stresses responsible for deformation to be due to westward underthrusting of the Sevier belt by the relatively rigid Precambrian basement, bringing rocks of low metamorphic grade into contact with those of higher grade.

Armstrong and Hansen (1966) and Price and Mountjoy (1970) propose an *Abscherungszone*, or zone of tectonic adjustment, separating metamorphosed rocks of the infrastructure from unmetamorphosed rocks of the suprastructure in the belt. This is considered to be a zone of sharp metamorphic gradient. In places the zone is a low-angle fault. Metamorphism and thrusting are treated as a single-stage or continuous process which affected rocks above and below the *Abscherungszone* in different ways.

Price and Mountjoy (1970) infer that deformation and metamorphism took place at depth and that diapiric flow brought these metamorphic rocks upward while they were still hot and ductile. Biotite that cuts earlier structures formed during penetrative deformation has been dated at 111 million years; this is considered to place an upper limit on the time of penetrative deformation.

Scholten and Ramspott (1968) and Scholten (1968) also recognize infrastructural and suprastructural levels in the Beaverhead Range, Idaho–Montana, which are separated by subhorizontal detachment faults. In this more marginal part of the orogenic belt, detachment generally took place in middle Paleozoic shaly rocks; the deeper structural levels did not undergo the penetrative deformation and metamorphism seen farther west and southwest, but they do show strong folding and extensive thrusting. Scholten (1968, 1971, 1973) suggests that regional uplift in the orogenic belt may be related to the emplacement of the Idaho batholith.

Although some differences in interpretations are apparent, most workers agree that the hinterland in early stages of orogeny was a zone of deep-seated deformation, metamorphism, and plutonism from Middle Jurassic through Early Cretaceous time. Uplift can only have been moderate, however, inasmuch as only fine clastics were shed eastward into the Rocky Mountain trough.

Late-Stage Deformation—Suprastructure. Late-stage deformation in the Sevier belt was characterized by large-scale uplift and eastward movement of gravity plates, resulting in attenuation and partial denudation of the hinterland. At first relief was low, and flysch-like sediments, such as the Lower Cretaceous Mancos shale, were transported into eastern Utah, but by Late Cretaceous uplift reached a maximum as tremendous thicknesses of coarse clastics caused the shoreline to prograde far eastward locally into the central part of the United States. Concomitant with maximum uplift and transport of debris, plates moved eastward under the influence of gravity, ultimately reaching the craton.

Gilluly and others (1970) and Eardley (1968) have queried the great transport of these plates, and instead they favor derivation from local uplifts in western Utah. Woodward (1970) and Crittenden (1972) have examined two of Eardley's "uplifts" and clearly show that travel of the plates was consistently eastward rather than radial. Both recumbent folds and displaced sedimentary facies indicate origin from a distant uplifted zone to the west, rather than from local sources.

The uplift that caused gravity movement is visualized as broad regional arching along the Sevier belt (Roberts and others, 1965; Scholten and Ramspott, 1968; Crosby, 1969; Scholten, 1970). However, the local strong folding of the suprastructure in addition to the complex nature of the infrastructure makes it clear that this was not simple static uplift (Mudge, 1970, 1971) but was a continuation of early-stage deformation that involved marked horizontal transport. Within the hinterland, as the following examples show, the rocks of the suprastructure characteristically show younger-on-older thrusts and steep faults accompanied by marked vertical attenuation and horizontal extension.

Whitebread's (1969) cross sections of the Snake Range, Nevada (Fig. 7A), show that more than 12,000 ft of beds are cut out in places; beds as young as Pennsylvanian are

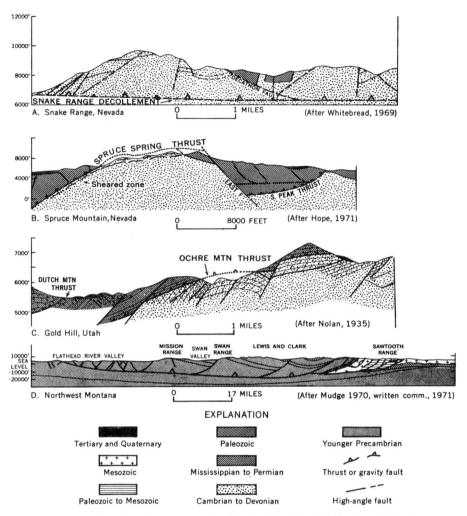

FIG. 7 Sections showing attenuation, Sevier belt, Nevada, Utah, and Montana.

brought down against Cambrian limestone on a series of high-angle faults that end downward at the décollement and are inferred to have formed during eastward movement of the upper plate.

In Spruce Mountain, northeastern Nevada (Fig. 7B), Hope (1971) shows similar low-angle faults that bring late Paleozoic rocks into contact with Ordovician rocks. Similar stratigraphic attenuation is shown by Nolan (1935) in the Gold Hill mining district, Utah (Fig. 7C), where the Ochre Mountain thrust cuts across Paleozoic rocks, locally juxtaposing Carboniferous units on Cambrian.

Mudge (1970, 1971) has discussed the origin of the disturbed belt in northwestern Montana, where vertical uplift leading to eastward gravitation gliding is postulated. He also shows attenuation in the hinterland similar to that farther south (Fig. 7D).

The cross sections drawn by most workers indicate that deformation in the hinterland resulted in significant thinning in this zone during eastward movement of glide plates. A regional section across the Sevier belt (Fig. 8), whose position is indicated on Fig. 1,

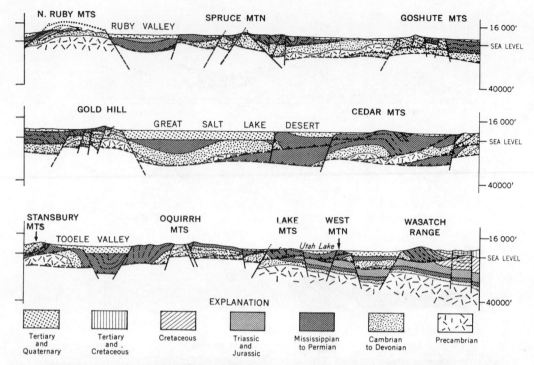

FIG. 8 Cross section from the Ruby Range, Nevada, to the Wasatch Mountains, Utah (compiled from Gilluly, 1932; Stokes, 1963; Hope, 1971; Snelson, 1955; Bissell, 1964; Howard, 1966; Cook and others, 1966; Walton, 1964; Cook and Berg, 1958; Bullock, 1951; R. B. Maurer, written communication, 1970). Location of section shown in Fig. 1.

shows attenuation in the hinterland, imbrication in its middle course, and folding and imbrication in the frontal zone.

Relations to Cordilleran Fold Belt

The structural coherence of the continent-long Cordilleran thrust and fold belt, of which the Sevier orogenic belt is but a part, has been brought into clear focus by the tectonic map of North America and accompanying discussion (King, 1969a, 1969b, p. 64–74). It is shown to be a nearly continuous belt of Alpine-type deformation, extending from southern Nevada through the Brooks range of Alaska. It is characterized throughout by eastward or northeastward transport —a feature that distinguishes it from the belt of Laramide (*Germanotype*) block uplifts which, in the Rocky Mountains of the United States, adjoins it on the east, and which, in part, overlaps it in time. The geology of individual segments of the Cordilleran fold belt were described during symposia of the Geological Society of America (Riverside, 1971; Calgary, 1971). These papers reveal differences in structural detail, which we conclude are the result of variations in sedimentary rock involved or in the depth of exposure rather than indications of basic differences in structural style (Scholten, 1970). This basic structural unity implies that the same underlying tectonic forces are involved throughout the belt. In the following discussion, therefore, the tectonic models and history of the entire Cordillera will be brought into view.

Although deformation in the Cordilleran fold belt has been discussed in terms of many different orogenic models, most can be summarized as variations on only a few basic themes: (1) underthrusting of craton from the east; (2) underthrusting of oceanic plates from the west; (3) development of compressive thrusts by large-scale crustal shortening of unspecified origin; and (4) at least two kinds of gravity-thrust models, (4a) some *with* crustal shortening and deep-seated deformation and (4b) others essentially *without* crustal shortening (Table 2).

Proposed Model of Orogenic Evolution

Although there is wide agreement that low-angle gravity gliding has played a large part in the late-stage deformation within the suprastructure, the nature of processes within the infrastructure that gave rise to the gravitational potential is subject to much greater uncertainty.

Mechanism of Early-Stage Deformation. Mechanisms that have been proposed to explain the deformation within the infrastructure commonly include some form of compressive deformation. Crustal shortening is commonly assumed, though the driving force is often unspecified or speculative (Table 2). In recent years, however, the rapidly evolving concepts of ocean-floor spreading and plate

TABLE 2 Orogenic Models and Timing, Cordilleran Thrust and Fold Belt

MECHANICS AND TIMING		REFERENCE
EARLY STAGES	LATE STAGES	
Underthrusting of Craton Middle Mesozoic	Middle Mesozoic	Misch (1957, 1960); Misch and Hazzard (1962); Thorman (1970)
Oceanic Underthrusting Jurassic and Early Cretaceous	Late Cretaceous	Burchfiel and Davis (1968); Wilson (1968); Burchfiel and others (1970); Gilluly (1971)
Cratonic and/or Oceanic Underthrusting (Listric thrusts in detached suprastructure) Jurassic and Cretaceous	Jurassic and Cretaceous	Bally, Gordy, and Stewart (1966)
Large Single-Stage Crustal Shortening (Origin unspecified) Mobilization of infrastructure Jurassic–Early Cretaceous	Deformation of detached suprastructure Late Cretaceous–Paleocene	Armstrong and Hansen (1966); Armstrong (1968a, 1968c); Fleck (1970)
Large Crustal Shortening (Origin unspecified) Mobilization of infrastructure Jurassic and Cretaceous	*Gravity Spreading of Plates* Deformation of suprastructure Jurassic and Cretaceous	Price and Mountjoy (1970); Price (1971a, 1971b, 1973); Rubey and Hubbert (1959)
Moderate Crustal Shortening (Coupled with oceanic underthrusting) Jurassic–Early Cretaceous	Uplift Gravity Plates Late Cretaceous–Early Tertiary	Roberts and Crittenden (this paper)
Minor Crustal Shortening (Origin unspecified) Jurassic and Cretaceous	Gravity Plates Late Cretaceous–Early Tertiary	Roberts and others (1965); Hose and Daneš (1967); Hose and Blake (1969); Tooker (1970); Rubey and Hubbert (1959)
Uplift with No or Unspecified Crustal Shortening (Infrastructure) Late Mesozoic	*Gravity Gliding of Plates* (Suprastructure) Late Mesozoic–Early Tertiary	Crosby (1969); Eardley (1969); Gilluly, Reed, and Cady (1970); Hose and Daneš (1973); Mudge (1970); Scholten and Ramspott (1968); Scholten (1970, 1971, 1973)

tectonics have provided a source of compressive stresses of appropriate scope and duration. Coney (1971), for example, relates the tectonic history of the Cordillera to events in the North Atlantic; Burchfiel and Davis (1971) have suggested that early Mesozoic plutonism, volcanism, and thrusting in eastern California and western Nevada were synchronous with underthrusting of the North American continent by the Pacific plate and infer that these features are genetically related.

This relationship seems reasonable for the Coast Ranges and possibly the Sierra Nevada, but its application to the Sevier belt has appeared unsatisfactory to Roberts, because it requires that the stresses be transmitted through western and central Nevada (the root zone of the Antler and Sonoma belts) where there is little evidence of deformation during middle Mesozoic time; although a few small intrusive bodies of Jurassic and Cretaceous age have been mapped in this region, the typical polymetamorphic, folded terranes of the Sevier belt are absent. It is possible, however, that the prolonged and areally extensive stress fields resulting from plate interaction may be transmitted through such areas without observable effect, and that the deformation may be localized within orogenic belts by processes (discussed further below) initiated during the depositional phase of geosynclinal history.

Miogeosynclinal sedimentation in the site of the Sevier belt during early and middle Paleozoic time totaled 15,000–25,000 ft (4600–7700 m); this load was accommodated without significant disturbance. Late Paleozoic and early Mesozoic sedimentation resulted in addition of 7000–25,000 ft (2200–7700 m). Doubling the subsidence by early Mesozoic time caused the late Precambrian surface to be depressed into higher temperature regimes at depths of 4.4–10 miles (7–16 km) below its original position. This culminated in the development of mobile zones, which responded to the regional stress by plastic flow, now expressed as regional metamorphism, plutonism, and penetrative deformation. We infer that the infrastructure now exposed in the Ruby, Snake, and Raft River Ranges was generated by this process. Our model for this stage resembles that evolved by Price and Mountjoy (1970) for the hinterland areas of the Canadian Rockies.

The precise nature of these processes of deep-seated deformation are not clearly understood, but they appear to involve a well-recognized sequence of events: (1) subsidence; (2) deformation; and (3) uplift and magmatism. Inasmuch as the areas of deepest subsidence appear to be the sites of maximum uplift, the two halves of the process must be tied to a series of closely linked mechanisms within the subcrust or mantle. These mechanisms may include thermal blanketing, phase changes, and partial melting, which result in expansion and uplift (Kennedy, 1959; MacDonald, 1963; Birch, 1965). Roberts (1968) has suggested that subcrustal processes such as changes from dense to less dense mineral phases (e.g., eclogite → gabbro) in zones of deep subsidence may lead to magmatism and uplift demanded by the observed sequence of events. Loading and subsidence appear to provide a built-in mechanism for uplift and orogeny which operates whenever critical limits are reached—here 4.4–10 miles of sediment accumulation.

Compression and related crustal shortening during the early orogenic phase have been emphasized by Bally and others (1966), Armstrong and Hansen (1966), and Price and Mountjoy (1970). As indicated previously, recumbent folds and imbricate thrusts within the infrastructure suggest that compressive tectonics were involved. Nevertheless, our only available calibration of the *amount* of crustal shortening comes from the foreland belt where the leading edges of the suprastructure impinge upon the craton. Therefore, until the effects of horizontal extension within the hinterland portion of the suprastructure can be evaluated independently, the net crustal shortening remains indeterminate. We are impressed by the dominant eastward transport across the entire eastern Cordilleran belt and by the fact that the effects of Antler, Sonoma, and Sevier orogenies add up to some 200 miles of apparent shortening. Crittenden feels that net crustal shortening may be at least half that amount; Roberts suggests it might be less.

The problem is further complicated by the probability that vertical attenuation of essentially plastic character, as indicated by steep apparent σ_1 inclinations (Compton, 1971; Moench, 1970), has been superimposed on the earlier thrust tectonics within the infrastructure, probably lasting into Tertiary time, to judge from K-Ar dates (Armstrong, 1963).

Mechanism of Late-Stage Deformation. Late-stage deformation was characterized by broad arching within the belt and eastward movement of glide plates. Uplift reached its culmination during this stage, judging by the coarseness and volume of orogenic sediments that resulted (Mudge and Sheppard, 1968). This marked uplift resulted from expansion in the subcrust genetically related to the mechanism of the early-stage deformation, and it was triggered by processes initiated during the geosynclinal phase. As unloading in the orogenic belt proceeded, isostatic response caused renewed uplift, thus maintaining the emergent zone. These later processes are epeirogenic rather than orogenic.

The amount of subsidence, height of uplift, and distance of gravity gliding are thus genetically related, as Roberts (1968) and Crosby (1969) have suggested, and as indicated in Fig. 4. The eastward bulges in the frontal thrust zone lie directly east of prominent deeps on both sides of the Cortez–Uinta axis, indicating that greater accumulation resulted in greater distance of travel of glide plates.

Price (1971a, 1971b) has criticized simplistic gravity models of late-stage deformation on two counts: (1) that an *eastward slope* on the basement beneath the foreland is not supported by available data, and (2) that a tectonic gap or pull-apart has *nowhere* been observed. Both criticisms appear valid at least in part. The basement under the leading edge of the thrust belt now slopes westward everywhere any information is available—a feature common to both the Price–Mountjoy (1970) and Mudge (1970) models. The difference relates to the basement slope farther west between the hinterland and the westernmost points from which direct evidence is available (e.g., Rocky Mountain trench, Willard thrust). This part of Mudge's model shows an east-sloping basement and is obviously diagrammatic. It should be noted, however, that neither Mudge's nor the Hose–Daneš model (1973) requires an east slope of *basement*; it merely requires an eastward slope of the *surface* on which detachment and sliding develops. In the Sevier belt, this may be well above basement.

The second criticism appears valid only in regard to highly simplified models (Mudge, 1970, 1971). In the Sevier orogenic belt, tectonic denudation, vertical attenuation, and horizontal elongation of the suprastructure have been demonstrated in many areas in eastern Nevada, western Utah, and parts of Idaho and Montana; this is believed to be characteristic of the Sevier belt as a whole.

Conclusions

Orogeny in the western Cordillera has recently been ascribed exclusively to ocean-floor spreading and related processes. These mechanisms seem readily applicable to the Coast Range and possibly the Nevadan orogenic belts, but they are not directly applicable to the Sevier belt, 100 miles (160 km) to the east. Here early stages of the orogeny include deep-seated deformation, metamorphism, and plutonism followed by late-stage uplift, shallow folding, and eastward gliding.

The Sevier orogenic belt was coincident with a trough that lay between the Antler orogenic belt and the craton, and it is concluded that loading and subsidence in the trough triggered subcrustal expansion and magmatism that led to orogeny.

Large-scale basement uplift in the Sevier hinterland is clearly indicated by the enormous amounts of debris shed eastward, especially during the late stage. No tectonic gap in the hinterland remains, but attenuation and thinning of glide plates have been demonstrated in many areas. These features of late-stage orogeny in the Sevier belt support gravity-sliding as a major mechanism.

The tectonic development of the Sevier belt parallels that of the Alps in many significant ways. Van Bemmelen (1966) has recognized deep-seated early-stage deformation of the Alpine basement complex, followed by late-stage uplift and gravity tectonics. This parallel sequence of events indicates that the mechanisms of orogeny are not dependent on local and accidental events but are fundamental processes of world-wide extent. Mesozoic orogeny in the Cordilleran fold belt is a further expression of these processes.

No single orogenic model is capable of performing all the tasks of deformation, metamorphism, plutonism, and uplift in the Sevier orogenic belt. We believe that a fuller understanding of these related processes can be achieved if consideration is given to mechanisms related not only to oceanic processes operating along the continental margins, but also to mechanisms related to loading and subsidence in troughs and basins within the continents.

References

Adams, J. A. S., Burchfiel, B. C., Davis, G. A., and Sutter, J. F., 1966, Absolute geochronology of Mesozoic orogenies, southeastern California (abs.): *Geol. Soc. Am. Spec. Paper 101*, p. 1.

Armstrong, R. L., 1963, Geochronology and geology of the eastern Great Basin in Nevada and Utah: Yale Univ., New Haven, Conn., Ph.D. thesis, 202 p.

———, 1968a, Sevier orogenic belt in Nevada and Utah: *Geol. Soc. Am. Bull.*, v. 79, p. 429–458.

———, 1968b, The Cordilleran miogeosyncline in Nevada and Utah: *Utah Geol. Mineral. Survey Bull. 78*, 58 p.

———, 1968c, Mantled gneiss domes in the Albion Range, southern Idaho: *Geol. Soc. Am. Bull.*, v. 79, p. 1295–1314.

———, 1970, Mantled gneiss domes in the Albion Range, southern Idaho: A revision: *Geol. Soc. Am. Bull.*, v. 81, p. 909–910.

———, 1971, Tectonic complexity of the Sevier orogenic belt (abs.): *Geol. Soc. Am. Abs. with Programs*, v. 3, p. 73–74.

———, and Hansen, E., 1966, Cordilleran infrastructure in the eastern Great Basin: *Am. Jour. Sci.*, v. 264, p. 112–127.

Bailey, E. H., and Blake, M. C., 1969, Late Mesozoic tectonic development of western California: *Geotectonics*, no. 3, p. 148–154.

Bailey, E. H., Irwin, W. P., and Jones, D. L., 1964, Franciscan and related rocks and their significance in the geology of western California: *Calif. Div. Mines Geol. Bull. 183*, 177 p.

Bally, A. W., Gordy, P. L., and Stewart, G. A., 1966, Structure, seismic data, and orogenic evolution of southern Canadian Rocky Mountains: *Can. Petrol. Geol.*, v. 14, p. 337–381.

Bateman, P. C., and Eaton, J. P., 1967, Sierra Nevada batholith: *Science*, v. 158, p. 1407–1417.

Birch, F., 1965, Speculations on the Earth's thermal history: *Geol. Soc. Am. Bull.*, v. 76, p. 133–154.

Bissell, H. J., 1964, Chaos near Ferguson Mountain, Elko County, Nevada (abs.): *Geol. Soc. Am. Spec. Paper 76*, p. 265–266.

Blue, D. M., 1960, Geology and ore deposits of the Lucin mining district, Box Elder County, Utah, and Elko County, Nevada: Univ. Utah, Salt Lake City, Utah, M.S. thesis, 101 p.

Brew, D. A., and Gordon, M., Jr., 1971, Mississippian stratigraphy of the Diamond Peak area, Eureka County, Nevada: *U.S. Geol. Survey Prof. Paper 661*, 81 p.

Bullock, K. C., 1951, Geology of Lake Mountain, Utah: *Utah Geol. Mineral. Survey Bull. 41*, p. 1–46.

Burchfiel, B. C., and Davis, G. A., 1968, Two-sided nature of the Cordilleran orogen and its tectonic implications: *Int. Geol. Cong., 23rd, Prague 1968, Proc.*, v. 3, p. 175–184.

———, 1971, Nature of Paleozoic and Mesozoic thrust faulting in the Great Basin area of Nevada, Utah, and southeastern California (abs.): *Geol. Soc. Am. Abst. with Programs*, v. 3, p. 88–90.

Burchfiel, B. C., Pelton, P. J., and Sutter, J., 1970, An early Mesozoic deformation belt in south-central

Nevada-southeastern California: *Geol. Soc. Am. Bull.*, v. 81, p. 211–216.

Clark, L. D., 1964, Stratigraphy and structure of part of the western Sierra Nevada metamorphic belt, California: *U.S. Geol. Survey Prof. Paper 410*, 70 p.

Coats, R. R., Marvin, R. F., and Stern, T. W., 1965, Reconnaissance of mineral ages of plutons in Elko County, Nevada, and vicinity, *in* Geological Survey Research 1965: *U.S. Geol. Survey Prof. Paper 525-D*, p. D11–D15.

Compton, R. R., 1971, Geologic map of the Yost quadrangle, Box Elder County, Utah, and Cassia County, Idaho: *U.S. Geol. Survey Misc. Geol. Inv. Map I-672*, scale 1:31,680.

Coney, P. J., 1970, The geotectonic cycle and the new global tectonics: *Geol. Soc. Am. Bull.*, v. 81, p. 739–747.

———, 1971, Cordilleran tectonic transitions and motion of the North American plate: *Nature*, v. 233, p. 462–465.

Cook, K. L., and Berg, J. W., Jr., 1958, Regional gravity survey along the central and southern Wasatch Front, Utah: *U.S. Geol. Survey Prof. Paper 316-A*, p. A75–A88.

Cook, K. L., Berg, J. W., Jr., Johnson, W. W., and Novotny, R. T., 1966, Some Cenozoic structural basins in the Great Salt Lake area, Utah, indicated by regional gravity surveys: *Utah Geol. Soc. Guidebook to the Geology of Utah*, no. 20, p. 57–76.

Crittenden, M. D., Jr., 1969, Interaction between Sevier orogenic belt and Uinta structures near Salt Lake City, Utah (abs.): *Geol. Soc. Am. Abst. with Programs*, pt. 5, p. 18.

———, 1972, Willard thrust and the Cache allochthon: *Geol. Soc. Am. Bull.*, v. 83, p. 2871–2880.

Crosby, G. W., 1969, Radial movements in the western Wyoming salient of the Cordilleran overthrust belt: *Geol. Soc. Am. Bull.*, v. 80, p. 1061–1078.

Dewey, J. F., and Bird, J. M., 1970, Mountain belts and the new global tectonics: *Jour. Geophys. Research*, v. 75, p. 2625–2647.

Dickinson, W. R., 1969, Evolution of calc-alkaline rocks in the geosynclinal system of California and Oregon, *in* Proceedings of the Andesite Conference, Oregon: *Oregon Dept. Geology and Mineral Industries Bull. 65*, p. 151.

———, 1970a, Second Penrose Conference—The New Global Tectonics: *Geotimes*, April 1970, p. 18–22.

———, 1970b, Relations of andesites, granites, and derivative sandstones to arc-trench tectonics: *Jour. Geophys. Research*, v. 75, p. 813–842.

Dott, R. H., Jr., 1955, Pennsylvanian stratigraphy of Elko and northern Diamond Ranges, northeastern Nevada: *Am. Assoc. Petrol. Geol. Bull.*, v. 39, p. 2211–2305.

Dover, J. H., 1969, Bedrock geology of the Pioneer Mountains, Blaine and Custer Counties, central Idaho: *Idaho Bur. Mines Geol. Pamph. 142*, 66 p.

Eardley, A. J., 1968, Major structures of the Rocky Mountains of Colorado and Utah: *UMR Jour.*, no. 1, p. 71–99.

———, 1969, Willard thrust and the Cache uplift: *Geol. Soc. Am. Bull.*, v. 80, p. 669–680.

Evernden, J. F., and Kistler, R. W., 1970, Chronology of emplacement of Mesozoic batholithic complexes in California and western Nevada: *U.S. Geol. Survey Prof. Paper 623*, 42 p.

Ferguson, H. G., and Muller, S. W., 1949, Structural geology of the Hawthorne and Tonopah quadrangles, Nevada: *U.S. Geol. Survey Prof. Paper 216*, 55 p.

Fleck, R. J., 1970, Tectonic style, magnitude, and age of deformation in the Sevier orogenic belt in southern Nevada and eastern California: *Geol. Soc. Am. Bull.*, v. 81, p. 1705–1720.

Gilluly, J. 1932, Geology and ore deposits of the Stockton and Fairfield quadrangles, Utah: *U.S. Geol. Survey Prof. Paper 173*, p. 1–165.

———, 1971, Plate tectonics and magmatic evolution: *Geol. Soc. Am. Bull.*, v. 82, p. 2383–2396.

———, Reed, J. C., Jr., and Cady, W. M., 1970, Sedimentary volumes and their significance: *Geol. Soc. Am. Bull.*, v. 81, p. 353–376.

Hamilton, W., 1969, Mesozoic California and the underflow of Pacific mantle: *Geol. Soc. Am. Bull.*, v. 80, p. 2409–2429.

———, 1970, The Uralides and the motion of the Russian and Siberian platforms: *Geol. Soc. Am. Bull.*, v. 81, p. 2553–2576.

Harris, H. D., 1959, Late Mesozoic positive area in western Utah: *Am. Assoc. Petrol. Geol. Bull.*, v. 43, p. 2636–2652.

Hope, R. A., 1971, Geologic map of the Spruce Mountain quadrangle, Elko County, Nevada: *U.S. Geol. Survey Geol. Quad. Map GQ-942*, scale 1:62,500.

Hose, R. K., and Blake, M. C., Jr., 1969, Structural development of the Eastern Great Basin during the Mesozoic (abs.): *Geol. Soc. Am. Abs. with Programs*, v. 1, pt. 5, p. 34.

Hose, R. K., and Daneš, Z. F., 1967, Late Mesozoic structural evolution of the eastern Great Basin (abs.): *Geol. Soc. Am. Ann. Mtg.*, New Orleans 1967, Programs, p. 102.

———, 1973, Development of the late Mesozoic to early Cenozoic structures of the eastern Great Basin (this volume).

Hose, R. K., and Repenning, C. A., 1959, Stratigraphy of Pennsylvanian, Permian, and Lower Triassic rocks of Confusion Range, west-central Utah: Am. Ass. Petrol. Geol. Bull., v. 43, p. 2167–2196.

Hotz, P. E., and Willden, R., 1964, Geology and mineral deposits of the Osgood Mountains quadrangle, Humboldt County, Nevada: U.S. Geol. Survey Prof. Paper 431, 128 p.

Howard, K. A., 1966, Structure of the metamorphic rocks of the northern Ruby Mountains, Nevada: Yale Univ., New Haven, Conn., Ph.D. thesis, 170 p.

———, 1971, Paleozoic metasediments in the northern Ruby Mountains, Nevada: Geol. Soc. Am. Bull., v. 82, p. 259–264.

Hsü, K. J., 1968, The principles of mélanges and their bearing on the Franciscan–Knoxville paradox: Geol. Soc. Am. Bull., v. 79, p. 1063–1074.

Irwin, W. P., 1964, Late Mesozoic orogenies in the ultramafic belts of northwestern California and southwestern Oregon in Geological Survey Research, 1964: U.S. Geol. Survey Prof. Paper 501-C, p. C1–C9.

Johnson, M. S., and Hibbard, D. E., 1957, Geology of the Atomic Energy Commission Nevada proving grounds area, Nevada: U.S. Geol. Survey Bull. 1021-K, p. 333–384.

Kennedy, G. C., 1959, The origin of the continents, mountain ranges, and ocean basins: Am. Scientist, v. 47, p. 491–504.

King, P. B., compiler, 1969a, Tectonic map of North America: U.S. Geol. Survey, scale 1:5,000,000.

———, 1969b, The tectonics of North America—A discussion to accompany the Tectonic Map of North America scale 1:5,000,000: U.S. Geol. Survey Prof. Paper 628, 95 p.

Kistler, R. W., Evernden, J. F., and Shaw, H. R., 1971, Sierra Nevada plutonic cycle—Part I, Origin of composite granitic batholiths: Geol. Soc. Am. Bull., v. 82, p. 853–868.

Kistler, R. W., and Willden, R., 1969, Age of thrusting in the Ruby Mountains, Nevada (abs.): Geol. Soc. Am. Abs. with Programs, v. 1, pt. 5, p. 40–41.

Lanphere, M. A., Irwin, W. P., and Hotz, P. E., 1968, Isotopic age of the Nevadan orogeny and older plutonic and metamorphic events in the Klamath Mountains, California: Geol. Soc. Am. Bull., v. 79, p. 1027–1057.

Longwell, C. R., Pampeyan, E. H., Bowyer, B., and Roberts, R. J., 1965, Geology and mineral deposits of Clark County, Nevada: Nevada Bur. Mines Bull. 62, p. 1–203.

MacDonald, G. J. F., 1963, The deep structure of continents: Rev. Geophys., v. 1, p. 587–665.

McKee, E. D., Oriel, S. S., and others, 1967, Paleotectonic maps of the Permian System: U.S. Geol. Survey Misc. Geol. Inv. Map I–450, scale 1:5,000,000.

Misch, P., 1957, Magnitude and interpretation of some thrusts in northeast Nevada (abs.): Geol. Soc. Am. Bull., v. 68, p. 1854.

———, 1960, Regional structural reconnaissance in central-northeast Nevada and some adjacent areas—observations and interpretations in Guidebook to geology of east central Nevada: Intermountain Ass. Petrol. Geol. 11th Ann. Field Conf., p. 17–42.

———, and Hazzard, J. C., 1962, Stratigraphy and metamorphism of late Precambrian rocks in central northeastern Nevada and adjacent Utah: Am. Assoc. Petrol. Geol. Bull., v. 46, p. 289–343.

Moench, R. II., 1970, Premetamorphic down-to-basin faulting, folding, and tectonic dewatering, Rangeley area, western Maine: Geol. Soc. Am. Bull., v. 81, p. 1463–1496.

Morris, H. T., and Lovering, T. S., 1961, Stratigraphy of the East Tintic Mountains, Utah: U.S. Geol. Survey Prof. Paper 361, 145 p.

Mudge, M. R., 1970, Origin of the disturbed belt in northwestern Montana: Geol. Soc. Am. Bull., v. 81, p. 377–392.

———, 1971, Gravitational sliding and the foreland thrust and fold belt of the North American Cordillera —Reply: Geol. Soc. Am. Bull., v. 82, p. 1139–1140.

———, and Sheppard, R. A., 1968, Provenance of igneous rocks in Cretaceous conglomerates in northwestern Montana in Geological Survey Research, 1968: U.S. Geol. Survey Prof. Paper 600-D, p. D137–D146.

Nolan, T. B., 1935, The Gold Hill mining district, Utah: U.S. Geol. Survey Prof. Paper 177, 172 p.

Oriel, S., and Armstrong, F. C., 1966, Times of thrusting in the Idaho–Wyoming thrust belt— Reply: Am. Assoc. Petrol. Geol. Bull., v. 50, p. 2614–2621.

Paddock, R. E., 1956, Geology of the Newfoundland Mountains, Box Elder County, Utah: Univ. Utah, Salt Lake City, Utah, M.S. thesis, 101 p.

Price, R. A., 1971a, The Cordilleran foreland thrust and fold belt in the southern Canadian Rockies (abs.): Geol. Soc. Am. Abs. with Programs, v. 3, p. 404–405.

———, 1971b, Gravitational sliding and the foreland thrust and fold belt of the North American Cordillera —Discussion: Geol. Soc. Am. Bull., v. 82, p. 1133–1138.

———, 1973, Large-scale gravitational flow of supracrustal rocks, southern Canadian Rockies (this volume).

———, and Mountjoy, E. W., 1970, Geologic structure of the Canadian Rocky Mountains between Bow and

Athabasca Rivers—A progress report: *Geol. Assoc. Can. Spec. Paper 6*, p. 7–25.

Rigby, J. K., ed., 1958, Geology of the Stansbury Mountains, Tooele County, Utah: *Utah Geol. Soc. Guidebook to the Geology of Utah*, no. 13, 150 p.

Riva, J., 1970, Thrusted Paleozoic rocks in the northern and central HD Range, northeastern Nevada: *Geol. Soc. Am. Bull.*, v. 81, p. 2689–2716.

Roberts, R. J., 1964, Stratigraphy and structure of the Antler Peak quadrangle, Humboldt and Lander Counties, Nevada: *U.S. Geol. Survey Prof. Paper 459-A*, p. A1–A93.

———, 1968, Tectonic framework of the Great Basin: *UMR Jour.* no. 1, p. 101–119.

———, 1969, The Cordilleran continental margin—Continental collisions vs. geotectonic cycles [Summ.]: *Geol. Soc. Am. Abs. with Programs*, pt. 7, p. 286–288.

———, 1972, Evolution of the Cordilleran fold belt: *Geol. Soc. Am. Bull.*, v. 83, p. 1989–2004.

———, Crittenden, M. D., Jr., Tooker, E. W., Morris, H. T., Hose, R. K., and Cheney, T. M., 1965, Pennsylvanian and Permian basins in northwestern Utah, northeastern Nevada, and south-central Idaho: *Am. Assoc. Petrol. Geol. Bull.*, v. 49, p. 1926–1956.

Roberts, R. J., Hotz, P. E., Gilluly, J., and Ferguson, H. G., 1958, Paleozoic rocks of north-central Nevada: *Am. Assoc. Petrol. Geol. Bull.*, v. 42, p. 2813–2857.

Roberts, R. J., Radtke, A. S., and Coats, R. R., 1971, Gold-bearing deposits in north-central Nevada and southwestern Idaho, *with a section on* Periods of plutonism in north-central Nevada (*by* M. L. Silberman and E. H. McKee): *Econ. Geol.*, v. 66, p. 14–33.

Roeder, D. H., and Nelson, T. H., 1970, Fossil subduction zones in the North American Cordillera (abs.): *EOS (Am. Geophys. Union Trans.) Ann. Mtg. 51st, Washington, D.C.*, p. 421.

Rubey, W. W., and Hubbert, M. K., 1959, Role of fluid pressure in mechanics of overthrust faulting, II. Overthrust belt in geosynclinal area of western Wyoming in light of fluid-pressure hypothesis: *Geol. Soc. Am. Bull.*, v. 70, p. 167–206.

Schaeffer, F. E., and Anderson, W. L., 1960, Geology of the Silver Island Mountains, Box Elder and Tooele Counties, Utah, and Elko County, Nevada: *Utah Geol. Soc. Guidebook to the Geology of Utah*, no. 15, 185 p.

Scholten, R., 1968, Model for evolution of Rocky Mountains east of Idaho batholith: *Tectonophysics*, v. 6, p. 109–126.

———, 1970, Origin of the disturbed belt in northwestern Montana—Discussion: *Geol. Soc. Am. Bull.*, v. 81, p. 3789–3792.

———, 1971, The overthrust belt of southwest Montana and east-central Idaho (abs.): *Geol. Soc. Am. Abs. with Programs*, v. 3, p. 411.

———, 1973, Gravitational mechanisms in the northern Rocky Mountains of the United States (this volume).

———, and Ramspott, L. D., 1968, Tectonic mechanisms indicated by structural framework of central Beaverhead Range, Idaho-Montana: *Geol. Soc. Am. Spec. Paper 104*, 71 p.

Shaw, H. R., Kistler, R. W., and Evernden, J. F., 1971, Sierra Nevada plutonic cycle: Part II, Tidal energy and a hypothesis for orogenic-epeirogenic periodicities: *Geol. Soc. Am. Bull.*, v. 82, p. 869–896.

Silberling, N. J., and Roberts, R. J., 1962, Pre-Tertiary stratigraphy and structure of northwestern Nevada: *Geol. Soc. Am. Spec. Paper 72*, 58 p.

Smith, F., and Ketner, K. B., 1968, Devonian and Mississippian rocks and the date of the Roberts Mountains thrust in the Carlin-Pinon Range area, Nevada: *U.S. Geol. Survey Bull. 1251-I*, p. 1–19.

Snelson, S., 1955, The geology of the southern Pequop Mountains, Elko County, northeastern Nevada: Univ. Wash., Seattle, Wash., M.S. thesis.

Spieker, E. M., 1946, Late Mesozoic and early Cenozoic history of central Utah: *U.S. Geol. Survey Prof. Paper 205-D*, p. D117–D161.

Stanley, K. O., Jordan, W. M., and Dott, R. H., Jr., 1971, New hypothesis of Early Jurassic paleogeography and sediment dispersal for western United States: *Am. Assoc. Petrol. Geol. Bull.*, v. 55, p. 10–19.

Stevens, C. H., 1969, Middle to Late Triassic deformation in the Inyo, White and northern Argus Mountains, California (abs.): *Geol. Soc. Am. Abs. with Programs*, pt. 5, p. 78.

Stifel, P. B., 1964, Geology of the Terrace and Hogup Mountains, Box Elder County, Utah: Univ. Utah, Salt Lake City, Utah, Ph.D. thesis, 248 p.

Stokes, W. L., 1963, Geologic map of northwestern Utah: *Utah Geol. Mineral. Survey Map*, scale 1:250,000.

Thorman, C. H., 1970, Metamorphosed and non-metamorphosed Paleozoic rocks in the Wood Hills and Pequop Mountains, northeast Nevada: *Geol. Soc. Am. Bull.*, v. 81, p. 2417–2448.

Tooker, E. W., 1970, Radial movements in the western Wyoming salient of the Cordilleran overthrust belt: *Geol. Soc. Am. Bull.*, v. 81, p. 3503–3506.

———, and Roberts, R. J., 1970, Upper Paleozoic rocks in the Oquirrh Mountains and Bingham mining district, Utah: *U.S. Geol. Survey Prof. Paper 629-A*, p. A1–A76.

———, 1971, Structures related to thrust faults in the Stansbury Mountains, Utah, *in* Geological Survey Research, 1971: *U.S. Geol. Survey Prof. Paper 750-B*, p. B1–B12.

Tschanz, C. M., and Pampeyan, E. H., 1970, Geology and mineral deposits of Lincoln County, Nevada: *Nevada Bur. Mines Bull. 73*, p. 1–183.

Van Bemmelen, R. W., 1966, The structural evolution of the southern Alps: *Geol. Mijnb.*, v. 45, p. 405–444.

Walton, P. T., 1964, Late Cretaceous and early Paleocene conglomerates along the western margin of the Uinta Basin, *in* Guidebook to the geology and mineral resources of the Uinta Basin: *Intermountain Assoc. Petrol. Geol., 13th Ann. Field Conf.*, p. 139–143.

Whitebread, D. H., 1969, Geologic map of the Wheeler Peak and Garrison quadrangles, Nevada and Utah: *U.S. Geol. Survey Misc. Geol. Inv. Map I-578*, scale 1:48,000.

Willden, R., and Kistler, R. W., 1969, Geologic map of the Jiggs quadrangle, Elko County, Nevada: *U.S. Geol. Survey Map GQ-859*, scale 1:62,500.

Wilson, J. T., 1968, Static or mobile earth: The current scientific revolution, *in* Gondwanaland revisited: New evidence for continental drift: *Am. Philos. Soc. Proc.*, v. 112, p. 309–320.

Woodward, L. A., 1970, Tectonic implications of structure of Beaver and northern San Francisco Mountains, Utah: *Geol. Soc. Am. Bull.*, v. 81, p. 1577–1584.

Young, J. C., 1955, Geology of the southern Lakeside Mountains, Utah: *Utah Geol. Mineral. Survey Bull. 56*, 116 p.

RICHARD K. HOSE
U.S. Geological Survey
Menlo Park, California[1]

ZDENKO F. DANEŠ
University of Puget Sound
Tacoma, Washington

Development of the Late Mesozoic to Early Cenozoic Structures of the Eastern Great Basin[2]

Low-angle faults of large displacement are an important structural element of the eastern Great Basin under discussion here (Fig. 1). These faults and related structures were developed mainly during the late Mesozoic to early Cenozoic in marine and continental strata from 30,000 to 40,000 ft (9000 to 12,000 m) thick. Subsequent block-faulting (Basin and Range orogeny), erosion, sedimentation, and vulcanism obscured or removed these structures in many places. Enough is exposed, however, to enable us to formulate ideas of how the low-angle faults may have formed. We suggest that they were produced mainly by differential uplift and gravity sliding, and we also proffer a tentative model to explain the unique style of displacement of the low-angle faults in the western

---

[1] Publication authorized by the Director, U.S. Geological Survey.

[2] We are very grateful to the following people who helped with various phases of the manuscript preparation: M. C. Blake, Jr., M. D. Crittenden, Jr., A. M. Johnson, D. L. Jones, P. B. King, M. Meier, H. J. Moore II, H. T. Morris, W. G. Pierce, G. Plafker, R. J. Roberts, E. W. Tooker, D. J. Varnes, and D. H. Whitebread. We are particularly grateful to Crittenden, Morris, Tooker, and Whitebread for sharing with us their vast knowledge of the geology of various parts of the subject area and to Arvid Johnson for his counsel on our analysis. We shoulder full responsibility for the conceptual aspects of this paper except as credited within the text.

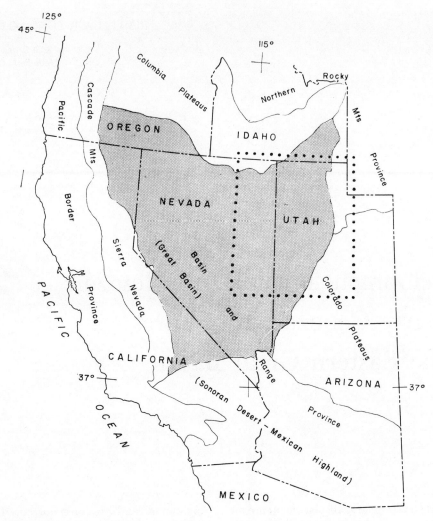

FIG. 1 Index map showing physical divisions of the western United States and outline of the subject area (dotted). Great Basin shown by gray area.

part of the area where younger or structurally higher strata are emplaced over older or structurally lower strata.

Regional Structural Geology [3]

The part of the eastern Great Basin that is considered here (Fig. 1) can be divided into an eastern and western part on the basis of structural differences (Fig. 2). The two parts are separated approximately by a line that extends from the eastern edge of the Raft River Range to the southeastern edge of the Snake Range. This line is "the eastern border of known occurrences of décollement thrusting" (Misch, 1960, p. 18), and it is also the eastern boundary of Misch's "northeastern

---

[3] The reader who is interested in the geologic history of the Great Basin before the late Mesozoic to early Cenozoic orogenesis should refer to the paper in this book by Roberts and Crittenden.

Nevada structural province." The eastern part of the area, which is unnamed, extends eastward to the eastern limit of thrusting (Fig. 2).

Low-angle faults of large displacement are the most striking pre-mid-Cenozoic structural features of this region. They were first described by Blackwelder (1910) in the Wasatch Mountains, but because they could not be traced beyond the individual mountain ranges, due to the disruption of the older terrane by the Basin and Range orogeny, their continuity along the eastern edge of the Great Basin went unrecognized for many years. Billingsley and Locke (1933) were the first to recognize the continuity of the thrust structures in the various ranges along the eastern edge of the Great Basin, and they named the whole structure the "Rocky Mountain–Great Basin thrust arc." Later it was called "the zone of Laramide thrusting" by Harris (1959) and, more recently, "Sevier orogenic belt" by Armstrong (1968).

Low-angle faults in the western part of the region were first discovered by Nolan (1935) at Gold Hill, Utah. Later, Hazzard and others (1953) described a large low-angle fault in the Snake Range, Nevada. Discovery of this fault was followed by similar finds (Misch and Easton, 1954) in the Schell Creek Range, and later Misch (1957) suggested that these faults also extended to the northern Egan Range, Cherry Creek Range, and Ruby Mountains, all in Nevada. Hazzard and Turner (1957) suggested that this fault or fault system was also present in the East Humboldt and Pequop Ranges, Nevada, the Albion Range, Idaho, and the Raft River and Grouse Creek Ranges, Utah.

Misch (1960) made the point that all of his eastern Nevada structural province was probably underlain by a décollement-type fault or fault complex, and that it was probably coextensive with and correlative with the thrust faults of the southern Wah Wah Range described by Miller (1966) and San Francisco Mountain area, Utah, described by East (1966).

Eastern Terrane

Low-angle faults of the eastern part of the region generally have large displacement, and most of them emplace older rock over younger rock. These faults have moved a thick miogeosynclinal sequence over a thin cratonic sequence of different facies. Estimates of separation vary from as little as 12 miles (19 km) (Hintze, 1960) to as much as 75 miles (120 km) (Eardley, 1951, p. 330). Crittenden (1961) has estimated eastward movement of 40 miles (65 km) along a fault in the Wasatch Mountains. Low-angle faults with the same sense of displacement occur structurally higher within the allochthon, and most of these also emplace older rock on younger, although there are a few exceptions in the western part of the eastern area.

The allochthonous rocks of the eastern part of the area are broadly folded and may show steep dips near the frontal thrusts. Crittenden (oral communication, 1970) has mapped a large recumbent fold overturned to the east in the autochthonous rock in the Wasatch Mountains. High-angle faults within the allochthon are mainly tears associated with the higher level imbricate thrusting. In the Wasatch Mountains, rocks of presumed Paleocene age depositionally overlap a large thrust; in the Canyon Range the fault at the base of the allochthon cuts rocks that are of probable Cretaceous age—relations that seem to place the orogeny within the Cretaceous to early Paleocene.

Western Terrane

Structures of the western part differ from those to the east. Both regions contain low-angle faults of large displacement, but most of those on the west emplace younger or structurally higher rock over older or structurally lower rock. Among the few exceptions, one in the Antelope Range developed earlier than a large younger-on-older fault in

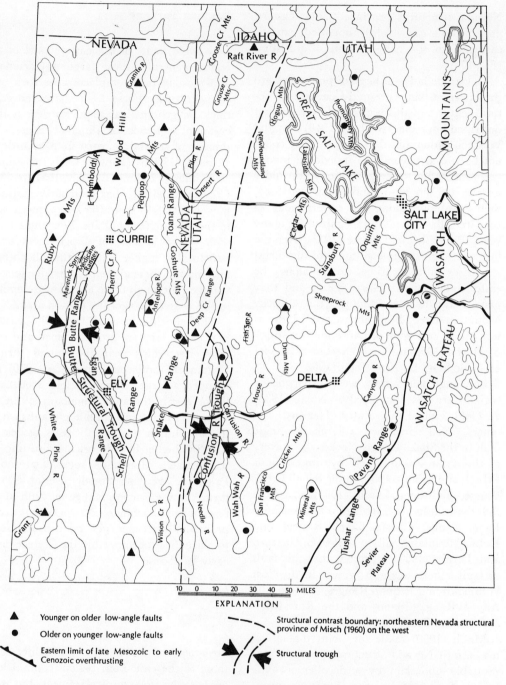

FIG. 2 Subject area showing distribution of structural elements.

the same range. Where the major or basal décollement is exposed, as in the Snake Range, structurally higher low-angle faults are also mostly of the younger-on-older type. The allochthonous units contain normal faults, in great abundance in places, and these form series of steps, or horsts and grabens, which are truncated by the low-angle faults.

Another important feature of the western part is the general homoclinal structure of many of the mountain ranges, even those cut by low-angle faults. Young (1960, p. 166) noted that in the north-central Schell Creek Range "Folds are surprisingly scarce in the map area considering the prevalence of thrust faults," but these low-angle thrust faults are of the younger-on-older type. The only conspicuous folds are along the axial part of the Butte structural trough (Fig. 2), and these are of moderate to large wavelength and moderate amplitude.

Autochthonous rocks are present in several of the ranges in the western part of the area. The rocks exhibit a different structural style from that of the allochthon, and all are metamorphosed to some extent. In the Snake Range, Nevada, strongly rodded autochthonous quartzite and deformed marble form a gentle arch. Autochthonous Precambrian and Paleozoic strata in the Ruby Mountains are greatly attenuated, metamorphosed, and tightly folded (Howard, 1966). Similar features occur in autochthonous rocks of the Raft River Range, Grouse Creek Range, and Deep Creek Range, Utah, and in the Wood Hills, Pequop Mountains, Cherry Creek Range, and Schell Creek Range, Nevada. Metamorphism and deformation of these rocks occurred before movement on the décollement, as unmetamorphosed rock rests tectonically on metamorphic rock. We regard the folding and gentle arching of the décollement of the Snake Range reported by Hazzard and others (1953) as a late phase of the late Mesozoic to early Cenozoic orogeny.

Extent and Correlation of Low-Angle Faults

Misch (1960) inferred that the allochthonous rocks of his eastern Nevada structural province were coextensive with the allochthonous rocks of a part of the Sevier orogenic belt. It is our intent to strengthen this inference by summarizing briefly some of the geological work that has become available since Misch's (1960) paper was published. The work of Whitebread (1969) in the southern Snake Range, Nevada, of Misch and Hazzard (*in* Hose and Blake, 1970) in the central Snake Range, and of Hose and Blake (1970) in the northern Snake Range, has established essential structural and stratigraphic differences between autochthonous and allochthonous rocks in the Snake Range and immediately adjacent areas.

First, except for intrusive rocks which are present in abundance, autochthonous rocks are early Middle Cambrian or older and, second, they are generally metamorphosed and internally deformed. Additionally, high-angle faults are much less common in the autochthon than in the allochthon. On the other hand, rocks of the allochthon range from Middle Cambrian to Permian in age and are essentially unmetamorphosed but severely broken by both high- and low-angle faults. Furthermore, no intrusive rocks have been reported from upper plate rocks of the Snake Range.

In view of the profound stratigraphic and structural differences between autochthon and allochthon observed along the entire 50-mile (80 km) length and 18-mile (29 km) width of the Snake Range, we feel that the differences should extend at least a short distance beyond the range. We contend also that these differences would enable us to resolve the very important question whether the décollement of the Snake Range extends eastward beneath the Burbank Hills (a southwestern appendage to the Confusion Range, Utah) or emerges in the intervening area, which is 5 miles (8 km) wide and almost entirely covered by younger deposits. Whitebread's (1969) detailed map shows that the single isolated outcrop of Paleozoic in the covered area, as well as the westernmost edge of the Burbank Hills, contain Devonian rock assigned to the same formation found in the easternmost part of the southern Snake Range where it is clearly part of the allochthon. Therefore it seems most probable that the décollement does indeed extend eastward beneath the Burbank Hills, for if it had emerged somewhere beneath the covered interval, the Burbank Hills would have been autochthonous and therefore most likely made up of early Middle Cambrian or older strata (Fig. 3).

Another problem of importance is whether the décollement of the Snake Range is coextensive with the thrust faults of the Sevier orogenic belt. Resolution of this problem is wholly dependent upon the conclusion just reached that the décollement extends beneath the Burbank Hills. Figure 3 shows that the allochthonous rocks of the Snake Range, the Burbank Hills along with the Confusion Range proper, and the Wah Wah Range form a structurally discrete and continuous unit. This unit, although cut by high-angle faults, contains only one low-angle fault that could compare in magnitude of displacement with the décollement. Miller (1966) shows that the Lower Cambrian rocks of the southern Wah Wah Range are thrust relatively eastward over strata at least as young as Middle and possibly Late Jurassic.

This thrust fault, plus two mapped in the vicinity of the San Francisco Mountains

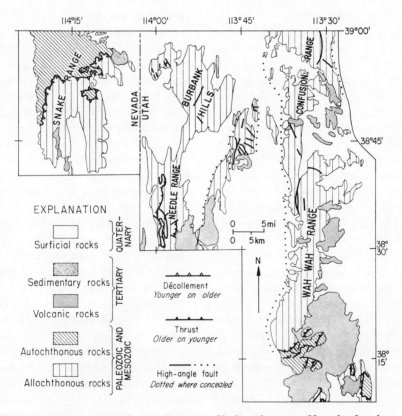

FIG. 3 Generalized geologic map of part of western Utah and eastern Nevada showing continuity of the allochthon and its relation to the autochthon.

(East, 1966), are part of the Sevier orogenic belt, and they juxtapose quite different facies of Paleozoic rocks, as is the case with thrusts to the northeast in the Wasatch Mountains. Their displacement must be considerable; furthermore, since the Wah Wah Range is in the upper plate with respect to the thrusts of the Sevier orogenic belt and, by virtue of its structural continuity with the Burbank Hills, is almost certainly upper plate with respect to the décollement, these low-angle fault surfaces must be coextensive and correlative (Misch, 1960; Hose and Daneš, 1968) (Fig. 3).

Interpretation of Contrasts in the Allochthonous Terrane

If the correlations are valid, we are faced with the problem of explaining contrasting structural styles in adjacent regions. Folding, thrusting, and general thickening of the section are characteristic of the eastern part of the allochthon and this could result only from horizontal compression as idealized in Fig. 4. On the west, the original section is attenuated as suggested by Fig. 5, particularly III and IV, a situation that could result only from extension. This combination of structures in one allochthonous complex is precisely what would result if the region were uparched on the west or downwarped on the east, so that an east-facing slope was formed. Under the circumstances defined by Hubbert and Rubey (1959) in which a fluid zone of greater than hydrostatic pressure is present near the base of a thick tabular mass (the developing allochthon) so that it could glide eastward, extension would prevail in the vicinity of the uparched area whereas compression would prevail at the toe. Such a combination of conditions should result in attenuation of the section on the west and thickening on the east.

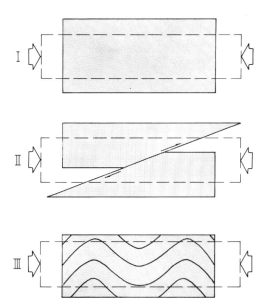

FIG. 4 Diagrams showing the ideal consequences of compressive forces applied to a slab as indicated by arrows. (I) Elastic-plastic deformation; (II) thrust-reverse fault deformation; (III) fold deformation (not to scale). (------- = original shape; ——— = final shape.)

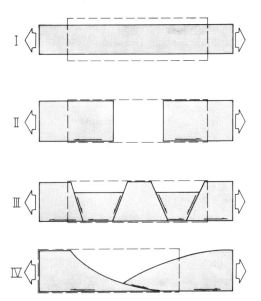

FIG. 5 Schematic diagram of effects of tensile forces applied to a slab as shown by arrows. (I) Elastic-plastic deformation; (II) pull-apart; (III) horst and graben on a basal shear surface; (IV) cycloidal shear surfaces. (------ = original shape; ——— = final shape.)

The combined width of the two parts of the allochthon at the present is 200 miles (320 km). If we accept 40 miles (65 km) of movement (Crittenden, 1961) on an important thrust of the frontal edge, then the original width of the developing allochthon would have been about 160 miles (260 km) or less. The thickness of the allochthon at the onset of tectonism was greater than 18,000 ft (5500 m) and perhaps locally as much as 35,000 ft (10,700 m) or even 40,000 ft (12,200 m). In view of the many lines of evidence for uniform relative eastward movement of the allochthon, a completely compressive origin would require that a push be applied to the sides as in Fig. 4 (II and III). However, application of tectonic compressive forces to this prism of strata without deformation of the autochthon is mechanically difficult to envision. One could, of course, visualize crustal (plus mantle) convergence as conventionally portrayed in sketches of tectogenes or zwischengebirge, and argue on this basis for compressive deformation and thickening of the rocks, which in turn would exert horizontal components of force against the rearward part of a prism of strata. Such an explanation would require severely deformed terrane in the rearward part of the allochthon—a situation that does not exist.

If, on the other hand, the thrust complex were compressive and rooted, uplift and movement along the root zone would be at least as much as the movement along the leading edge of the allochthon; that is, 40 miles (65 km). This situation would require that mantle rock or at least low crustal rock be exposed in the western part of the allochthon—also a circumstance not borne out in the subject area. The mechanics of underthrusting of a nonyielding foreland as suggested by Misch (1960, p. 41) imply that compression would have prevailed along the entire width of the allochthon. By any of these explanations, the entire allochthon should have developed internal structures that either resulted from compression or at least contained few, if any, that resulted from extension. Certainly most of the structures of the western part of the subject area could not possibly have been formed by compression; on the contrary, the only reasonable conclusion is that they were produced by extension.

Role of Gravity

In the past, suggestions that the allochthon could have been emplaced by gravity were rejected, at least partly because no satisfactory mechanism was known that would allow gravitational movement on a small gradient. The analysis of Hubbert and Rubey (1959) circumvented that objection. But Misch (1960, p. 40) objected to a gravitational origin because he felt that tectonic denudation of the basement should have resulted. We feel, on the other hand, that total denudation would be possible only where an allochthon was so thin as to be able to support its own weight in a near-vertical wall. Such a situation is obtained in the breakaway zone of the Heart Mountain fault (Pierce, 1960, 1973) where the glide block was about 2000 ft (610 m) thick.

In a somewhat thicker allochthon, the result of extension in the western part of the allochthon would be development of horsts and grabens, rather similar to Fig. 5, III. This is the situation depicted by Rubey and Hubbert (1959, Fig. 9). Our interpretation of the structures of the allochthon in the eastern Great Basin requires that those in the western part of that area, including especially low-angle faults, resulted from horizontally applied tensile force and vertical gravitational compression. We suggest that the great thickness of the allochthon in the western part of the area (>18,000 ft or 5500 m) must have been one of the critical factors in the evolution of the low-angle, younger-on-older faults.

A mathematical–physical explanation of the genesis of low-angle faults of the younger-on-older variety was first suggested to us by Varnes' (1962) work in the south Silverton area, Colorado, where his theoretically derived shear trajectories compared remarkably well with the fault pattern he had mapped in the field. Varnes used a modification of Prandtl's compressed cell to derive the pattern of shear trajectories. We later found that Nye (1951) had published an analysis of shear trajectories in glaciers, very similar to the one we are presenting to explain the younger-on-older type low-angle faults of eastern Nevada and western Utah.

Our explanation is based upon the assumption that the entire allochthon was initially a large slab or prism (Fig. 6A) no more than 160 miles (260 km) across (E–W), from 18,000 to 40,000 ft (5500 to 12,200 m) thick, and at least 200 to 250 miles (320 to 400 km) long. The solution to the problem of development of low-angle faults is, however, independent of the length. We shall arbitrarily assume an initial thickness of the allochthon of 30,000 ft (9000 m).

The prism is then transformed into something similar to Fig. 6B by some indeterminate tectonic process. In a relative sense the western edge is elevated and the eastern edge depressed so that an east-facing slope is produced. Since the western edge is approximately aligned with the axis of Nolan's (1943) early Mesozoic geanticline, the presence of an east-facing slope seems to be reasonably well established.

As the gradient is developed, if there exists a zone of abnormally high fluid pressure close to the base of the slab, the maximum compressive stress in the western part of the area is vertical, whereas to the east it is horizontal. In response to the vertical compression, which was induced by gravity, cycloidal shear trajectories come into being (Figs 6C and 7) The vertical compression, coupled with the tendency of the entire prism to move down gradient, causes movement along the trajectories of maximum shear, as illustrated in the left part of Fig. 6D. Faults would develop first on the west.

The mathematical analysis of the problem closely parallels that of Nye (1951), Varnes (1962), and Kanizay (1962). It is based on the following assumptions (Fig. 7):

1. The crustal block analyzed here may be approximated by an infinite slab of homogeneous, isotropic material acted upon by its own weight only.
2. Failure of the material occurs when the

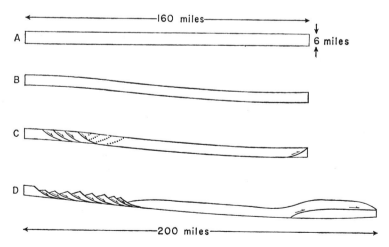

FIG. 6 Diagrammatic cross section of the eastern Great Basin showing development of younger-on-older faults. (A) Configuration of slab at outset; (B) development of gradient; (C) development of shear trajectories of cycloidal form in the rearward or active part of the allochthon; (D) final configuration showing younger-on-older faults on the west and older-on-younger faults on the east.

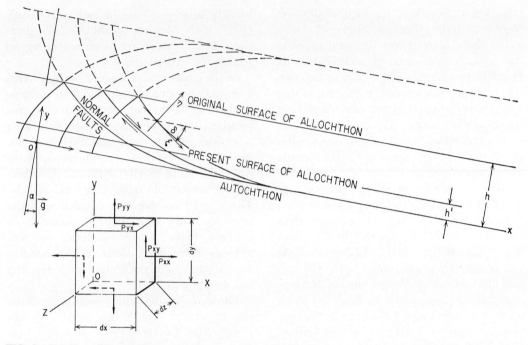

FIG. 7 Principal shear-stress trajectories within the rearward part of the postulated allochthon, and stress on a cubical element.

maximum shear stress reaches a constant value k characteristic of the material (Von Mise's criterion).
3. Failure follows a trajectory of maximum shearing stress.
4. The problem is independent of the coordinate Z.

The only body force acting on a particle is its weight; therefore

$$\frac{\partial Pxx}{\partial x} + \frac{\partial Pyx}{\partial y} = -\rho g \sin \alpha \quad (1)$$

$$\frac{\partial Pxy}{\partial x} + \frac{\partial Pyy}{\partial y} = \rho g \cos \alpha \quad (2)$$

where ρ is the density of the material, g the gravity acceleration, x the cartesian coordinate along the slab, y at right angles to the slab (positive sense up), and α the dip of the slab.

Solving Eqs. 1 and 2, setting the stress components equal to zero at the free surface, $y = h$; and setting $Pxy = k$ at the bottom of the slab, $y = 0$ yields

$$Pxx = a(y - h) + f(y) - f(h) \quad (3)$$

$$Pxy = -\rho g \sin \alpha \, (y - h) = k\left(1 - \frac{y}{h}\right) \quad (4)$$

$$Pyy = \rho g \cos \alpha \, (y - h) = \frac{k}{\tan \alpha}\left(\frac{y}{h} - 1\right) \quad (5)$$

where a is an arbitrary constant and f is an arbitrary function.

Applying assumption 2 and the rules of tensor transformation, we are now able to obtain via a series of equations (5a to 5j) given in the Appendix the following mathematical expressions for the shape of the fault surfaces:

$$\frac{dy}{dx} = \frac{2 - y/h}{\sqrt{(y/h)(2 - y/h)}} \quad (6)$$

or

$$\frac{dy}{dx} = \frac{-y/h}{\sqrt{(y/h)(2 - y/h)}} \quad (7)$$

Both Eq. 6 and Eq. 7 represent families of cycloids. One family has the cusps at $y = 0$, pointing downward; the other, at $y = 2h$, pointing upward. Every trajectory of one of the families intersects every trajectory of the other family at right angles, and both intersect the free surface, $y = h$, at an angle of 45°.

The foregoing analysis shows how cycloidal shear trajectories could indeed be generated and that, in view of the assumed plastic nature of the prism and the east-sloping gradient, movement by body forces should take place. In the western part of the allochthon this movement would be of the type that would move structurally higher (younger) rock over structurally lower (older) rock. For the case of the eastern Great Basin we conclude that most of the upper portion of the allochthon has already been eroded since the average observed dip of the fault surfaces is gentle to flat.

Movement is possible only on a slope, and the minimum gradient, according to the work of Hubbert and Rubey (1959), is a function of the ratio λ of fluid pressure at the base of the allochthon to the weight of rock above. As the fluid pressure approaches the geostatic pressure ($\lambda \to 1$), the required gradient becomes less. Raleigh and Griggs (1963) have pointed out that it is necessary to consider the fact that energy has to be expended to force the leading edge of an allochthon up a ramp or rise, a problem not considered by Hubbert and Rubey (1959). This requires a slight increase in the gradient, or an increase in λ, or a slight push from the rear.

In the westernmost part of the developing allochthon the segment of rock resting on a hemicycloidal shear trajectory is resting on an average slope of about 22°. If λ along this shear surface is reasonably high, the block of rock resting on the cycloidal surface would exert a considerable eastward force on the back end of the allochthon, probably not enough to fully account for the extra force needed to push the leading edge up the ramp but certainly enough to minimize the increase in gradient called for by Raleigh and Griggs.

As the allochthon moves downslope it creates a mass deficiency in the hinterland and mass surplus along the leading edge as it overrides the autochthon. The net effect in the rear would be uplift of the autochthon to provide a slight local gradient increase. Along the leading edge the increased mass would cause isostatic adjustment that would have the effect of reducing the angle of the slope against which the front part of the allochthon would move. The combined effects would make movement of the allochthon easier.

Appendix

According to the rules of transformation of components of a tensor,

$$P\xi\eta \equiv \cos(\xi, x)\, Pxx \cos(x, \eta)$$
$$+ \cos(\xi, x)\, Pxy \cos(y, \eta)$$
$$+ \cos(\xi, y)\, Pyx \cos(x, \eta)$$
$$+ \cos(\xi, y)\, Pyy \cos(y, \eta) \quad (5a)$$

where ξ, η are an arbitrary (local) orthogonal coordinate system. Let the angle between x and ξ be γ; then

$$P\xi\eta = -Pxx \cos\gamma \sin\gamma$$
$$+ Pxy (\cos^2\gamma - \sin^2\gamma)$$
$$+ Pyy \sin\gamma \cos\gamma$$
$$= -\frac{Pxx - Pyy}{2} \sin 2\gamma + Pxy \cos 2\gamma \quad (5b)$$

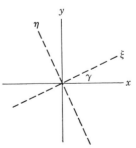

If $P\xi\eta$ along such a direction is to reach a maximum value, then it must vanish along a plane inclined 45° to the former:

$$0 = -\frac{P_{xx} - P_{yy}}{2}\sin 2(\gamma + 45°)$$
$$+ P_{xy}\cos 2(\gamma + 45°)$$
$$= \frac{P_{xx} - P_{yy}}{2}\cos 2\gamma + P_{xy}\sin 2\gamma \quad (5c)$$

so that

$$\tan 2\gamma = -\frac{P_{xx} - P_{yy}}{2P_{xy}} \quad (5d)$$

Therefore

$$\sin 2\gamma = -\frac{P_{xx} - P_{yy}}{\sqrt{4P_{xy}^2 + (P_{xx} - P_{yy})^2}} \quad (5e)$$

and

$$\cos 2\gamma = \frac{2P_{xy}}{\sqrt{4P_{xy}^2 + (P_{xx} - P_{yy})^2}} \quad (5f)$$

We substitute (5e) and (5f) in (5b) and use assumption 2:

$$P\xi\eta = \frac{1}{2}\frac{(P_{xx} - P_{yy})^2}{\sqrt{4P_{xy}^2 + (P_{xx} - P_{yy})^2}}$$
$$+ \frac{2P_{xy}^2}{\sqrt{4P_{xy}^2 + (P_{xx} - P_{yy})^2}}$$
$$= \frac{1}{2}\frac{4P_{xy}^2 + (P_{xx} - P_{yy})^2}{\sqrt{4P_{xy}^2 + (P_{xx} - P_{yy})^2}} = k \quad (5g)$$

or

$$(P_{xx} - P_{yy})^2 + 4P_{xy}^2 = 4k^2 \quad (5h)$$

Next, we transform (5d):

$$\tan 2\gamma = \frac{2\tan\gamma}{1 - \tan^2\gamma}$$
$$= -\frac{P_{xx} - P_{yy}}{\sqrt{4k^2 - (P_{xx} - P_{yy})^2}} \quad (5i)$$

or

$$\tan\gamma = \frac{\sqrt{4k^2 - (P_{xx} - P_{yy})^2} \pm 2k}{P_{xx} - P_{yy}}$$
$$= \frac{\partial y}{\partial x} \quad (5j)$$

References

Armstrong, R. L., 1968, Sevier orogenic belt in Nevada and Utah: *Geol. Soc. Am. Bull.*, v. 79, p. 429–458.

Benioff, H., and Gutenberg, B., 1951, Strain characteristics of the Earth's interior, *in* Gutenberg, B., ed., *Internal constitution of the Earth*: New York, Dover Publications, p. 382–407.

Billingsley, P. R., and Locke, A., 1933, Tectonic position of ore districts in the Rocky Mountain region: *Am. Inst. Mining Metall. Engineers Tech. Pub. 501*, 12 p.

Blackwelder, E., 1910, New light on the geology of the Wasatch Mountains, Utah: *Geol. Soc. Am. Bull.*, v. 21, p. 517–542.

Crittenden, M. D., 1961, Magnitude of thrust faulting in northern Utah, *in* Geological Survey research 1961: *U.S. Geol. Survey Prof. Paper 424-D*, p. D128–D131.

Eardley, A. J., 1951, Structural geology of North America: New York, Harper and Bros., 624 p.

East, E. H., 1966, Structure and stratigraphy of San Francisco Mountains, western Utah: *Am. Assoc. Petrol. Geol. Bull.*, v. 50, p. 901–920.

Harris, H. D., 1959, Late Mesozoic positive area in western Utah: *Am. Assoc. Petrol. Geol. Bull.*, v. 43, p. 2636–2652.

Hazzard, J. C., Misch, P., Wiese, J. H., and Bishop, W. C., 1953, Large-scale thrusting in northern Snake Range, White Pine County, northeastern Nevada (abs.): *Geol. Soc. Am. Bull.*, v. 64, p. 1507–1508.

Hazzard, J. C., and Turner, E. F., 1957, Décollement-type overthrusting in south-central Idaho, northwestern Utah, and northeastern Nevada (abs.): *Geol. Soc. Am. Bull.*, v. 68, p. 1829.

Hintze, L., 1960, Thrust-faulting limits in western Utah (abs.): *Geol. Soc. Am. Bull.*, v. 71, p. 2062.

———, 1963, *Geologic map of southwestern Utah*: Utah State Land Board.

Hose, R. K., and Blake, M. C., Jr., 1970, *Geologic map of White Pine County, Nevada*: U.S. Geol. Survey open-file map.

Hose, R. K., and Daneš, Z. F., 1968, Late Mesozoic structural evolution of the eastern Great Basin (abs.): *Geol. Soc. Am. Spec. Paper 115*, p. 102.

Howard, K. A., 1966, Structure of metamorphic rocks of the northern Ruby Mountains, Nevada: Yale University, Ph.D. dissertation, New Haven, Conn.

Hubbert, M. K., and Rubey, W. W., 1959, Role of fluid pressure in mechanics of overthrust faulting. I—Mechanics of fluid-filled porous solids and its application to overthrust faulting: *Geol. Soc. Am. Bull.*, v. 70, p. 115–166.

Kanizay, S. P., 1962, Mohr's theory of strength and Prandtl's compressed cell in relation to vertical tectonics: *U.S. Geol. Survey Prof. Paper 414-B*, p. B1–B16.

Miller, G. M., 1966, Structure and stratigraphy of southern part of Wah Wah Mountains, southwest Utah: *Am. Assoc. Petrol. Geol. Bull.*, v. 50, p. 858–900.

Misch, P., 1957, Magnitude and interpretation of some thrusts in northeastern Nevada (abs.): *Geol. Soc. Am. Bull.*, v. 68, p. 1854–1855.

———, 1960, Regional structural reconnaissance in central-northeast Nevada and some adjacent areas: Observations and interpretations, *in* Geology of east-central Nevada: *Intermountain Assoc. Petrol. Geol. 11th Ann. Field Conf. Guidebook*, p. 17–42.

Misch, P., and Easton, W. H., 1954, Large overthrusts near Connors Pass in the southern Schell Creek Range, White Pine County, eastern Nevada (abs.): *Geol. Soc. Am. Bull.*, v. 65, p. 1347.

Nolan, T. B., 1935, The Gold Hill mining district, Utah: *U.S. Geol. Survey Prof. Paper 177*, 172 p.

———, 1943, The Basin and Range province in Utah, Nevada, and California: *U.S. Geol. Survey Prof. Paper 197-D*, p. 141–196.

Nye, J. F., 1951, The flow of glaciers and ice-sheets as a problem in plasticity: *Proc. Royal Soc. London*, ser. A, v. 207, p. 554–572.

Pierce, W. G., 1960, The "break-away" point of the Heart Mountain detachment fault in northwestern Wyoming, *in* Geological Survey research 1960: *U.S. Geol. Survey Prof. Paper 400-B*, p. B236–B237.

———, 1973, Principal features of the Heart Mountain fault and the mechanism problem (this volume).

Raleigh, C. B., and Griggs, D. T., 1963, Effect of the toe in the mechanics of overthrust faulting: *Geol. Soc. Am. Bull.*, v. 74, p. 819–830.

Rubey, W. W., and Hubbert, M. K., 1959, Role of fluid pressure in mechanics of overthrust faulting. II—Overthrust belt in geosynclinal area of western Wyoming in light of fluid-pressure hypothesis: *Geol. Soc. Am. Bull.*, v. 70, p. 167–206.

Varnes, D. J., 1962, Analysis of plastic deformation according to von Mise's theory with application to the South Silverton area, San Juan County, Colorado: *U.S. Geol. Survey Prof. Paper 378-B*, p. B1–B49.

Whitebread, D., 1969, Geologic map of the Wheeler Peak and Garrison quadrangles, Nevada and Utah: *U.S. Geol. Survey Misc. Geol. Inv. Map I-578*.

Young, J. C., 1960, Structure and stratigraphy in north-central Schell Creek Range: *Intermountain Assoc. Petroleum Geologists 11th Ann. Field Conf. Guidebook*, p. 158–172.

RICHARD M. FOOSE

Department of Geology
Amherst College
Amherst, Massachusetts

Vertical Tectonism and Gravity in the Big Horn Basin and Surrounding Ranges of the Middle Rocky Mountains[1]

The purpose of this paper is to show in summary the sequence of tectonic events that have led to the development of the major crustal features in the area of the Bighorn Basin and surrounding ranges of the Middle Rocky Mountains. This is an area in which large blocks of the crust have been vertically raised or have subsided. Coincidentally, one may observe the importance of gravity as a force that has altered the vertically displaced crustal blocks and contributed its own style to the final deformational picture.

Regional Description

General Setting

The Bighorn Basin with its surrounding ranges is located in northwestern Wyoming and part of south-central Montana (Fig. 1). The area comprises a large part of the so-called Middle Rocky Mountains. The Middle Rockies are characterized by large mountain

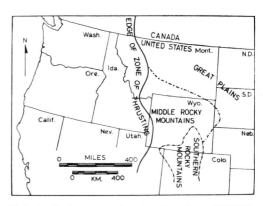

FIG. 1 Index map showing location of Middle Rocky Mountains.

[1] I acknowledge with pleasure the critical suggestions for improvement of the paper by Donald U. Wise.

ranges with more or less rectangular shapes, alternating with large, flat-floored basins filled with Tertiary sedimentary rocks. The ranges are commonly 30–70 km wide and 120–180 km long. The range fronts are commonly very steep features with a relief of 1000–2000 m reflecting a strong structural control. Some of the ranges are relatively flat-topped, such as the Beartooth and the Bighorn Mountains, and reveal vast areas of Precambrian crystalline rocks stripped of their Paleozoic and Mesozoic cover.

Tectonic Setting

Middle Rocky Mountains. The Middle Rocky Mountains are part of the Wyoming foreland shelf east of the Cordilleran fold and thrust belt of geosynclinal sediments that locally are more than 16,000 m thick (Rubey and Hubbert, 1959) (Fig. 1). By contrast, slightly less than 1000 m of shallow-water Paleozoic sediments, largely carbonate, and approximately 2000 m of Mesozoic rocks, mostly clastic, were deposited on the crystalline basement of the Wyoming shelf. Only in Cretaceous time did the cratonic shelf begin to subside strongly, resulting in increasingly thicker deposits of clastic sediments accompanied by interbedded tuffs and other volcanic products from the vulcanism that characterized the beginning of Laramide deformation. During Laramide deformation and after its termination in the Eocene, as much as 3000 m of Tertiary sediments accumulated in some of the subsiding basins. Before Laramide deformation, however, the basement was near the surface throughout the region, and it played a determining role in the style of deformation that occurred in the Laramide orogeny.

Bighorn Basin and Surrounding Ranges. The Bighorn Basin is almost completely surrounded by uplifted basement blocks (Fig. 2). To the west, south, and east of the basin these blocks form great mountain masses. By contrast, there is little or no topographic expression to reveal the sharp vertical displacement of the basement that marks the northern structural boundary of the basin, the Nye-Bowler fault zone, along which both vertical and horizontal movements have occurred. On the west side a great mass of Eocene volcanic tuffs and agglomerates, constituting the Yellowstone Plateau and Absaroka Mountains, accumulated during the later part of the Laramide orogeny.

The basin boundaries are characterized by a large structural relief between the subsided and uplifted blocks within a horizontal distance of a few kilometers. The uppermost surface of basement rocks in the Bighorn Basin is as deep as 6300 m *below* sea level, whereas the exposed surface of the same units at the top of the adjacent mountains is as much as 4000 m *above* sea level. Thus the structural relief has an order of magnitude of more than 10 km (Love, 1960), exceeding 12 km in some places (Berg, 1963).

The boundaries between the basin and the surrounding mountain blocks are of two types. Many are marked by sharply delineated basement faults, whereas others involve steep tilting of the basement without faulting. Where faults are absent or confined to the basement, the overlying sediments were draped into great monoclinal ramps or were strongly folded (locally into overturned positions) in response to basement movement (Foose and others, 1961; Prucha and others, 1965). In places they became detached and glided into lower topographic positions along the mountain fronts. The geometry of their occurrence helps reveal much of the influence of gravity on the style of the vertical uplifts produced during the Laramide orogeny. In the following sections characteristic structural features illustrating the tectonic evolution and mode of deformation of several ranges in the Bighorn Basin area will be examined.

Beartooth Mountain Block

Structural Features

The Beartooth crustal block (Fig. 2), approximately 130 by 65 km in area, is described and illustrated in detail by Foose and

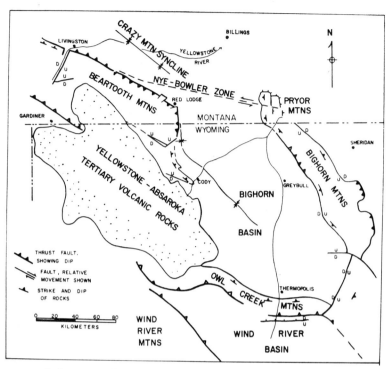

FIG. 2 Major tectonic features of the Bighorn Basin and surrounding mountain ranges.

others (1961). Only the salient features are reviewed here.

Except for a small area in the southwest part of the crustal block all of the structural boundaries between the block and adjacent basin areas are major faults. The largest of these, the Beartooth fault which bounds the block along its eastern edge, is vertical at the southeast corner of the block. At this point, it meets the similarly vertical Clarks Fork fault at almost a perfect right angle. Neither fault cuts the other, for there is no trace of either fault past their junction. The Precambrian crystalline rocks stand at an elevation of nearly 4000 m above sea level at the corner. About 1 km to the east they are at a depth of more than 1000 m below sea level, buried under the Tertiary sediments of the Bighorn Basin, thus making a structural relief of approximately 5 km (Fig. 3).

Northward along the Beartooth fault its dip progressively flattens toward the massif (west),

attaining an angle of 50° 6 km from the Clarks Fork corner, and a 20° angle 11 km farther north (Fig. 4). Where it is possible to observe the fault throughout a considerable vertical distance, one can see it steepening toward the bottom of the canyon. Thus the geometry does not fit the concave-upward, "sled-runner" profile postulated by Thom (1923) and other earlier workers for the range faults in the Central Rockies. Berg's (1963) projection of the frontal fault of the Wind River Mountains to a 45° dip at great depth does, however, remain a possibility.

The Beartooth block was not raised uniformly. Flueckinger (1970) has shown that clasts of crystalline rocks were accumulating along the front 10–15 km north of the southeast corner at the same time that Paleozoic carbonate clasts accumulated near the corner. Evidently, the Paleozoic section had been more rapidly eroded in the northern area due to greater uplift and exposure, and there was

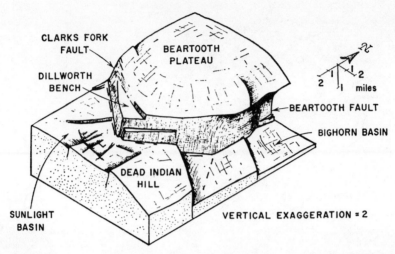

FIG. 3 Structural configuration of the basement at the southeast corner of the Beartooth crustal block.

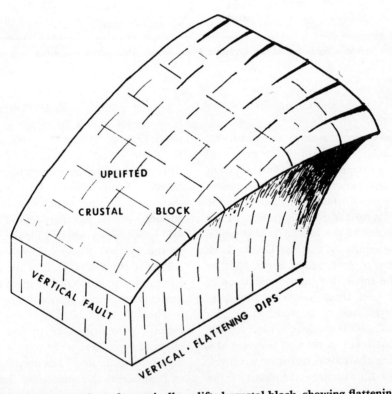

FIG. 4 Attitude of fault surface of a vertically uplifted crustal block, showing flattening of dip with increasing distance from the block corner.

a southward tilt, or plunge, to the rising frontal edge of the block, in spite of the fact that the flat attitude of the fault in the north would by itself tend to minimize the vertical throw.

The presence of two intersecting vertical faults at the southeast corner of the Beartooth block precludes the possibility of any horizontal movement of the massif at that place. The northwest corner of the Beartooth block has been upraised in an almost identical way. Thus there are two corners of the block where the limiting geometric relationship of intersecting vertical faults prevents any significant horizontal transport of the upraised block over the adjacent subsided basins.

At the northeast corner of the Beartooth block, in the vicinity of Red Lodge, Montana, however, there has been horizontal transport of the block out across the basin over a distance of approximately 3 km. Three prominent structural features provide mechanisms for horizontal displacement of the frontal edge of the block (Foose, 1960): (1) the outward curvature of the Beartooth fault into the attitude of a thrust; (2) a series of nearly parallel, imbricate thrusts in those parts of the upraised block that have had the least movement toward the basin; and (3) a series of tear faults that broke the block corner, allowing some sections of the block to move farther outward over the basin. Imbrication and tear faulting are characteristic only of the northeast corner of the Beartooth block (Fig. 5).

One other structural feature of importance may be observed at two places, each located approximately 16 km from the northeast block corner. Northwest of the corner, in the vicinity of Red Lodge Creek, and south of the corner, at Line Creek, the vertical Paleozoic and Mesozoic strata, already deformed along subhorizontal axes by draping over the edge of the raised crystalline block, have been laterally sheared, faulted, and folded along steeply plunging axes (Figs. 5 and 6). The

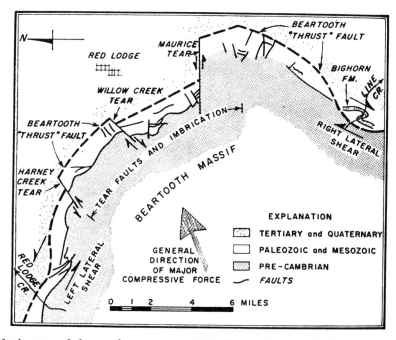

FIG. 5 Geologic map of the northeast corner of the Beartooth crustal block, showing structural features involved in gravitational breakup of the block corner.

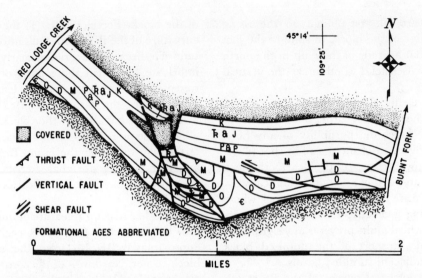

FIG. 6 Thrust and shear structure in draped sedimentary beds, north side of Beartooth Mountains.

motion was left lateral in the first and right lateral in the second locality, and it caused a horizontal shortening of at least 1 km and possibly as much as $2\frac{1}{2}$ km. These two major structural features indicate that the upraised Beartooth block has moved relatively toward the northeast in response to compressional forces.

Tectonic Forces and Mechanisms

Three major forces were involved in the deformation of the Beartooth crustal block: (1) regional west-east crustal compression; (2) vertical forces, either as a component of horizontal crustal compression or by vertical movements within the mantle; and (3) gravity. While recognizing that there is cause for debate about the relative importance of these forces, the author wishes to emphasize the force of gravity, which played an increasingly important role in Laramide deformation as the basement blocks were elevated into ever higher and less confined positions with respect to the adjacent subsided basins. Without either confirming or denying a possible role of regional crustal compression, the following discussion examines the manner in which gravity was effective, through a variety of structural mechanisms, in deforming the edges of the rising Beartooth block.

Beartooth Fault. One of the most important structural features that was influenced by gravity and which, in turn, changed the structural relations and style of much of the uplifted Beartooth block was the Beartooth fault. Experimental work by Hafner (1951), Sanford (1959) and others with models has clearly shown that vertical uplift of blocks by faulting will occur along concave-downward fractures. The spreading pattern of stress trajectories and some actual expansion of the rock itself cause this fault geometry, so characteristic of the Beartooth fault. Obviously, the potential effect of gravity is most important in the highest part of the section where there has been less structural support, or confinement, of the uplifted block.

That gravity has indeed played a role is suggested initially by the relation between the amount of uplift along the Beartooth fault and the dip of the fault, as discussed previously. Furthermore, where the crystalline rocks of the basement are examined in close juxtaposition to the steeply dipping Paleozoic sedimentary rocks along the Beartooth Mountain front there is good evidence of large-scale eastward basement rotation or flexing. For example, the Stillwater Complex of ultramafic rocks in the northwestern Beartooth Mountains has clearly been rotated and overturned (Butler, 1966). In addition, there are many thousands of joints in the crystalline basement rocks. The orientation of these joints includes major sets that are almost exactly parallel to the structurally controlled edges of the block (Fig. 4) (Spencer, 1959). These joints have widened as a consequence both of weathering and of expansion due to block uplift. Although block uplift is difficult to document precisely, Wise (1964) has shown that an extensive network of microjoints developed in late and post-Laramide time in the crystalline rocks of all of the uplifted crustal blocks in the Bighorn Basin area. The cause of microjoint formation was the horizontal expansion of the blocks due to gravity acting on their unsupported boundaries. The orientation of the microjoint sets coincides closely with the macrojoints. The combination of local basement block rotation along fractures with the expansion of both macrojoints and microjoints provided an excellent mechanism for block expansion, which explains why there is bending of the frontal faults along most of the uplifted basement blocks.

The presence of low-angle reverse faults along many of the range fronts in the Middle Rockies indicates local horizontal compression. The features described here neither preclude nor support the thought by some geologists that regional horizontal compression in the crust was a very important force in the Middle Rockies. The amount of horizontal transport along low-angle reverse faults, however, is not nearly as great as the amount of vertical uplift along those same faults where they are steep or vertical at greater depth. Low-angle reverse faults should not be interpreted automatically as indicating great regional compression. In the Middle Rockies they are a relatively minor feature directly associated with vertical uplift, and they are due largely or wholly to a local compressional component of the powerful drag by gravity on the poorly supported or unconfined range fronts that rose far above the crystalline basement of the adjacent basins.

Tear Faults. One may assume that the northeast corner of the block was initially lifted along uncomplicated faults similar to those forming the southeast and northwest corners of the block. However, because the northeast corner rose $1-1\frac{1}{2}$ km higher than the southeast corner, the likelihood was much greater that the northeast corner would rupture in response to the gravitational force acting upon it. This force, possibly combined with regional compressive stress acting on the rising block, was sufficient to cause a series of breaks in the uplifted and exposed northeast corner of the block so that three prominent tear faults enabled portions of the block to be moved outward separately over the adjacent basin (Fig. 7). The apparent horizontal displacements range from 1 to 3 km.

The actual "outward" movement of the Beartooth block in the corner area probably occurred in exactly the same way as along the eastern side of the block, that is, primarily by gravitational block expansion and bending of the Beartooth fault. The tear faults accommodated the extensional stress due to the generally radial movement away from the corner. It seems likely that there was also a relative upward component of movement along the tear faults in those block segments that moved less far over the sediment-filled basin. In these segments a lesser amount of the total displacement along the frontal fault would have been horizontal, and a greater amount vertical, in comparison to the adjacent segments. Because of post-Laramide erosion and the difficulty of establishing the exact sense of displacement within the crystalline rocks, the amount of relative vertical movement between adjacent blocks is unknown.

Imbrication. At the same time that tear faults were occurring, imbrication developed in the northeast corner of the Beartooth block. Numerous small, low-angle reverse faults, most of them parallel to each other and presumably also to the deeper, alluvium-buried

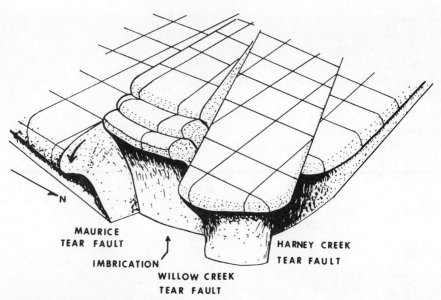

FIG. 7 Block diagram showing relationship of tear faults and imbrication at the northeast corner of the Beartooth crustal block.

Beartooth fault, broke the recessed segments of the northeast corner (Fig. 7). The uppermost fault in one of the pie-shaped corner segments caused Precambrian crystalline rocks to lie upon all of the Paleozoic formations from Cambrian to Mississippian.

Imbrication was caused by a combination of gravitational bending of the Beartooth fault followed by creation of a younger, nearly vertical fault in response to the continued vertical uplift of the crustal block. This new fault, in turn, also bent outward, and another new vertical fault was created (Fig. 8). In this manner imbrication in the recessed segments

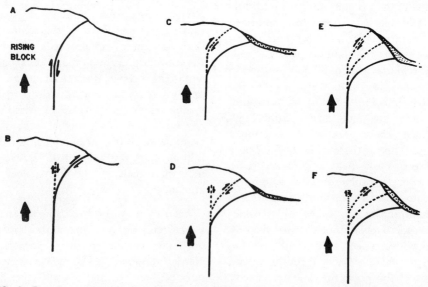

FIG. 8 Development of imbrication by combined effect of vertical uplift and gravity.

provided a mechanism for the release of gravitative stress produced continually by the continued vertical rise of the block. The segments between the recessed ones continued to move forward along the first frontal fault, rising to a lesser extent and thus failing to develop imbrication.

Lateral Shearing. The tectonic significance of the previously described left and right lateral fold and shear structures at distances of 16 km from the northeast corner (see Figs. 5 and 6) is that they are both younger than the initial uplift of the massif, as shown by the fact that they displace and modify the already folded and imbricated sedimentary rocks along the frontal edge. Evidently, they developed late in the Laramide orogeny, when the Beartooth block had been uplifted far above the adjacent basins and the northeast corner had already begun to break along tear faults and by imbrication. These structures may be interpreted in terms of the local gravitational model. The northeastward movement at the block corner, and the resulting decrease of confinement at the sides of the corner, caused lateral shearing stresses between the massif and the adjacent basins. The complexity of these features may be seen as the result of the local effect of gravity on already steeply dipping and overturned beds along the mountain front.

Bighorn Mountain Block

Structural Features

Many of the same features that have been described for the Beartooth Mountains may also be seen in the Bighorn Mountains. Like the Beartooths, the Bighorn Mountains were uplifted vertically along faults during the Laramide orogeny. Features particularly characteristic of the Bighorns are marginal basement rotation and drape-folding (in addition to faulting) of the sedimentary veneer over the edges of the crystalline block.

Across most of the Bighorn Range the Precambrian crystalline basement is nearly flat. Locally it is stripped of its sedimentary veneer. Within 500 ft of the western range front at Five Springs Canyon the crystalline basement abruptly becomes very steep. Wise (1964) has plotted the structural orientation of 23 pegmatite dikelets and many hundreds of microjoints in this zone. These features show almost constant dip in the rotated zone, thus "indicating fault block rotation, possibly of an overhanging basement lip." The age of the basement rotation must be very late in the Laramide uplift of the block.

At the same locality the Five Springs Fault, a low-angle reverse fault dipping toward the Range displaces the steeply overturned Paleozoic and Mesozoic sedimentary rocks that have been drape-folded over the crystalline block edge. Both north and south of the fault the section is steeply monoclinal above a basement fault which must be nearly vertical, because the structural relief between the basin and the range exceeds 6 km within a distance of 1–2 km. Drape-folding is also excellently displayed along the eastern boundary of the Bighorn Range.

Tectonic Forces and Mechanisms

The sharp basement rotation, extensive microjointing, and low-angle reverse faulting on the west side of the Bighorn Range are further examples of features resulting from local horizontal components of gravity acting on the unsupported edge of a block (compare Osterwald, 1961, Fig. 4) and giving rise to basinward movement of the range and extension of the basement in the range itself. The passive deformation of the sedimentary rock into steeply monoclinal or overturned drape folds over basement faults implies stretching of the beds, which was accomplished by a combination of plastic thinning and dilation along fractures. The precise mechanisms and the time relations of these structural developments in relation to the stage of basement deformation are subjects requiring further study.

Owl Creek Mountain Block

The crystalline basement rises gradually from the Bighorn Basin southward toward the Owl Creek Mountains in a ramp on top of which the Paleozoic sedimentary rocks lie as a monoclinal veneer. At the south front of the range there is about 10 km of structural relief (Love, 1960) with respect to the Wind River Basin. Earlier interpretations of the structural boundary between the range and the Wind River Basin by Fanshawe (1939), Tourtelot and Thompson (1948), and Berg (1962) postulated a large thrust of the range southward. This was in keeping with similar ideas for most of the ranges in the Middle Rockies. Wise (1963) showed that the structural relation between the Owl Creek Range and the Wind River Basin to the south are similar to those described previously in the Beartooth and Bighorn Mountain blocks, although the amount of horizontal transport is greater. In addition, along the outer edge of the range the steeply dipping carbonate rocks in the Paleozoic section have locally glided from their original position outward onto the Tertiary sedimentary fill of the basin in a series of tectonic "landslides" (Fig. 9), perhaps similar on a small scale to the Heart Mountain "thrust" (Pierce, 1973).

A third structural phenomenon attributable to gravitational forces consists of a series of graben along a zone near and parallel to the range crest and the highest point of vertical uplift (Fig. 10). None of the other block uplifts in the Bighorn Basin area exhibit this type of collapse feature. Their presence in the Owl Creek Range probably indicates a somewhat broader arching of the outer part of the block edge, which is compatible with a somewhat greater horizontal transport of the block over the basin.[2]

Evolution of Gravity-Created Structures

A brief summary is presented here of the gravity-created structural features and their

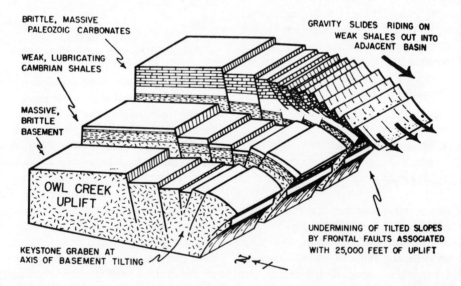

FIG. 9 Relationship of basement uplift to gravity gliding, Wind River Canyon area, Wyoming (from Wise, 1963).

---

[2] Wise (1971, personal communication) interprets the graben as phenomena associated with the outer edge of a compressionally bent "basement beam," thereby assigning more importance to crustal compression.

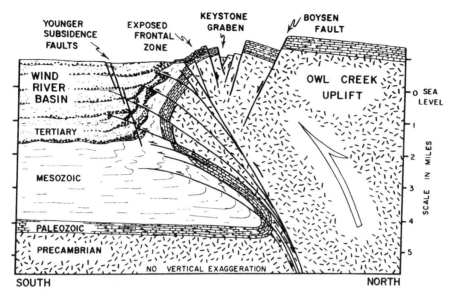

FIG. 10 Suggested structure of frontal zone of Owl Creek uplift, Wyoming (from Wise, 1963).

probable time of creation in relation to stages of block uplift in the Middle Rockies.

Initial Breaking of Crustal Blocks

1. Beginning of uplift along steeply dipping or vertical faults.
2. Beginning of gentle drape-folding and formation of monoclines in sedimentary veneer above the block boundaries.

Low Elevation of Crustal Block

1. Bending of vertical faults at a distance from the range corners into the attitude of high-angle reverse faults (Fig. 4).
2. Dilation of macrojoints and major fractures within the crustal block, particularly near its edge (Figs. 4 and 7).
3. Sharp drape-folding and thinning of sedimentary beds above block boundaries.

Medium Elevation of Crustal Block

1. Continued bending of frontal faults into attitudes of low-angle reverse faults.
2. Initial breaking of certain block corners under great stress by tear faults. Development of a pattern of tear faults around the corner (Figs. 5 and 7).
3. Local rotation of the upper edge of some basement blocks along low-angle reverse faults.
4. Dilation of joints and fractures within a larger area of the basement block (Fig. 7).
5. Extreme drape-folding and steep monoclinal development above the block boundaries; extension of some faults through the sedimentary cover.
6. Local removal of the sedimentary veneer from most highly tilted parts of block, with resulting development of truncated sedimentary bed edges; beginning of tectonic gliding of small, detached sedimentary rock masses toward basins.

Highest Elevation of Crustal Block

1. Maximum development of low-angle reverse faulting; order of magnitude of horizontal transport by this mechanism is 2–3 km (Figs. 4 and 7).
2. Continued tear faulting around broken block corners with apparent horizontal

transport in orders of magnitude up to 3 km (Fig. 7).
3. Imbrication in recessed segments of broken corner areas due to combination of continued block uplift and creation of steeper branch faults above older, low-angle reverse faults; horizontal transport by this mechanism in order of 1–2 km (Fig. 8).
4. Development of microjoints, enhancing volume expansion of the blocks.
5. Creation of lateral shear structures along block margins near laterally expanding block corners that are moving basinward along already existing faults.
6. Creation of graben along range crests in response to tensional stresses caused by broad arching of the basement and basinward transport of the crystalline block edge along high- and low-angle imbricated reverse faults (Fig. 10).
7. Extensive denudation of range tops, resulting in stripped surfaces and irregular erosional edges of sedimentary rock veneer now standing vertically or overturned along range fronts (Figs. 3 and 7).
8. Widespread detachment of sedimentary rock masses along bedding and other structural planes with consequent gravity gliding to lower topographic positions (Fig. 9).

Conclusions

Throughout the Middle Rocky Mountains a tectonic style may be observed that emphasizes the role of two major forces that acted on the crust during the Laramide orogeny, partly in sequence and partly concurrently. The primary and earliest force was that of vertical tectonism, with or without some horizontal compression. The secondary force that began during crustal block uplift and continued after its completion was that of gravity, which extensively remodeled the basically simple geometry of the initial blocks in the Middle Rockies by creating a variety of structural features that provided for the release of stress within the blocks. Release of stress was accompanied by movements of parts of the block along the newly created structures in directions outward and downward toward the adjacent basins. The tectonic evolution of the Middle Rockies followed this bicausal pattern of tectonism in the same general way that has so clearly been discussed by Van Bemmelen (1966).

In shaping the structures observed in the field, the importance of gravity as a local tectonic force was of course secondary to the major forces responsible for the initial vertical uplift of the range. However, it was not insignificant! Many new structural features owed their origin directly to it, and others were modified by it. Had this been an area of thick geosynclinal sediment accumulation similar to that of Idaho and western Montana (up to 16 km thick) instead of one of thin shelf-type sediments (average of 3 km), there would have been much more extensive lateral redistribution of the upper crust by gravity.

How does one relate the large tectonic province of the Middle Rockies, in which vertical tectonism and gravity appear to dominate as forces, to some adjacent tectonic provinces west of it where there is good evidence for strong regional horizontal compression, or to the western continental margin where many data support the hypothesis of a convergent encounter between oceanic and continental crustal plates? The question is important. Although the answer is not yet clear, it must be one that provides an understanding of compatibility between structural styles in large crustal blocks that differ widely. The validity of a tectonic interpretation for any specific crustal region need not obviate the validity of a totally different interpretation in another crustal region. Final answers to global-scale tectonic problems will be derived not by blurring the facts concerning tectonic features and forces in smaller crustal units but by sharpening them.

References

Berg, R. R., 1962, Mountain flank thrusting in Rocky Mountain foreland, Wyoming and Colorado: *Am. Assoc. Petrol. Geol. Bull.*, v. 46, p. 2019–2032.

———, 1963, Laramide sediments along Wind River thrust, Wyoming, *in* Childs, O. E., and Beebe, B. W., eds., Backbone of the Americas: *Am. Assoc. Petrol. Geol. Mem. 2*, p. 220–230.

Butler, J. R., 1966, Geologic evolution of the Beartooth Mountains, Montana and Wyoming, Part 6: Cathedral Peak Area, Montana: *Geol. Soc. Am. Bull.*, v. 77, p. 45–64.

Fanshawe, J. R., 1939, Geology of the Wind River Canyon area: *Am. Assoc. Petrol. Geol. Bull.*, v. 23, p. 1439–1492.

Flueckinger, L., 1970, Stratigraphy, petrography, and origin of Tertiary sediments off the front of the Beartooth Mountains, Montana–Wyoming: Ph.D. thesis, The Pennyslvania State University, University Park, Pennsylvania, 249 p.

Foose, R. M., 1960, Secondary structures associated with vertical uplift in the Beartooth Mountains, Montana: *Proc. 21st Int. Geol. Cong.*, v. 18, p. 53–61.

———, Wise, D. U., and Garbarini, G., 1961, Structural geology of the Beartooth Mountains, Montana and Wyoming: *Geol. Soc. Am. Bull.*, v. 72, p. 1143–1172.

Hafner, W., 1951, Stress distribution and faulting: *Geol. Soc. Am. Bull.*, v. 62, p. 373–398.

Love, J. D., 1960, Cenozoic sedimentation and crustal movement in Wyoming: *Am. Jour. Sci.*, v. 258-A, p. 204–214.

Osterwald, F. W., 1961, Critical review of some tectonic problems in the Cordilleran foreland: *Am. Assoc. Petrol. Geol. Bull.*, v. 45, p. 219–237.

Pierce, W. G., 1973, Principal features of the Heart Mountain fault and the mechanism problem (this volume).

Prucha, J. J., Graham. J. A., and Nickelsen, R. P., 1965, Basement-controlled deformation in Wyoming province of Rocky Mountains foreland: *Am. Assoc. Petrol. Geol. Bull.*, v. 49, p. 966–992.

Rubey, W. W., and Hubbert, M. K., 1959, Role of fluid pressure in mechanics of overthrust faulting, II: Overthrust belt in geosynclinal area of western Wyoming in light of fluid pressure hypothesis: *Geol. Soc. Am. Bull.*, v. 70, p. 115–206.

Sanford, A. R., 1959, Analytical and experimental study of simple geologic structures: *Geol. Soc. Am. Bull.*, v. 70, p. 19–52.

Spencer, E. W., 1959, Fracture patterns in the Beartooth Mountains, Montana and Wyoming: *Geol. Soc. Am. Bull.*, v. 70, p. 467–508.

Thom, W. T., 1923, The relation of deep seated faults to the surface structural features of central Montana: *Am. Assoc. Petrol. Geol. Bull.*, v. 7, p. 1–14.

Tourtelot, H. A., and Thompson, R. M., 1948, Geology of the Boysen area, central Wyoming: *U.S. Geol. Survey Prelim. Map 91*, Oil and Gas Inv. Ser.

Van Bemmelen, R. W., 1966, The structural evolution of the southern Alps: *Geol. Mijnb.*, v. 45, p. 405–444.

Wise, D. U., 1963, Keystone faulting and gravity sliding driven by basement uplift of Owl Creek Mountains, Wyoming: *Am. Assoc. Petrol. Geol. Bull.*, v. 47, p. 586–598.

———, 1964, Microjointing in basement, Middle Rocky Mountains of Montana and Wyoming: *Geol. Soc. Am. Bull.*, v. 75, p. 287–306.

WILLIAM G. PIERCE

U.S. Geological Survey
Menlo Park, California[1]

Principal Features of the Heart Mountain Fault and the Mechanism Problem

The author has been studying the Heart Mountain fault over a period of years and reporting on it as information became available. The reports, in several different journals, have grown so in number that it now seems desirable to bring together in a single brief paper the salient features of this uniquely preserved low-angle fault along with new, unpublished information. The most perplexing aspect of the Heart Mountain fault is a mechanism for emplacement of the upper plate in accordance with the observed features of the fault. Other geologists have also been considering the problem, and their suggestions, as well as the mechanism preferred by the author, will be discussed.

Regional Setting

The Heart Mountain detachment or décollement fault is in northwestern Wyoming, adjoining and east of Yellowstone National Park. It lies on the east flank of the Rocky Mountain uplift and on the western part of the Bighorn Basin, which is both a structural and topographic basin. The Rocky Mountain uplift began in the later part of the Late Cretaceous and reached its climax during the Eocene. The Bighorn structural basin developed concomitantly with the Rocky Mountain uplift. The rocks transported by the Heart Mountain fault became detached from the eastern border of the uplift area and moved southeastward into the adjoining Bighorn Basin near the close of the early Eocene.

Two smaller detachment faults, the South Fork (Dake, 1918; Pierce, 1941, 1957; Bucher, 1947) and Reef Creek (Pierce, 1963), which preceded the Heart Mountain fault, discussed only briefly here, are mentioned because they are similar in movement and closely related in time to the Heart Mountain fault (Fig. 1, left). They are also on the east

[1] Publication authorized by the Director, U.S. Geological Survey.

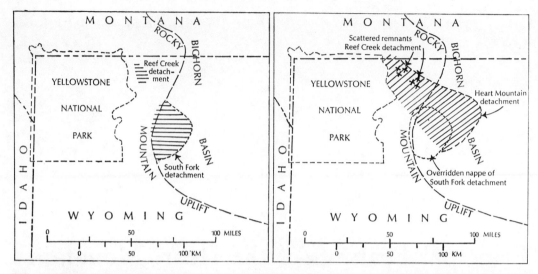

FIG. 1 Sketch maps showing the location of the Reef Creek, South Fork, and Heart Mountain detachment faults. (left) Location of the Reef Creek and South Fork detachment faults which preceded the Heart Mountain fault. (right) Location of Heart Mountain detachment fault, and its relation to the preceding Reef Creek and South Fork detachments.

flank of the Rocky Mountain uplift, and their movement also was to the southeast. The upper plate of the South Fork detachment was thicker and more extensive than the Reef Creek. The Heart Mountain detachment took place below the Reef Creek fault, and the scattered upper plate blocks of that fault were carried in piggy-back fashion by the Heart Mountain fault (Fig. 1, right). The Heart Mountain fault rode out over the upper plate of the South Fork detachment, however, and left upper plate blocks of the Heart Mountain fault on top of the South Fork detachment. The Heart Mountain fault is the youngest and least deformed of the three faults and is the most extensive, best exposed, and best preserved.

Nature and Extent

The upper plate of the Heart Mountain fault is composed of about 1600 ft (490 m) of predominantly carbonate rocks of Ordovician to Mississippian age and from 0 to 900 ft (270 m) or more of volcanic rocks (Cathedral Cliffs formation) of Eocene age (Fig. 2). Underlying the detachment horizon are about 1200 ft (365 m) of Cambrian shales with some limestone beds and at the base a sandstone which rests on Precambrian granitic rocks. Volcanic rocks of the Wapiti formation, previously called the "early basic breccia," deposited very soon after the emplacement of the Heart Mountain fault blocks, have preserved many of the structural features that would otherwise have been lost.

The Heart Mountain fault is a composite of four types of faults, illustrated in Fig. 3. They are (1) a high-angle breakaway fault, trending north-south for at least 7 miles (11 km) at the northwestern limit of the detachment fault, which developed as the detached mass moved away from strata that remained in place, (2) a bedding plane fault, 35 miles (56 km) long and 15 miles (24 km) or more wide, stratigraphically located a few feet above the base of the Bighorn dolomite, (3) a transgressive fault, more than 9 miles (15 km) long and about 2 miles (3 km) across, which cut obliquely up-section or transgressed the strata between the bedding fault and the

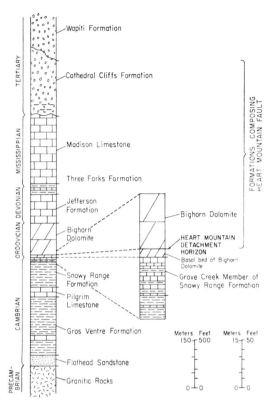

FIG. 2 Generalized stratigraphic section showing the formations associated with the Heart Mountain fault.

ground surface, and (4) a fault along which the upper plate blocks moved up to 30 miles (50 km) over a former land surface. Most of the remnants on the former land surface were recognized and described by Dake in 1918 as the Heart Mountain fault. In 1920 Hewett added the remnants at McCulloch Peaks, which earlier had been ascribed a glacial origin by Granger and Sinclair (1912). The writer recognized the transgressive and bedding parts of the fault (Pierce, 1941, 1957) and the breakaway fault (Pierce, 1960). Detailed geologic maps of most of the fault-covered area have now been published as U.S. Geological Survey Geologic Quadrangle maps GQ–477, GQ–478, GQ–542, GQ–755, GQ–778, GQ–817, and GQ–935.

There are more than 50 upper plate remnants of the Heart Mountain fault that range in size from a few hundred feet to 5 miles (8 km) across (Fig. 3) and many more that are from a few feet to a few tens of feet across. These pieces of upper plate, which before faulting covered an area of about 500 square miles (1300 km^2) comprising the bedding and transgressive parts of the fault, are now scattered over a pie-shaped area of 1300 square miles (3400 km^2) which is about 65 miles (105 km) long in the direction of movement, and 8–30 miles (13–50 km) wide. The fault blocks that moved over the former land surface were widely scattered over an area on the order of 800 square miles (2100 km^2). Extensive pediment deposits composed of carbonate fragments derived from upper plate blocks attest to the former presence of larger and more numerous fault blocks on the former land surface part of the fault in the Heart Mountain and McCulloch Peaks areas (Fig. 3).

Although the upper plate has suffered some erosion, particularly the former land surface phase, which lacked the protective armor of volcanic rocks that covered the bedding part of the fault, the present distribution of the upper plate into scattered blocks is not due to subsequent erosion of a once-continuous sheet. As indicated in Fig. 3, the upper plate originally occupied the 500-square-mile (1300 km^2) area of the bedding and transgressive faults. Approximately half of that area, or about 250 square miles, was tectonically denuded and almost immediately covered by volcanic rocks. Thus there was available only about 250 square miles of upper plate to move onto the former land surface part of the fault. The area of the land surface part of the fault, however, is about 800 square miles (2000 km^2), or more than three times the area of upper plate available to cover it. To scatter 250 square miles of upper plate over an 800-square-mile area requires that the upper plate be broken up into a number of smaller segments. Moreover, the lack of erosion of the surface of tectonic denudation and the preservation of carbonate fault breccia on it indicate that the upper plate broke up and

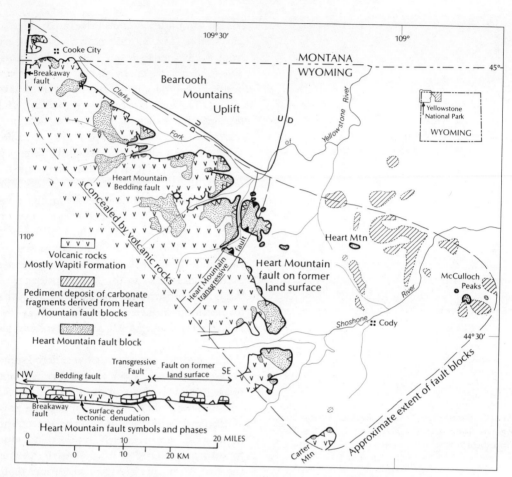

FIG. 3 Sketch map showing exposed remnant blocks of the Heart Mountain fault plate. Movement was to the southeast. Reef Creek and South Fork detachments are not shown.

separated into scattered blocks while it was being emplaced. The breakup, separation, and scattering of the upper plate as it was being emplaced must be taken into account when considering the emplacement mechanism.

Slope of Fault Surface

The slope on which the Heart Mountain fault blocks moved can be determined fairly closely inasmuch as the elevations and distance between the highest and lowest points on the fault can be determined fairly accurately.

The southeasternmost remnants moved over a former land surface that was underlain by Eocene continental beds of the Willwood formation. The Willwood does not contain marine sedimentary rocks, so it follows that the elevation at places such as McCulloch Peaks was at least above sea level. The distance back to the highest point, the breakaway fault, is 65 miles (105 km). If a slope of 2° is projected from sea level for that distance, the elevation at the breakaway would be at least 12,000 ft (3700 m) above sea level. Volcanic rocks at the breakaway fault, some slightly older and some slightly younger than the faulting, contain petrified wood and fossil forests. Dorf (1964, p. 9) believes these plants grew at an elevation of no more than 3000 ft (900 m) above sea level. Consequently, a slope of less than 2° seems to be indicated.

Information is lacking as to whether the slope of the bedding fault and the fault on the former land surface differed significantly. Inasmuch as the average of the two presumably was less than 2°, and an increase in one would require an equivalent decrease in the other, it seems likely that there was little difference.

The transgressive fault now slopes about 10° to the northwest, as can be observed for a distance of 1½ miles (2.4 km) between its connection with the bedding fault on the northwest and the fault on the former land surface on the southeast. However, the dip of the transgressive fault originally may have been less than this, because there has been some postfaulting subsidence in the region northwest of the transgressive fault, which could have increased the northwest dip of the transgressive fault.

Observations on the Fault Itself

Throughout the 35-mile (56 km) southeastward extent of the bedding plane phase, the Heart Mountain fault occurs at precisely the same stratigraphic horizon, with only a few very minor exceptions. This horizon is at the top of a dolomitic limestone bed, generally about 8 ft (2½ m) thick, that is the basal unit of the Bighorn dolomite (Fig. 2). This is an unexpected place for a detachment or gravity type of fault to occur. The horizon at which the faulting occurred can be examined in unfaulted beds west of the breakaway fault, east of the transgressive fault, and north of the bedding fault on the flank of the Beartooth uplift. At these places the fault horizon is nothing more than an inconspicuous bedding plane in a carbonate sequence. The faulting mechanism did not require the presence of fissile or lubricated rock—neither shale nor evaporite occur at this horizon. The mechanism did not even favor a fissile or lubricated horizon, for if it did the fault would have been slightly lower in the section, in the Snowy Range formation or its Grove Creek member.

Although the basal bed of the Bighorn formation has some irregular vertical fractures, it is not brecciated (even immediately below the fault), unlike the beds above the fault. The strata beneath the basal bed likewise do not appear to be fractured or deformed.

Carbonate fault breccia, cemented into a hard, dense rock, rests on the fault surface at a number of places. It ranges from less than 1 in. to 2 ft in thickness and exceptionally is as much as 12 ft (3.6 m) thick. Commonly more than 50 percent of the breccia is matrix, but the matrix is also carbonate breccia of microscopic to submicroscopic size. The megascopic breccia fragments range from specks to 2 mm or more in size; most are angular, some are subangular, and some rounded. Where the Bighorn dolomite rests on the Heart Mountain fault, the lowermost few inches is commonly a breccia in which the fragments range in size from a few millimeters to rock flour. The breccia has the appearance of being thoroughly mixed but actually it may not be, for it begins to grade upward within inches into shattered dolomite in which the dolomite is broken into fragments similar to those below but have not rotated. Shattering has been observed to extend 50 ft (15 m) or more above the base of the upper plate. Breccia blocks become increasingly larger upward, and the brecciation gradually dies out. Although there is some variation in detail in the fault breccia, it is either a carbonate breccia or minutely shattered carbonate rock; it is not a mylonite. Where Madison limestone rests directly on the fault, fault breccia is less common and slickensides may be present.

In places, the fault breccia has been injected upward as clastic carbonate dikes (Pierce, 1968a). A few of these clastic dikes have been noted in the carbonate rocks of the upper plate, but more of them have been found in the volcanic rocks. They are less than 1 in. to 2 ft (0.6 m) wide and some extend upward as much as 100 ft (30 m) above the Heart Mountain fault.

Time and Nature of Movement

Some of the fault blocks rest on rocks as young as early Eocene, and the fault is overlain in places by rocks that are either very late early Eocene or early middle Eocene. Emplacement must have occurred, therefore, near the close of the early Eocene.

Some seemingly conflicting evidence as to the age relationship between the Heart Mountain fault and a stream channel deposit, the Crandall conglomerate (late lower Eocene or early middle Eocene), was noted at the time the deposit was named and described (Pierce, 1957, p. 609–613), and at that time it was suggested that perhaps the conglomerate developed more or less contemporaneously with the Heart Mountain fault. Further study of the deposit indicates that it is related to a preliminary Heart Mountain fault movement, during which roughly the northeastern one-third of the bedding and adjoining transgressive part of the fault moved less than 1 mile southeastward. This preliminary movement of part of the Heart Mountain upper plate occurred only a very short time before the main movement of the entire upper plate.

As the upper plate moved, it broke up into many blocks, which separated as they moved, and as movement progressed they became further separated. The amount of horizontal movement covers a wide range. The fragments of the upper plate nearest to the breakaway fault have moved from less than a mile to a few miles. The most southeasterly remnants of the plate have moved some 30 miles (50 km), at least, from the transgressive fault. Nowhere did the upper plate imbricate or ride up on itself. The fault blocks were most widely separated on the former land surface, as indicated in Fig. 3. In the transgressive fault and adjoining area to the northwest, the fault blocks are less scattered. Presumably, the scattered blocks that were moving along the bedding fault were retarded as they started moving up along the transgressive fault. As additional blocks moved down the bedding fault they would abut against the slower moving blocks, and some of their forward momentum would be transferred to the blocks moving up the transgressive fault.

The main movement occurred during a brief interval of time, as shown by (1) the absence of erosion on the surface of tectonic denudation in the bedding fault area, which was exposed during faulting, and by (2) the clastic carbonate dikes. The carbonate dikes are composed of Heart Mountain fault breccia that was left on the surface of tectonic denudation as the fault blocks separated; the carbonate fault breccia was injected into volcanic rocks of the Wapiti formation that subsequently filled gaps between the blocks. This finely divided carbonate material could not have survived lengthy exposure to erosion. Consequently, the Heart Mountain fault movement is believed to have been very rapid, probably cataclysmic.

The Emplacement Mechanism

Statement of the Problem

Bucher (1933) first recognized the unusual character of the Heart Mountain fault and suggested that the limestone plates constituting the thrust (as it was then known) were emplaced by the horizontal component of the force of a large volcanic explosion. Later (1947) he suggested that uptilting of the region resulted in shearing and sliding by gravity, probably aided by frequent earthquake shocks that preceded the outbreak of volcanic activity. The only remnants that he recognized as belonging to the Heart Mountain fault, however, were those that were on the former land surface (Bucher, 1947, Fig. 1). He did not accept as part of the Heart Mountain fault the newly recognized remnants mapped 6 years earlier (Pierce, 1941, Fig. 2), which extended the fault from the former land surface through the transgressive fault and into the bedding fault. However, from his study of the allochthonous masses on the former land surface, he correctly concluded that they "do not represent erosion of a once continuous 'thrust sheet,' but are independent fragments that came into existence by a process radically different from that of normal orogeny and at a rate very much greater than that of normal orogenic processes" (Bucher, 1947, p. 194–196).

The presence of a breakaway fault rather than a root zone and the separation of the upper plate into numerous blocks clearly indicate that the Heart Mountain fault was not an overthrust emplaced by a push from the rear. As stated earlier (Pierce, 1957, p. 615), once the detachment or shearing off had started along the bedding and transgressive faults, the force of gravity could have accounted in considerable part for the continuing movement of the upper plate, but some factor other than gravity definitely seems to be needed to explain each of the following characteristics of the fault:

1. The slope of less than 2° on which the fault blocks moved seems to be much too low for movement by gravity alone, particularly on a fault that lacks any sign of being lubricated. In some places today, as along the transgressive fault, upper plate blocks rest on the Heart Mountain fault where it slopes as much as 10°, but they are not now sliding under the influence of gravity.
2. The horizon along which the Heart Mountain bedding fault occurred was in the most viscous rather than in the least viscous strata. Kehle (1970, p. 1642) suggests that in gravity sliding, deformation will be concentrated in the strata with the lowest viscosity. If the bedding fault was due primarily to gravity sliding, however, then it should have been in the soft Cambrian shales less than 50 ft (15 m) below the detachment horizon, for they have a lower viscosity than the dolomite and dolomitic limestone in which the deformation occurred. Indications are lacking in the Cambrian shale of movement within them, or of stratigraphic displacement or angular discordance, whereas at the fault horizon between the basal Bighorn bed and the Bighorn dolomite (Fig. 2) evidence of stratigraphic displacement, angular discordance, and fault breccia are common features.
3. The fact that the entire area of the bedding fault is at the same stratigraphic horizon, and also that the surface of tectonic denudation exposed by the bedding fault was not eroded, seems to indicate that the initial movement on it occurred essentially simultaneously over either the entire fault surface or over a large part of it, a conclusion that in either case carries important implications regarding the amount of shear stress that is required. According to Hubbert and Rubey (1959, p. 125), a fault produced by stress applied to the rearward edge of the upper plate most likely would be propagated as a dislocation in which the stress involved at any one time affects only a minute fraction of the total area of the fault surface, and therefore can be assumed to be negligible. In the Heart Mountain bedding fault, however, deformation did not take place as a result of stress applied to the rearward edge, so the foregoing conclusion is not applicable. If shearing simultaneously occurred over all parts of a bedding fault area of several hundred square miles, a great deal of shearing stress would be required. That shearing stress would be in addition to the force needed to keep the plate moving once motion had started.

The unspecified factor that seems to be indicated by the three items just discussed would be the one that operated in combination with the constant force of gravity to produce both lateral transport and dispersal of the parts of the upper plate. In particular, the need for a mechanism that would have the effect of reducing friction along the fault by at least several orders of magnitude has led to five suggested mechanisms for the Heart Mountain fault. Three of them are similar in one respect: friction is greatly reduced owing to the presence along the fault of fluid, either liquid or gas, under pressure approaching the lithostatic pressure of the upper plate. These are the fluid pressure mechanism of Hubbert and Rubey (1959; also Rubey and Hubbert, 1959), the air cushion-landslide mechanism suggested by Hsü (1969), and the hovercraft (volcanic gas) mechanism of Hughes (1970a). Gravity sliding by shear in low-viscosity strata has been proposed by Kehle (1970). Earthquake oscillations are favored by the author.

Fluid Pressure

High fluid pressure as an emplacement mechanism for low-angle thrust faults of large lateral displacement, which was proposed by Hubbert and Rubey (1959) and Rubey and Hubbert (1959), has won wide acceptance among geologists. It does not seem applicable for aiding movement on the Heart Mountain fault, for as previously stated (Pierce, 1963, p. 1234) the fault is at shallow depth where high fluid pressure would not be expected, but if there was an abnormal pressure, it would be lost as soon as the fault mass began to break up. Davis (1965) discussed the Heart Mountain fault as one of several low-angle faults which he considers inadequately explained by the high fluid pressure mechanism, and he observed that "abnormal fluid pressure at the base of isolated blocks moving along a bedding or erosional surface is an improbable phenomenon to facilitate such movements." In reply to Davis, Rubey and Hubbert (1965) wondered whether any of the field evidence clearly disproves the possibility that movement on the Heart Mountain fault, which began before the accumulation of the Wapiti formation, may also have continued during its accumulation, and if so, whether it could support a broadened interpretation that part of the fault movement was the result of relatively local movements within the volcanic breccia and part was the result of larger-scale movements of an extensive thrust sheet composed of limestone blocks plus volcanic breccia. In regard to this suggestion evidence was presented (Pierce, 1966) showing that throughout the bedding-plane area of the fault, the emplacement of the fault masses definitely occurred before the volcanism that resulted in deposition of the Wapiti formation. The Heart Mountain faulting, however, occurred between what seems to be a more or less continuing sequence of deposition of volcanic rocks from Cathedral Cliffs to Wapiti formations, which suggests there may be a relationship between the faulting and some mechanism associated with volcanism.

Goguel (1969) has discussed the conditions at a depth of less than 1 km under which the heat produced by friction and earthquakes, by vaporizing the water and dehydrating minerals such as gypsum, might elevate the water pressure sufficiently to allow sliding. Among several possible examples he mentions are the Heart Mountain fault in Wyoming and the "dislocations of the summits" or "Gipfelfaltungen" in the Dolomites.

Landslides

Hsü (1969) has suggested the possibility that the Heart Mountain fault might be a landslide like the Flims slide in the Rhine Valley, southwest of Chur, Switzerland. He states that although he has not had an opportunity to visit the Heart Mountain thrust he was impressed by some of the phrases used in describing the thrust, which he says are also applicable to the Flims landslide. The similarities he mentions are misleading, however, and there are very significant differences which suggest that they do not have a common origin. One of the similarities he notes is that "even the scale of the Flims slide is comparable to the Heart Mountain thrust." A size or scale comparison that was not made is the areal extent of each, and areal extent is an important measure of scale, especially when considering the mechanism of emplacement. The Flims landslide was spread over an area of 25 square miles (65 km^2), whereas the Heart Mountain fault masses were scattered over an area on the order of 1300 square miles (3400 km^2), that is, 50 times greater. In fact, the area of the Heart Mountain fault is many times greater than the area of any known landslide.

But more significant than the difference in areal extent between landslides and the Heart Mountain fault is their difference in texture. The "shattered and broken" brecciation which Hsü (1969, p. 944) quotes from my description of the Heart Mountain fault is applicable only to the basal part of the fault blocks, and not to the blocks as a whole. He also comments that Bucher (1933, p. 57) emphasized the "extremely shattered" nature of the Heart Mountain thrust masses, but here, too, it should be noted that Bucher does not say that extreme shattering is a general characteristic of the thrust masses. To put his statement into context, Bucher (1933, p. 57) said that "three smaller masses (east and south of the main sheets) are extremely shattered." One of these masses, however, is the deposit north of Carter Mountain (which for simplicity I have grouped with the carbonate pediment deposits in Fig. 3), and this deposit is extremely shattered only because it was let down 2100 ft (640 m) vertically and 3 miles (5 km) horizontally on the Carter Mountain landslide in Quaternary time (Pierce, 1968b, p. D237-D238). The other two small masses which Bucher referred to probably are the McCulloch Peaks remnants. These are scattered erosional vestiges of a once larger thrust mass which has been modified and broken up by erosional processes and likewise are unlike remnants that have not been disturbed by erosional undercutting or landsliding.

A comparison of the texture of the material comprising cataclysmic landslide masses such as Flims (Heim, 1932), Blackhawk (Woodford and Harriss, 1928; Shreve, 1968) and Sherman (Shreve, 1966) and that of the Heart Mountain fault masses shows that they are very dissimilar. For example, both Woodford and Harriss (1928) and Shreve (1968) mapped the Blackhawk landslide as a breccia unit, the Blackhawk breccia, which is described as consisting "almost entirely of gray, unsorted and unstratified breccia of fresh angular fragments of Furnace marble ranging in size from powder up to 25 cm in diameter, with most clasts being about 2.5 cm in diameter" (Shreve, 1968, p. 17). The Furnace limestone is of Carboniferous age and the Blackhawk breccia is Quaternary. The formations that were transported in this and similar landslides have been destroyed as formation units. The material making up the landslides is not mapped as transported formations but has become a new mappable unit or formation, and its age is no longer that of the formations from which it was derived. Maps and descriptions of Heart Mountain fault blocks (Stevens, 1938; Bucher, 1947; Pierce, 1941, 1963), on the other hand, show that the fault blocks are composed of formations which are identical in physical characteristics and stratigraphic sequence with their unfaulted counterparts. The basal part of a fault block may be shattered and broken for some tens of feet upward but it gradually grades upward into unbroken rock. Brecciation—a few inches to a few feet of comminuted fault breccia—is confined to the lowermost part of the Heart Mountain fault masses and in many places is absent entirely. It is difficult to believe that the mechanism which transported these masses for many miles with so little deformation that their allochthonous character went undetected for many years is the same mechanism that transformed the parent rocks of the Flims, Blackhawk, and similar landslides into breccia.

Furthermore, the initial steep slope essential for attaining the high velocity of cataclysmic landslides such as Flims and Blackhawk cannot be called on as part of the

Heart Mountain fault mechanism. The initial movement on the fault was along the bedding plane part, which probably had an average southeastward slope of less than 2° for a distance of 35 miles. Those conditions are incompatible with an initial movement on a slope that is steep enough to impart an initial high velocity.

The distribution of the Heart Mountain fault blocks is also unlike that of landslide deposits. The Heart Mountain fault blocks were widely scattered when they came to rest (see Fig. 3). The material making up a cataclysmic landslide is deposited essentially as a single contiguous sheet (see Shreve, 1968, Fig. 2).

For all of the foregoing reasons, landsliding appears to be far from adequate as a mechanism to explain emplacement of the Heart Mountain fault blocks. The great differences between cataclysmic landslides and the Heart Mountain fault blocks indicate that they are not products of the same kind of transporting process.

Hovercraft

A mechanism that is dependent on volcanic gas under pressure has been suggested by Hughes (1970a). He proposed the term "hovercraft tectonics" for a mechanism whereby "voluminous volcanic gas was introduced laterally at a favored stratigraphic horizon below extremely massive limestone, effectively reducing friction so that an extensive but thin sheet of this limestone and overlying rocks were able to slide along a very gentle slope." He proposed that the gas "was intruded laterally at the base of the Bighorn dolomite to produce a fluidized aggregate of fluid and solid which acted as a lubricant." The evidence of fluidization that he determined from "an examination of only one exposure and one hand specimen" of the fault breccia at White Mountain included (1) igneous rock fragments, of which some have alteration rims and some are embayed; (2) vesicles; (3) rims of finely comminuted matrix material plastered around embayed larger fragments; and (4) ovoid-shaped matrix material.

Hughes suggested that the volcanic gas was intruded much like some well-known lateral intrusions of diabase sills in flat-lying strata. It should be noted, however, that the base of the Bighorn dolomite is an uncommon horizon for intrusive injection in this area. Numerous sills and tabular bodies have been mapped in the Cooke City area (Lovering, 1930, pl. 1), but they are in the Cambrian shales below the Bighorn Dolomite.

In a paper on the geology of White Mountain (Nelson, Pierce, Parsons, and Brophy (1972)) evidence is presented showing that the igneous rock fragments in the fault breccia, which Hughes implied are indicative of fluidization, came from nereby allochthonous dikes and sills in the basal part of the upper plate. Therefore the presence of igneous rock fragments in the fault breccia at White Mountain is not evidence of introduction of volcanic gas along the fault. It has been pointed out (Pierce and Nelson, 1970) that at 29 localities where the fault breccia has been examined, volcanic residue is lacking. Hughes replied (1970b) that it never was his contention that igneous rock fragments should become incorporated in a fault breccia that is fluidized by igneous gases. Although he does not believe that the lack of volcanic residues negates his theory, many of the more widely accepted examples of fluidized rock (Reynolds, 1954) contain abundant fragments of igneous rocks that seem to have been formed by rapid emission of dissolved volcanic gas from a parent magma. If the Heart Mountain fault blocks were transported on magma-derived gases, I would expect igneous rock fragments to be present in some of the 29 fault breccia localities.

If fluid is to facilitate movement of blocks along a low-angle fault, it must be confined under high pressure. Consequently, the fault blocks must be large enough so that the fluid supporting them cannot escape from beneath them more rapidly than it is being replaced.

Breakup of the upper plate into numerous blocks would allow the fluid pressure to drop, except possibly over the one or more vents that were assumed to furnish the fluid, and as Hughes notes (1970a, p. 114) loss of fluid pressure would cause the blocks to come to rest. As indicated previously, a breakup into many smaller segments is precisely what must have happened. The distribution shown in Fig. 3 indicates that this breakup must have occurred before the southeasternmost blocks came to rest, so some mechanism other than high fluid pressure is needed to transport them.

The inadequacy of the hovercraft mechanism as applied to the former land surface part of the fault is apparent for yet another reason. The upper plate of the Heart Mountain fault can be traced as a continuous sheet from the bedding fault across the transgressive fault and into the fault on the former land surface (Fig. 3), clearly showing that the latter fault is an integral part of the Heart Mountain fault. Hughes (1970a, p. 629) apparently agrees that the bedding and land surface faults are part of the same tectonic event. There is no reason to assume that the fault blocks which moved over the land surface did so by a mechanism basically different from those in the bedding fault area. There is little evidence of the presence of high-pressure gas in the bedding fault area, but there is none at all in the land surface fault area, so the upper plate presumably moved as a solid against a solid, not on a fluidized bed. Recognizing this, Hughes suggested as a modification of the mechanism, "that once virtually frictionless motion was initiated, however slowly, upon the bedding plane fault, the sliding blocks would tend to travel onward for a considerable distance, not by simple gravity sliding as I stated, but by their own momentum, thus accounting for the scattered klippen-like outlying blocks, including Heart Mountain."

This modification implies that the upper plate blocks which moved the farthest on the land surface needed the greatest momentum. These blocks can reasonably be assumed to have come from the southeasternmost part of the bedding plane fault, and they were the first to move up the transgressive fault, onto the land surface, and away from any possible sources of volcanic gas. Their momentum would be slight to begin with, and for it to have increased the blocks would have to be pushed by the coherent upper plate moving on the bedding fault. As has been pointed out earlier in the paper, however, the whole style of the Heart Mountain fault movement is a separation or pulling apart of the upper plate by body forces, which seems to have begun almost as soon as lateral movement began. An upper plate that is being pulled apart by the body forces that are acting on it cannot at the same time be applying a push from the rear sufficient to impart a high momentum to the forward part. Therefore, Hughes' suggestion (1970a, p. 629) that the 15 miles (25 km) of movement of Heart Mountain over a former land surface is due to its own momentum seems to be completely inadequate, and the 30 miles (50 km) of movement of the McCulloch Peaks mass on its own momentum is even less likely.

Gravity Sliding by Shear in Low-Viscosity Strata

Kehle (1970, p. 1658) has suggested:

Gravity sliding and orogenic translation are accomplished by the deformation of the least competent (lowest viscosity) strata within a rock sequence. The mode of deformation is primarily simple (rectilinear) shear. ... This process, dominated by the fluid-like properties of rocks, requires less work to achieve a given displacement than does translation across a sole fault.... Thus, rather than bedding plane faults, entire zones of rock undergo massive shear to accomplish transport of overlying rock masses.

Kehle apparently interprets the Heart Mountain fault as due to gravity sliding inasmuch as a computation is made on the possible rate of movement of the upper plate. As noted earlier in the discussion of the mechanism problem, however, if movement was due entirely to gravity sliding, deformation should be concentrated in the strata with the lowest viscosity, the Cambrian shales, the top of which is less than 50 ft (15 m) below the fault horizon. But the fault plane is in dolomite, which has a much higher viscosity.

There are several other features of the Heart Mountain fault that do not appear compatible with gravity sliding by fluidlike shear of the décollement zone: (1) movement is concentrated along a sharp, narrow stratigraphic horizon, rather than being distributed through a massive, thick, shear zone; (2) the upper plate not only broke into numerous blocks which separated as they moved but many of them are very small with dimensions of a few feet to a few tens of feet; (3) the fault breccia consists of angular fragments and shows no evidence of flow or shear structure; and (4) there is no zone of shear flow deformation in the rocks either below, above, or near the fault.

Earthquake Oscillation

A combination of gravity and earthquake oscillations is the only mechanism that has been suggested (Bucher, 1947, p. 196; Pierce, 1963, p. 1234) which is compatible with the observed features of the Heart Mountain fault. The common association of volcanism and earthquakes, and the close association in time between the extrusion of the Wapiti volcanics and the faulting, point up the likelihood of there having been earthquakes at that time. The vertical displacement of 20,000 ft (6100 m) along the east front of the Beartooth uplift, most of which occurred during Early Tertiary time, no doubt was accompanied by earthquakes (Pierce, 1963, p. 1234). The cataclysmic nature of the Heart Mountain fault movement, which is suggested by the short time involved in emplacement, is consistent with an earthquake-associated mechanism.

If under certain unusual conditions a unit of gently inclined rocks received slightly more upward acceleration than a unit of rocks on which it rested, the frictional resistance to movement between them would be lessened and the units would tend to separate along an incipient fault. If, during the upward cycles of oscillatory seismic motion, accelerations approaching 1 g were imparted to the rocks above the incipient fault, then, at the moments that the upward accelerations ceased, the stress normal to the fault as a result of gravity would approach zero and separation would tend to occur at the base of the rock unit with higher acceleration. In effect the upper plate would be almost weightless. With innumerable repetitions of upward acceleration, the upper plate would be intermittently nearly unrestrained by friction and free to move laterally on a very low slope. Moreover, perhaps a forward impetus might be imparted to the upper plate by seismic waves that emanated from beneath the upper plate and moved tangentially upward to the southeast.

It has been stated that the maximum acceleration of the ground within 5 miles (8 km) of the epicenter of a magnitude 5.0 shock may be 5–10 percent of gravity, and that a magnitude 6.0 shock can have a maximum ground acceleration on the order of 20 percent of gravity (Natl. Research Council, 1969). During the San Fernando earthquake of February 9, 1971, accelerations much higher than these were recorded. Thus in at least some instances, previous ideas as to the maximum acceleration of earthquakes may be revised upward. The San Fernando earthquake, of moderate magnitude 6.6, has, according to Maley and Cloud (1971, p. 165), provided the largest collection of significant strong-motion data ever obtained. They report (p. 163):

At Pacoima Dam, 5 miles (8 km) south of the epicenter, the earthquake accelerations were the highest ever recorded, i.e. in the 0.5 to 0.75 g range with several high-frequency peaks to 1 g. The initiation of strongest shaking began 1 to 2 seconds after instrumental triggering and lasted for approximately 12 seconds.

Degenkolb (1971, p. 133) comments in regard to apparently the same record:

> It is reported that the closest strong motion record of this earthquake, taken at or near the summit of a fractured rock formation, indicated ground accelerations of more than 100 percent g horizontal and 70 percent g vertical. For 12 seconds the recurring motions were recorded at about 50 percent g. This would indicate that ground motions for even a moderate earthquake may be much greater than assumed by many engineers.... It is interesting to note that these high concentrations of force, motion, or energy are on some form of rock rather than on saturated alluvium.

In discussing records obtained in the upper levels of buildings, Maley and Cloud (1971, p. 164) report that "Peak values in excess of 0.4 g were recorded as distant as downtown Los Angeles, nominally 26 miles (42 km) [from the epicenters], although the vast majority of measurements were in the 0.2- to 0.4-g range.... The highest top-floor accelerations are approximately two to three times those measured at the ground level in the same structure."

In discussing the vertical accelerations at Kagel Canyon, Morrill (1971, p. 177) reports that the San Fernando earthquake "produced evidence of vertical accelerations far in excess of any ever recorded instrumentally." Evidence at the Los Angeles County Fire Station No. 74, 8 miles (13 km) from the epicenter,

> ... suggests that the vertical acceleration exceeded that of gravity and may have equaled or exceeded that of the great Assam earthquake of June 12, 1897 (Oldham, 1897) in very local situations.... A 20-ton fire truck enclosed in a garage moved 6 to 8 feet fore and aft, 2 to 3 feet sideways without leaving visible skid marks on the garage floor. The truck was in gear, and the brakes were set.... Marks which appear to have been made by the right rear tire were found on the door frame 3 feet above the floor, while the metal fender extending several inches out beyond the upper portion of the tire was not damaged. Four feet above the floor the hose rack was broken by the rear step of the truck. The step was bent up while the hose rack was broken downward.

The living quarter's building has shingle siding which overlapped the foundation $4\frac{3}{8}$ in. After the quake a corner of the building was displaced 11 in. laterally from its former position. To reach that position the building must have been elevated more than $4\frac{3}{8}$ in. relative to its foundation before it was displaced laterally. This "suggests that the building accelerated upward, with respect to the ground, at a rate of at least 1 g for about 0.1 second; that is, if one assumes only 1 g and the given distance."

An unusual effect of the San Fernando, California, earthquake of February 9, 1971 is the shattered earth at Wallaby Street, Sylmar, which furnishes some information regarding the effects of seismic waves (Nason, 1971). Some of the special features of the shattered earth site which Nason (p. 97–98) has described in detail are

> ... the tossed-earth phenomenon of overturned soil, which may indicate a seismic shaking exceeding 1.0 g; the pulverization of the soil may mean that the high acceleration occurred over many wave cycles.... The very localized nature of the high acceleration indicated by the tossed-earth may result from unknown effects of constructive interference of seismic waves arriving along different paths and possible focusing or concentration of seismic wave energy at the narrowed-down ridge-top. This suggests that seismic wave interference effects are probably very important to the detailed distribution of strong shaking and earthquake damage.... The fact that the soil was thoroughly overturned is an indication of the strength of the shaking. It would appear that possible seismic wave interference and focusing patterns may be very important to the local distribution of strong seismic shaking and related damage.

My purpose in citing this unusual effect is to suggest that some as yet unknown kind of focusing and combining of seismic waves may have occurred in the region of the Heart Mountain fault so as to produce the initial shearing along the bedding fault and aid in the lateral transport of the upper plate. In this connection it is noted that the Heart

Mountain fault is essentially at a seismic velocity discontinuity since the velocity of the compressional or P wave in the Bighorn dolomite is considerably higher (possibly 50 percent or more) than in the underlying Cambrian shale.

The preceding discussion of mechanism has been restricted to the Heart Mountain fault, but a more exhaustive study of proposed mechanisms should consider their applicability to the Reef Creek and South Fork faults. As has been mentioned, these faults are closely related to the Heart Mountain fault in time and direction of movement; they too are of the décollement type that initially broke along a bedding surface. A similar mechanism seems to be indicated for all three faults. The features of the Reef Creek and South Fork faults are compatible with an earthquake-associated mechanism as far as is known, but not with the other mechanisms that have been proposed for the Heart Mountain fault.

The earthquakes that have occurred in historic time do not appear to have been of sufficient intensity or duration to have operated in conjunction with gravity as an emplacement mechanism. If earthquakes were a part of the mechanism, presumably they would have to be of greater intensity and either of longer duration or recurring at shorter intervals than those known today. Consequently, an earthquake-related mechanism must be regarded as little more than a suggestion bolstered by circumstantial evidence until more is known about how seismic waves might interact at shallow depth in a specific geologic setting, such as for the Heart Mountain fault, so as to both produce a widespread bedding plane rupture and aid in many miles of lateral transport of fault blocks over several kinds of surfaces.

References

Bucher, W. H., 1933, Volcanic explosions and overthrusts: *Am. Geophys. Union Trans. 14th Ann. Mtg.*, p. 238–242.

———, 1947, Heart Mountain problem, *in Wyoming Geol. Assoc. Guidebook 2nd Ann. Field Conf., Bighorn Basin, 1947*: p. 189–197.

Dake, C. L., 1918, The Heart Mountain overthrust and associated structures in Park County, Wyoming: *Jour. Geol.*, v. 26, p. 45–55.

Davis, G. A., 1965, Role of fluid pressure in mechanics of overthrust faulting: Discussion: *Geol. Soc. Am. Bull.*, v. 76, p. 463–468.

Degenkolb, H. J., 1971, Preliminary structural lessons from the earthquake, *in* The San Fernando, California, earthquake of February 9, 1971: *U.S. Geol. Survey Prof. Paper 733*, p. 133–134.

Dorf, E., 1964, *The petrified forests of Yellowstone National Park*: Washington, D.C., Govt. Printing Office, 12 p.

Goguel, J., 1969, Le rôle de l'eau et de la chaleur dans les phénomènes tectoniques: *Rev. Géogr. Phys. Géol. Dynam.*, v. 11, p. 153–164.

Granger. W. and Sinclair, W. J., 1912, Notes on the Tertiary deposits of the Bighorn Basin: *Am. Mus. Nat. History Bull. 32*, p. 64–66.

Heim, A., 1932, *Bergsturz und Menschenleben*: Zürich, Fretz and Wasmuth Verlag, 218 p.

Hewett, D. F., 1920, The Heart Mountain overthrust, Wyoming: *Jour. Geol.*, v. 28, p. 536–557.

Hsü, K. J., 1969, Role of cohesive strength in the mechanics of overthrust faulting and of landsliding: *Geol. Soc. Am. Bull.*, v. 80, p. 927–952.

Hubbert, M. K., and Rubey, W. W., 1959, Role of fluid pressure in mechanics of overthrust faulting, I. Mechanics of fluid-filled porous solids and its application to overthrust faulting: *Geol. Soc. Am. Bull.*, v. 70, p. 115–166.

Hughes, C. J., 1970a, The Heart Mountain detachment fault—A volcanic phenomenon?: *Jour. Geol.*, v. 78, p. 107–116.

———, 1970b, The Heart Mountain detachment fault—A volcanic phenomenon? A reply: *Jour. Geol.*, v. 78, p. 629–630.

Kehle, R. O., 1970, Analysis of gravity sliding and orogenic translation: *Geol. Soc. Am. Bull.*, v. 81, p. 1641–1664.

Lovering, T. S., 1930, The New World or Cooke City mining district, Park County, Montana: *U.S. Geol. Survey Bull. 811*, p. 1–87.

Maley, R. P., and Cloud, W. K., 1971, Preliminary strong-motion results from the San Fernando earthquake of February 9, 1971, *in* The San Fernando, California, earthquake of February 9, 1971: *U.S. Geol. Survey Prof. Paper 733*, p. 163–176.

Morrill, B. J., 1971, Evidence of record vertical accelerations at Kagel Canyon during the earthquake, *in* The San Fernando, California, earthquake of February 9, 1971: *U.S. Geol. Survey Prof. Paper 733*, p. 177–181.

Nason, R. D., 1971, Shattered earth at Wallaby Street, Sylmar, *in* The San Fernando, California, earthquake of February 9, 1971: *U.S. Geol. Survey Prof. Paper 733*, p. 97–98.

National Research Council, Committee on Earthquake Engineering Research, 1969, *Earthquake Engineering Research*: Natl. Acad. Sci., p. 5 (available as PB 188 636 from Natl. Tech. Inf. Service, U.S. Dept. Commerce, Springfield, Va., 22151).

Nelson, W. H., Pierce, W. G., Parsons, W. H., and Brophy, G. P., 1972, Igneous activity, metamorphism, and Heart Mountain faulting at White Mountain, northwestern Wyoming: *Geol. Soc. Am. Bull.*, v. 83, p. 2607–2620.

Oldham, R. D., 1897, Report on the Great Earthquake of 12th June, 1897: *India Geol. Survey Mem.*, v. 29, p. 1–379.

Pierce, W. G., 1941, Heart Mountain and South Fork thrusts, Park County, Wyoming: *Am. Assoc. Petrol. Geol. Bull.*, v. 25, p. 2021–2045.

———, 1957, Heart Mountain and South Fork detachment thrusts of Wyoming: *Am. Assoc. Petrol. Geol. Bull.*, v. 41, no. 4, p. 591–626.

———, 1960, The "break-away" point of the Heart Mountain detachment fault in northwestern Wyoming, *in* Geological Survey research 1960: *U.S. Geol. Survey Prof. Paper 400-B*, p. B236–B237.

———, 1963, Reef Creek detachment fault, northwestern Wyoming: *Geol. Soc. Am. Bull.*, v. 74, p. 1225–1236.

———, 1966, Role of fluid pressure in mechanics of overthrust faulting: Discussion: *Geol. Soc. Am. Bull.*, v. 77, p. 565–568.

———, 1968a, Tectonic denudation as exemplified by the Heart Mountain fault, Wyoming, *in Orogenic Belts*: Internat. Geol. Cong., 23rd, Prague, 1968, Repts., Sec. 3, Proc., p. 191–197.

———, 1968b, The Carter Mountain landslide area, northwest Wyoming, *in* Geological Survey research 1968: *U.S. Geol. Survey Prof. Paper 600-D*, p. D235–D241.

———, and Nelson, W. H., 1970, The Heart Mountain detachment fault—A volcanic phenomenon? A discussion: *Jour. Geol.*, v. 78, p. 116–123.

Reynolds, D. L., 1954, Fluidization as a geologic process, and its bearing on the problem of intrusive granites: *Am. Jour. Sci.*, v. 252, p. 577–613.

Rubey, W. W., and Hubbert, M. K., 1959, Role of fluid pressure in mechanics of overthrust faulting, II. Overthrust belt in geosynclinal area of western Wyoming in light of fluid-pressure hypothesis: *Geol. Soc. Am. Bull.*, v. 70, p. 167–205.

———, 1965, Role of fluid pressure in mechanics of overthrust faulting: Reply: *Geol. Soc. Am. Bull.*, v. 76, p. 469–474.

Shreve, R. L., 1966, Sherman landslide, Alaska: *Science*, v. 154, p. 1639–1643.

———, 1968, The Blackhawk landslide: *Geol. Soc. Am. Spec. Paper 108*, 47 p.

Stevens, E. H., 1938, Geology of the Sheep Mountain remnant of the Heart Mountain thrust sheet, Park County, Wyoming: *Geol. Soc. Am. Bull.*, v. 49, p. 1233–1266.

Woodford, A. O., and Harriss, T. F., 1928, Geology of Blackhawk Canyon, San Bernardino Mountains, California: *California Univ. Pubs. Geol. Sci.*, v. 17, p. 265–304.

ROBERT SCHOLTEN

Department of Geosciences
Pennsylvania State University
University Park, Pennsylvania

Gravitational Mechanisms in the Northern Rocky Mountains of the United States

The area discussed in this paper is that part of the Rocky Mountain fold and thrust belt to the east and north of the Idaho batholith (Fig. 1). As the figure shows, this batholith is the only major intrusive body in the eastern half of the Cordilleran geosyncline, which was deformed in Cretaceous and Tertiary time into the Sevier orogenic belt (see Roberts and Crittenden, 1973, Fig. 2). This unique position of the batholith strongly suggests a genetic relation to the development of a structural style east of it which is distinctively different from the style farther north (Scholten, 1970). In this paper the area north of the batholith is called the "northern province," and the area of the batholith and the deformed belt to the east is termed the "southern province."

Regional Framework and Structural Style

Distribution of Rock Units

As shown in Fig. 2, much of the area north of the Idaho batholith is underlain by quartzites, pelites, and their metamorphosed equivalents of the Precambrian Belt group (or supergroup), which attains thicknesses of 10–20 km toward the west (Sloss, 1950; Harrison and Campbell, 1963; Harrison, 1972). Regional metamorphism increases with depth and toward the west (Maxwell and Hower, 1967) but remains low-grade except near Precambrian intrusives and in the vicinity of the Idaho batholith, where gneisses and

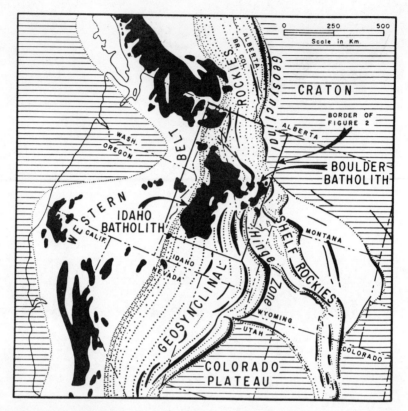

FIG. 1 Tectonic map of the Cordilleran region of western North America, showing deformed western geosynclinal belt and eastern "Geosynclinal Rockies," separated from craton and shelf by geosynclinal hinge zone. Dotted and bold lines indicate structural trends (major fold axes and thrust zones). Batholiths shown in black. Rectangle outlines area of this report (compare Fig. 2). (Slightly modified after Eardley, 1962, Fig. 21.1.)

FIG. 2 Tectonic map of the northern Rocky Mountains of the United States (modified after Tectonic Map of the United States, U.S.G.S.-A.A.P.G., 1962, with data added). (1) Precambrian crystalline basement; (2) Late Precambrian Belt supergroup (quartzites, pelites, and metamorphic equivalents near Idaho batholith); (3) Paleozoic and Mesozoic rocks of Cordilleran geosyncline and adjacent cratonic shelf (pre- to syntectonic); (4) Early Cretaceous to Paleocene (and Eocene?) rocks on cratonic shelf (syntectonic) [(A) Adel Mountain and Elkhorn volcanics near Boulder Batholith; (B) Beaverhead conglomerate in extreme southwest Montana and west of Boulder batholith]; (5) Syntectonic granitoid batholiths and stocks [(A) Jurassic (?) and Cretaceous; (B) Early Tertiary (perhaps more abundant than shown along Idaho Batholith margin)]; (6) Eocene to Recent sediments (s) and volcanics (v) (post-tectonic); (a) unclassified faults; (b) upthrusts, overthrusts, and gliding sheets; (c) normal faults (largely or wholly post-tectonic). EE' = cross section of Fig. 3; AA', BB', CC', DD' = cross sections of Fig. 4.

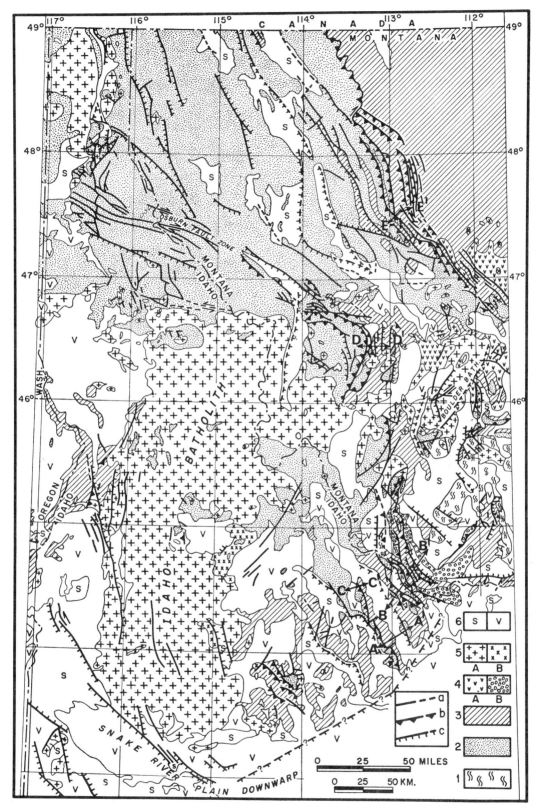

related rocks of amphibolite facies represent equivalents of the lower Belt (Hietanen, 1961; Reid and Greenwood, 1968; Reid and others, 1970). In the southern province the batholith is in contact relation not only with the Belt, but also with Paleozoic rocks. Tertiary sediments and volcanic rocks overlap the batholith.

The overthrust belt contains Paleozoic and Belt rocks in both provinces, and Mesozoic rocks are exposed in the frontal thrust sheets. The pre-Beltian basement emerges in the southern province only: in a mantled gneiss dome east of the southern Idaho batholith (Dover, 1969), and in the area east of the thrust belt, where its emergence is due partly to the absence of a Belt cover and partly to upthrusting and major arching in Cretaceous and Tertiary time (Scholten and others, 1955; M'Gonigle, 1965; Scholten, 1967). This basement outcrop area represents the edge of a westward salient of the North American stable interior.

Structural Pattern

The Belt is only gently deformed north of the Idaho batholith behind the overthrust belt, except for zones of cleavage and flow folding within 40 km of the pluton (Hietanen, 1961). In both the northern and southern provinces Belt strata do participate in the deformation within the thrust belt, but the absence of pre-Beltian crystalline rocks at the surface strongly suggests that the basement was not involved in the overthrusting. Geophysical data in Canada confirm that the thrusts converge downward and westward into a major décollement horizon above the crystalline basement (Bally and others, 1966).

In the northern province overthrusting advanced some 125 km farther eastward than in the southern province (Fig. 2). Between the two provinces a complex east-west zone of lateral shear known as the Lewis and Clark "line" or lineament (Billingsley and Locke, 1965) extends from a series of faults north of the Idaho batholith for some 800 km into eastern Montana. The offset of the thrust belt, however, cannot be attributed mainly to right-lateral displacement along this shear zone. Field work in the major Osburn fault zone north of the batholith has demonstrated a right-lateral displacement of only 20–25 km (Wallace and others, 1960), and Smith (1965) argues in favor of an overall left-lateral movement along the Lewis and Clark zone. Moreover, the shear zone does not perfectly separate the northern from the southern geosynclinal thrust front: east of the Boulder batholith (and east of the area of Fig. 2) the northern front continues southward for some distance across the lineament.

The most striking difference between the two provinces lies in their style of deformation. Northwestern Montana and Canada are characterized by numerous subparallel thrusts that tend to follow bedding and dip almost uniformly westward, separating west-dipping, essentially homoclinal panels of older rocks thrust on younger rocks. Figure 3 (after Mudge, 1966a, 1966b) provides a typical example in Montana, and the generalized cross section by Price (1973, Fig. 7) shows that the same style persists into Canada. Although local complexities do occur, particularly where thrusts cut from a lower to a higher bedding plane and perhaps also as a result of drag exerted by a higher thrust on a lower one, the overall regional aspect is one of remarkable structural tranquillity by comparison to the southern province. Along strike, the thrusts overlap where they die out (Price, 1973, Fig. 4), so that the total shortening is about the same along any given section.

Figure 4 illustrates the style of deformation in the southern province. Typical features are the following: (1) thrusts involving the Belt and lower Paleozoic quartzites are commonly folded and dip eastward over many miles (cross sections AA', CC' no. 2, and DD'); (2) many thrusts are not clearly related to bedding and to a single frontal fold but truncate at the base folds to which they are not geometrically related (CC' no. 1, DD'); (3)

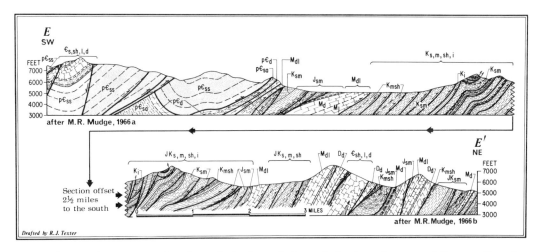

FIG. 3 Composite cross section through the frontal thrust belt of northwestern Montana. Location shown in Fig. 2. Lower half is slightly offset continuation of upper half. Symbols for this figure and Fig. 4: (p Є, C, O, D, M, P, P, J, K, T, Q) Precambrian, Cambrian, Ordovician, Devonian, Mississippian, Pennsylvanian, Permian, Jurassic, Cretaceous, Tertiary, and Quaternary sedimentary rocks; (L) Lower; (M) Middle; (U) Upper; (q, k) quartzite; (sh) shale; (m) mudstone; (s) sandstone; (ss) sandstone and siltstone; (sa) siltstone and argillite; (sl) shaly limestone; (l) limestone; (d) dolomite; (dl) dolomite, limestone, and dolomitic limestone; (gs) gneiss; (i) intrusive rocks; (v) volcanic rocks.

strong folding of the beds within or below the thrust sheets is not local but occurs throughout, in places to the point of overturning into antiformal synclines and synformal anticlines (see fold involving Belt on left side of CC' no. 1, and fold not involving Belt on right side of AA'); (4) recumbent folds or cascade-type flap structures similar to those described by Harrison and Falcon (1936) in Iran are common in middle to upper Paleozoic carbonates, generally verging northeastward, but in places radially away from domal uplifts (left side of BB'); (5) near the eastern edge of the deformed belt rootless epidermal allochthonous sheets of upper Paleozoic carbonate rest on previously faulted or folded terrane of older and younger rocks covered in one instance by a syntectonic conglomerate (BB').

The markedly different structural style east of the Idaho batholith and south of the Lewis and Clark lineament as compared to that north of the lineament strongly suggests tectonic mechanisms that were at least in part different.

Syntectonic Deposits

Late Jurassic to Paleocene syntectonic deposits derived from the west occur throughout the North American Rockies and are important in revealing structural events in the more central parts of the deforming geosyncline. In the southern province of this report extensive remains of the subaerial Beaverhead conglomerate occur at the eastern margin of the overthrust belt (Fig. 2). This conglomerate attains thicknesses of at least 5000 m, and a large part is composed of highly rounded pebbles and cobbles of lower Paleozoic and Belt quartzite, deposited continuously from Early Cretaceous (Albian) to Early Tertiary (Paleocene or early Eocene) times, as shown by pollen (Ryder and Ames, 1970). Farther east, a large part of the Late Cretaceous Harebell and Paleocene Pinyon conglomerates of western Wyoming is similarly composed of Belt quartzite clasts, probably deposited by recycling of the earlier Beaverhead. In between, the Plio-Pleistocene Snake

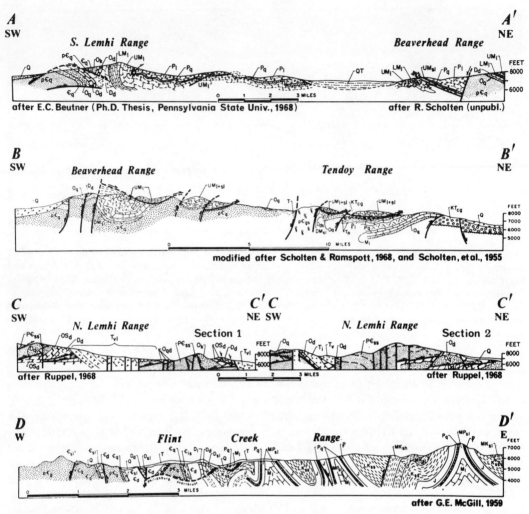

FIG. 4 Cross sections through mountain ranges east of the Idaho batholith. Locations shown in Fig. 2. Section 1, CC' is approximately 3 miles (5 km) south of section 2, CC'. Symbols see Fig. 3, caption. Note differences in structural style as compared to Fig. 3.

River lava undoubtedly covers additional occurrences. The only possible source for this huge volume of quartzite clasts lies in the extensive terrain of Belt rocks exposed today in the vicinity of the Idaho batholith (Fig. 2), some 50–75 km west of the Beaverhead conglomerate area. Detailed sedimentological studies by Ryder confirm a general southeastward transport (Ryder and Scholten, 1973). Isolated remains of conglomerate in Idaho, close to the batholith, contain many clasts of granite and have been tentatively equated with the Beaverhead by Shockey (1957) and Anderson (1959). The Beaverhead conglomerate is underlain by latest Jurassic to Early Cretaceous shales, siltstones, and sandstones containing abundant chert grains from chert-bearing upper Paleozoic and Triassic (?) rocks.

Thus it appears that the batholith was being uplifted and vigorously eroded as far back as latest Jurassic time, and that large areas had been highly elevated and unroofed well down into the Precambrian by Albian time, and probably even earlier. To explain a continuous transport of cobbles over 50 km or more during the Cretaceous and Paleocene it must be assumed that active uplift persisted and probably expanded laterally throughout this period, thus maintaining a high elevation of the source area and permitting repeated eastward recycling of the conglomerates on steep surface slopes until they arrived in their present positions.

Armstrong (1968), in discussing the Sevier orogenic belt south of the Snake River plain (Fig. 2), states that Eocambrian clasts first appear in early Upper Cretaceous time and that their abundant appearance in late Upper Cretaceous time was almost coincident throughout the region, suggesting "that evolutionary stages of the orogenic belt were similar over the entire region and that structures within the belt are of similar age and may be extrapolated and correlated along strike." Clearly, however, this extrapolation cannot be carried northward in all its particulars into the southern province of this report where the Precambrian was vigorously eroded as far back as Early Cretaceous time. Similarly, the conclusion of Roberts and Crittenden (1973) that, from Middle Jurassic through Early Cretaceous time, uplift in the western zone of the Sevier belt "can only have been moderate . . . inasmuch as only fine clastics were shed eastward into the Rocky Mountain trough" does not apply to the northern continuation of this zone where it crosses the Idaho batholith.

In the northern province, local conglomerate lenses containing some rounded igneous pebbles occur in the Lower Cretaceous mudstones and sandstones of northwestern Montana, and shales and sandstones with local conglomerate lenses mark the Upper Cretaceous (e.g., see Mudge, 1966a, 1966b; Mudge and Sheppard, 1968). In southern Canada a conglomerate at the base of the Lower Cretaceous Blairmore formation locally attains more than 1000 ft (300 m) in thickness; interdigitated lenses of "rock-fragment sandstone" and conglomerate occur in shale and sandstone both higher up in the Blairmore and in the underlying Upper Jurassic or Lower Cretaceous Kootenay formation, aggregating several hundred feet (Rapson, 1965). According to Rapson, lithic clasts include mostly chert, limestone, and phosphatic and volcanic fragments derived from upper Paleozoic rocks; less than 25 percent of the lithic clasts are sandstone and siltstone whose source could have been in the upper, lower, or pre-Paleozoic, and metamorphic fragments are rare. The Upper Jurassic to Lower Cretaceous clastic wedge deposits of the Southern Canadian Rockies lie beneath overthrusts far back in the deformed belt and probably were at least partly derived from the leading edges of emerging thrusts (Price and Mountjoy, 1970).

In summary, it appears that the syntectonic deposits in the southern province of this report are significantly different from those farther north with respect to the volume of coarse conglomerate in general, and the volume of Precambrian pebbles and cobbles in particular. They differ from those farther south in that very large quantities of Precambrian clasts appear in late Lower Cretaceous rather than late Upper Cretaceous time. In light of this, the inference seems justified that, as far back as Early Cretaceous or even Late Jurassic time, the roof of the Idaho batholith was raised more highly and eroded more deeply and extensively than elsewhere in the Rocky Mountains. The fact that the metamorphic assemblages in premetamorphic or symmetamorphic imbrications of Beltian (?) metasediments close to the eastern edge of the batholith indicate a relatively low-pressure environment (Dover, 1969) also suggests that much of the overburden had already been removed during the early stages of orogenesis. Even by Eocene time the batholith area, though considerably lowered in elevation, still stood out above the surrounding regions, as shown by paleotopographic contours drawn by Axelrod (1968) on the

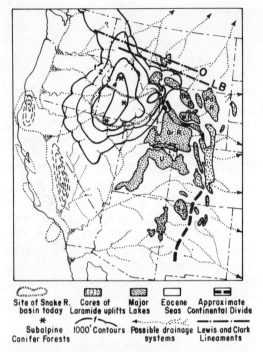

FIG. 5 Eocene paleogeography of the western United States as interpreted from paleobotanical data by Axelrod (1968, Fig. 10), with main lineaments of Lewis and Clark zone added (O = Osburn lineament; LB = Lake Basin lineament). Paleotopographic contour interval 1000 ft (305 m). Heavy dashed line shows present-day continental divide.

basis of the distribution of fossil remains of conifers and deciduous trees (Fig. 5). Today, the average elevation of the source area is approximately the same as that of the clastic wedge deposits.

Dynamic Models

In reviewing the structure of the Canadian Rocky Mountains, Price (1973) points out that, viewed on the scale of the entire mountain chain, the interlocked system of subparallel slip surfaces (overthrusts) so characteristic of that region represents a type of penetrative plastic flow within the chain without loss of cohesion. He attributes this remarkable structural pattern to gravitational spreading of uplifted supracrustal rocks under their own weight, away from a core of buoyant upwelling in the infrastructure to the west of the thrust belt, which created the topographic uplift and resulting gravitational potential. The décollement surface, in this view, continually sloped westward and even increased its dip, so that the gravitational thrust movement was entirely upslope, much as in a continental glacier. The similar structure of northwestern Montana north of the Lewis and Clark lineament would seem to imply an origin similar to that of the Canadian Rockies. Mesozoic (?) orthogneisses reported by Reid and Greenwood (1968) north of the Idaho batholith may indicate a core of buoyant upwelling similar to that in Canada.

By contrast, the deformed belt of the southern province presents no coherent system of interlocked slip surfaces. At least, the cascade and flap folds and slip sheets of upper Paleozoic carbonates indicate loss of cohesion within the belt at that level and cannot be described in terms of penetrative plastic flow on the scale of the whole chain. They create the impression of downslope gliding rather than upslope spreading features. More specific evidence for such a process includes transport directions radially away from a denuded central uplift in the Beaverhead Range (Scholten and Ramspott, 1968), westward transport away from a lately tilted Lemhi Range (Beutner, 1972), and the lack of root zones for these subhorizontal faults. The allochthonous sheet in the Tendoy Range, for example (Fig. 4, cross section BB', right side) can only have been derived by eastward gliding from the area of Precambrian rocks directly to the west, before postorogenic subsidence of that area along a major normal fault. The sheet overlies the Beaverhead conglomerate, which, in addition to rounded quartzite cobbles, contains subangular clasts of Mesozoic and Paleozoic limestone derived from nearby. The allochthonous sheet is itself overlain by ancient landslide deposits with Precambrian boulders that came from the immediate west. This shows that the now downfaulted area was topographically high during the Early Tertiary and was unroofed successively by erosion, tectonic gliding, and landsliding, with the resulting conglomerate, slip sheet, and landslide deposits reflecting the stratigraphy of the source area in reverse order. The slip sheet must have been epiglyptic, traveling essentially across the earth's surface while bulldozing away some conglomerate in the process.

As to the more extensive and deep-seated, folded, and rooted thrust sheets involving the Precambrian Belt, it seems plausible to explain their geographic relation to the Idaho batholith in terms of a genetic relation. One such model would be that buoyant upwelling and lateral spreading, being more pronounced here than in the northern province, provided a direct push from the rear, as advocated by Langton (1935), thus causing the eastward thrusting and the development of a distinctively different structural style in the southern province. Such a model would also account for the wrapping of thrusts around the northeastern corner of the batholith (Fig. 2). The difficulty with this model is that K/Ar and lead-alpha data, though as yet not conclusive, suggest that the primary igneous activity was Early Cretaceous or even earlier (Larsen and others, 1958; McDowell and Kulp, 1969); igneous and structural relationships near the contact point to the same conclusion (Hietanen, 1961). An additional major event, involving partial reheating or additional intrusion, or both, occurred in late Eocene time. Since much or most of the thrusting in the eastern part of the Sevier belt occurred between Early Cretaceous and Eocene times, a model relating all or most thrusting and folding to lateral mushrooming of the pluton does not seem satisfactory.

On the same grounds a model of fundamental, Greenland-type *Abscherung* of supracrustal rocks over a hot, highly mobile infrastructure, plausibly applied by Armstrong and Hansen (1966) to Late Jurassic (?) deformation in the eastern Great Basin, fails for later deformation in the Sevier belt. Finally, regional Cordilleran compression alone, whether connected with plate tectonics or other processes, fails to account for the structural differences between the northern and southern provinces of this report.

An alternative model linking the batholith with the deformation to the east postulates that deformation was the result of large-scale gravitational gliding away from a major uplift in the intruded area (Scholten, 1968, 1970), with the uplift representing a type of "undation" in the sense of Van Bemmelen (1964, 1966, 1967). A downslope gliding origin has also been proposed for the "disturbed belt" of northwestern Montana, in the northern province of this report (Mudge, 1970a), It has been questioned by Scholten (1970) and Price (1971a) and further defended in Mudge's replies to these discussions. Mudge (1970b) discounts the contrast in structural style across the Lewis and Clark lineament, but it appears so striking and fundamental that it must be accounted for by different dynamic models. The downslope gliding model is more likely to apply to the more complex and less coherent structural framework south of the Lewis and Clark lineament.

Evidence for the actual existence of an undation in the Idaho batholith area before and during the period of overthrusting lies in the syntectonic deposits discussed previously. According to the proposed model, and employing the theory of gravity gliding offered by Kehle (1970), a thick, low-viscosity "zone" (rather than a discrete surface), situated at a depth of many kilometers, attained an eastward slope away from the undation and underwent fluidlike internal deformation under the influence of the downslope gravitational component. The overlying rocks, disengaged in the rear by erosion or structural pull-apart, or both, were carried along eastward on top of the low-viscosity zone. Downslope buttressing gave rise in part to internal folding and in part to thrusting along faults that cut across bedding. As older thrusts were folded, motion along them became impossible and new ones formed farther east. Strongly arched areas became the loci of superficial gravitational cascade and flap folds and slip sheets of mid- to late Paleozoic carbonate rocks.

The depth and location of the décollement zone within the sequence are not known. As pointed out by Mudge (1971), there is no need to postulate that it lies at the base of the Belt. In southern Canada, Bally and others (1966, Pl. 12) infer it to lie above the base of the Belt (or equivalent Purcell Series) in the interior part of the geosyncline. An interpretive restoration by Price and Mountjoy (1970, Fig. 2–3) places it at a depth of about 14 km (46,000 ft) below the base of the Upper Jurassic in the area adjacent to the future core of buoyant upwelling. East of the Idaho batholith it is likely to follow a deeply buried pelitic unit. Above the batholith itself it must have been at least partly uncontrolled by stratigraphy. Since the southern half of the batholith intrudes the Paleozoic, detachment could have occurred here only within the Paleozoic or at the intrusive contact, cutting downward across bedding toward the east to meet the pelitic unit in the Belt. The northern half of the batholith may or may not have intruded the Paleozoic, and a detachment zone within the Belt is possible there.

If this model is juxtaposed against Price's model for the Canadian Rockies (also applicable to the U.S. Rockies north of the Lewis and Clark lineament), the result is the composite model diagrammatically represented in Fig. 6. The view here is southward, and the diagram shows only the northern portion of the Idaho batholith. The upper portion of the pluton is represented as concordant and laccolithic, as is, in effect, probable. Uplift is shown as accomplished in part by vertical movement along the western end of the Lewis and Clark lineament, shown schematically as a single fault. The probability of such a genetic relation is strongly suggested by the fact that the lineament lies at the northern border of the Eocene remnant uplift as reconstructed by Axelrod (1968) (see Fig. 5). Tectogenesis

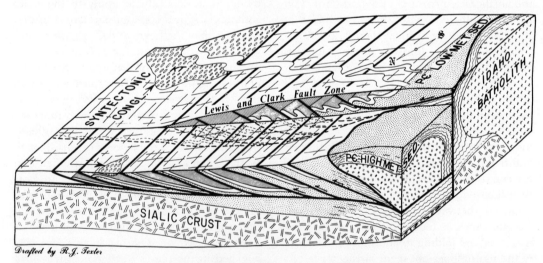

FIG. 6 Schematic block diagram, looking south, showing uplift along the Lewis and Clark zone. Northern (frontal) block shows tectonic evolution due to buoyant upwelling and gravitational spreading (compare Fig. 3, and Price, 1973). Southern (rear) block shows gravitational gliding and cascading down the temporarily reversed slope of a décollement zone (compare Fig. 4 and Scholten, 1968). High uplift of southern block gives rise to thick and extensive syntectonic conglomerate with abundant Precambrian clasts. (p C = Precambrian, KT = Cretaceous to Tertiary.)

in the front (northern) half of the diagram is portrayed as a process of buoyant upwelling and thrusting in the core and lateral upslope spreading and thrusting according to the hypothesis of Price and Mountjoy (1970). High uplift in the southern half is shown as resulting in extensive conglomerate deposition and in folding and thrusting due to downslope gliding across a décollement zone hypothetically assumed to lie near the base of low-metamorphic upper Belt rocks above the batholith and continuing eastward along a stratigraphic horizon in the Belt. Thus the composite model accounts for the contrasts in both structural style and syntectonic deposits.

It may be noted here that the gliding model for the southern province is capable of being tested, particularly near the northeastern corner of the Idaho batholith, where much detailed field work remains to be done. If it is correct, a décollement zone in the Belt should be evident, wrapping around the batholith along with the thrusts (Fig. 2) and rising to the rear (though very likely at an angle less than the original gliding angle, owing to postorogenic subsidence of the batholith area). Ideally, the zone should project westward over the batholith, but in view of the possibility of a widespread second intrusive event in the late Eocene along the margin of the pluton (McDowell and Kulp, 1969) there could also be a truncating relationship.[1] It should further be noted that the composite model for the two provinces is not incompatible with crustal compression, early tectonic fundamental *Abscherung*, or eastward or westward movement of plates. To the extent in which plate movement played a role in the evolution of the Cordilleran belt, it must have affected the entire belt and the Idaho batholith area cannot have been immune to it. The model does, however, emphasize the significant role gravity appears to have played in the specific development of the supracrustal fold and thrust belt in the northern U.S. Rockies.

Finally, it should be emphasized that, if gravitational spreading (as opposed to gliding) occurred in the north, it would be illogical to assume it did not occur in the southern province. The fact that the frontal thrust belt of the northern province extends for some distance across the lineament suggests that the same mechanism operated southward as far as the northern edge of the cratonic salient of southwest Montana, beyond which point the process was no longer operative. Gliding, however, appears to have been the dominant process east of the Idaho batholith, placing its characteristic stamp on the resulting structural style.

The Undation Model: Analysis and Ultimate causes

The question arises whether it is realistic to postulate that a deep-seated décollement zone was sufficiently raised to initiate gliding by gravity, without at the same time creating topographic elevations of improbable or impossible magnitudes. A further question relates to the possible fundamental causes of such an undation. To examine these questions, two cross sections have been constructed to scale (Fig. 7), taking into account approximate thicknesses of sediments in the geosyncline and showing the hypothetical growth of an undation according to the geothermal model proposed by Schuiling (1973). The upper section applies to the northern half of the Idaho batholith and the lower one to the southern half.

In the upper section initial depths to the pre-Beltian basement and the top of the Belt

---

[1] Recent work by R. B. Chase and J. L. Talbot (Geol. Soc. Am., Abs. with Progr, p. 470–471, 1973; Talbot, pers. comm., April, 1973) reveals a prominent, one-half mile thick detachment zone dipping about 30° away from the northeast corner of the batholith, and suggests gravitationally induced movement of the suprastructure. (Footnote added during impression).

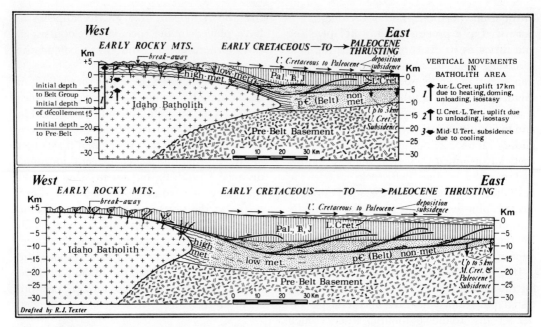

FIG. 7 Schematic cross sections across the northern (top section) and southern (bottom section) parts of the Idaho batholith and the area to the east, showing eastward gliding of sedimentary rocks away from uplifted batholith area and eastward progression of thrusting during Cretaceous and Paleocene. Initial depths of Belt strata and pre-Beltian basement at left side of upper section are estimated; initial depth of décollement zone is assumed. Gliding is initiated after calculated uplift of 17 km in batholith area (arrow 1), and is followed by additional isostatic uplift (arrow 2) and post-tectonic subsidence (arrow 3). Area to the east subsides due to loading by sedimentation and thrusting. Gravitative potential is equivalent to a gliding slope of about 1.5° in the southern section and slightly over 2° in the northern section. Gliding may have been facilitated by abnormal pore pressures and lowering of the décollement zone viscosity due to heat, penetration of hydrothermal fluids and (momentarily) earthquakes. Updip attachment resistance is reduced by deep erosion in breakaway area.

prior to uplift are shown at 21 and 6 km, respectively, implying a thickness of 15 km for the Belt sediments in the future area of intrusion. Harrison and Campbell (1963) report approximately 42,000 ft (almost 13 km) of Belt in the Pend Oreille area, and Harrison (1972, Fig. 3) shows 52,000 ft (almost 16 km) near Missoula, Montana, both north of the batholith. Since the bottom is not exposed, the total thickness must have been greater than that. The thickness of post-Precambrian rocks varied greatly in the geosyncline, increasing from north to south (Sloss, 1950, Fig. 1). A total of 6 km is reasonable for the northern section. This implies a net uplift of somewhere near 20 km between

the time when the area began to rise and the present time, when the surface stands well above 1 km. It may safely be assumed that at least 8 km of this uplift, and possibly much more, had occurred by Albian time, when voluminous Belt pebbles and cobbles were arriving at the present site of the Beaverhead conglomerate through repeated recycling.

The geothermal model of Schuiling (1973) starts with the assumption that a 30-km-thick crust in a linear belt is thickened by 10 km. Schuiling calculates that radiogenic heat will then cause a primary uplift of nearly 3.5 km by thermal expansion and phase changes. Isostatic rise directly due to a 10-km thickening may be calculated at 1.5 km, for crust and mantle densities of 2.8 and 3.3, respectively. This adds up to a 5-km primary uplift.

Crustal thickening may be the result of plate underthrusting, and the position of the Idaho batholith behind the intersection of two great subarcs (Fig. 1) may suggest that two subplates within a major underthrusting plate were colliding, causing correspondingly greater thickening, heating, and uplift by expansion, phase changes, and isostasy. Direct evidence of a greater amount of thickening lies in seismic data indicating a present crustal thickness between 30 and 40 km in this area (Kanasewich, 1966). In view of the conclusion that nearly 20 km of rock has been removed, the crustal thickness would appear to have been more than 50 km as a result of the thickening process. Since the continental interior crust is around 40 km thick according to Kanasewich, it is highly plausible that thickening in the batholith area exceeded 10 km. Exceptionally great thickening and heat development at the subarc intersection would at the same time account well for the melting that gave rise to the Idaho batholith, which represents a local "hot spot" in the Sevier belt. To this effect may be added the doming of the Belt and younger rocks over the laccolithic top of a pluton, which continued to be replenished at depth by molten material moving in laterally toward the central core of upwelling. If these two effects (thickening greater than 10 km and laccolithic doming) may be assumed to add another 2 km to the primary uplift, the result would be a total of 7-km primary uplift. The rise of the land surface was less in the northern province, but the steepest topographic gradient was not from south to north but from the ancestral uplift in the west to the subsiding foredeep trough in the east; the same was true for the main direction of sedimentary transport.

It may further be calculated that each 1 km of uplift creates concurrent erosion and isostatic rebound amounting to approximately 1750 and 1450 m, respectively, and that the topography will be raised by increments of a little over 700 m in the process. Thus a primary uplift of 7 km would cause about 12 km of erosion and about 10 km of isostatic rebound, while the surface would gradually reach a height of about +5 km. At this stage, crustal uplift due to primary causes (7 km) and secondary causes (10 km) would amount to 17 km. This situation is portrayed in the upper section of Fig. 7 (arrow 1).

If, in central Idaho, a pelitic, low-viscosity, potential décollement zone existed somewhere near the middle of the Belt at around 13 km below sea level prior to deformation, a 17-km uplift would raise it to +4 km, or about 1 km below the earth's surface. The gravitative energy potential now available to initiate gliding is given by the difference in surface height between the eastern limit of the thrust belt (assumed to remain at +1 km with concurrent thrusting, subaerial sedimentation, and isostatic subsidence) and a breakaway point some 100 km to the west over the roof of the batholith at +5 km. This is equivalent to an eastward potentiometric gradient of 4 percent, or somewhat over 2°.

In the southern (lower) section of Fig. 7 the décollement zone is portrayed as rising westward into the Paleozoic rocks that are intruded by the batholith, and to lie at, or near, the top of the batholith. Since thrusting progressed farther eastward in this section, the distance between the breakaway point and the eastern limit of thrusting may be taken at about 160 km, giving a gravitative energy potential represented by a potentiometric gradient of about 2.5 percent, or 1.5°. In both sections the eastward slope of the décollement zone itself, between the breakaway point and the point of greatest depth, would of course be many times greater, but the potentiometric gradient is used to take into account the resistance to be overcome where the décollement (as well as the thrusts cutting up through the section) revert to a westward slope (downslope buttressing effect).

In assessing whether gliding is possible under gravitative energy potentials of this magnitude, the analysis by Kehle (1970) is useful. Kehle states that gravity gliding above a zone of distributed fluidlike shear is favored by (1) low viscosity in the potential décollement zone, as in the case of shale interbedded between rocks of much higher viscosity, (2) increased thickness of the potential low-viscosity zone, (3) increased dip of the zone, and (4) increased depth of the zone. For a model in which a 1-km-thick shale unit underlies a 1-km-thick "stiff" unit, Kehle predicts a gliding velocity of almost 80 km/m.y. for Newtonian, and 35 km/m.y. for non-Newtonian viscosity, on a slope of 2 percent, or a little over 1°. End-effects (updip attachment and downdip buttressing) are neglected. If the shale is $\frac{1}{3}$ km thick and is overlain by $1\frac{1}{3}$ km of "stiff" rock, velocities are reduced to one-third the foregoing values. Increased water content in the shale, on the other hand, will increase the gliding velocity by enhancing the mobility of the décollement zone.

In Fig. 7 the average depth of burial of the presumed décollement zone is much greater than 1 km. The thickness of the zone is unknown, but the values assumed by Kehle are entirely acceptable for the stratigraphic sequence of the Belt. The slope in the upper (northern) section is twice that assumed by Kehle, and the downdip buttressing effect can be eliminated from consideration because the slope of the potentiometric gradient rather than the slope of the décollement zone was used. The updip attachment effect may be considered small where the rear of the plate is dissected by deep valleys such as must have existed in a mountain range of this height. Finally, high water content, increased pore pressures, lowered viscosities, and enhanced mobility may be expected at shallow depths above the batholith as the result of continued heat flow and rising hydrothermal fluids (Goguel, 1969). Earthquakes may well have contributed to high pore pressures during ephemeral moments, and if this was an area of plate and subplate collision they may have been exceptionally frequent.

On the basis of these considerations it may be concluded that, under the conditions shown in the upper section of Fig. 7, gliding velocities should be more than sufficient to account for observed displacements along thrusts in the roughly 50 million year time span available. For the lower (southern) section, where the décollement zone is portrayed as cutting across bedding and rising westward into the Paleozoic above the pluton, it would have to be assumed that rock viscosities were lowered in the zone as a result of heat and high water content and pore pressures created by the intrusion. East of the pluton the zone is buried more deeply in the southern than in the northern section owing to the greater thickness of post-Precambrian rocks, thus further enhancing the possibility of fluidlike shear. The potentiometric slope is still one and one-half times greater than the slope of the décollement zone in Kehle's model.

It should be emphasized that no pretense is intended that the parameters given here are quantitatively known, but only that the assumptions seem reasonable. Neither the depth and thickness of the décollement zone, nor the amount and velocity of primary uplift are known. A shallower depth of the zone would require less primary uplift without necessarily decreasing the gravitational gradient. In any case, considerable latitude is possible, for even average velocities one order of magnitude lower than those obtained by Kehle would be adequate. All that the foregoing analysis is intended to show is that the model is realistic in terms of known geologic data and current understanding of the mechanics of orogenic translation, geothermal expansion, and isostasy.

As gliding and thrusting progress, the crust in the batholith area continues to rise by isostatic rebound upon tectonic and erosional unloading. The net effect will be a gradual lowering of the earth's surface, which now lies at 2–3 km above sea level. This could be the end result of an additional 13 km of unloading and 11 km of rebound (still assuming specific gravities of 2.8 and 3.3 for crust and mantle). Although the lowering of the surface may have been partly due to the postthermal foundering that occurs in the cooling phase of the process envisioned by Schuiling (1973), it is nevertheless safe to assume that uplift was sufficiently great, and unloading sufficiently deep, to remove the entire roof and upper portion of the pluton. Schuiling also points out that the foundering stage is likely to be accompanied by late, shallow intrusion and volcanism. The group of late Eocene K/Ar ages in the Idaho batholith established by McDowell and Kulp (1969) as well as the occurrence of the Oligocene (?) Challis volcanics in the region may constitute a confirmation of this theoretical deduction.

Finally, Schuiling's model predicts that unloading in an area of exceptional heating and uplift should in the end exceed the initial amount of crustal thickening, resulting in a crust that is thinner than it was initially. Crustal thickness contours for North America based on explosion seismic data (Kanasewich, 1966) show, in effect, that the northern Rockies do not have a root but, on the contrary, the crust is thinner than in the continental interior. The data are not yet sufficiently detailed to show whether the thinning is greatest in the Idaho batholith area in particular, and a further means of testing the model lies, therefore, in the performance of future seismic work.

Conclusions

The conclusion that adjacent segments of a mountain chain are the product of different tectogenic processes is distasteful to many geologists. However, one need but examine the summaries of the various segments of the entire Rocky Mountain fold and thrust belt from Mexico to Alaska as presented in a recent symposium (Price, 1971b) to find not only that different models have been proposed for different segments, but that they are probably necessary in view of major differences in geosynclinal evolution, rock types, plutonism, and structural style between the segments.

Such a conclusion, however, does not imply chaos or mutual independence between segments. Better than other physical sciences, geology shows us that end results may be the product of many interacting parameters and processes, and that their final form depends on which process or processes were dominant in a given place or time. This holds true no less for tectonics, but it does not mean that there is no common underlying master process or controlling factor. All it implies is that the *manifestations* of the fundamental process are likely to be different in different areas and at different times, depending on variations in regional conditions of a lesser scale.

Ultimately, the only sources of tectonic energy are heat and gravity. At the level of the crust, the most fundamental manifestation of these energies may be, in the currently popular view, the drifting of "plates" driven by thermal convection, gravity or both. Second-order manifestations include the formation and subsequent deformation of the great geosynclinal belts of the earth, perhaps in response to subduction of plates, whether continental and oceanic, or both continental. Depending on the nature, thickness, shape, motion, and number of colliding plates and subplates, degree of underthrusting, distance from the subduction zone, and other factors, the type and magnitude of the tectonic response will differ between belts and from place to place within a single belt. Deformation may be linked *directly* to the subduction process in some cases or, in other cases, mostly to thermal and vertical gravitational effects that were *indirectly* triggered by subduction. Local "hot spots" may arise within a belt, causing gravitational buoyant upwelling and diapiric "mushrooming" of gneisses or batholithic intrusion of molten matter, and undations or linear uplifts may form as a result of thermal expansion, doming, and isostasy. Third-order responses include gravitational spreading or gliding (or both) of supracrustal rocks, both leading to thrusting. Still lower order responses include the formation of cascade and flap folds and the emplacement of shallow slip sheets off local uplifts, and, finally, the emplacement of landslides.

The view presented in this paper is that the area of the Idaho batholith represents an unusually thickened, heated, and highly uplifted portion of the crust in the eastern part of the North American Cordillera, and that the characteristic structural manifestations to the east can be attributed to this.

References

Anderson, A. L., 1959, Geology and mineral resources of the North Fork quadrangle, Lemhi County, Idaho: *Idaho Bur. Mines Geol. Pamphlet 118*, 92 p.

Armstrong, R. L., 1968, Sevier orogenic belt in Nevada and Utah, *Geol. Soc. Am. Bull.*, v. 79, p. 429–458.

———, and E. Hansen, 1966, Cordilleran infrastructure in the eastern Great Basin: *Am. J. Sci.*, v. 264, p. 112–127.

Axelrod, D. I., 1968, Tertiary floras and topographic history of the Snake River Basin, Idaho: *Geol. Soc. Am. Bull.*, v. 79, p. 713–734.

Bally, A. W., Gordy, P. L., and Stewart, G. A., 1966, Structure, seismic data and orogenic evolution of southern Canadian Rocky Mountains: *Can. Petrol. Geol. Bull.*, v. 14, p. 337–381.

Beutner, E. C., 1972, Reverse gravitative movement on earlier overthrusts, Lemhi Range, Idaho: *Geol. Soc. Am. Bull.*, v. 83, p. 839–846.

Billingsley, P., and Locke, A., 1939, *Structure of ore districts in the continental framework*: New York, Am. Inst. Mining Metall. Engs., 51 p.

Dover, J. H., 1969, Bedrock geology of the Pioneer Mountains, Blaine and Custer Counties, central Idaho: *Idaho Bur. Mines Geol. Pamphlet 142*, 66 p.

Eardley, A. J., 1962, *Structural geology of North America*: New York, Harper and Row, 743 p.

Goguel, J., 1969, Le rôle de l'eau et de la chaleur dans les phénomènes tectoniques: *Rev. Géogr. Phys. Géol. Dyn.*, v. 11, p. 153–164.

Harrison, J. E., 1972, Precambrian Belt basin of northwestern United States: Its geometry, sedimentation, and copper occurrences: *Geol. Soc. Am. Bull.*, v. 83, p. 1215–1240.

———, and Campbell, A. B., 1963, Correlations and problems in Belt Series stratigraphy, northern Idaho and western Montana: *Geol. Soc. Am. Bull.*, v. 74, p. 1413–1428.

Hietanen, A., 1961, Metamorphic facies and style of folding in the Belt Series northwest of the Idaho Batholith: *Bull. Comm. Géol. Finlande*, no. 196, p. 73–103.

Kanasewich, E. R., 1966, Deep crustal structure under the Plains and Rocky Mountains: *Can. J. Earth Sci.*, v. 3, p. 937–945.

Kehle, R. O., 1970, Analysis of gravity sliding and orogenic translation: *Geol. Soc. Am. Bull.*, v. 81, p. 1641–1664.

Langton, C. M., 1935, Geology of the northeast part of the Idaho Batholith and adjacent region in Montana: *J. Geol.*, v. 43, p. 27–60.

Larsen, E. S., Jr., Gottfried, D., Jaffe, H. W., and Waring, C. L., 1958, Lead-alpha ages of the Mesozoic batholiths of western North America: *U.S. Geol. Survey Bull. 1070-B*, p. 35–62.

M'Gonigle, J. W., 1965, Structure of the Maiden Peak area, Montana–Idaho: Ph.D. thesis, Pennsylvania State University, University Park, 146 p.

McDowell, F. W., and Kulp, J. L., 1969, Potassium-argon dating of the Idaho Batholith: *Geol. Soc. Am. Bull.*, v. 80, p. 2379–2382.

McGill, G. E., 1959, Geologic map of the northwest flank of the Flint Creek Range, western Montana: *Montana Bur. Mines. Geol. Spec. Publ. 8* (map sheet).

Maxwell, D. T., and Hower, J., 1967, High-grade diagenesis and low-grade metamorphism of illite in the Precambrian Belt Series: *Am. Mineralogist*, v. 52, p. 843–857.

Mudge, M. R., 1966a, Geologic map of the Pretty Prairie quadrangle, Lewis and Clark County, Idaho: *U.S. Geol. Survey Geol. Quad. Map GQ-454*.

———, 1966b, Geologic Map of the Patricks Basin Quadrangle, Teton, and Lewis and Clark Counties, Montana: *U.S. Geol. Survey Geol. Quad. Map GQ-453*.

———, 1970a, Origin of the disturbed belt in northwestern Montana: *Geol. Soc. Am. Bull.*, v. 81, p. 377–392.

———, 1970b, Origin of the disturbed belt in northwestern Montana: A reply: *Geol. Soc. Am. Bull.*, v. 81, p. 3793–3794.

———, and Sheppard, R. A., 1968, Provenance of igneous rocks in Cretaceous conglomerates in northwestern Montana, *in* Geological Survey Research 1968: *U.S. Geol. Survey Prof. Paper 600-D*, p. D137–D146.

Price, R. A., 1971a, Gravitational sliding and the foreland thrust and fold belt of the North American Cordillera: Discussion: *Geol. Soc. Am. Bull.*, v. 82, p. 1133–1138.

———, convenor, 1971b, The foreland fold and thrust belt of the North American Cordillera: Symposium: *Geol. Soc. Am. Progr. 24th Ann. Meeting Rocky Mountain*

Section, p. 370–371, 373–375, 377–378, 398–399, 400–401, 404–405, 408–409, 411, 415, 416.

———, 1973, Large-scale gravitational flow of supracrustal rocks, southern Canadian Rockies (this volume).

———, and E. W. Mountjoy, 1970, Geologic structure of the Canadian Rock Mountains between Bow and Athabasca Rivers—A progress report: *Geol. Assoc. Can. Spec. Paper 6*, p. 7–25.

Rapson, J. E., 1965, Petrography and derivation of Jurassic–Cretaceous clastic rocks, southern Rocky Mountains, Canada: *Am. Assoc. Petrol. Geol. Bull.*, v. 49, p. 1426–1452.

Reid, R. R., and Greenwood, W. R., 1968, Multiple deformation and associated progressive metamorphism in the Beltian rocks north of the Idaho Batholith, U.S.A.: *Int. Geol. Cong., Rept. 23rd Session (Prague)*, p. 122–123.

Reid, R. R., Greenwood, W. R., and Morrison, D. A., 1970, Precambrian metamorphism of the Belt Supergroup in Idaho: *Geol. Soc. Am. Bull.*, v. 81, p. 915–918.

Roberts, R. J., and Crittenden, M. D., 1973, Orogenic mechanisms, Sevier orogenic belt, Nevada and Utah (this volume).

Ryder, R. T., and Ames, H. T., 1970, The palynology and age of the Beaverhead Formation, Montana–Idaho, and their paleotectonic implications in the Lima region, Montana–Idaho: *Am. Assoc. Petrol. Geol. Bull.*, v. 54, p. 1155–1171.

Ryder, R. T., and Scholten, R. (1973), Syntectonic conglomerates in southwestern Montana: Their nature, origin, and tectonic significance: *Geol. Soc. Am. Bull.*, v. 84, p. 773–796.

Scholten, R., 1967, Structural framework and oil potential of extreme southwestern Montana: *Montana Geol. Soc. Guidebook No. 18*, p. 7–19.

———, 1968, Model for evolution of Rocky Mountains east of the Idaho Batholith: *Tectonophysics*, v. 6, p. 109–126.

———, 1970, Origin of the disturbed belt in northwestern Montana: A discussion: *Geol. Soc. Am. Bull.*, v. 81, p. 3789–3792.

———, Keenmon, K. A., and Kupsch, W. O., 1955, Geology of the Lima region, southwestern Montana and adjacent Idaho: *Geol. Soc. Am. Bull.*, v. 66, p. 345–404.

Scholten, R., and Ramspott, L. D., 1968, Tectonic mechanisms indicated by structural framework of central Beaverhead Range, Idaho–Montana: *Geol. Soc. Am. Spec. Paper 104*, 71 p.

Schuiling, R. D., 1973, Active role of continents in tectonic evolution—Geothermal models (this volume).

Shockey, P. N., 1957, Reconnaissance geology of the Leesburg quadrangle, Lemhi County, Idaho: *Idaho Bur. Mines Geol. Pamphlet 113*, 42 p.

Sloss, L. L., 1950, Paleozoic sedimentation in Montana area: *Am. Assoc. Petrol. Geol. Bull.*, v. 34, p. 423–451.

Smith, J. G., 1965, Fundamental transcurrent faulting in northern Rocky Mountains: *Am. Assoc. Petrol. Geol. Bull.*, v. 49, p. 1398–1409.

Van Bemmelen, R. W., 1964, Der gegenwärtige Stand der Undationstheorie: *Mitt. Geol. Gesellsch. Wien*, v. 57, p. 379–399.

———, 1966, Stockwerktektonik *sensu lato*, in Etages Tectoniques: Wegmann Symposium, Univ. Neuchâtel, p. 19–40.

———, 1967, The importance of the geonomic dimensions for geodynamic concepts: *Earth Sci. Rev.*, v. 3, p. 79–110.

Wallace, R. E., Griggs, A. B., and Hobbs, S. W., 1960, Tectonic setting of the Coeur d'Alene district, Idaho: *U.S. Geol. Survey Prof. Paper 400B*, p. 25–27.

RAYMOND A. PRICE

Department of Geological Sciences
Queen's University
Kingston, Ontario

Large-Scale Gravitational Flow of Supracrustal Rocks, Southern Canadian Rockies[1]

The Canadian Rocky Mountains, a series of linear mountain ranges that rises above the high western Interior Plains, is a distinctive but relatively narrow physiographic province within the Cordilleran mountain system, northeast of the Rocky Mountain trench (Fig. 1). As a physiographic province it corresponds in a general way with a distinctive structural province that is characterized by low-angle thrust faulting and shallow folding in distinctly layered but essentially unmetamorphosed sedimentary rocks of platform or "miogeosynclinal" aspect. This structural province forms the foreland fold and thrust zone along the external margin of the Eastern Cordilleran fold belt (Fig. 2), one of two severely deformed belts within the southern Canadian Cordillera that are marked by mountainous core zones of metamorphic and plutonic rocks and are separated by a less deformed region of generally lower elevation —the Intermontane zone (Wheeler, 1970).

Although the southern Canadian Rockies only comprise the external part of one of these two adjacent fold belts, it is nevertheless a representative segment of a more or less continuous belt of thrust faulting and folding that stretches the length of the Cordillera for some

[1] This study is an outgrowth of a regional reconnaissance investigation of bedrock geology over a large part of the southern Canadian Rockies for the Geological Survey of Canada. It reflects the influence of discussions with many collaborators, particularly R. J. W. Douglas, G. B. Leech, E. W. Mountjoy, and D. K. Norris; however, final responsibility for any errors of fact or interpretation rests with me. This study has been supported by research grants from the National Research Council of Canada and the Geological Survey of Canada.

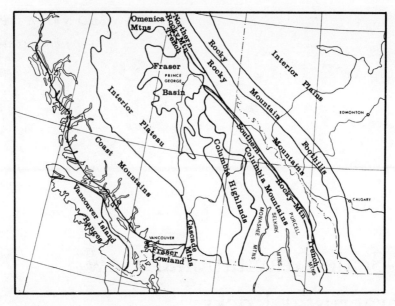

FIG. 1 Physiographic subdivisions of the southern Canadian Cordillera.

4000 miles (6500 km) from northern Alaska through western Canada and the northwestern United States into southeastern California (King, 1969). Moreover, because its origin is inextricably bound to that of the rest of the southern Canadian Cordillera, the structural evolution of the southern Canadian Rocky Mountains provides some insight on the tectonic evolution of the entire system (Price and Mountjoy, 1970) as well as on the origin of the belt of thrust faulting and folding in other parts of the North American Cordillera.

The essential features of the geologic structure of the southern Canadian Rockies have been attributed to "a series of gigantic thrust faults, which have carried the older formations forward, and placed them in a number of places above the highest beds of the series.... The region has been broken by a number of parallel or nearly parallel fractures into series of oblong orographic blocks, and these tilted and shoved one over the other into the form of a westerly-dipping compound homocline.... Overturned folds (occur) along the courses of some of the faults, but they are usually small, and are of minor importance as a structural feature, and the great earth rents of the district seem to have been produced without much preliminary bending" (McConnell, 1887, p. 31D–33D).

In the 85 years following this perceptive outline of the geological structure of the mountains west of Calgary, there has been a phenomenal growth of new information and concepts which is proceeding at an ever increasing rate. Systematic regional studies within even the most remote parts of the mountains have provided geologic maps with new and more detailed information on near-surface structures, and there has been an extensive program of deep drilling and seismic exploration for hydrocarbons, which has added new dimensions to this picture and has made the eastern part of the Canadian Rockies one of the best known foreland zones of thrust faulting and folding in the world (Bally and others, 1966).

Many new concepts have been adopted. The deformation is "thin-skinned." It involves displacements on southwest-dipping listric thrust faults that flatten above the

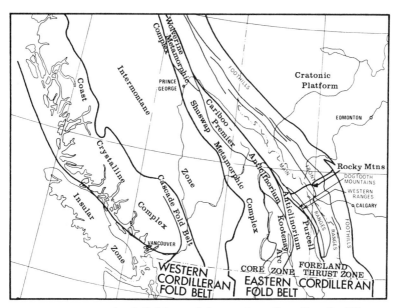

FIG. 2 Structural subdivisions of the southern Canadian Cordillera.

basement of Precambrian metamorphic and plutonic rocks which slopes under the Canadian Rockies from the Churchill province of the Canadian Shield (Shaw, 1963, Keating, 1966; Bally and others, 1966). Folds have developed within the anisotropic mass of layered sedimentary rocks as a direct consequence of displacements on "stepped" thrust faults that alternately follow the sedimentary layering and then cut across it (Douglass, 1950). Some of the thrust faults themselves have been folded as a result of displacements along other thrust faults lower in the sedimentary pile (Douglas, 1950, 1958). Thrust faulting and the development of parallel (concentric) folds have proceeded conjointly, side by side, as two different but equivalent manifestations of the same basic process of horizontal shortening and vertical thickening within the layered mass (Price, 1967; Dahlstrom, 1970). The novelty of much of this contemporary fabric of concepts and information on the geologic structure of the southern Canadian Rockies tends to obscure the fact that almost all of it has been woven over the conceptual framework which McConnell (1887) established.

Scale of Observation

McConnell's basic notion of the form of the structures and the mechanical nature of the deformation remains as the essence of contemporary views on structural evolution within the southern Canadian Rockies. The deformation is viewed in terms of brittle failure in an anisotropic layered mass, and of simple translations or rotations of discrete, relatively rigid blocks or slabs of rock that are bounded by the thrust faults along which they have been stacked one upon the other like shingles on a roof. The idea that the beds between the faults are undistorted except for the modest strains associated with the development of parallel folds is widely held and has been adopted as a basic premise for the construction of balanced cross sections in which the surface area of every bed (and its length in a cross section) is the same after the deformation as before (Dahlstrom, 1969). These ideas arise from the fact that the rocks, when viewed on the scale of a hand specimen or a thin section, generally show little or no evidence of distortion due to penetrative flow.

Over much of the southern Canadian

Rockies, even the finest details of the primary sedimentary features and the most delicate of organic structures have been preserved within a mass of rock that obviously has been deformed on a larger scale. At the scale of an individual outcrop, fractures are ubiquitous and faults occur as distinct surfaces of slip rather than zones of distributed shear. Obviously, *at this scale of observation* the deformation is a result of brittle failure and simple rigid body translations and rotations. But this is not the scale of observation from which the whole of the southern Rocky Mountains, or even the whole of one thrust sheet, can be encompassed in a single view. To perceive more, it is necessary to adopt a new perspective at a different scale of observation. However, changes in the scale of our observations can have important consequences in shaping our impressions of the fundamental nature of the deformation in rocks.

Our thinking about deformed rocks is done with models. We use precise and elegant mathematical models to represent the "elastic" and "fluid" behavior of rocks, rigid geometric models to portray the shape of cylindrical folds, and rather vague conceptual models to convey the notion of brittle failure or plastic flow in rock deformation. Our choice of models, whether conceptual and vague or mathematical and precise, must be consistent with the particular scale of our deliberations at the moment. A model eminently suited to represent the deformation of rock at one scale of observation may be entirely inadequate as a representation of the same deformation viewed at some different scale (Carey, 1962). The main purpose of this paper is to assess the effects of changes in scale of observation on what we perceive of the nature of the geologic structure within the whole of the southern Canadian Rockies, and then to consider the nature and implications of the role of gravity in the evolution of this structure. To do this, it is necessary to abandon the traditional perspective which has been focused, since McConnell's day, on the scale of the outcrop and the individual thrust sheet, and to "stand back" in order to view the whole of the southern Canadian Rockies as the single entity that it is. The emphasis must be shifted from models based on individual small bodies of rock and on parts of specific thrust faults to models that represent large masses of rock and reflect the group behavior of large numbers of individual thrust faults.

Geologic Structure

The sedimentary suprastructure of the southern Canadian Rockies is a northeasterly tapering wedge of distinctly layered and strongly anisotropic supracrustal rocks that has been stripped from a passive, unfaulted infrastructure of metamorphic and plutonic basement rocks and displaced northeastward up the basement slope, and on the flank of the craton. The whole mass has been foreshortened by more than 200 km and thickened commensurately as a result of displacements that range up to several tens of kilometers on distinct listric thrust faults which are generally southwest-dipping (Shaw, 1963; Bally and others, 1966; Price and Mountjoy, 1970). The upper sides of these faults have been displaced relatively northeastward and upward, and the faults gradually cut up through the stratigraphic layering northeastward, commonly following the layering over large areas in response to the anisotropy inherent in the layering (Price, 1965). Displacements along the thrust faults consistently involved a translation of older rocks over younger, repetitions in the vertical succession of stratigraphic units, and a thickening of the wedge of supracrustal rocks.

At the scale of the individual thrust fault the most conspicuous aspect of the deformation is the fact that a thoroughly fractured mass of otherwise essentially unstrained layered sedimentary rocks has remained intact while having undergone large-scale translations on a distinct shear surface. Fractures that reflect a loss of cohesion within the mass are ubiquitous. Blocks bounded by fractures have moved with respect to each other in orderly patterns involving a combination of extension marked by veins, compression marked by stylolites, and simple shear marked by slickensided fractures (Price, 1967). Brittle fracture is the dominant mode of deformation.

At the scale of a group of larger thrust faults (Fig. 3), the most conspicuous aspect of the deformation is the manner in which the displacements vary along and among the components of the group. The faults interfinger and overlap with one another along strike to form a mechanically interlocked system across which the net displacement is distributed among the components of the group in a pattern that changes from one locality to the next (Douglas, 1958; Price, 1967). Although the displacement along one individual thrust fault may decrease rapidly along strike, that along other adjacent thrust faults increases proportionately, so that the total displacement distributed over the group as a whole remains relatively constant. All of the thrust faults terminate within the rock mass. The mass is physically continuous around the ends of each and every one of them. The thrust faulting clearly did not involve a complete loss of cohesion across the mass. At this scale the fractures separating individual unstrained blocks are too small to be seen, but their cumulative effects are still evident. Displacements along and across the fractures provide mechanical adjustments for folding and other forms of bulk strain between the thrust faults. The deformation is a kind of inhomogeneous ductile shear failure, which combines the characteristic features of penetrative flow with those of relatively brittle faulting.

At the scale of the whole of the southern Canadian Rockies (Fig. 4), the most conspicuous aspects of the deformation are the overall change in the shape of the northeasterly tapering wedge of supracrustal rock, and the fact that the area occupied by even the largest of the thrust faults is small compared with the total area of the southern Canadian Rocky Mountains. A wedge of supracrustal rocks more than 370 km wide and up to 15 km thick was compressed horizontally by more than 50 percent (to 170 km) and thickened proportionately (Price and Mountjoy, 1970). This large-scale distortion, achieved without loss of cohesion across the mass, represents a type of penetrative plastic flow. Large net displacements across the belt

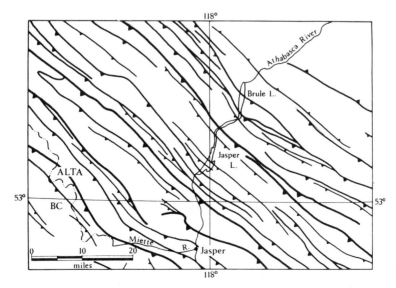

FIG. 3 Relationships among thrust faults in the vicinity of Jasper, Alberta. Triangles mark overthrust side. Main faults shown as heavy lines.

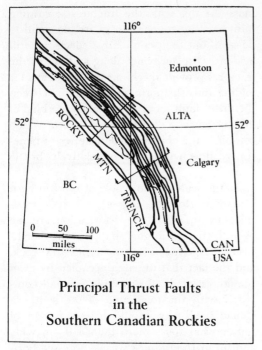

FIG. 4 Relationships among principal thrust faults in the southern Canadian Rockies. Lines with arrows mark the locations of structure sections shown in Fig. 7.

are the cumulative effect of smaller componental displacements distributed over a penetrative array of discrete, discontinuous slip surfaces. These slip surfaces are the mechanically interlocked system of thrust faults that overlap and interfinger with one another along strike, and are contained entirely within the mass of supracrustal rocks, or else merge with a zone of detachment near the basement surface beneath it. The supracrustal rocks have flowed northeastward, up the slope of the basement, out of the Cordillera, and on to the flank of the North American craton.

Gravitational Flow of Supracrustal Rocks

The idea that thrust faulting can be an integral part of large-scale lateral flow in supracrustal rocks sheds a different light on the enigma posed by the notion that there must be a simple basic relationship between gravity and low-angle thrust faulting. Gravitational gliding, the most popular paradigm for such a relationship, arises from our familiar conceptual model based on the frictional resistance to sliding of a rigid rectangular block on an inclined plane (Fig. 5). Elegant theoretical analyses have established the basic principles behind this process (Hafner, 1951; Hubbert and Rubey, 1959; Hsü, 1968) and there are some obvious natural examples of it; but the critical data of regional geology indicate that most low-angle thrust faults have always sloped in the wrong direction for gravitational sliding. This enigma can be ascribed to an inappropriate choice of conceptual models for the analyses, rather than to misguided intuition. A model based on the dynamic gravitational equilibrium in a plastic solid which flows in response to its own weight is more realistic at this scale of observation than one based on the limiting static equilibrium of a perfectly rigid block of rock of rectangular cross section on an inclined plane. This concept of lateral gravitational spreading is the very essence of Van Bemmelen's (1960) model for "secondary tectogenesis," and it was advocated by Bucher (1956) to account for the décollement structures in the foreland zones of mountain belts (Fig. 6).

Viewed from this perspective, the southern Canadian Rockies are analogous to the peripheral zone of a large ice sheet. There has been large-scale plastic flow and lateral gravitational spreading from a zone of buoyant upwelling in a hot mobile infrastructure of metamorphic and plutonic rocks which is exposed in the core zone of the Eastern Cordilleran fold belt (Price and Mountjoy, 1970). The gravitational potential for the flow can be attributed to the buoyant upwelling of these metamorphic and plutonic rocks. It resulted from the northeasterly slope of the surface of the mass of supracrustal rocks, not from the slope of the basement beneath it. The gravity-induced displacements occurred in spite of the slope of the underlying basement

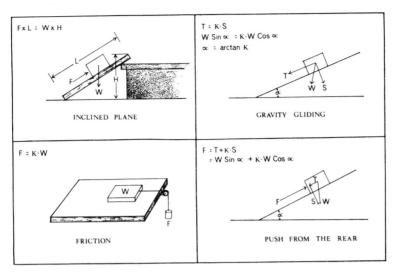

FIG. 5 Gravitational gliding model for thrust faulting. The model is based on the concept of the inclined plane (upper left) and the concept of frictional resistance to sliding (lower left) and involves establishing the limiting static equilibrium of a rigid rectangular block that slides down the plane in response to body forces (upper right) or moves up the plane in response to surface forces (lower right).

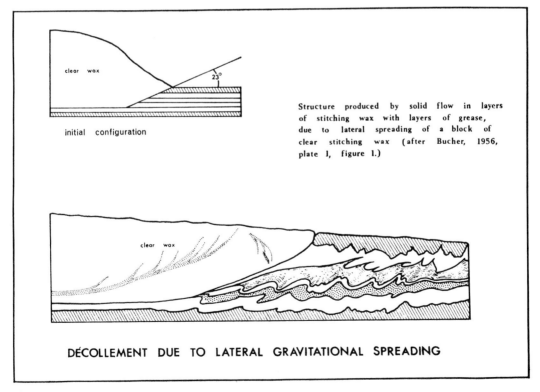

FIG. 6 A model for horizontal compression of supracrustal rocks due to lateral gravitational spreading.

surface, not because of it. The situation is analogous to the flow of the Pleistocene ice sheets up the bedrock surface, from near sea level in the vicinity of Hudson Bay, to the headwaters of the Mississippi River. Just as in the case of the Pleistocene ice sheets, this process of lateral gravitational flow imposed a load on the crust, in response to which it subsided isostatically.

The amount of isostatic subsidence beneath the southern Canadian Rockies can be estimated by comparing restored sections, in which there has been compensation for the effects of the deformation in the supracrustal rocks, with structure sections along the same profile that show the supracrustal rocks in their present deformed state. Two restored sections across the southern Canadian Rockies have been drawn with respect to a horizontal datum corresponding to the transition from marine to nonmarine beds in the Upper Jurassic strata of the Fernie group and the Kootenay and Nikanassin formations (Fig. 7). This is an essentially contemporaneous sea level datum over the whole of the region, and the aggregate thickness of supracrustal rocks below this datum at any point is a measure of the depth to the basement as it existed just prior to the onset of the thrusting and folding (Price and Mountjoy, 1970). The difference between the depth to the basement at any point as it existed at the end of the Jurassic and the depth to the basement at the same point now is a measure of the net amount of isostatic subsidence since the deformation of the supracrustal rocks began.

The net subsidence of the basement surface since the onset of the thrust faulting is about 8 km beneath the western Rockies and about 2 km beneath the western Interior Plains. In the western Rockies the average elevation of the bedrock surface is about 2 km above sea level. Since the thrusting began, the basement has subsided 8 km and the bedrock surface has risen 2 km. Accordingly, there has been a net increase of about 10 km in the total thickness of the continental crust in this area, and this has occurred by accretion at the top. The subsidence amounts to about 80 percent of the total increase in thickness, as might be expected if it were an isostatic effect dependent upon a ratio in specific gravity between the supracrustal rocks and the mantle of about 0.8.

Isostatic subsidence, which extends well beyond the actual limits of any large load imposed on the Earth's crust, can be ascribed to an elastic warping of the lithosphere beneath the load; the width and depth of the peripheral depression can be related to the magnitude of the load and the effective flexural parameter of the lithosphere. Thus, even for relatively modest loads due to Pleistocene ice sheets, subsidence may extend several hundred kilometers beyond the edge of the actual supracrustal load (Walcott, 1970). Accordingly, the foredeep trough (exogeosyncline) that migrated northeastward in advance of the deformation in the eastern Cordillera (Bally and others, 1966) must have been an isostatically induced peripheral depression that developed in response to the load imposed on the lithosphere by the northeasterly flow of supracrustal rocks up on to the flank of the craton (Fig. 8). This has important geotectonic implications. This moatlike depression was an efficient sediment trap for the detrital outwash that was discharged intermittently from the Cordillera; thus the subsidence and filling of the foredeep was an indirect result of the same buoyant upwelling and lateral gravitational spreading

---

FIG. 7 Relationships between tectonic thickening of supracrustal rocks and basement subsidence in the southern Canadian Rockies. The upper and lower sections are drawn along the northern and southern lines, respectively, in Fig. 4. The restored sections are referred to a horizontal pretectonic sea level datum marked by the transition from marine to nonmarine rocks in the Late Jurassic. The net subsidence of the unfaulted basement surface is the difference between the present depth to the basement and the depth in the Late Jurassic.

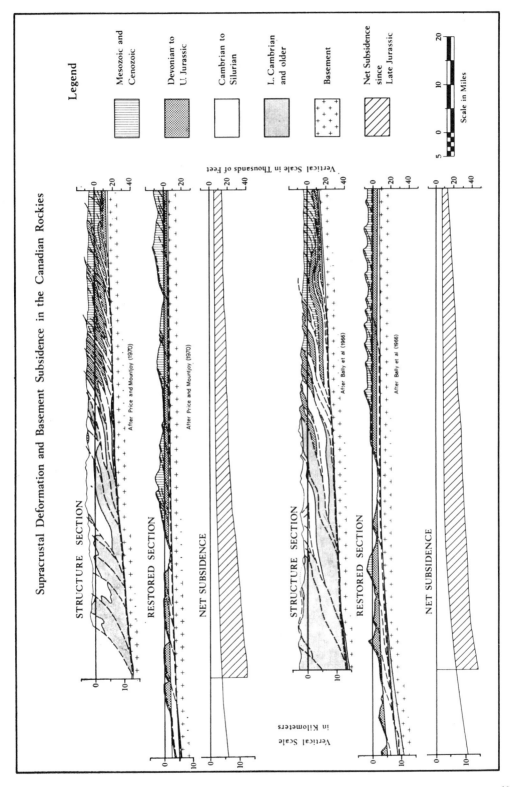

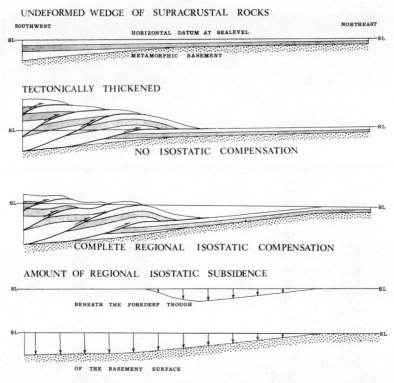

FIG. 8 Tectonic thickening of supracrustal rocks, isostasy, and the origin of a migrating foredeep trough. The relationships shown schematically are based on the assumption that, because of flexural rigidity in the lithosphere, isostatic subsidence extends beyond the actual limits of the supracrustal load.

that produced the clastic detritus within the Cordillera. A sensitive chain of reaction from buoyant upwelling in the metamorphic and plutonic infrastructure of the core zone, to lateral gravitational spreading involving adjacent supracrustal rocks, isostatic subsidence of the cratonic margin beneath this load, and settling of the clastic detritus eroded from the internal zones in this external sediment trap, underscores the basic underlying role of gravity at all stages in the evolution of orogenic belts of this type.

The record of fluctuating rates of sedimentation and of intermittent erosion in the clastic wedge deposits of the foredeep is a sensitive gauge of variations in the intensity of deformation within the whole of the orogenic belt. The time delay involved in isostatic adjustment to supracrustal loads is geologically very short. The lateral gravitational spreading that occurred in conjunction with each pulse of upwelling in the core zone of the orogenic belt led to a commensurate and essentially contemporaneous subsidence of the foredeep, where part of the newly generated outwash of clastic detritus was immediately trapped. Thus intervals of rapid subsidence and sedimentation in the foredeep are evidence of concurrent intense deformation in the orogenic belt; and, conversely, lulls in the sedimentation in the foredeep are evidence of relative tectonic quiescence in the orogenic belt. This means that, contrary to popular belief, unconformities in the clastic wedge sequence of the foredeep do not mark the times of most intense deformation within the orogenic belt but instead reflect anorogenic intervals. It is noteworthy in this context that the times of maximum uplift in the plutonic and metamorphic core zone of the Eastern Cordilleran fold belt, as defined by the frequency of occurrence of potassium-argon isotope dates from these rocks, correlate very closely with the times of maximum deposition in the foredeep (Bally and others, 1966).

The origin of the gravitational instability that drives the buoyant upwelling in the core zone must be sought at some larger scale of observation. If it is a consequence of a relative convergence between plates of lithosphere (Dewey and Bird, 1970), the record of the interaction between the converging plates is written in the clastic wedge deposits of the foredeep trough.

Conclusions

Although the characteristic *structures in* the southern Canadian Rockies can be attributed to brittle failure in an anisotropic layered mass of supracrustal rocks, and to simple translations or rotations of discrete, relatively rigid blocks or plates of rock on thrust faults, the *structure of* the southern Canadian Rockies cannot. At a scale of observation which encompasses the whole of the southern Canadian Rockies in a single view, the deformation involves large strain without loss of cohesion. It is a type of plastic flow in which displacements are distributed among a myriad array of discrete, interleaved, and overlapping slip surfaces, all of which terminate within the deformed mass.

Gravity has been the dominant factor at all stages in the deformation. Northeasterly flow of the supracrustal rocks up the flank of the craton is a result of gravitational spreading from a zone of buoyant upwelling in the metamorphic and plutonic rocks of the infrastructure in the core of the Eastern Cordilleran fold belt. The basement was depressed under the added load of the northeasterly-flowing supracrustal rocks, and a migrating foredeep trough (exogeosyncline) in front of the advancing deformation was an isostatically induced moat in which the detrital outwash from the Cordillera was trapped. The record of sedimentation in the subsiding foredeep is a record of fluctuations in orogenesis in the whole of the Eastern Cordilleran fold belt.

References

Bally, A. W., Gordy, P. L., and Stewart, G. A., 1966, Structure, seismic data, and orogenic evolution of southern Canadian Rocky Mountains: *Bull. Can. Petrol. Geol.*, v. 14, p. 337–381.

Bucher, W. H., 1956, The role of gravity in orogenesis: *Geol. Soc. Am. Bull.*, v. 67, p. 1295–1318.

Carey, S. W., 1962, Scale of geotectonic phenomena: *Jour. Geol. Soc. India*, v. 3, p. 95–105.

Dahlstrom, C. D. A., 1969, Balanced cross sections: *Can. Jour. Earth Sci.*, v. 6, p. 743–757.

―――, 1970, Structural geology in the eastern margin of the Canadian Rocky Mountains: *Bull. Can. Petrol. Geol.*, v. 18, p. 332–406.

Dewey, J. F., and Bird, J. M., 1970, Mountain belts and the new global tectonics: *Jour. Geophys. Res.*, v. 75, p. 2625–2647.

Douglas, R. J. W., 1950, Callum Creek, Langford Creek, and Gap map-areas, Alberta: *Geol. Survey Can. Mem. 255*, 124 p.

―――, 1958, Mount Head map-area, Alberta: *Geol. Survey Can. Mem. 291*, 241 p.

Hafner, W., 1951, Stress distribution and faulting: *Geol. Soc. Am. Bull.*, v. 62, p. 373–398.

Hsü, K. J., 1969, Role of cohesive strength in the mechanics of overthrust faulting and landsliding: *Geol. Soc. Am. Bull.*, v. 80, p. 927–952.

Hubbert, M. K., and Rubey, W. W., 1959, Role of fluid pressure in mechanics of overthrust faulting, I. Mechanics of fluid-filled porous solids and its application to overthrust faulting: *Geol. Soc. Am. Bull.*, v. 70, p. 115–166.

Keating, L. F., 1966, Exploration in the Canadian Rockies and Foothills: *Can. Jour. Earth Sci.*, v. 3, p. 713–723.

King, P. B., 1969, *Tectonic map of North America*: Washington, D.C., *U.S. Geological Survey*.

McConnell, R. G., 1887, Report on the geological features of a portion of the Rocky Mountains, accompanied by a section measured near the 51st Parallel: *Geol. Survey Can. Ann. Rpt.*, N.S., v. II, 1886, Report D, p. 1D–41D.

Price, R. A., 1965, Flathead map-area, British Columbia and Alberta: *Geol. Survey Can. Mem. 336*, 221 p.

——, 1967, The tectonic significance of mesoscopic subfabrics in the southern Canadian Rocky Mountains of Alberta and British Columbia: *Can. Jour. Earth Sci.*, v. 4, p. 39–70.

——, and Mountjoy, E. W., 1970, Geologic structure of the Canadian Rocky Mountains between Bow and Athabasca Rivers—A progress report, *in* Wheeler, J. O., ed., *Structure of the Southern Canadian Cordillera*. Geol. Assoc. Can. Spec. Paper no. 6, p. 8–25.

Shaw, E. W., 1963, Canadian Rockies in time and space, *in* Childs, O. E., and Beebe, B. W., eds., *Backbone of the Americas*: Am. Assoc. Petrol. Geol. Mem. 2, p. 231–242.

Van Bemmelen, R. W., 1960, New views on East-Alpine orogenesis: *Int. Geol. Cong., Rept. 21st Session (Norden)*, Pt. XVII, p. 99–116.

Walcott, R. I., 1970, Isostatic response to loading of the crust in Canada: *Can. Jour. Earth Sci.*, v. 7, p. 716–727.

Wheeler, J. O., 1970, Introduction, *in* Wheeler, J. O., ed., *Structure of the Southern Canadian Cordillera*: Geol. Assoc. Can. Spec. Paper no. 6, p. 1–5.